PROJECT

MANAGEMENT CASE STUDIES

PROJECT
MANAGEMENT
CASE STUDIES

FIFTH EDITION

HAROLD KERZNER, PH.D.

Senior Executive Director for Project Management
The International Institute for Learning
New York, New York

WILEY

Published by John Wiley & Sons, Inc., Hoboken, New Jersey
Published simultaneously in Canada
PMI, CAPM, PMBOK, PMP and Project Management Professional are registered marks of the
Project Management Institute, Inc.

For general information about our other products and services, please contact our Customer Care
Department within the United States at (800) 762-2974, outside the United States at (317) 572-3993
or fax (317) 572-4002.

Wiley publishes in a variety of print and electronic formats and by print-on-demand. Some material
included with standard print versions of this book may not be included in e-books or in print-on-
demand. If this book refers to media such as a CD or DVD that is not included in the version you
purchased, you may download this material at http://booksupport.wiley.com. For more information
about Wiley products, visit www.wiley.com.

Cover image: © Aeriform/Getty Images, Inc.
Cover design: Wiley

Library of Congress Cataloging-in-Publication Data:

Names: Kerzner, Harold, author.
Title: Project management: case studies/Harold Kerzner, Ph.D.
Other titles: Project management (Case studies)
Description: Fifth edition. | Hoboken, New Jersey: John Wiley & Sons, Inc.,
    [2017] | Includes index. | Description based on print version record and
    CIP data provided by publisher; resource not viewed.
Identifiers: LCCN 2016056664 (print) | LCCN 2016057574 (ebook) | ISBN
    9781119389156 (pdf) | ISBN 9781119389163 (epub) | ISBN 9781119385974
    (paperback: acid-free paper)
Subjects: LCSH: Project management–Case studies.
Classification: LCC HD69.P75 (print) | LCC HD69.P75 K472 2017 (ebook) | DDC
    658.4/04–dc23
LC record available at https://lccn.loc.gov/2016057574

Printed in the United States of America

10 9 8 7 6 5 4 3 2 1

To Andrea and Jeremy
For successfully managing the "miracle" project:
Our grandson, Asher Kaiden Thompson

Contents

Preface

Other than on-the-job training, case studies and situations are perhaps the best way to learn project management. Project managers pride themselves on finding solutions to problems, and case studies are an excellent way for this to happen. Case studies require that students investigate what went right in the case, what went wrong, and what recommendations should be made to prevent these problems from recurring in the future. The use of case studies is applicable both to undergraduate- and graduate-level project management courses as well as training programs to pass various certification examinations in project management.

Situations are smaller case studies that focus on one or two points that need to be addressed, whereas case studies can focus on a multitude of interrelated issues. The table of contents identifies several broad categories for the cases and situations, but keep in mind that the larger case studies, such as "Corwin Corporation," "The Blue Spider Project," or "The Rise, Fall, and Resurrection of Iridium," could have been listed under several topics. Some of the case studies, such as "The Need for Metrics" and "The Singapore Software Group," are well suited for group exercises. Other smaller or minicases can be covered during the class period.

Several smaller cases or situations are included in this edition at the request of faculty members who asked for cases that could be discussed in class and worked on in a team environment. These smaller cases can be used as in-class assignments or take-home assignments.

Almost all of the cases and situations have seed questions either in the case itself or in the instructor's teaching notes on the case to assist the reader in the analysis of the case. The seed questions from the instructor's manual will be provided by the instructor. An instructor's manual is available from John Wiley & Sons only to faculty members who adopt the book for classroom use.

Almost all of the case studies are factual. In most circumstances, the cases and situations have been taken from the author's consulting practice. The names of many of the companies and the people in the companies have been disguised for obvious reasons.

Some educators prefer not to use case studies that are more than 10 or 20 years old. However, the circumstances surrounding many of these older cases and situations are the same today as they were years ago. Unfortunately, we seem to be repeating several of the mistakes made previously.

Eighteen new cases have been added to this edition and some existing cases have been updated. Seed questions in the case studies reflect on some of the issues that project managers might face. The new cases are:

- Disney (A): Case study discusses how Disney's Imagineering project managers may need a different set of skills from those possessed by most traditional project managers.
- Disney (B): Case study discusses some of the challenges Disney faced with the Haunted Mansion Project.
- Disney (C): Case study discusses how the enterprise environmental factors impacted Disney's decisions to build new theme parks.
- Disney (D): Case study discusses the contractual decisions that Disney faces with some of its partners in the construction of worldwide theme parks.
- Disney (E): Case study discusses the challenges faced by an established theme park in Hong Kong when Disney announced it would build a Disney theme park nearby.
- Olympics (A): Case study shows the complexities and enterprise environmental factors that impact the decision to host the Olympic Games.
- Olympics (B): Case study shows the complexities of following the PMI Code of Ethics and Professional Conduct when managing Olympic projects that involve billions of dollars and often-greedy contractors.
- Olympics (C): Case study shows what is involved with managing a project designed to feed 20,000 Olympic athletes and staff at the Olympic Village when they come from almost every country in the world and may have different nutritional needs.
- Olympics (D): Case study discusses the some of the health and safety risks that the Olympic athletes faced in the Rio Olympic Games
- The Project Audit: Case study discusses a company's recognition that it needed a process in place to audit projects, but it was unsure about how to do it, when to do it, or who should do it.
- Trade-off Decisions (A): Case study discusses the challenges that a company faces when having to make critical trade-off decisions.
- Trade-off Decisions (B): Case study discusses the options that a company faces with regard to making a critical decision.
- The Executive Director: Case study discusses how a newly appointed executive director at a government agency got immersed in political gamesmanship to protect his image.

- Boeing 787 Dreamliner Battery Problems: Case study illustrates the importance of safety as a project management constraint when designing a commercial aircraft.
- Airbus A380: Case study focuses on some of the business-related decisions that project managers must make in the commercial aircraft industry.
- Agile (A): Case study focuses on some of the strategic business decisions that may be impacted when converting to agile or Scrum, especially when your business survives on competitive bidding and your clients may not understand or allow you to use agile or Scrum.
- Agile (B): Case study describes some of the operational issues facing project managers when they must manage a project in an agile environment rather than in a traditional project management environment.
- Agile (C): Case study illustrates how reporting project status in an agile environment may be different from status reporting in a traditional project management environment.

Most of the case studies are factual, but the names of the companies, the names of the individuals involved, and other identifying details have been changed (with the exception of Disney, Boeing, and Iridium, and the case studies of the 2016 Olympics and the Challenger space shuttle disaster).

Part 1

PROJECT MANAGEMENT METHODOLOGIES

As companies approach some degree of maturity in project management, it becomes readily apparent to all that some sort of standardization approach is necessary for the way that projects are managed. The ideal solution might be to have a singular methodology for all projects, whether they are for new product development, information systems, or client services. Some organizations may find it necessary to maintain more than one methodology, however, such as one methodology for information systems and a second methodology for new product development.

The implementation and acceptance of a project management methodology can be difficult if the organization's culture provides a great deal of resistance toward the change. Strong executive leadership may be necessary such that the barriers to change can be overcome quickly. These barriers can exist at all levels of management as well as at the worker level. The changes may require that workers give up their comfort zones and seek out new social groups.

Lakes Automotive

Lakes Automotive is a Detroit-based tier-one supplier to the auto industry. Between 1995 and 1999, Lakes Automotive installed a project management methodology based on nine life-cycle phases. For the next 10 years, all 60,000 employees worldwide accepted the methodology and used it. Management was pleased with the results. Also, Lakes Automotive's customer base was pleased with the methodology and provided Lakes Automotive with quality award recognition that everyone attributed to how well the project management methodology was executed.

In February 2015, Lakes Automotive decided to offer additional products to its customers. Lakes Automotive bought out another tier-one supplier, Pelex Automotive Products (PAP). PAP also had a good project management reputation and also provided quality products. Many of its products were similar to those provided by Lakes Automotive.

Because the employees from both companies would be working together closely, a single project management methodology would be required that would be acceptable to both companies. PAP had a good methodology based on five life-cycle phases. Both methodologies had advantages and disadvantages, and both were well liked by their customers.

QUESTIONS

1. How do companies combine methodologies?
2. How do you get employees to change work habits that have proven to be successful?
3. What influence should a customer have in redesigning a methodology that has proven to be successful?
4. What if the customers want the existing methodologies left intact?
5. What if the customers are unhappy with the new combined methodology?

Ferris HealthCare, Inc.

In July of 2014, senior management at Ferris recognized that its future growth could very well be determined by how quickly and how well it implemented project management. For the past several years, line managers had been functioning as project managers while still managing their line groups. The projects came out with the short end of the stick, most often late and over budget, because managers focused on line activities rather than project work. Everyone recognized that project management needed to be an established career path position and that some structured process had to be implemented for project management.

A consultant was brought into Ferris to provide initial project management training for 50 out of the 300 employees targeted for eventual project management training. Several of the employees thus trained were then placed on a committee with senior management to design a project management stage-gate model for Ferris.

After two months of meetings, the committee identified the need for three different stage-gate models: one for information systems, one for new products/ services provided, and one for bringing on board new corporate clients. There were several similarities among the three models. However, personal interests dictated the need for three methodologies, all based on rigid policies and procedures.

After a year of using three models, the company recognized it had a problem deciding how to assign the right project manager to the right project. Project managers had to be familiar with all three methodologies. The alternative, considered impractical, was to assign only those project managers familiar with that specific methodology.

After six months of meetings, the company consolidated the three methodologies into a single methodology, focusing more on guidelines than on policies and procedures. The entire organization appeared to support the new single methodology. A consultant was brought in to conduct the first three days of a four-day training program for employees not yet trained in project management. The fourth day was taught by internal personnel with a focus on how to use the new methodology. The success to failure ratio on projects increased dramatically.

QUESTIONS

1. Why was it so difficult to develop a single methodology from the start?
2. Why were all three initial methodologies based on policies and procedures?
3. Why do you believe the organization later was willing to accept a single methodology?
4. Why was the single methodology based on guidelines rather than policies and procedures?
5. Did it make sense to have the fourth day of the training program devoted to the methodology and immediately attached to the end of the three-day program?
6. Why was the consultant not allowed to teach the methodology?

Clark Faucet Company

BACKGROUND

By 2010, Clark Faucet Company had grown into the third largest supplier of faucets for both commercial and home use. Competition was fierce. Consumers would evaluate faucets on artistic design and quality. Each faucet had to be available in at least 25 different colors. Commercial buyers seemed more interested in the cost than the average consumer, who viewed the faucet as an object of art, irrespective of price.

Clark Faucet Company did not spend a great deal of money advertising on the radio, television, or Internet. Some money was allocated for ads in professional journals. Most of Clark's advertising and marketing funds were allocated to the two semiannual home and garden trade shows and the annual builders' trade show. One large builder could purchase more than 5,000 components for the furnishing of one newly constructed hotel or one apartment complex. Missing an opportunity to display the new products at these trade shows could easily result in a six- to 12-month window of lost revenue.

CULTURE

Clark Faucet had a noncooperative culture. Marketing and engineering would never talk to one another. Engineering wanted the freedom to design new products, whereas marketing wanted final approval to make sure that what was designed could be sold.

The conflict between marketing and engineering became so fierce that early attempts to implement project management failed. Nobody wanted to be the project manager. Functional team members refused to attend team meetings and spent most of their time working on their own pet projects rather than on the required work. Their line managers also showed little interest in supporting project management.

Project management became so disliked that the procurement manager refused to assign any of his employees to project teams. Instead, he mandated that all project work come through him. He eventually built a virtual brick wall around his employees. He claimed that this would protect them from the continuous conflicts between engineering and marketing.

THE EXECUTIVE DECISION

The executive council mandated that another attempt to implement good project management practices must occur quickly. Project management would be needed not only for new product development but also for specialty products and enhancements. The vice presidents for marketing and engineering reluctantly agreed to try to patch up their differences but did not appear confident that any changes would take place.

Strange as it may seem, no one could identify the initial cause of the conflicts or how the trouble actually began. Senior management hired an external consultant to identify the problems, provide recommendations and alternatives, and act as a mediator. The consultant's process would have to begin with interviews.

ENGINEERING INTERVIEWS

The following comments were made during engineering interviews:

- "We are loaded down with work. If marketing would stay out of engineering, we could get our job done."
- "Marketing doesn't understand that there's more work for us to do other than just new product development."
- "Marketing personnel should spend their time at the country club and in bar rooms. This will allow us in engineering to finish our work uninterrupted!"
- "Marketing expects everyone in engineering to stop what they are doing in order to put out marketing fires. I believe that most of the time the problem is that marketing doesn't know what they want up front. This leads to change after change. Why can't we get a good definition at the beginning of each project?"

MARKETING INTERVIEWS

These comments were made during marketing interviews:

- "Our livelihood rests on income generated from trade shows. Since new product development is four to six months in duration, we have to beat up on engineering to make sure that our marketing schedules are met. Why can't engineering understand the importance of these trade shows?"
- "Because of the time required to develop new products [four–six months], we sometimes have to rush into projects without having a good definition of what is required. When a customer at a trade show gives us an idea for a new product, we rush to get the project under way for introduction at the next trade show. We then go back to the customer and ask for more clarification and/or specifications. Sometimes we must work with the customer for months to get the information we need. I know that this is a problem for engineering, but it cannot be helped."

The consultant wrestled with the comments but was still somewhat perplexed. "Why doesn't engineering understand marketing's problems?" pondered the consultant. In a follow-up interview with an engineering manager, the following comment was made: "We are currently working on 375 different projects in engineering, and that includes those that marketing requested. Why can't marketing understand our problems?"

QUESTIONS

1. What is the critical issue?
2. What can be done about it?
3. Can excellence in project management still be achieved and, if so, how? What steps would you recommend?
4. Given the current noncooperative culture, how long will it take to achieve a good cooperative project management culture and even excellence?
5. What obstacles exist in getting marketing and engineering to agree to a single methodology for project management?
6. What might happen if benchmarking studies indicate that either marketing or engineering are at fault?
7. Should a single methodology for project management have a process for the prioritization of projects, or should some committee external to the methodology accomplish this?

Creating a Methodology

BACKGROUND

John Compton, the president of the company, expressed his feelings quite bluntly at the executive staff meeting. He said:

> We are no longer competitive in the marketplace. Almost all of the requests for proposal that we want to bid on have a requirement that we must identify in the proposal the project management methodology we will use on the contract should we be awarded the contract. We have no project management methodology. We have just a few templates we use based upon the *PMBOK® Guide*. All of our competitors have methodologies, but not us.
>
> I have been asking for a methodology to be developed for more than a year now, and all I get are excuses. Some of you are obviously afraid that you might lose power and authority once the methodology is up and running. That may be true, but losing some power and authority is obviously better than losing your job. In six months I want to see a methodology in use on all projects or I will handle the situation myself. I simply cannot believe that my executive staff is afraid to develop a project management methodology.

CRITICAL ISSUES

The executive staff knew this day was inevitable; they had to take the initiative in the implementation of a project management methodology. Last year, a consultant was brought in to conduct a morning three-hour session on the benefits of project management and the value of an enterprise project management (EPM) methodology. As part of the session, the consultant explained that the time needed to develop and implement an EPM system can be shortened if the company has a project management office (PMO) in place to take the lead role. The consultant also explained that whichever executive gets control of the PMO may become more powerful than other executives because he or she now controls all of the project management intellectual property. The executive staff fully understood the implication of this and therefore were reluctant to visibly support project management until they could see how their organization would be affected. In the meantime, project management suffered.

Reluctantly, a PMO was formed reporting to the chief information officer. The PMO was comprised of a handful of experienced project managers that could, it was hoped, take the lead in the development of a methodology. The PMO concluded that five steps had to be done initially. After the five steps were done, the executive committee would receive a final briefing on what had been accomplished. The final briefing would be in addition to the monthly updates and progress reports. The PMO believed that getting executive support and sign-offs in a timely manner would be difficult.

The first step that needed to be done was the establishment of the number of life-cycle phases. Some people interviewed wanted 10 to 12 life-cycle phases. That meant that there would be 10 to 12 gate-review meetings, and the project managers would spend a great deal of time preparing paperwork for the gate-review meetings rather than managing the project. The decision was then made to have no more than six life-cycle phases.

The second step was to decide whether the methodology should be designed around rigid policies and procedures or go the more informal route of using forms, guidelines, checklists, and templates. The PMO felt that project managers needed some degree of freedom in dealing with clients and therefore the more informal approach would work best. Also, clients were asking to have the methodology designed around client business needs, and the more informal approach would provide the flexibility to do this.

The third step was to see what could be salvaged from the existing templates and checklists. The company had a few templates and checklists but not all project managers used them. The decision was made to develop a standardized set of documents in accordance with the information in the *PMBOK® Guide*. The project managers could then select whatever forms, guidelines, templates, and checklists were appropriate for a particular project and client.

The fourth step would be to develop a means for capturing best practices using the EPM system. Clients were now requiring in their requests for proposal that best practices on a project must be captured and shared with the client prior to the close out of the project. Most of the people in the PMO believed that this could be done using forms or checklists at the final project debriefing meeting.

The fifth step involved education and training. The project managers and functional organizations that would staff the projects would need to be trained in the use of the new methodology. The PMO believed that a one-day training program would suffice and the functional organizations could easily release their people for a one-day training session.

QUESTIONS

1. What can you determine about the corporate culture from the fact that they waited this long to consider the development of an EPM system?
2. Can a PMO accelerate the implementation process?
3. Is it acceptable for the PMO to report to the chief information officer, or should it report to someone else?
4. Why is it best to have six or fewer life-cycle phases in an EPM system?
5. Is it best to design an EPM system around flexible or inflexible elements? Generally, when first developing an EPM system, do companies prefer to use formal or informal designs?
6. Should an EPM system have the capability of capturing best practices?

Honicker
Corporation

BACKGROUND

Honicker Corporation was well recognized as a high-quality manufacturer of dashboards for automobiles and trucks. Although it serviced mainly U.S. automotive and truck manufacturers, the opportunity to expand to a worldwide supplier was quite apparent. The company's reputation was well known worldwide, but it was plagued for years with ultraconservative senior management leadership that prevented growth into the international marketplace.

When the new management team came on board in 2009, the conservatism disappeared. Honicker was cash rich, had large borrowing power and lines of credit with financial institutions, and received an AA-quality rating on its small amount of corporate debt. Rather than expand by building manufacturing facilities in various countries, Honicker decided to go the fast route by acquiring four companies around the world: Alpha, Beta, Gamma, and Delta Companies.

Each of the four acquired companies serviced mainly its own geographic area. The senior management team in each of the four companies knew the culture in their geographic area and had a good reputation with their clients and local stakeholders. The decision was made by Honicker to leave each company's senior management teams intact, provided that the necessary changes, as established by corporate, could be implemented.

Honicker wanted each company to have the manufacturing capability to supply parts to any Honicker client worldwide. But doing this was easier said than done. Honicker had an EPM methodology that worked well. Honicker understood project management and so did the majority of Honicker's clients and stakeholders in the United States. Honicker recognized that the biggest challenge would be to get all of the divisions at the same level of project management maturity and using the same corporate-wide EPM system or a modified version of it. It was expected that each of the four acquired companies might want some changes to be made.

The four acquired divisions were all at different levels of project management maturity. Alpha did have an EPM system and believed that its approach to project management was superior to the one that Honicker was using. Beta Company was just beginning to learn project management but did not have any formal EPM system, although it did have a few project management templates that were being used for status reporting to its customers. Gamma and Delta Companies were clueless about project management.

To make matters worse, laws in each of the countries where the acquired companies were located created other stakeholders that had to be serviced, and all of these stakeholders were at different levels of project management maturity. In some countries government stakeholders were actively involved because of employment procurement laws; in other countries government stakeholders were passive participants unless health, safety, or environmental laws were broken.

It would certainly be a formidable task to develop an EPM system that would satisfy all of the newly acquired companies, their clients, and their stakeholders.

ESTABLISHING THE TEAM

Honicker knew that there would be significant challenges in getting a project management agreement in a short amount of time. Honicker also knew that there is never an acquisition of equals; there is always a "landlord" and "tenants," and Honicker is the landlord. But acting as a landlord and exerting influence in the process could alienate some of the acquired companies and do more harm than good. Honicker's approach was to treat this as a project and to treat each company, along with its clients and local stakeholders, as project stakeholders. Using stakeholder relations management practices would be essential to getting an agreement on the project management approach.

Honicker requested that each company assign three people to the project management implementation team that would be headed up by Honicker personnel. The ideal team member, as suggested by Honicker, would have some knowledge and/or experience in project management and be authorized by their senior levels of management to make decisions for their company. The representatives should also understand the stakeholder needs from their clients and local stakeholders. Honicker wanted an understanding to be reached as early as possible that each company would agree to use the methodology that was finally decided on by the team.

Senior management in each of the four companies sent a letter of understanding to Honicker promising to assign the most qualified personnel and agreeing to use the methodology that was agreed on. Each stated that its company understood the importance of this project.

The first part of the project would be to come to an agreement on the methodology. The second part of the project would be to invite clients and stakeholders to see the methodology and provide feedback. This was essential since the clients and stakeholders eventually would be interfacing with the methodology.

KICKOFF MEETING

Honicker had hoped that the team could come to an agreement on a companywide EPM system within six months. But after the kickoff meeting was over, Honicker realized that it would probably be two years before an agreement would be reached on the EPM system. Several issues became apparent at the first meeting:

- Each company had different time requirements for the project.
- Each company saw the importance of the project differently.
- Each company had its own culture and wanted to be sure that the final design was a good fit with that culture.
- Each company saw the status and power of the project manager differently.
- Despite the letters of understanding, two of the companies, Gamma and Delta, did not understand their role and relationship with Honicker on this project.
- Alpha wanted to micromanage the project, believing that everyone should use its methodology.

Senior management at Honicker asked the Honicker representatives at the kickoff meeting to prepare a confidential memo on their opinion of the first meeting with the team. The Honicker personnel prepared a memo including the following comments:

- Not all of the representatives at the meeting openly expressed their true feelings about the project.
- It was quite apparent that some of the companies would like to see the project fail.
- Some of the companies were afraid that the implementation of the new EPM system would result in a shift in power and authority.
- Some people were afraid that the new EPM system would show that fewer resources were needed in the functional organization, thus causing a downsizing of personnel and a reduction in bonuses that were currently based on headcount in functional groups.

- Some seemed apprehensive that the implementation of the new system would cause a change in the company's culture and working relationships with their clients.
- Some seemed afraid of learning a new system and being pressured into using it.

It was obvious that this would be no easy task. Honicker had to get to know all companies better and understand their needs and expectations. Honicker management had to show them that their opinions were of value and find ways to win their support.

QUESTIONS

1. What are Honicker's options now?
2. What would you recommend that Honicker do first?
3. What if, after all attempts, Gamma and Delta companies refuse to come on board?
4. What if Alpha Company is adamant that its approach is best and refuses to budge?
5. What if Gamma and Delta Companies argue that their clients and stakeholders have not readily accepted the project management approach and they wish to be left alone with regard to dealing with their clients?
6. Under what conditions would Honicker decide to back away and let each company do its own thing?
7. How easy or difficult is it to get several geographically dispersed companies to agree to the same culture and methodology?
8. If all four companies were willing to cooperate with one another, how long do you think it would take for an agreement on and acceptance to use the new EPM system?
9. Which stakeholders may be powerful and which are not?
10. Which stakeholder(s) may have the power to kill this project?
11. What can Honicker do to win their support?
12. If Honicker cannot win their support, then how should Honicker manage the opposition?
13. What if all four companies agree to the project management methodology and then some client stakeholders show a lack of support for use of the methodology?

Acquisition Problem

BACKGROUND

All companies strive for growth. Strategic plans are prepared identifying new products and services to be developed and new markets to be penetrated. Many of these plans require mergers and acquisitions to obtain the strategic goals and objectives rapidly. Yet often even the best-prepared strategic plans fail when based on mergers and acquisitions. Too many executives view strategic planning for a merger or acquisition as planning only and often give little consideration to implementation, which takes place when both companies are actually combined. Implementation success is vital during any merger and acquisition process.

PLANNING FOR GROWTH

Companies can grow in two ways—internally or externally. With internal growth, companies cultivate their resources from within and may spend years attaining their strategic targets and marketplace positioning. Since time may be an unavailable luxury, meticulous care must be given to make sure that all new developments fit the corporate project management methodology and culture.

External growth is significantly more complex. External growth can be obtained through mergers, acquisitions, and joint ventures. Companies can purchase the expertise they need very quickly through mergers and acquisitions. Some companies execute occasional acquisitions while other companies have sufficient access to capital such that they can perform continuous acquisitions. However, once again, companies often neglect to consider the impact on project management after the acquisition is made. Best practices in project management may not be transferable from one company to another. The impact on project management systems resulting from mergers and acquisitions is often irreversible, whereas joint ventures can be terminated.

Project management often suffers after the actual merger or acquisition. Mergers and acquisitions allow companies to achieve strategic targets at a speed not easily achievable through internal growth, provided the sharing or combining of assets and capabilities can be done quickly and effectively. This synergistic effect can produce opportunities that a firm might be hard-pressed to develop by itself.

Mergers and acquisitions focus on two components: preacquisition decision making and postacquisition integration of processes. Wall Street and financial institutions appear to be interested more in the near-term financial impact of the acquisition rather than the long-term value that can be achieved through combined or better project management and integrated processes. During the mid-1990s, companies rushed into acquisitions in less time than the company required for a capital expenditure approval. Virtually no consideration was given to the impact on project management and on whether project management knowledge and best practices would be transferable. The result appears to have been more failures than successes.

When a firm rushes into an acquisition, often very little time and effort are spent on postacquisition integration. Yet this is where the real impact of the acquisition is felt. Immediately after an acquisition, each firm markets and sells products to each other's customers. This may appease the stockholders, but only in the short term. In the long term, new products and services will need to be developed to satisfy both markets. Without an integrated project management system where both parties can share the same intellectual property and work together, this may be difficult to achieve.

When sufficient time is spent on preacquisition decision making, both firms look at combining processes, sharing resources, transferring intellectual property, and the overall management of combined operations. If these issues are not addressed in the preacquisition phase, then the unrealistic expectations may lead to unwanted results during the postacquisition integration phase.

STRATEGIC TIMING ISSUE

Lenore Industries had been in existence for more than 50 years and served as a strategic supplier of parts to the automobile industry. Lenore's market share was second only to its largest competitor, Belle Manufacturing. Lenore believed that the economic woes of the U.S. automobile industry between 2008 and 2010 would reverse themselves by the middle of the next decade and that strategic opportunities for growth were at hand.

The stock prices of almost all of the automotive suppliers were grossly depressed. Lenore's stock price was also near a 10-year low. But Lenore had rather large cash reserves and believed that the timing was right to make one or more strategic acquisitions before the market place turned around. With this in mind, Lenore decided to purchase its largest competitor, Belle Manufacturing.

PREACQUISITION DECISION MAKING

Senior management at Lenore fully understood that the reason for most acquisitions is to satisfy strategic and/or financial objectives. Table I shows the six reasons identified by senior management at Lenore for the acquisition of Belle Manufacturing and the most likely impact on Lenore's strategic and financial objectives. The strategic objectives are somewhat longer term than the financial objectives, which are under pressure from stockholders and creditors for quick returns.

Lenore's senior management fully understood the long-term benefits of the acquisition, which were:

- Economies of combined operations
- Assured supply or demand for products and services
- Additional intellectual property, which may have been impossible to obtain otherwise
- Direct control over cost, quality, and schedule rather than being at the mercy of a supplier or distributor
- Creation of new products and services
- Putting pressure on competitors by creating synergies
- Cutting costs by eliminating duplicated steps

TABLE I   ACQUISITION OBJECTIVES

Reason for Acquisitions	Strategic Objective	Financial Objective
Increase customer base	Bigger market share	Bigger cash flow
Increase capabilities	Become a business solution provider	Larger profit margins
Increase competitiveness	Eliminate costly steps and redundancy	Stable earnings
Decrease time-to-market for new products	Market leadership	Rapid earnings growth
Decrease time to market for enhancements	Broad product lines	Stable earnings
Closer to customers	Better price–quality–service mix	Sole-source or single-source procurement

Lenore submitted an offer to purchase Belle Manufacturing. After several rounds of negotiations, Belle's board of directors and Belle's stockholders agreed to the acquisition. Three months later, the acquisition was completed.

POSTACQUISITION INTEGRATION

The essential purpose of any merger or acquisition is to create lasting value and value that would not exist had the companies remained separate. The achievement of these benefits, as well as attaining the strategic and financial objectives, could rest on how well the project management value-added chains of both firms are integrated, especially the methodologies within their chains. Unless the methodologies and cultures of both firms can be integrated, and reasonably fast, the objectives may not be achieved as planned.

Lenore's decision to purchase Belle Manufacturing never considered the compatibility of their respective project management approaches. Project management integration failures occurred soon after the acquisition happened. Lenore had established an integration team and asked the integration team for a briefing on what critical issues were preventing successful integration.

The integration team identified five serious problems that were preventing successful integration of their project management approaches:

1. Lenore and Belle have different project management methodologies.
2. Lenore and Belle have different cultures and integration is complex.
3. There are wage and salary disparities.
4. Lenore overestimated the project management capability of Belle's personnel.
5. There are significant differences in functional and project management leadership.

It was now apparent to Lenore that these common failures resulted because the acquisition simply cannot occur without organizational and cultural changes that are often disruptive in nature. Lenore had rushed into the acquisition with lightning speed but with little regard for how the project management value-added chains would be combined.

The first common problem area was inability to combine project management methodologies within the project management value-added chains. This occurred for four reasons:

1. A poor understanding of each other's project management practices prior to the acquisition
2. No clear direction during the preacquisition phase on how the integration would take place

3. Unproven project management leadership in one or both firms
4. The existence of a persistent attitude of "we–them"

Some methodologies may be so complex that a great amount of time is needed for integration to occur, especially if each organization has a different set of clients and different types of projects. As an example, a company developed a project management methodology to provide products and services for large publicly held companies. The company then acquired a small firm that sold exclusively to government agencies. The company realized too late that integration of the methodologies would be almost impossible because of requirements imposed by government agencies for doing business with the government. The methodologies were never integrated and the firm servicing government clients was allowed to function as a subsidiary, with its own specialized products and services. The expected synergy never took place.

Some methodologies simply cannot be integrated. It may be more prudent to allow the organizations to function separately than to miss windows of opportunity in the marketplace. In such cases, pockets of project management may exist as separate entities throughout a large corporation.

Lenore knew that Belle Manufacturing services many clients outside of the United States but did not realize that Belle maintained a different methodology for those clients. Lenore was hoping to establish just one methodology to service all clients.

The second major problem area was the existence of differing cultures. Although project management can be viewed as a series of related processes, it is the working culture of the organization that must eventually execute these processes. Resistance by the corporate culture to effectively support project management can cause the best plans to fail. Sources for the problems with differing cultures include a culture that:

- Has limited project management expertise (i.e., missing competencies) in one or both firms
- Is resistant to change
- Is resistant to technology transfer
- Is resistant to transfer of any type of intellectual property
- Will not allow for a reduction in cycle time
- Will not allow for the elimination of costly steps
- Must reinvent the wheel
- Views project criticism as personal criticism

Integrating two cultures can be equally difficult during favorable and unfavorable economic times. People may resist any changes to their work habits or comfort zones, even though they recognize that the company will benefit by the changes.

Multinational mergers and acquisitions are equally difficult to integrate because of cultural differences. Several years ago, an American automotive supplier acquired a European firm. The American company supported project management vigorously and encouraged its employees to become certified in project management. The European firm provided very little support for project management and discouraged its workers from becoming certified, arguing that its European clients do not regard project management as highly as do General Motors, Ford, and Chrysler. The European subsidiary saw no need for project management. Unable to combine the methodologies, the American parent company slowly replaced the European executives with American executives to drive home the need for a single project management approach across all divisions. It took almost five years for the complete transformation to take place. The American parent company believed that the resistance in the European division was more of a fear of change in its comfort zone than a lack of interest by its European customers.

Planning for cultural integration can also produce favorable results. Most banks grow through mergers and acquisitions. The general practice in the banking industry is to grow or be acquired. One Midwest bank recognized this and developed project management systems that allowed it to acquire other banks and integrate the acquired banks into its culture in less time than other banks allowed for mergers and acquisitions. The company viewed project management as an asset that had a very positive effect on the corporate bottom line. Many banks today have manuals for managing merger and acquisition projects.

The third problem area Lenore discovered was the impact on the wage and salary administration program. The common causes of the problems with wage and salary administration included:

- Fear of downsizing
- Disparity in salaries
- Disparity in responsibilities
- Disparity in career path opportunities
- Differing policies and procedures
- Differing evaluation mechanisms

When a company is acquired and integration of methodologies is necessary, the impact on wage and salary administration can be profound. When an acquisition takes place, people want to know how they will be affected individually, even though they know that the acquisition is in the best interests of the company.

The company being acquired often has the greatest apprehension about being lured into a false sense of security. Acquired organizations can become resentful to the point of trying to subvert the acquirer. This will result in value destruction where self-preservation becomes paramount, often at the expense of project management systems.

Consider the following situation. Company A decides to acquire company B. Company A has a relatively poor project management system, where project management is a part-time activity and not regarded as a profession. Company B, in contrast, promotes project management certification and recognizes the project manager as a full-time, dedicated position. The salary structure for the project managers in Company B was significantly higher than for their counterparts in Company A. The workers in Company B expressed concern that "We don't want to be like them," and self-preservation led to value destruction.

Because of the wage and salary problems, Company A tried to treat Company B as a separate subsidiary. But when the differences became apparent, project managers in Company A tried to migrate to Company B for better recognition and higher pay. Eventually, the pay scale for project managers in Company B became the norm for the integrated organization.

When people are concerned with self-preservation, the short-term impact on the combined value-added project management chain can be severe. Project management employees must have at least the same, if not better, opportunities after acquisition integration as they did prior to the acquisition.

The problem area that the integration team discovered was the overestimation of capabilities after acquisition integration. Included in this category were:

- Missing technical competencies
- Inability to innovate
- Speed of innovation
- Lack of synergy
- Existence of excessive capability
- Inability to integrate best practices

Project managers and those individuals actively involved in the project management value-added chain rarely participate in preacquisition decision making. As a result, decisions are made by managers who may be far removed from the project management value-added chain and whose estimates of postacquisition synergy are overly optimistic.

The president of a relatively large company held a news conference announcing that his company was about to acquire another firm. To appease the financial analysts attending the news conference, he meticulously identified the synergies expected from the combined operations and provided a timeline for new products to appear on the marketplace. This announcement did not sit well with the workforce, who knew that the capabilities were overestimated and the dates were unrealistic. When the product launch dates were missed, the stock price plunged and blame was erroneously placed on the failure of the integrated project management value-added chain.

In this case the problem area identified was leadership failure during postacquisition integration. Included in this category were:

- Leadership failure in managing change
- Leadership failure in combining methodologies
- Leadership failure in project sponsorship
- Overall leadership failure
- Invisible leadership
- Micromanagement leadership
- Believing that mergers and acquisitions must be accompanied by major restructuring

Managed change works significantly better than unmanaged change. Managed change requires strong leadership, especially with personnel experienced in managing change during acquisitions.

Company A acquires Company B. Company B has a reasonably good project management system, but it has significant differences from Company A's system. Company A then decides, "We should manage them like us," and nothing should change. Company A then replaces several Company B managers with experienced Company A managers, a change that took place with little regard for the project management value-added chain in Company B. Employees within the chain in Company B were receiving calls from different people, most of whom were unknown to them and were not told whom to contact when problems arose.

As the leadership problem grew, Company A kept transferring managers back and forth. This resulted in smothering the project management value-added chain with bureaucracy. As expected, performance was diminished rather than enhanced, and the strategic objectives were never attained.

Transferring managers back and forth to enhance vertical interactions is an acceptable practice after an acquisition. However, it should be restricted to the vertical chain of command. In the project management value-added chain, the main communication flow is lateral, not vertical. Adding layers of bureaucracy and replacing experienced chain managers with personnel inexperienced in lateral communications can create severe roadblocks in the performance of the chain.

The integration team then concluded that any of the problem areas, either individually or in combination, could cause the project management value chain to have problem areas, such as:

- Poor deliverables
- Inability to maintain schedules
- Lack of faith in the chain

- Poor morale
- Trial by fire for all new personnel
- High employee turnover
- No transfer of project management intellectual property

Company A now realized that it may have bitten off more than it could chew. The problem was how to correct these issues in the shortest amount of time without sacrificing its objectives for the acquisition.

QUESTIONS

1. Why is it so difficult to get senior management to consider the impact on project management during preacquisition decision making?
2. Are the acquisition objectives in Table I realistic?
3. How much time is really needed to get economies of combined operations?
4. How should Lenore handle differences in the project management approach if Lenore has the better approach?
5. How should Lenore handle differences in the project management approach if Belle has the better approach?
6. How should Lenore handle differences in the project management approach if neither Lenore nor Belle has any project management?
7. How should Lenore handle differences in the culture if Lenore has the better culture?
8. How should Lenore handle differences in the culture if Belle has the better culture?
9. How should Lenore handle differences in the wage and salary administration program?
10. Is it possible to prevent an overoptimistic view of the project management capability of the company being acquired?
11. How should Lenore handle disparities in leadership styles?

Part 2

IMPLEMENTATION OF PROJECT MANAGEMENT

The first step in the implementation of project management is to recognize the true benefits that can be achieved from using project management. These benefits can be recognized at all levels of the organization. However, each part of the organization can focus on a different benefit and want the project management methodology to be designed for its particular benefit.

Another critical issue is that the entire organization may not end up providing the same level of support for project management. This could delay the final implementation of project management. In addition, there may be some pockets within the organization that are primarily project-driven and will give immediate support to project management, whereas other pockets, which are primarily non–project-driven, may be slow in their acceptance.

Kombs Engineering

In June 2013, Kombs Engineering had grown from just a few employees to a company with $250 million in sales. The business base consisted of two contracts with the U.S. Department of Energy (DOE), one for $150 million and one for $80 million. The remaining $20 million consisted of a variety of smaller jobs for $150,000 to $500,000 each. Kombs expected the growth in smaller jobs to exceed $100 million in a few years.

The larger contract with DOE was a five-year contract for $150 million per year. The contract was awarded in 2008 and was up for renewal in 2013. DOE had made it clear that, although it was very pleased with the technical performance of Kombs, the follow-on contract must go through competitive bidding by law. Marketing intelligence indicated that DOE intended to spend $100 million per year for five years on the follow-on contract with a tentative award date of October 2013.

On June 21, 2013, the solicitation for proposal was received at Kombs. The technical requirements of the proposal request were not considered to be a problem for Kombs. There was no question in anyone's mind that on technical merit alone, Kombs would win the contract. The more serious problem was that DOE required a separate section in the proposal on how Kombs would manage the $100 million/year project as well as a complete description of how the project management system at Kombs functioned.

When Kombs won the original bid in 2008, there was no project management requirement. All projects at Kombs were accomplished through the traditional organizational structure. Line managers acted as project leaders.

In July 2013, Kombs hired a consultant to train the entire organization in project management. The consultant also worked closely with the proposal team in responding to the DOE project management requirements. The proposal was submitted to DOE during the second week of August. In September 2013, DOE provided Kombs with a list of questions concerning its proposal. More than 95 percent of the questions involved project management. Kombs responded to all questions.

In October 2013, Kombs received notification that it would not be granted the contract. During a postaward conference, DOE stated that it had no "faith" in the Kombs project management system. Kombs Engineering is no longer in business.

QUESTIONS

1. What was the reason for the loss of the contract?
2. Could it have been averted?
3. Does it seem realistic that proposal evaluation committees could consider project management expertise to be as important as technical ability?

Williams Machine Tool Company

For 85 years, the Williams Machine Tool Company had provided quality products to its clients, becoming the third largest U.S.-based machine tool company by 1990. The company was highly profitable and had an extremely low employee turnover rate. Pay and benefits were excellent.

Between 1980 and 1990, the company's profits soared to record levels. The company's success was due to one product line of standard manufacturing machine tools. Williams spent most of its time and effort looking for ways to improve its bread-and-butter product line rather than to develop new products. The product line was so successful that companies were willing to modify their production lines around these machine tools rather than asking Williams for major modifications to the machine tools.

By 1980, Williams Company was extremely complacent, expecting this phenomenal success with one product line to continue for 20 more years. The recession of the early 1990s forced management to realign its thinking. Cutbacks in production had decreased the demand for the standard machine tools. More and more customers were asking for either major modifications to the standard machine tools or a completely new product design.

The marketplace was changing, and senior management recognized that a new strategic focus was necessary. However, lower-level management and the workforce, especially engineering, were strongly resisting a change. The employees, many of them with over 20 years of employment at Williams Company, refused to recognize the need for this change in the belief that the glory days of yore would return at the end of the recession.

By 1995, the recession had been over for at least two years, yet Williams Company had no new product lines. Revenue was down, sales for the standard product (with and without modifications) were decreasing, and the employees were still resisting change. Layoffs were imminent.

In 1996, the company was sold to Crock Engineering. Crock had an experienced machine tool division of its own and understood the machine tool business. Williams Company was allowed to operate as a separate entity from 1995 to 1996. By 1996, red ink had appeared on the Williams Company balance sheet. Crock replaced all of the Williams senior managers with its own personnel. Crock then announced to all employees that Williams would become a specialty machine tool manufacturer and that the "good old days" would never return. Customer demand for specialty products had increased threefold in just the last 12 months alone. Crock made it clear that employees who would not support this new direction would be replaced.

The new senior management at Williams Company recognized that 85 years of traditional management had come to an end for a company now committed to specialty products. The company culture was about to change, spearheaded by project management, concurrent engineering, and total quality management.

Senior management's commitment to product management was evident from the time and money spent in educating the employees. Unfortunately, the seasoned 20-year-plus veterans still would not support the new culture. Recognizing the problems, management provided continuous and visible support for project management, in addition to hiring a project management consultant to work with the people. The consultant worked with Williams from 1996 to 2001. From 1996 to 2001, the Williams Division of Crock Engineering experienced losses in 24 consecutive quarters. The quarter ending March 31, 2002, was the first profitable quarter in over six years. Much of the credit was given to the performance and maturity of the project management system. In May 2002, the Williams Division was sold. More than 80 percent of the employees lost their jobs when the company was relocated over 1,500 miles away.

QUESTIONS

1. Could project management have been used for those projects involving modifications to the standard machine tool line?
2. How should you handle experienced employees who live in the past and refuse to accept project management?
3. Who should be blamed for the failure of Williams Company to react quickly to changes in the marketplace?

4. Is project management the real reason why the company was sold and employees lost their jobs?

5. Should Crock Engineering have held onto the Williams Company? If so, how?

6. Could there have been a hidden agenda behind the sale of Williams Company (such as poor prognosis for the future of the machine tool industry)?

The Reluctant
Workers

Tim Aston had changed employers three months earlier. His new position was project manager. At first he had stars in his eyes about becoming the best project manager his company had ever seen. Now he wasn't sure if project management was worth the effort. He made an appointment to see Phil Davies, director of project management.

Tim Aston: "Phil, I'm a little unhappy about the way things are going. I just can't seem to motivate my people. Every day, at 4:30 p.m., all of my people clean off their desks and go home. I've had people walk out of late-afternoon team meetings because they were afraid that they'd miss their car pool. I have to schedule morning team meetings."

Phil Davies: "Look, Tim. You're going to have to realize that in a project environment, people think that they come first and the project is second. This is a way of life in our organizational form."

Tim Aston: "I've continually asked my people to come to me if they have problems. I find that the people do not think that they need help and, therefore, do not want it. I just can't get my people to communicate more."

Phil Davies: "The average age of our employees is about 46. Most of our people have been here for 20 years. They're set in their ways. You're the first person that we've hired in the past three years. Some of our people may just resent seeing a 30-year-old project manager."

Tim Aston: "I found one guy in the accounting department who has an excellent head on his shoulders. He's very interested in project management. I asked his boss if he'd release him for a position in project management, and his boss just laughed at me, saying something to the effect that as long as that guy is doing a good job for him, he'll never be released for an assignment elsewhere in the company. His boss seems more worried about his personal empire than he does about what's best for the company.

"We had a test scheduled for last week. The customer's top management was planning on flying in for firsthand observations. Two of my people said that they had scheduled vacation days coming and that they would not change under any conditions. One guy was going fishing and the other guy was planning to spend a few days working with fatherless children in our community. Surely these guys could change their plans for the test."

Phil Davies: "Many of our people have social responsibilities and outside interests. We encourage social responsibilities and only hope that the outside interests do not interfere with their jobs.

"There's one thing you should understand about our people. With an average age of 46, many of our people are at the top of their pay grades and have no place to go. They must look elsewhere for interests. These are the people you have to work with and motivate. Perhaps you should do some reading on human behavior."

QUESTIONS

1. How has Tim Aston handled these situations?
2. Can the company help Tim?
3. What are your suggested solutions?

Macon, Inc.

Macon was a 50-year-old company in the business of developing test equipment for the tire industry. The company had a history of segregated departments with very focused functional line managers. The company had two major technical departments: mechanical engineering and electrical engineering. Both departments reported to a vice president for engineering, whose background was always mechanical engineering. For this reason, the company focused all projects from a mechanical engineering perspective. The significance of the test equipment's electrical control system was often minimized when, in reality, the electrical control systems were what made Macon's equipment outperform that of the competition.

Because of the strong autonomy of the departments, internal competition existed. Line managers were frequently competing with one another rather than focusing on the best interest of Macon. Each would hope the other would be the cause for project delays instead of working together to avoid project delays altogether. Once dates slipped, fingers were pointed and the problem would worsen over time.

One of Macon's customers had a service department that always blamed engineering for all of its problems. If the machine was not assembled correctly, it was engineering's fault for not documenting it clearly enough. If a component failed, it was engineering's fault for not designing it correctly. No matter what problem occurred in the field, customer service would always put the blame on engineering.

As might be expected, engineering would blame most problems on production, claiming that production did not assemble the equipment correctly and did not maintain the proper level of quality. Engineering would design a product and then throw it over the fence to production without ever going down to the manufacturing floor to help with its assembly. Errors or suggestions reported from production to engineering were being ignored. Engineers often perceived the assemblers as incapable of improving the design.

Production ultimately assembled the product and shipped it out to the customer. Oftentimes during assembly, the production people would change the design as they saw fit without involving engineering. This would cause severe problems with documentation. Customer service would later inform engineering that the documentation was incorrect, once again causing conflict among all departments.

The president of Macon was a strong believer in project management. Unfortunately, his preaching fell upon deaf ears. The culture was just too strong. Projects were failing miserably. Some failures were attributed to the lack of sponsorship or commitment from line managers. One project failed as the result of a project leader who failed to control scope. Each day the project would fall farther behind because work was being added with very little regard for the project's completion date. Project estimates were based on a gut feel rather than on sound quantitative data.

The delay in shipping dates was creating more and more frustration for the customers. The customers began assigning their own project managers as watchdogs to look out for their companies' best interests. The primary function of these watchdog project managers was to ensure that the equipment purchased would be delivered on time and complete. This involvement by the customers was becoming more prominent than ever before.

The president decided that action was needed to achieve some degree of excellence in project management. The question was what action to take, and when.

QUESTIONS

1. Where will the greatest resistance for excellence in project management come from?
2. What plan should be developed for achieving excellence in project management?
3. How long will it take to achieve some degree of excellence?
4. Explain the potential risks to Macon if the customer's experience with project management increases while Macon's knowledge remains stagnant.

Cordova Research Group

Cordova Research Group spent more than 30 years conducting pure and applied research for a variety of external customers. With the reduction, however, in research and development (R&D) funding, Cordova decided that the survival of the firm would be based on becoming a manufacturing firm as well as performing R&D. The R&D culture was close to informal project management with the majority of the personnel holding advanced degrees in technical disciplines. To enter the manufacturing arena would require hiring hundreds of new employees, mostly non-degreed.

QUESTIONS

1. What strategic problems must be solved?
2. What project management problems must be solved?
3. What time frame is reasonable?
4. If excellence can be achieved, would it more likely be achieved using formal or informal project management?

Cortez Plastics

Cortez Plastics was having growing pains. As the business base of the company began to increase, more and more paperwork began to flow through the organization. The "informal" project management culture that had worked so well in the past was beginning to deteriorate and was being replaced by a more formal project management approach. Recognizing the cost implications of a more formal project management approach, senior management at Cortez Plastics decided to take some action.

QUESTIONS

1. How can a company maintain informal project management during periods of corporate growth?
2. If the organization persists in creeping toward formal project management, what can be done to return to a more informal approach?
3. How would you handle a situation where only a few managers or employees are promoting the more formal approach?

The Enterprise Resource Planning Project

"What have I gotten myself into?" remarked Jerry as he looked at himself in the mirror. "I must have been crazy to volunteer for this project. Although I consider myself a good project manager, having been in project management for more than 25 years, I know very little about how to recover a failing project. This may be more than I can handle. I certainly do not want to end up in the hospital like the two previous project managers!"

BACKGROUND

Most of Jerry's project management career had been working for Mannix Corporation, a company that provides information technology (IT) business solutions to companies around the world. In the past 10 years, Mannix had developed expertise in enterprise resource planning (ERP) systems. ERP is an enterprisewide information system designed to coordinate all resources, information, and activities needed to complete business solutions. ERP generally focuses on a single database that is common to all departments. Information can be stored and retrieved on a real-time basis. However, some companies maintain an ERP in modules. The modular software design means that a business can select the modules as needed, mix and match modules from different vendors, and add new modules of its own to improve business performance.

More than a year ago, Mannix Corporation won a contract from the Prylon Company to create and install an ERP system using a single database. Prylon did not have a complete ERP system but had various modules that were purchased from a variety of vendors. Prylon tried desperately on its own to coordinate all modules into one database but failed. Prylon then hired Mannix Corporation to try to unite all of the modules into one package. If that did not work, then Mannix Corporation would have the right to remove the modular design and start over with a single database design concept.

The first project manager failed terribly. Several of Prylon's business systems were shut down temporarily while the project manager tried to coordinate the modules. Functional managers were furious that they did not have access to the business systems they needed, and the daily operations of Prylon's business suffered. Prylon asked Mannix Corporation to remove the first project manager and replace him.

The second project manager who was assigned was like a bull in a china shop. He did not understand Prylon's business, refused to learn and understand Prylon's business needs and culture, and made unrealistic demands on Prylon for additional support. The second project manager alienated senior management at Prylon to the point where they were willing to cancel the contract with Mannix Corporation and go out for competitive bidding again.

EXECUTIVE DECISION

Executives from Mannix Corporation met with senior management at Prylon Corporation and asked for one more chance. The original contract schedule was 18 months. It was now one year into the project, and it appeared that at least one more year would be needed to finish the job. Prylon did not want to let Mannix continue on with the contract. But going out for competitive bidding again and having another contractor come back with an 18-month schedule would mean that the ERP system would not be operational for at least another two years. If Mannix Corporation could succeed, the ERP system could be operational in less than a year. Mannix Corporation was given a third chance but was told that a new project manager must be assigned.

Immediately after receiving the news that the project would continue, Jerry was asked to attend a meeting with senior management. One of the executives spoke up:

> Jerry, I guess you know about the problems we are having with Prylon Corporation. They have been one of our best clients over the years and we do not want to lose their business.
>
> You are one of our best project managers and I am asking you to volunteer to become the third project manager and finish the project successfully. The choice is yours.
>
> The first two project managers never looked at the early warning signs indicating that the project was getting into trouble. Projects do not go from "green" to "red" overnight. The early warning signs were either misunderstood or overlooked. In either event, we have a displeased client. We need a project manager who can reverse a possibly failing project. I know you have never been asked before today to take over a distressed project.

Jerry thought about it for a couple of minutes and then agreed to become the next project manager. One of the executives then stated:

> When a project gets way off track, the cost of recovery is huge and vast resources are often required for corrections. We cannot give you any more resources and the contract is a firm-fixed-price effort. We will have to absorb the cost overruns.
>
> I expect that some of the requirements will have to change during recovery. The ultimate goal of a recovery project is not to finish on time, but to finish with reasonable value and benefits for Prylon Corporation. The longer you wait to make the necessary repairs on the contract, the more costly the repairs will be.
>
> As I see it, your biggest challenge will be the team. You cannot recover a distressed project in isolation. You need the team, and their morale is quite low at the moment. The team has been through two project managers already. Not all project managers have the ability to recover a failing project. But I think you can handle it.

MEETING THE TEAM

Jerry understood that his first concern had to be the morale of his team. He knew many of the team members personally through socialization and having worked with them on previous projects. At the meeting, the team stated that they felt that they were on a death spiral. The previous project managers had created unnecessary additional work, causing the team to work excessive hours on overtime. This placed increased stress and pressure on the team. Several team members were replaced, but at inopportune times. A consultant was hired to support the team, and they felt that it made matters worse.

It was pretty obvious now what Jerry had inherited:

- A burned-out team
- An emotionally drained team
- A team with poor morale
- An exodus of the talented team members who might be in high demand elsewhere
- A team with a lack of faith in the recovery process
- Furious customers
- Nervous management
- Invisible sponsorship
- Either invisible or highly active stakeholders

Jerry told the team that there were six life-cycle phases that must be accomplished to recover the distressed project. Jerry drew Figure I on the board and said that this would be his approach.

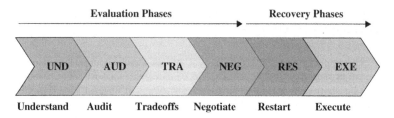

FIGURE I   Recovery life-cycle phases

He also told the team that all overtime was canceled and that they were not to work on this project for a few days until he reviewed the project and all of the facts. He said that they should reestablish their work-life balance and that this project was not the end of the world. He also said that he would reestablish an incentive program aligned with the successful completion of the project. Jerry knew there was a risk in asking team members not to work on the project for a few days because they might find a home on another project. But he felt comfortable about the first meeting with the team and believed that they would help him recover the project.

UNDERSTANDING PHASE

Jerry collected all of the files, reports, memos, and letters that were part of the project. He reviewed the history of the project and had meetings with those senior managers at Mannix who had information on Prylon Corporation. He reviewed the business case for the project, the expected benefits, the assumptions, and the project's objectives.

He also had to evaluate the enterprise environmental factors to see if they were still valid. To do this, he would need to talk to people at Prylon Corporation. This would be essential. They would have to get to know him and trust him, and he would have to understand their needs and sensitivities.

The meeting with Prylon personnel went well. Prylon still wanted the entire ERP system as promised and was willing to accept the fact that the project would probably be six months late. However, Prylon still wanted to see the recover plan and what trade-offs, if any, needed to be made before agreeing to a continuation of the project.

AUDIT PHASE

Having completed the understanding phase, it was now time to reconvene the team and begin the audit phase. First, Jerry informed the team of his meeting with Prylon and stated that:

- The project is still considered to be of value to Prylon.
- The project is still aligned to Prylon's strategy.
- Mannix Corporation is still committed to completing the project successfully for Prylon.

- All of the stakeholders are still committed but want to see the final recovery plan.
- Both Prylon and Mannix are motivated toward the rescue of this project.

The next item on Jerry's agenda during the audit phase was to improve morale. Jerry had already asked the team to stop working on the project for a few days and rest up. The next step was to make team members aware of Jerry's desire to listen to their concerns by allowing them to vent their issues. First, Jerry asked the team to look at the good things that happened on the project. The intent was to build morale.

Jerry then asked three questions:

1. Was the original plan overly optimistic?
2. Were there political problems that led to active or passive resistance by the team?
3. Were the work hours and workloads demoralizing?

The answers came quickly and to the point: The plan was overly optimistic; the requirements package was incomplete, resulting in numerous changes; and the previous two project managers assumed that the client was always right and agreed to all of the changes, thus resulting in increased workloads. The team felt that many of the changes were not necessary. To make matters worse, political infighting at Prylon impacted the project team. Both senior and middle managers at Prylon were interfacing directly with the Mannix project team members and asking them to do things that were not part of the original statement of work.

Jerry told the team that he would personally insulate them from interference from Prylon. From this point forth, any and all interfacing, questions, requests, or scope changes by Prylon personnel had to go through Jerry. The team seemed quite pleased with this change of events.

The next step in the audit phase would be most important. The team had to critically assess performance to date. As part of the audit, the team performed a root-cause analysis to identify the surface and hidden failure points.

Once the failure points were identified, the team had to determine what could be done within the original time frame established by the contract and what could be done if the project were allowed to slip by six months. The team listed all of the critical deliverables and beside each one indicated what was "a must-have," "nice to have," "can wait," and "not needed." The information was then drawn on the board. (See Figure II.) The team was told simply to list the issues on Figure II but not to analyze them yet.

TRADE-OFF PHASE

With all of the issues now listed, Jerry asked the team to see where trade-offs could be made. Jerry wrote the following questions on the board beside Figure II:

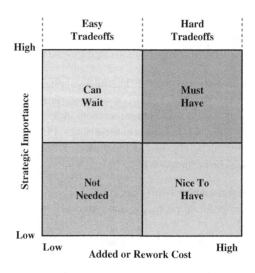

FIGURE II   Trade-off categories

- Where are the trade-offs?
- What are the expected casualties?
- What can and cannot be done?
- What must be fixed first?
- Can we stop the bleeding?
- Have the priorities of the competing constraints changed?
- Have the features changed?
- What are the risks?

Jerry and the team then reviewed all of the opportunities for trade-offs and came up with a recommended game plan to be presented to senior management at Prylon.

NEGOTIATION PHASE

The team came up with a recovery plan, including various options. Now Jerry had to present the recovery plan to Prylon. Although it was common practice to ask some of the team members to accompany the project manager, Jerry decided to do it alone, in keeping with his promise of insulating the team from management at Prylon. Jerry knew there might be questions he could not answer but believed this to be the best approach.

Jerry started his presentation with an explanation of why he was there alone and that he would now be the only go-between from Prylon and Mannix. He explained his reason for doing this and that this was the only way he could have confidence in the execution of the project according to the proposed new schedule. Jerry presented the team's recommended recovery plan and the various options based on what would be most important to Prylon right now—for example, time, cost value, and scope. Jerry appeared quite honest in his beliefs for recovery and continuously asserted that he was not giving them any unrealistic expectations. The project would be no more than six months late, and he would do everything possible to accelerate the schedule. Jerry also asserted that he needed effective governance from Prylon to make this work, and he was now asking for their buy-in for the recovery plan.

Jerry had expected Prylon to ask for a few days for them to discuss his recovery plan, but much to his surprise, they took an immediate vote with Jerry present in the board room and gave him the authorization to proceed. They were a little unhappy that he was severing the interfacing between Prylon and the Mannix team and that everything had to go through him, but they understood his reasons for doing this and accepted his approach.

RESTART PHASE

Returning to Mannix Corporation, Jerry called a meeting of the team to discuss the good news and the small changes that Prylon made to the recovery plan. Jerry knew that there were three options to restart a failing project:

1. Full anesthetic: Bring all work to a standstill until a recovery plan is finalized.
2. Partial anesthetic: Bring some work to a standstill until scope is stabilized.
3. Scope modification: Continue work but with modifications as necessary.

Prylon's rapid approval of the team's recovery plan made the third option a reality. Jerry commended the team for doing a good job. When projects get into trouble, it is customary to bring new team members on board with new ideas. However, Jerry felt reasonably comfortable with the assigned team members.

EXECUTION PHASE

Execution was now under way. Jerry prepared a memo and sent it out to all team members. The memo stated Jerry's expectations for recovery and included the following points:

- We must learn from past mistakes; making the same mistake twice is unacceptable.
- We must stabilize scope.
- We must rigidly enforce the scope change control process.

- It may be necessary to perform critical health checks.
- Effective communication is essential.
- We must maintain positive morale.
- We must adopt proactive stakeholder management and I will be responsible for this.
- Do not rely on the company's project management methodology system to save us; the team, not the methodology, is responsible for the recovery.
- Do not allow unwanted stakeholder intervention, which increases pressure.
- I will carefully manage stakeholder expectations.
- I will try to insulate the team from politics.

QUESTIONS

1. Why did Prylon give Mannix Corporation a third chance?
2. Do projects go from green to red overnight? If they do, then what is the most likely cause?
3. Should a firm-fixed-price contract have been awarded from the ERP effort?
4. Is it reasonable to expect that requirements will change during recovery?
5. What is the ultimate goal of a recovery project?
6. Do stakeholders expect trade-offs during recovery?
7. What generally happens to constraints such as time and cost during recovery?
8. Why was morale low when Jerry first took over the project?
9. What are the characteristics of a death spiral on a failing project?
10. What was Jerry's intent in canceling overtime and asking the team to stop working on the project for a few days?
11. What were the risks in Question 10?
12. As identified in the case, what were the life-cycle phases for recovery, and what is accomplished in each phase?
13. Suppose that during the audit phase, Jerry discovered that one of the team members, who was a close friend of his, was the cause of most of the issues. How should Jerry handle the situation?
14. What should Jerry do during the negotiation phase if Prylon Corporation comes up with its own recovery plan and the plan is unacceptable to Mannix?

The Prioritization
of Projects

BACKGROUND

The directorates of Engineering, Marketing, Manufacturing, and R&D all had projects that they were working on, and each directorate established its own priorities for the projects. The problem was that the employees were working on multiple projects and had to deal with competing priorities.

PRIORITIZATION ISSUES

Lynx Manufacturing was a low-cost producer of cables and wires. The industry itself was considered a low-technology industry, and some of its products had been manufactured the same way for decades. There were some projects to improve the manufacturing processes, but they were few and far between.

Each of the four directorates—Engineering, Marketing, Manufacturing, and R&D—had projects, but the projects were generally quite small and used resources from only each individual directorate.

By the turn of the twenty-first century, manufacturing technologies began to grow. Lynx had to prepare for the technology revolution that was about to impact its business. Each directorate began preparing lists of projects that it would need to work on, and some lists contained as many as 200 projects. These projects were more complex than projects worked on previously, and project team members from all directorates were assigned on either a full-time or part-time basis.

Each directorate chief officer would establish the priorities for the projects originating in his or her directorate even though the projects required resources from other directorates. This created significant staffing issues and numerous conflicts:

- Each directorate would hoard its best project resources even though some projects outside of the directorate were deemed more important to the overall success of the company.
- Each directorate would put out fires by using people who were assigned to projects outside of its directorate rather than using people who were working on internal projects.
- Each directorate seemed to have little concern about any projects done in other directorates.
- Project priorities within each directorate could change on a daily basis because of the personal whims of the chief of that directorate.
- The only costs and schedules that were important were those related to projects that originated within the directorate.
- Senior management at the corporate level refused to get involved in the resolution of conflicts between directorates.

The working relationships between the directorates deteriorated to the point where senior management reluctantly agreed to step in. The total number of projects that the four directorates wanted to complete over the next few years exceeded 350, most of which required a team with members coming from more than one division.

QUESTIONS

1. Why is it necessary for senior management to step in rather than let the chiefs of the directorates handle the conflicts?
2. What should the senior management team do to resolve the problem?
3. Let's assume that the decision was to create a list that included all of the projects from the four directorates. How many of the projects on the list should have a priority number or priority code?
4. Can the directorate chiefs assign the priority or must it be done with the involvement of senior management?
5. How often should the list of prioritized projects be reviewed, and who should be in attendance at the review meetings?
6. Suppose that some of the directorate chiefs refuse to assign resources according to the prioritized list and still remain focused on their own pet projects. How should this issue now be resolved?

Selling Executives on Project Management

BACKGROUND

The executives at Levon Corporation watched as their revenue stream diminished and refused to listen to their own employees who were arguing that project management implementation was necessary for growth. Finally, the executives agreed to listen to a presentation by a project management consultant.

NEED FOR PROJECT MANAGEMENT

Levon Corporation had been reasonably successful for almost 20 years as an electronics component manufacturer. The company was a hybrid between project-driven and non–project-driven businesses. A large portion of its business came from development of customized products for government agencies and private-sector companies around the world.

The customized or project-driven portion of the business was beginning to erode. Even though Levon's reputation was good, the majority of these contracts were awarded through competitive bidding. Every customer's request for proposal asked for a section on the contractor's project management capability. Levon had no real project management capability. Since most of the contracts were awarded on points rather than going to the lowest bidder, Levon was constantly downgraded in the evaluation of the proposals because it had no project management capability.

The sales and marketing personnel continuously expressed their concerns to senior management, but the concerns fell on deaf ears. Management was afraid that support of project management could result in a shift in the balance of power in the company. Also, whatever executive ended up with control of the project management function could become more powerful than the other executives.

GAP ANALYSIS

Reluctantly, the executives agreed to hire a project management consultant. The consultant was asked to identify the gaps between Levon and the rest of the industry and to show how project management could benefit the company. The consultant was also asked to identify the responsibilities of senior management once project management is implemented.

After a few weeks of research, the consultant was ready to make his presentation before the senior staff. The first slide that the consultant presented was Figure I, which showed that Levon's revenue stream was not as good as managers thought. Levon was certainly lagging the industry average, and distance between Levon and the industry leader was getting larger.

The consultant then showed Figure II. The consultant had developed a project management maturity factor based on such elements as time, cost, meeting scope, ability to handle risks, providing quality products, and customer interfacing and reporting. Using the project management maturity factor, the consultant showed that Levon's understanding and use of project management were lagging the industry trend.

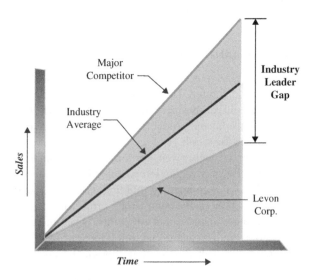

FIGURE I Levon's gap analysis

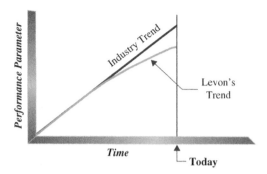

FIGURE II   Project management performance trend

The consultant then showed Figure III, which clearly illustrated that, unless Levon takes decisive action to improve its project management capability, the gap will certainly increase. The executives seemed to understand this, but the consultant could still see their apprehension in supporting project management.

QUESTIONS

1. Why did the executives refuse to listen to their own employees but were willing to listen to a consultant?
2. Was the consultant correct in beginning the presentation by showing the gap between Levon and the rest of the industry?
3. Why did the executives still seem apprehensive even after the consultant's presentation?
4. What should the consultant say next to get the executives to understand and support project management?

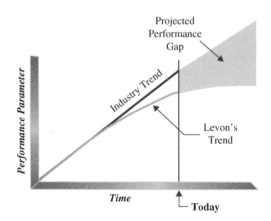

FIGURE III   Increasing performance gap

The New CIO

BACKGROUND

Rose Industries was a manufacturer of electrical products for the home. A large portion of its business base was devoted to the design, development, and manufacturing of specialized electronic components for public and private-sector clients.

Ralph Williams had been with Rose Industries for more than 45 years, beginning in the mail room and working himself up to president and chief executive officer (CEO) of Rose. He was now beginning his tenth year as president and CEO.

Rose Industries believed in inbreeding. All promotions were from within the ranks. Rose often had trouble attracting talented people, especially people with MBA degrees, because the company's conservative policy dictated that all new employees begin at the bottom of the company and work their way up. Every senior manager at Rose had been with the company for at least 30 years. Rose Industries discouraged employees from taking outside seminars and courses. If you wanted to attend a conference or symposium, the policy was "Take vacation and pay your own way." There were several training programs available to the workforce, but they were all taught by internal personnel and covered only the skills needed to do each job more effectively or to become qualified for a promotion to the next pay grade. Each employee was allowed a maximum of seven days off a year to attend internal training programs.

The company did not have any tuition reimbursement policy. Numerous colleges and universities in the surrounding area provided a variety of evening programs leading to various certifications as well as undergraduate and graduate degrees, but employees had to pay all expenses out of pocket. In order to satisfy some of the needs of the employees at Rose Industries, many of the professional societies in the surrounding area held conferences, symposiums, and professional meetings on weekends rather than weekdays.

A TIME FOR CHANGE

By 2003, the ultraconservative nature of Rose Industries began to affect growth. Rose was falling farther behind its competitors, and gross sales were declining rather than increasing. Although Rose was considered a low-cost manufacturer, it was losing business to some higher-priced competitors whose advertising campaigns attacked Rose's weak project management capability. For years, Rose could not see any value in using project management and seemed to discourage its personnel from becoming a Project Management Professional (PMP)®. Project management was never identified as one of Rose's strengths in its proposals during competitive bidding. Rose did use a few templates when managing projects, but no formal project management methodology existed.

Rose's information systems were somewhat outdated. When software was needed, especially for more sophisticated business requirements, Rose would look for off-the-shelf products even though the products were not 100 percent applicable to or did not satisfy all of the company's needs. Rose did maintain an Information Technology Department, which would create software for smaller requirements, but no systems development methodology was used.

HIRING THE CIO

Ralph Williams, president and CEO of Rose Industries, understood quite well the seriousness of the situation. The company must become good at project management, improve its information system, and develop methodologies for both project management and information systems. Mr. Williams decided to break with tradition and hire a new chief information officer (CIO) from outside the company.

After an extensive search and interviewing process, the company hired John Green, a 20-year veteran with one of the largest IT consulting companies in the world. There was no question about John's credentials and what he could bring to Rose Industries. The real issue was if and when he would be able to change the culture to accept his new ideas. Rose Industries had had the same culture for decades, and getting the seasoned veterans at the company to accept change would be difficult. John was told about the challenges before he was hired, and he felt that he could adequately handle the situation.

A BULL IN A CHINA SHOP

During the first two weeks on the job, John interviewed personnel from all levels of the organization to ascertain how difficult it would be to change the culture. The situation was worse than he had been led to believe.

John knew from decades of experience in project management that there are four characteristics of an effective project management culture: communications, cooperation, teamwork, and trust. The interviews made it quite apparent that there was no project management within the company and senior management had been reluctant to initiate improvements. Communications were quite poor because of the lack of a good information system. Cooperation and teamwork occurred only if people felt that they could benefit personally. There was more mistrust than trust. The building blocks for effective project management simply were not there.

John originally thought that he could make the necessary changes within two years. Now, after the interviews, it looked like five years would be closer to the truth. If this five-year time frame were allowed to happen, the health of Rose Industries could significantly degrade during that time period.

John came up with a four-step approach for implementing change:

Step 1: Hire several PMP credential holders quickly.

Step 2: Create two project management offices (PMOs); one would function as an IT PMO, and the other one would be a corporate or strategic PMO.

Step 3: The IT PMO would create an IT systems development methodology suitable for Rose Industries

Step 4: The corporate PMO would create an enterprise project management methodology for all of Rose's projects except the project in IT. The PMO would also participate in the portfolio selection of projects, strategic planning for project management, and project risk management activities.

John believed that this approach could accelerate the maturity in project management and some good practices could be in place in about two years. All he needed now was buy-in from the executive staff for his plan.

PRESENTATION OF THE PLAN

At the next executive staff meeting, John presented his plan. The responses were not what John had hoped for. The other executives in the room immediately attacked step 1, arguing that they had no intention of hiring additional people. John would have to get some of the existing labor pool personnel trained in project management, and there would be limited funding available to do this.

When step 2 was addressed, the executives argued that creating two PMOs would be the same as adding layers of management on top of the existing organizational structure. They simply could not see the need or value in having PMOs and viewed them as possible threats to their power and authority.

When step 3 was discussed, there were several questions as to why Rose Industries had to develop its own systems development methodology when several packages were commercially available. Some executives seemed to have no idea what a systems development methodology was or why it was needed at all. When step 4 was discussed, the executives became furious that John was recommending that someone other than the executive levels of management participate in the portfolio selection of projects, especially project managers who were not even on a management career path ladder. The portfolio selection of projects was obviously seen as a job done entirely by executives. Likewise, allowing anyone other than executives to be involved in strategic planning and risk management was as a serious threat to some executives who perceived that this could impact their power, authority, and bonuses.

John now saw quite clearly what he was up against and that all of the executive support that he was led to believe would be forthcoming would not happen. There was no way he could implement the necessary changes by himself, at least not in a reasonable time period.

Within two weeks after the meeting, John turned in his resignation. John believed that Rose Industries was doomed to failure, and John did not want his reputation to be tarnished by working for a company that failed.

QUESTIONS

1. Why was it so difficult for Rose Industries to implement project management prior to John Green coming on board?
2. Can inbreeding be detrimental to project management maturity?
3. Looking at Rose's current level of project management maturity, which is not much, how long might it take the company to see some reasonable project management maturity assuming John was not there?
4. Is it possible for an executive, or for anyone else for that matter, to determine the true challenges of the job at hand during hiring interviews? What questions, if any, should be asked?
5. How does one know during the job interview process if the promises made for support will be kept?
6. Was John correct in his four components of a good project management culture?
7. Was he too optimistic with his four-step approach?
8. Why were the other executives threatened by his four-step approach?
9. If John had decided to remain with Rose Industries, how might he change his four-step approach, given the responses by the other executives?
10. Was John correct in resigning from the company?
11. What is your prognosis on Rose's chances of remaining in business?

The Invisible Sponsor

BACKGROUND

Some executives prefer to micromanage projects whereas other executives are fearful of making a decision because, if they were to make the wrong decision, it could impact their career. In this case study, the president of the company assigned one of the vice presidents to act as the project sponsor on a project designed to build tooling for a client. The sponsor, however, was reluctant to make any decisions.

ASSIGNING THE VP

Moreland Company was well respected as a tooling design-and-build company. Moreland was project-driven because all of its income came from projects. Moreland was also reasonably mature in project management.

When the previous vice president (VP) for engineering retired, Moreland hired an executive from a manufacturing company to replace him. The new VP for engineering, Al Zink, had excellent engineering knowledge about tooling but had worked for companies that were not project-driven. Al had very little knowledge about project management and had never functioned as a project sponsor. Because of Al's lack of experience as a sponsor, the president decided that Al should get his feet wet as quickly as possible and assigned him as the project sponsor on a medium- size project. The project manager on this project was Fred Cutler. Fred was an engineer with more than 20 years of experience in tooling design and manufacturing. Fred reported directly to Al Zink administratively.

FRED'S DILEMMA

Fred understood the situation; he would have to train Al Zink on how to function as a project sponsor. This was a new experience for Fred because subordinates usually do not train senior personnel on how to do their job. Would Al Zink be receptive?

Fred explained the role of the sponsor and how there are certain project documents that require the signatures of both the project manager and the project sponsor. Everything seemed to be going well until Fred informed Al that the project sponsor is the person that the president eventually holds accountable for the success or failure of the project. Fred could tell that Al was quite upset over this statement. Al realized that the failure of a project where he was the sponsor could damage his reputation and career. Al was now uncomfortable about having to act as a sponsor but knew that he might eventually be assigned as a sponsor on other projects. Al also knew that this project was somewhat of a high risk. If he could function as an invisible sponsor, he could avoid making any critical decisions.

In the first meeting between Fred and Al where Al was the sponsor, Al asked Fred for a copy of the schedule for the project. Fred responded: "I'm working on the schedule right now. I cannot finish the schedule until you tell me whether you want me to lay out the schedule based upon best time, least cost, or least risk."

Al stated that he would think about it and get back to Fred as soon as possible. During the middle of the next week, Fred and Al met in the company's cafeteria. Al asked Fred again, "How is the schedule coming along?" and Fred responded as before: "I cannot finish the schedule until you tell me whether you want me to lay out the schedule based on best time, least cost, or least risk."

Al, furious, turned around and walked away from Fred. Fred was now getting nervous about how upset Al was and began worrying if Al might remove him as the project manager. But Fred decided to hold his ground and get Al to make a decision.

At the weekly sponsor meeting between Fred and Al, once again Al asked the same question, and once again Fred gave the same response as before. Al now became quite angry and yelled out: "Just give me a least time schedule."

Fred had gotten Al to make his first decision. Fred finalized his schedule and had it on Al's desk two days later awaiting Al's signature. Once again, Al procrastinated and refused to sign off on the schedule. Al believed that, if he delayed making the decision, Fred would take the initiative and begin working on the schedule without Al's signature.

Fred kept sending e-mails to Al asking when he intended to sign off on the schedule or, if something was not correct, what changes needed to be made. As expected, Al did not respond. Fred then decided that he had to pressure Al one way or another into making timely decisions as the project sponsor. Fred then sent Al this e-mail:

I sent you the project schedule last week. If the schedule is not signed by this Friday, there could be an impact on the end date of the project. If I do not hear from you, one way or another, by this Friday, I will assume you approve the schedule and I can begin implementation.

The president's e-mail address was also included in the CC location on the e-mail. The next morning, Fred found the schedule on his desk, signed by Al Zink.

QUESTIONS

1. Why do some executives refuse to function as project sponsors?
2. Can an executive be "forced" to function as a sponsor?
3. Is it right for the sponsor to be the ultimate person responsible for the success or failure of the project?
4. Were Al Zink's actions those of someone trying to be an invisible sponsor?
5. Did Fred Cutler act appropriately in trying to get Al Zink to act as a sponsor?
6. What is your best guess as to what happened to the working relationship between Al Zink and Fred Cutler?

The Trade-off
Decision (A)

Overview

Ellen's company, Wayne Corporation, was reasonably mature in project management. They maintained a PMO that was staffed with experienced project managers and kept up with the latest developments in project management.

In the past, senior management wanted all strategic projects to be managed by functional managers rather than project managers. Only the projects defined as operation projects were managed by project managers. But now senior managers had changed their minds and allowed project managers to manage the long-term strategic projects.

Wayne Corporation was a marketing-driven company. Project approvals were made by senior management; project managers did not participate in the approval process. Instead, project managers were brought on board after the project was approved, a schedule was established, and a cost baseline provided. Whatever assumptions and constraints were made in the project approval process were considered as gospel and never to be challenged. Many critical assumptions and other information were withheld from project managers, thus making their jobs more difficult.

Competing Constraints

For years, Ellen managed primarily operational projects that were much shorter in duration than strategic projects. On these operational projects, only three constraints were considered; time, cost, and scope. Whenever a trade-off was necessary, executives understood that they could maintain only two of three constraints and therefore found it relatively easy to make trade-off decisions because of the limited choices.

All of this was about to change. With the latest developments in project management, especially metric measurement techniques, Wayne's PMO promoted the use of competing constraints. There could be as many as 10 constraints on projects now, especially the strategic projects that were aligned to strategic objectives and impacted by changes in the enterprise environmental factors. Trade-offs would now be more complex.

Ellen's Dilemma

Ellen was placed in charge of a project to fill out the company's product line offerings. To achieve the sales target, marketing asked for a level of product quality that exceeded what was currently available in the marketplace. Furthermore, the product had to be developed by a certain date to meet strategic financial goals and expectations. Even though the desired level of quality was highly optimistic, the project team believed that this level of quality was achievable with existing technology and worker capability. When everyone agreed on the requirements, marketing announced to the general public that this product would be available on a certain date.

Ellen's critical milestone was the design freeze date. This was the date when the product design would be frozen and the blueprints would be turned over to manufacturing. There was no flexibility in the manufacturing schedule. If the blueprints were not available by the design freeze date, then manufacturing would not be able to meet marketing's announced launch date for the product.

Partway into the project, the optimism over meeting the desired quality dissipated. The project team did not believe that it could meet the desired quality level even if it was allowed to extend the schedule and push out the design freeze date. The team believed that it had achieved 90 percent of the desired quality, but the last 10 percent would be difficult. Even with more time, a technical breakthrough would be required to achieve the entire last 10 percent, and that appeared highly unlikely.

For operational projects with the three constraints of time, cost, and scope, Ellen had some degree of discretion in making trade-off decisions. But for this project, with competing constraints and information on enterprise environmental factors that was held by senior management, Ellen needed executive assistance in making trade-off decisions.

Ellen prepared Figure I for her discussions with senior management. The solid lines in the exhibit show that the project team had achieved less than the desired level of quality that marketing and senior management expected. If trade-offs take place, then the project team may be able to increase the level of quality to the dotted line in Exhibit I, but perhaps still not to the desired level of quality. However, any trade-offs would be accompanied by risks.

FIGURE I The quality gap

Ellen understood the project management risks but not the business risks that could affect trade-off decisions. She prepared Table I, which shows how the decision to perform a trade-off on quality could impact several other project management areas of knowledge. Ellen believed that this could help senior management make an informed decision on trade-offs.

TABLE I   POSSIBLE TRADE-OFF IMPACTS

Area of Knowledge	Possible Trade-off Impact
Scope Management	Increasing the quality further will require changes to the scope baseline. Additional scope changes may be necessary as the quality improvement initiative takes place. There is no guarantee that any significant quality improvement can take place in time to meet the design freeze date. The design freeze date must slip.
Time Management	Increasing quality further cannot be achieved without an elongation of the schedule. Time to market will increase based on how much time we are given to try to improve quality further.
Cost Management	Increasing quality further cannot be achieved without an increase in the project's cost baseline. The amount of additional funding needed is based on the additional time allowed.
Risk Management	There may be a high risk that any further improvements to quality cannot take place without a significant technical breakthrough. The team believes that, even with more time, the chances of a significant technical breakthrough are low to moderate.
Human Resource Management	The project team does not possess the skills needed for a technical breakthrough. The organization as a whole may lack available and/or qualified resources to improve quality further. Many of the existing team members are already committed to other projects when the existing design freeze date arrives. Delaying the design freeze date could impact the staffing of other projects.
Procurement Management	Further quality improvements will require outsourcing some of the work. This may require sharing intellectual property with the contractor. This is a legal risk that management must consider. At this late date, it is highly unlikely that any contractor could meet our original design freeze date.

TABLE II   OPTIONS CONSIDERED

Trade-off Option	Extension Needed	Expected Quality Level	Probability of Success
1	1 month	93%	30%
2	2 months	95%	20%
3	3 months	96%	5%

Ellen had high hopes at the start of the project that the expected quality could be achieved. Now she was feeling as if she had failed as a project manager. Her initial thoughts were to ask for more time and push out the design freeze date, if possible. But first she had to meet with the team to see what alternatives members felt were possible. The team presented Ellen with the options identified in Table II.

It was pretty obvious that achieving 100 percent of the desired quality would be impossible. Ellen knew that there were also enterprise environmental factors that had to be considered before deciding which trade-off option to recommend to senior management. She decided to meet with the vice presidents for marketing and sales individually before deciding which alternative would be best. The following information was obtained:

- The quality expectation that marketing had was highly optimistic, and marketing understood this during project approval.
- Marketing may be able to accept the achieved level of quality but that level would most likely result in lower sales expectations.
- Marketing would be willing to extend the launch date of the product provided that the gap between achieved and expected quality can be reduced by at least 25 percent.

Ellen now had to decide what recommendations, if any, to make to senior management.

QUESTIONS

1. Why do most projects eventually require some type of trade-off?
2. Why are strategic projects more prone to trade-offs than operational projects?
3. Why do projects that have more than three competing constraints generally require more trade-offs than projects based on just the traditional triple constraints?
4. What cultural issues at Wayne Corporation made it quite likely that trade-offs would be necessary?
5. Why are design freeze dates important?
6. Who would most likely make the decision for the trade-offs on Ellen's project?
7. What would Ellen most likely recommend to the executive team with regard to trade-offs?
8. What decision would the executives most likely make with regard to a trade-off?

The Trade-off Decision (B)

The Decision

Wayne Corporation wanted the best possible quality but was reluctant to allow the design freeze date to slip. A compromise was finally reached. The design freeze date would be allowed to slip by one month with the hope that Ellen could further improve the quality of the product. The executives knew that the desired quality level might not be reached but were hoping to get as close as possible. This would give Wayne Corporation a competitive edge.

Ellen was quite pleased with the decision because it gave her another chance to show that she was a good a project manager. Unfortunately, the ability to improve the quality further was in the hands of the technical team and beyond her control. However, she did have faith in the technical team's ability to improve the quality, even though it may be just a small change. Making significant improvements to quality would still require the use of external resources.

The Breakthrough

During the 30-day extension for the design freeze date, the technical team had some good news. The team believed that a technical breakthrough was possible and that it could not only meet the desired quality level requested by management but even exceed it significantly. However, this would require slippage of the design freeze date by at least another three months. The team could not guarantee that the breakthrough would take place but felt confident that it could be done.

TABLE I   OPTIONS

Option	Factors for Consideration	Trade-off
1	Launch the product as soon as possible. This requires maintaining the design freeze date with the 30-day extension. Revenue would be generated. Sales information has already been announced to the public. We will continue working on the possible technical breakthrough. Then later, if a technical breakthrough can be achieved, we can announce the next-generation product. If we do this, people who purchased the original product may be upset about the next-generation product being launched so quickly after the earlier version.	An increase in the cost baseline for continued R&D as well as some marketing costs for the next generation product.
2	Launch the product as soon as possible for revenue generation. However, if a technical breakthrough is achieved, delay the marketing and sales of the next-generation product for at least a year or two so as not to alienate existing customers.	An increase in the cost baseline for R&D and to keep much of the product team in place for launch of the next-generation product.
3	Delay the design freeze date another three months and see if the technical breakthrough actually takes place. This could result in lost revenue for the next four months or longer. There is also the risk of a competitor launching a similar product during this waiting period. If the technical breakthrough does not happen, several months have been wasted as well as additional cost for R&D.	An increase in the cost baseline as well as a schedule slippage of at least three to four months.
4	Same as Option #3 but begin aggressive marketing of the new product so that sales appear quickly when the product is launched. There is a risk that the technical breakthrough will not be achieved. People may be upset and confused if we tell them that the previous marketing information was incorrect and a newer product will be forthcoming.	A significant increase in the cost baseline as well as a schedule slippage.

Trade-off Options

Ellen met with the senior management team to discuss the trade-off options based on the good news. Table I lists the options considered.

QUESTION

1. All three options require some trade-offs on time and cost as well as possibly quality. Which option should Wayne Corporation select?

The Project Audit

Overview

The executive boardroom was uncharacteristically quiet. The six executives sitting around the table had been under the impression that their company was reasonably good at project management and better at it than their competitors. But according to the report of the external consultant they hired to perform an audit of several of their projects, project management was being performed poorly and most problems were caused by actions and inactions by the six people sitting around the table. This was certainly not what the executives had expected to hear.

History

In 20 years Spear Electronics had grown from a small electronics manufacturer into a manufacturer of custom-designed electronic components. Although Spear had several commercial product lines that were sold through various electronic outlets, the majority of its business was now from contracts for customized electronics. Spear had a global sales force, but all customization, engineering, and manufacturing was performed in corporate headquarters in the Midwest.

Until its customization business began to grow, Spear did not recognize the need for project management. All of that changed quickly because customization meant that Spear was now becoming a project-driven company where project management would evolve into a strategic or core competency.

Spear was not a big believer in education. It provided the minimum amount of education for its workers. Spear conducted a two-day training program on the basics of project management. Attendance was limited only to those people who were expected to work on customization projects. Workers who were interested in obtaining their certification in project management were expected to pay their own way and take evening or weekend courses to further their education. None of the executives attended any of the project management training courses because they believed that their "occasional" roles as project sponsors did not mandate any in-depth knowledge about project management.

Customers who were financing the customization projects began asking to see Spear's project management methodology. Senior management then established a PMO composed of two people who were given the mission of establishing an enterprise project management methodology for Spear. The workers were told that this would be a full-time assignment for a maximum of six months and that, at the end of that time, they would be reassigned to their previous functional managers. The PMO reported to a middle-level manager.

The Decision to Audit the Projects

Spear has more than 20 reasonably sized customization projects going on at the same time. Each of the six executives acted as the project sponsor for three or four projects, all possibly being performed at the same time. Because the sponsors were not sure about their roles as sponsors and had other duties that required much of their time, they assumed that, if they did not hear any bad news on the projects, things were going as planned. Unfortunately, a large percentage of the projects in the past came in late and over budget, and it appeared that this situation was continuing.

One executive decided to get more actively involved in one of the projects he was sponsoring and discovered that the project would be late. The project manager had been hoping to correct the situation before the project was scheduled to be completed and decided that it would not be prudent to inform the executives about the possible issues until at the very last moment. Learning that this trend of withholding information was happening, the executive became concerned that other project managers might also be withholding bad news from senior management. After talking with managers of other projects he was sponsoring, the executive concluded that many of the project managers were in fact suppressing bad news. The decision was then made by the executive to hire an outside consulting company to perform health checks on all of the projects requiring customization efforts.

The Health Check Report

The health check report made it quite clear that there were serious issues with the way that projects were being managed.

- Performance was not following the baselines. Most project managers seemed to believe that the baselines were created just to get the projects initiated and they could then manage any way they wished without adhering to the baselines.
- The baselines were created based on forecasts. Since the baselines were not being followed, most of the forecasts would not be achieved.
- The expected benefits at the end of the projects and value realization were also not being met.
- Clients and stakeholders were delaying making some critical decisions.
- Most of the projects did not have any risk mitigation plans.
- The projects were not using enough metrics, and the metrics being used did not clearly articulate the status of the projects. The consultant stated that the project managers may be fearful of using the correct metrics because doing so would bring to light the seriousness of some of their issues.

One of the more serious issues in the report was the criticism of senior management's lack of knowledge about project sponsorship and project governance. Some of the critical issues included:

- Sponsors do not understand the differences between their roles and responsibilities and those of the project managers. Many times they were both doing the same thing, and conflicts arose. There was no clear definition of each party's authority and what decisions each one should be making.
- Sponsors do not understand their own decision-making authority. Sponsors seem to be under the mistaken belief that project governance and corporate governance are the same. As a result, project managers seem to have insufficient authority to make some project decisions that they should have make.

Project staffing was also recognized as a serious problem. Some of the critical issues were:

- Project managers must rely on the functional managers for project staffing.
- Project managers lack information on the capabilities of the assigned personnel. The assumption is made that the functional managers have assigned qualified resources.

- Once the resources are assigned, they are still under the control of the functional managers. The project managers have no control over vacation time, training, firefighting on other projects, and the risk of early reassignment of resources.
- Once the resources are assigned, the project managers do not know if the resources are being used effectively.
- Functional managers are using their best resources on short-term projects that are under their direct control and affect their own year-end bonuses. Critical project resources, especially on long-term projects, are assigned at the latest possible moment.
- Overall manpower planning and project staffing is not done consistently. It appears to be just a best guess.
- The PMO is not authorized to perform capacity planning efforts and to ensure and validate that projects have the correct resources.
- Spear has too many projects in the queue and not enough resources.
- Critical resources are being consumed on too many non-value-added projects.
- Spear is bidding on projects without realizing that the existing labor force lacks some of the needed skills.

A Rude Awakening

The silence around the conference room table made it clear that the consultant's comments were correct. Obviously, something had to be done, but what? How should the company begin to address these issues?

Part 3

PROJECT MANAGEMENT CULTURES

Project management methodologies, regardless how good, are simply pieces of paper. What converts these pieces of paper into a world-class methodology is the culture of the organization and how quickly project management is accepted and used. Superior project management is attained when the organization has a culture based on effective trust, communication, cooperation, and teamwork.

A good culture cannot be created overnight. It may take years and strong executive leadership. Good project management cultures demonstrate leadership by example. Senior management must provide the same type of effective leadership that they wish to see implemented by the corporate culture. If roadblocks exist, then senior management must take the initiative in overcoming these barriers.

Como Tool and Die (A)

Como Tool and Die was a second-tier component supplier to the auto industry. Its largest customer was Ford Motor Company. Como had a reputation for delivering a quality product. During the 1980s and the early 1990s, Como's business grew because of its commitment to quality. Emphasis was on manufacturing operations, and few attempts were made to use project management. All work was controlled by line managers who, more often than not, were overburdened with work.

The culture at Como underwent a rude awakening in 1996. In the summer of that year, Ford Motor Company established four product development objectives for both tier 1 and tier 2 suppliers:

1. Lead time: 25–35 percent reduction
2. Internal resources: 30–40 percent reduction
3. Prototypes: 30–35 percent reduction (time and cost)
4. Continuous process improvement and cost reductions

The objectives were aimed at consolidation of the supply base with larger commitments to tier 1 suppliers, which would now have greater responsibility in vehicle development, launch, process improvement, and cost reduction. Ford had established a time frame of 24 months for achievement of the objectives. The ultimate goal for Ford would be the creation of one global, decentralized vehicle development system that would benefit from the efficiency and technical capabilities of the original equipment manufacturers and the subsupplier infrastructure.

STRATEGIC REDIRECTION: 1996

Como realized that it could no longer compete on quality alone. The market-place had changed. The strategic plan for Como was now based on maintaining an industry leadership position well into the twenty-first century. The four basic elements of the strategic plan were:

1. First to market (faster development and tooling of the right products)
2. Flexible processes (quickly adaptable to model changes)
3. Flexible products (multiple niche products from shared platforms and a quick-to-change methodology)
4. Lean manufacturing (low cost, high quality, speed, and global econo-mies of scale)

The implementation of the strategy mandated superior project management performance, but changing a 60-year culture to support project management would not be an easy task.

The president of the company established a task force to identify the cultural issues of converting over to an informal project management system. He believed that project management would eventually become the culture and, therefore, that the cultural issues must be addressed first. The following list of cultural issues was identified by the task force:

- Existing technical, functional departments currently do not adequately support the systemic nature of projects as departmental, and individual objectives are not consistent with those of the project and the customer.
- Senior management must acknowledge the movement away from tra-ditional, "over-the-fence" management and openly endorse the signifi-cance of project management, teamwork, and delegation of authority as the future.
- The company must establish a system of project sponsorship to support project managers by trusting them with responsibility and then empower-ing them to be successful.
- The company must educate managers in project and risk management and the cultural changes of cross-functional project support; it is in the manager's self-interest to support the project manager by providing nec-essary resources and negotiating for adequate time to complete the work.
- The company must enhance information systems to provide cost and schedule performance information for decision making and problem resolution.
- Existing informal culture can be maintained while utilizing project man-agement to monitor progress and review costs. Bureaucracy, red tape, and

lost time must be eliminated through project management's enhanced communications, standard practices, and goal congruence.

The task force, as a whole, supported the idea of informal project management and believed that all of the cultural issues could be overcome. The task force identified four critical risks and the method of resolution:

1. Trusting others and the system
   - *Resolution:* Training in the process of project management and understanding of the benefits. Interpersonal training to learn to trust in each other and in keeping commitments will begin the cultural change.
2. Transforming 60 years of tradition in vertical reporting into horizontal project management
   - *Resolution:* Senior management sponsors the implementation program, participates in training, and fully supports efforts to implement project management across functional lines with encouragement and patience as new organizational relationships are forged.
3. Capacity constraints and competition for resources
   - *Resolution:* Work with managers to understand constraints and to develop alternative plans for success. Develop alternative external capacity to support projects.
4. Inconsistency in application after introduction
   - *Resolution:* Set the clear expectation that project management is the operational culture and the responsibility of each manager. Set the implementation of project management as a key measurable for management incentive plans. Establish a model project and recognize the efforts and successes as they occur.

The president realized that project management and strategic planning were related. He wondered what would happen if the business base would grow as anticipated. Could project management excellence' enhance the business base even further? To answer this question, the president prepared a list of competitive advantages that could be achieved through superior project management performance:

- Project management techniques and skills must be enhanced, especially for the larger, complex projects.
- Development of broader component and tooling supply bases would provide for additional capacity.
- Enhanced profitability would be possible through economies of scale to utilize project managers and skilled trades resources more efficiently through balanced workloads and level production.

- Greater purchasing leverage would be possible through larger purchasing volume and sourcing opportunities.
- Disciplined coordination, reporting of project status, and proactive project management problem-solving must exist to meet timing schedules, budgets, and customer expectations.
- Effective project management of multitier supply base will support sales growth beyond existing, capital-intensive internal tooling and production capacities.

The wheels were set in motion. The president and his senior staff met with all of the employees of Como Tool and Die to discuss the implementation of project management. The president made it clear that he wanted a mature project management system in place within 36 months.

QUESTIONS

1. Does Como have a choice in whether to accept project management as a culture?
2. How much influence should a customer be able to exert on how the contractors manage projects?
3. Was Como correct in attacking the cultural issues first?
4. Does the time frame of 36 months seem practical?
5. What chance of success do you give Como?
6. What dangers exist when your customers are more knowledgeable than you are concerning project management?
7. Is it possible for your customers' knowledge of project management to influence the way that your organization performs strategic planning for project management?
8. Should your customer, especially if it is a powerful customer, have input in the way that your organization performs strategic planning for project management? If so, what type of input should the customer have and on what subject matter?

Como Tool and Die (B)

By 1997, Como had achieved partial success in implementing project management. Lead times were reduced by 10 percent rather than the target of 25 to 35 percent. Internal resources were reduced by only 5 percent. The reduction in prototype time and cost was 15 percent rather than the expected 30 to 35 percent.

Como's automotive customers were not pleased with the slow progress and relatively immature performance of the company's' project management system. Change was taking place, but not fast enough to placate the customers. Como was on target according to its 36-month schedule to achieve some degree of excellence in project management, but would its customers be willing to wait another two years for completion, or should Como try to accelerate the schedule?

FORD INTRODUCES "CHUNK" MANAGEMENT

In the summer of 1997, Ford announced to its suppliers that it was establishing a "chunk" management system. All new vehicle metal structures would be divided into three or four major portions with each chosen supplier (i.e., chunk manager) responsible for all components within that portion of the vehicle. To reduce lead time at Ford and to gain supplier commitment, Ford announced that advanced placement of new work (i.e., chunk managers) would take place without competitive bidding. Target agreements on piece price, tooling cost, and lead time would be established and equitably negotiated later with value engineering work acknowledged.

Chunk managers would be selected based on superior project management capability, including program management skills, coordination responsibility, design feasibility, prototypes, tooling, testing, process sampling, and start of production for components and subassemblies. Chunk managers would function as the second-tier component suppliers and coordinate vehicle build for multiple different vehicle projects at varied stages in the development–tool–launch process.

STRATEGIC REDIRECTION: 1997

Ford Motor Company stated that the selection of the chunk managers would not take place for another year. Unfortunately, Como's plan to achieve excellence would not have been completed by then, and its chances to be awarded a chunk management slot were slim.

The automotive division of Como was now at a critical junction. Como's management believed that the company could survive as a low-level supplier of parts but its growth potential would be questionable. Chunk managers might find it cost effective to become vertically integrated and produce for themselves the same components that Como manufactured. This could have devastating results for Como. This alternative was unacceptable.

The second alternative required that Como make it clear to Ford Motor Company that Como wished to be considered for a chunk manager contract. If Como were to be selected, then Como's project management systems would have to:

- Provide greater coordination activities than previously anticipated.
- Integrate concurrent engineering practices into the company's existing methodology for project management.
- Decentralize the organization so as to enhance the working relationship with the customers.
- Plan for better resource allocation so as to achieve a higher level of efficiency.
- Force proactive planning and decision making.
- Drive out waste and lower cost while improving on-time delivery.

There were also serious risks if Como were to become a chunk manager. The company would be under substantially more pressure to meet cost and delivery targets. Most of its resources would have to be committed to complex coordination activities rather than new product development. Therefore, value-added activities for its customers would be diminished. Finally, if Como failed to live up to its customers' expectations as a chunk manager, it might end up losing all automotive work.

The decision was made to inform Ford of Como's interest in chunk management. Now Como realized that its original three-year plan for excellence in project management would have to be completed in 18 months. The question on everyone's mind was: "How?"

QUESTIONS

1. What was the driving force for excellence before the announcement of chunk management, and what is it now?
2. How can Como accelerate the learning process to achieve excellence in project management? What steps should management take based on its learning so far?
3. What are Como's chances for success? Justify your answer.
4. Should Como compete to become a chunk manager?
5. Can the decision to become a chunk supplier change the way Como performs strategic planning for project management?
6. Can the decision to become a chunk supplier cause an immediate change in Como's single methodology for project management?
7. If a single methodology for project management already exists, then how difficult will it be to make major changes to the methodology and what type of resistance, if any, should management expect?

Apache Metals, Inc.

Apache Metals is an original equipment manufacturer of metalworking equipment. The majority of Apache's business is as a supplier to the automotive, appliance, and building products industries. Each production line is custom-designed according to application, industry, and customer requirements.

Project managers are assigned to each purchase order only after the sales department has a signed contract. The project managers can come from anywhere within the company. Basically, anyone can be assigned as a project leader. The assigned project leaders can be responsible for as many as 10 purchase orders at one time.

In the past, there has not been enough emphasis on project management. At one time, Apache even assigned trainees to perform project coordination. All failed miserably. At one point, sales dropped to an all-time low, and cost overruns averaged 20 to 25 percent per production line.

In January 2007, the board of directors appointed a new senior management team that would drive the organization to excellence in project management. Project managers were added through recruitment efforts and a close examination of existing personnel. Emphasis was on individuals with good people and communication skills.

The following steps were implemented to improve the quality and effectiveness of the project management system:

- Outside formal training for project managers
- Development of an apprenticeship program for future project managers
- Modification of the current methodology to put the project manager at the focal point
- Involvement of project managers to a greater extent with the customer

QUESTIONS

1. What problems can you see in the way project managers were assigned in the past?
2. Will the new approach taken in 2007 put the company on a path to excellence in project management?
3. What skill set would be ideal for the future project managers at Apache Metals?
4. What overall cultural issues must be considered in striving for excellence in project management?
5. What time frame would be appropriate to achieve excellence in project management? What assumptions must be made?

Haller Specialty Manufacturing

For the past several years, Haller has been marginally successful as a specialty manufacturer of metal components. Sales would quote a price to the customer. Upon contract award, engineering would design the product. Manufacturing had the responsibility to produce the product and ship the product to the customer. Manufacturing often changed the engineering design package to fit manufacturing capabilities.

The vice president of manufacturing was perhaps the most powerful position in the company next to the president. Manufacturing was considered to be the main contributor to corporate profits. Strategic planning was dominated by manufacturing.

To get closer to the customer, Haller implemented project management. Unfortunately, the vice president for manufacturing would not support project management for fear of a loss of power and authority.

QUESTIONS

1. If the vice president for manufacturing is a hindrance to excellence, how should this situation be handled?
2. Would your answer to the above question be different if the resistance came from middle or lower-level management?

Coronado Communications

BACKGROUND

Coronado Communications, Inc. (CCI) was a midsize consulting company with corporate headquarters in New York City and satellite divisions in more than 25 of the largest cities in the United States. CCI was primarily a consulting company for large and small firms that wished to improve their communication systems, including computer hardware and networking systems. Each of the 25 divisions serviced its own geographical areas. Whenever a request for proposal was sent to CCI, corporate decided which satellite office would bid on the job.

In 2009, Fred Morse took over as president and chief executive officer of CCI. Although CCI was successful and won a good portion of its contracts through competitive bidding, Morse felt that CCI could win more contracts if he created a climate of internal competition. Prior to Morse coming on board as the chief executive, CCI corporate would decide which satellite office would bid on the job. Morse decided that any and all CCI branches could bid on each and every contract. This process meant that each satellite office would be competing with other satellite offices.

COMPETITIVE SYSTEM

In the past, CCI encouraged the satellite office that would be bidding on the job to use internal resources whenever possible. If the office in Chicago was bidding on a contract and was awarded the contract, then the Chicago office could use resources from the Boston office to fulfill the contract. The workers in the Boston office would then bill the Chicago office a fully loaded or fully burdened hourly rate, but excluding profits. All profits would be shown on the financial statement of the office that won the contract. This technique fostered cooperation between the satellite offices because the Chicago office would get credit for all profits and the Boston office would be able to keep some of its employees on direct charges against contracts rather than on overhead account if they were between jobs.

With the new competitive system, Boston would have the right to charge Chicago a profit for each hour worked, and the profit on these hours would be credited to Boston's financial statement. In effect, Chicago would be treating Boston as if it were a contractor hired by Chicago. If Chicago felt that it could get resources at a cheaper rate by hiring resources from outside CCI, then it was allowed to do so.

The bonus system also changed. In the past, bonuses were paid out equally to each satellite office based on the total profitability to CCI. Now the bonuses paid to each satellite office would be based entirely on the profitability of each satellite office. Salary increases would also be heavily biased toward individual satellite office profitability.

Over the years, the company had developed an outstanding enterprise project management methodology with a proven record of success. Now each satellite office was still asked to use the methodology but could make its own modifications to satisfy its customer base.

TWO YEARS LATER

The following facts appeared after using the new competitive system for two years:

- The gross revenue to the corporation had increased by 40 percent, but the profit margin was only 9 percent, down from the 15 percent prior to the implementation of the new competitive system.
- Satellite offices were lowering their profit margins in order to win new business.
- Most satellite offices were outsourcing some of their work to low-cost suppliers rather than using available resources from other satellite offices.
- Some satellite offices had to lay off some of their talented people because of lack of work.
- Employees were asking for transfers to those satellite offices where greater opportunities existed.
- The cooperative working relationships that once existed between satellite offices was now a competitive relationship with hoarding of information and lack of communications.

- There was no longer a uniform process in place for promotions and awards; everything was based on yearly satellite office profitability.
- Each satellite office created its own project management methodology. The modifications were designed to reduce paperwork and lower the overall cost of using the methodology.
- Clients who had become accustomed to seeing the old methodology were somewhat unhappy with the changes because less information was being presented to them during status review meetings. The clients were also unhappy that updates and changes to the methodology were not being made as fast as necessary, and CCI appeared to be getting farther behind in project management capability.

QUESTIONS

1. Could these results have been anticipated?
2. What happened to the corporate culture?
3. Can project management practices be improved with a major repair to the corporate culture?
4. Is it realistic to expect each satellite office to have its own project management methodology? What happens when two or more satellite offices must work together?
5. Can CCI be fixed? If so, what would you do, and how long do you estimate it would take to make the repairs?

Radiance International

BACKGROUND

Radiance International (RI) had spent more than half a decade becoming a global leader in managing pollution, hazard, and environmental protection projects for its worldwide clients. It maintained 10 offices across the world with approximately 150 people in each office. Its projects ranged from a few hundred thousand dollars to a few million dollars and lasted from six months to two years.

When the downturn in corporate spending began in 2008, RI saw its growth stagnate. Line managers who previously spent most of their time interfacing with various project teams were now spending the preponderance of their time writing reports and memos trying to justify their position in case downsizing occurred. Project teams were asked to generate additional information that the line managers needed to justify their existence. This took a toll on the project teams and forced team members to do "busywork" that was sometimes unrelated to their project responsibilities.

REORGANIZATION PLAN

Management decided to reorganize the company primarily because of the maturity level of project management. Over the years, project management had matured to the point where senior management explicitly trusted the project managers to make both project-based and business-based decisions without continuous guidance from senior management or line management. The role of line management was simply to staff projects and then "get out of the way." Some line managers remained involved in some of the projects, but their interference actually did more harm than good. Executive sponsorship was also very weak because the project managers were trusted to make the right decisions.

The decision was made to eliminate all line management and go to the concept of pool management. One of the line managers was designated as the pool manager and administratively responsible for the 150 employees who were now assigned to the pool. Some of the previous line managers were let go while others became project managers or subject matter experts within the pool. Line managers who remained with the company were not asked to take a cut in pay.

In the center of the pool were the project managers. Whenever a new project came into the company, senior management and the pool manager would decide which project manager would be assigned to head up the new project. The project manager would then have the authority to talk to anyone in the pool who had the expertise needed on the project. If the person stated that he or she was available to work on the project, the project manager would provide that person with a charge number authorizing budgets and schedules for his or her work packages. If the person overran the budget or lengthened the schedule unnecessarily, project managers would not ask this person to work on his or her project again. Pool workers who ran out of charge numbers or were not being used by project managers were then terminated from the company. Project managers would fill out a performance review form on each worker at the end of the project and forward it to the pool manager. The pool manager would make the final decision concerning wage and salary administration but relied heavily on inputs from the project managers.

The culture fostered effective teamwork, communication, cooperation, and trust. Whenever a problem occurred on a project, the project manager would stand up in the middle of the pool and state his or her crisis, and 150 people would rush to his aid, asking what they could do to help. The organization prided itself on effective group thinking and group solutions to complex projects. The system worked so well that sponsorship was virtually eliminated. Once every week or two, a sponsor would walk into the office of a project manager and ask, "Are there any issues I need to know about?" If the project manager responded, "No," then the sponsor would say, "I'll talk to you in a week or two again" and leave.

TWO YEARS LATER

After two years, the concept of pool management was working better than expected. Projects were coming in ahead of schedule and under budget. Teamwork abounded throughout the organization and morale was at an all-time high in every RI location. Everyone embraced the new culture, and nobody was terminated from the company after the first year of the reorganization. Business was booming even though the economy was weak. There was no question that RI's approach to pool management had worked, and worked well.

By the middle of the third year, RI's success story was reported in business journals around the world. While all of the news was favorable and brought in more business, RI became a takeover target by large construction companies that saw the acquisition of RI as an opportunity. By the end of the third year, RI was acquired by a large construction firm. The construction company believed in strong line management with a span of control of approximately 10 employees per supervisor. The pool management concept at RI was eliminated; several line management positions were created in each RI location and staffed with employees from the construction company. Within a year, several RI employees left the company.

QUESTIONS

1. Is it a good idea to remove all of the line management slots?
2. If pool management does not work, can line management slots then be reinstated?
3. How important is the corporate culture to the pool management concept?
4. Are there project sponsors at RI?

The Executive Director

Background

Richard Damian was delighted that his political party had won the election. As a reward for his years of support, he was appointed executive director of this government agency, replacing a person from the other political party. Damian had been with the government for more than 30 years. This would be at least a four-year appointment, and, if his party was still in power after the elections four years from now, he could be the executive director for an additional four years or more.

Damian knew how to play political gamesmanship. He avoided anything that was considered controversial and voted with his party line on all issues even if he disagreed with his party's position. He knew how to get things done behind the scenes and without exposing himself to any risks or being scrutinized by the media. But now, as executive director, he realized that things might be different. He was now exposed to the media.

The Internet Security Project

Damian's predecessor had been plagued by Internet hackers who were getting access to some of the agency's proprietary information. The media was aware of this, and his predecessor had been engulfed with bad publicity. The media kept asking what Damian was planning to do to correct the situation, and Damian kept stating that he was not ready to discuss this until he had worked out a plan with his executive staff.

Damian's predecessor had tried unsuccessfully to correct the problems using the agency's internal information technology (IT) resources. Unfortunately, the internal resources had limited IT security knowledge. Budgetary cuts made it impossible to hire additional IT resources. Government hiring and firing practices also made it difficult to remove some poor performers and replace them with other workers trained in IT security practices. Damian's predecessor also tried to get support from other government agencies, but they had their own political agendas and could not or would not provide the needed support.

The project had to be outsourced. Damian instructed one of his direct reportees to assign someone as the project manager and begin by soliciting bids from at least three vendors. Since time was critical because of the pressure imposed by the media, Damian recommended that the quotes be obtained informally because of the time-consuming, rigid policies and procedures that must be followed for traditional government contracting.

Assigning the Project Manager

The person assigned as the project manager reported several levels below Damian. The project manager had been with the agency for less than two years and had a degree in information systems. In addition, the project manager, and the rest of the agency's IT group, had very little knowledge about IT security. The agency would have to rely on the expertise of contractors.

Damian believed that rank has its privilege. As such, he decided that he should not have to interface directly with people who report low on the organizational chart. He instructed one of his direct reportees to keep him informed about the status of the project.

A month later, Damian was informed that there were three qualified bidders. The bids ranged from $1.5 million to $1.75 million with a time frame of approximately three months. All three bidders said that their bid was just a rough estimate and that a final bid could not be made without a clear examination of the agency's existing hardware and software. Furthermore, all three bidders stated that they wanted a cost-reimbursable contract rather than a firm-fixed-price contract.

Damian was in a hurry to get the contract started. He instructed the procurement people to issue a cost-reimbursable contract to one of the vendors immediately. This required violation of traditional procurement policies and procedures, but Damian felt that this was an extraordinary situation that needed resolution quickly. Two weeks later, the contract was signed and the project had a go-ahead date of March 1.

The Work Begins

As soon as the work began, Damian held a news conference and announced that the security system was being modified and that all protocols for the new system would be operational within 90 days, the duration of the contract. Even though he had very limited knowledge about how the system would work, he still made promises concerning the system's capabilities. The media seemed somewhat skeptical about how quickly the changes would be made and Damian's promises and began asking questions. Damian knew how to play the political game. He certainly did not have enough information to answer the questions that might be forthcoming. He declined to answer any of the media's questions, stating that all questions would be answered at a future new conference.

Over the next month, the media kept asking why Damian's agency was not providing any information on the status of the new security system. It was rumored that the project would be coming in late and over budget. Damian was several layers of management removed from where the work was taking place and knew very little about the progress of the project'. Information on the status of the project, especially any bad news, was being filtered out as the information flowed up the organizational hierarchy to the point where it appeared that there were no issues. Unfortunately, that was not the case.

Damian learned that the project would be at least one month late and possibly over budget by $500,000. He called a news conference and informed the media about the schedule slippage and cost overrun. Trying to protect his image, Damian stated that he was never informed about the risks on the project. Furthermore, he said that he was forming a committee headed by one of his direct reportees to get to the bottom of the problem. Once again, Damian refused to answer any questions posed by the media.

The Problems Mount

Damian asked his direct reportee for a briefing on what was being done to correct the situation. The information provided by the contractor stated that his agency's hardware and existing software were outdated and needed to be replaced. The software and changes needed to enhance computer security could not run on the existing hardware. The contractors was now asking for an additional $4 million to update all of the hardware and software. The agency's IT personnel were all in favor of the upgrade. The entire changeover would add six months to the length of the schedule.

The $1.75 million project was now at $6 million and possibly increasing. Damian was convinced that the media would attack his credibility because of the security issues that still existed. Someone had to be blamed so that the "heat" would not be placed on Damian. His first thought was to blame the project manager, but everyone would know that this was not the case.

Playing the political game, Damian called another news conference and blamed his predecessor for all of the existing issues. He stated that these problems should have been addressed years ago and that his predecessor, who belonged to a different political party, failed to take the necessary steps to correct the situation. Furthermore, Damian asserted that one of his direct reportees was now the sponsor for the entire project and that Damian would receive weekly progress briefings.

Damian played the political game to the best of his ability. Not only did he blame the other political party and his predecessor for all of the problems, but he insulated himself from further disasters by stating that one of his direct reportees was now the sponsor. Damian now believed that he would be free from further criticism.

The Situation Worsens

Damian's agency did not have authorization for purchasing computer hardware and software without getting permission from the government's centralized procurement group. The hardware and software recommendations from Damian's contractor were not on the government's approved hardware and software list. The contractor either had to select hardware and software from the approved list or apply for add-ons to the list. Applying for add-ons to the approved list could take as much as three to six months, thus lengthening the existing contract.

The contractor reviewed the list and recommended purchasing hardware and software that would increase the project's cost by $5 million rather than $4 million. The length of the project was now estimated to be 1.5 years rather than three months—assuming, of course, that the hardware could be received within a reasonable amount of time. Trying to get additional hardware and software added to the approved list could have possibly saved $1 million but might have lengthened the project to two years.

The media became aware of the situation and began attacking Damian's credibility as an executive director. Once again Damian had to play the political game. He called a news conference and stated that, although time and cost were both constraints, time was prioritized as being more important than cost. Therefore, he accepted the cost overrun of $5 million. Furthermore, Damian stated that he would now be the sponsor for this project, although executive directors normally do not sponsor projects of this size.

Damian once again blamed his predecessor for all of the problems, stating that this problem could have been done years ago at a lower cost. He also left the media with the impression that his direct reportee who was previously the sponsor was being reassigned to another position. That move made it appear that his reportee was partially at fault.

The Project Approaches Completion

The original project, with a budget of $1.75 million and a duration of three months, was completed at the end of 2.5 years and at a cost of $9 million. Software bugs were found during testing that required overtime, software changes, and the purchase of some additional hardware.

Once again Damian played the political game. He informed the media that the project was completed and that security was now in place. Furthermore, he stated that he personally was in charge of the project all the way and took full credit for its successful completion.

QUESTIONS

1. Could the solution to the security problem have been managed internally, or was it necessary to outsource the work?
2. Because of the sensitivity of the problem, was it acceptable to get quotes from vendors on an informal basis, or should everything have been done through the formal channels for procurement?
3. Why did Damian want someone quite low in the organizational hierarchy to function as the project manager?
4. Why did Damian want one of his direct reportees to keep him informed about the status of the project rather than hearing it directly from the project manager?
5. Why did all three bidders want a cost-reimbursable contract rather than a fixed-price contract?
6. Why was Damian reluctant to address the media and answer their questions once the project started?
7. What was the driving force behind Damian's actions throughout the case?

Part 4

PROJECT MANAGEMENT ORGANIZATIONAL STRUCTURES

In the early days of project management, there existed a common belief that project management had to be accompanied by organizational restructuring. Project management practitioners argued that some organizational structures, such as matrix structures, were more conducive to good project management, while others were not quite effective. Every organizational structure comes with both advantages and disadvantages.

Today, we question whether organizational restructuring is necessary. Is it possible that project management can be implemented effectively in any organizational structure if the culture is cooperative? Restructuring is often accompanied by a shift in authority and the balance of power. Can effective project management occur at the same time that the organization undergoes restructuring?

Quasar Communications, Inc.

Quasar Communications, Inc. (QCI) is a 30-year-old, $350 million division of Communication Systems International, the world's largest communications company. QCI employs about 340 people, of whom more than 200 are engineers. Ever since the company was founded 30 years ago, engineers have held every major position within the company, including president and vice president. The vice president for accounting and finance, for example, has an electrical engineering degree from Purdue and a master's degree in business administration from Harvard.

Until 1996, QCI was a traditional organization where everything flowed up and down. In 1996, QCI hired a major consulting company to come in and train all personnel in project management. Because of the reluctance of the line managers to accept formalized project management, QCI adopted an informal, fragmented project management structure where the project managers had lots of responsibility but very little authority. The line managers were still running the show.

By 1999, QCI had grown to a point where the majority of its business base revolved around 12 large customers and 30 to 40 small customers. The time had come to create a separate line organization for project managers, where each individual could be shown a career path in the company and the company could benefit by creating a body of planners and managers dedicated to project completion.

The project management group was headed up by a vice president and included the following full-time personnel:

- Four individuals to handle the 12 large customers
- Five individuals for the 30 to 40 small customers
- Three individuals for research and development (R&D) projects
- One individual for capital equipment projects

The nine customer project managers were expected to handle two to three projects at one time if necessary. However, because customer requests usually did not come in at the same time, it was anticipated that each project manager usually would handle only one project at a time. The R&D and capital equipment project managers were expected to handle several projects at once.

In addition to the above personnel, the company also maintained a staff of four product managers who controlled the profitable off-the-shelf product lines. The product managers reported to the vice president of marketing and sales.

In October 1999, the vice president for project management decided to take a more active role in the problems that project managers were having and held counseling sessions with each project manager. The following major problem areas were discovered.

R&D PROJECT MANAGEMENT

Project manager: "My biggest problem is working with these diverse groups that aren't sure what they want. My job is to develop new products that can be introduced into the marketplace. I have to work with engineering, marketing, product management, manufacturing, quality assurance, finance, and accounting. Everyone wants a detailed schedule and product cost breakdown. How can I do that when we aren't even sure what the end item will look like or what materials are needed? Last month I prepared a detailed schedule for the development of a new product, assuming that everything would go according to the plan. I worked with the R&D engineering group to establish what we considered to be a realistic milestone. Marketing pushed the milestone to the left because they wanted the product to be introduced into the marketplace earlier. Manufacturing then pushed the milestone to the right, claiming that they would need more time to verify the engineering specifications. Finance and accounting then pushed the milestone to the left, asserting that management wanted a quicker return on investment. Now how can I make all of the groups happy?"

Vice president: "Whom do you have the biggest problems with?"

Project manager: "That's easy—marketing! Every week marketing gets a copy of the project status report and decides whether to cancel the project. Several times marketing has canceled projects without even discussing it with me, and I'm supposed to be the project leader."

Vice president: "Marketing is in the best position to cancel projects because they have inside information on profitability, risk, return on investment, and competitive environment."

Project manager: "The situation that we're in now makes it impossible for the project manager to be dedicated to a project where he does not have all of the information at hand. Perhaps we should either have the R&D project managers report to someone in marketing or have the marketing group provide additional information to the project managers."

SMALL-CUSTOMER PROJECT MANAGEMENT

Project manager: "I find it virtually impossible to be dedicated to and effectively manage three projects that have priorities that are not reasonably close. My low-priority customer always suffers. And even if I try to give all of my customers equal status, I do not know how to organize my work and have effective time management on several projects."

Project manager: "Why is it that the big projects carry all of the weight and the smaller ones suffer?"

Project manager: "Several of my projects are so small that they stay in one functional department. When that happens, the line manager feels that he is the true project manager operating in a vertical environment. On one of my projects, I found that a line manager had promised the customer that additional tests would be run. This additional testing was not priced out as part of the original statement of work. On another project, the line manager made certain remarks about the technical requirements of the project. The customer assumed that the line manager's remarks reflected company policy. Our line managers don't realize that only the project manager can make commitments on resources to the customer as well as on company policy. I know this can happen on large projects as well, but it is more pronounced on small projects."

LARGE-CUSTOMER PROJECT MANAGEMENT

Project manager: "Those of us who manage the large projects are also marketing personnel, and, occasionally, we are the ones who bring in the work. Yet everyone appears to be our superior. Marketing always looks down on us, and when we bring in a large contract, marketing just looks down on us as if we're riding their coattails or as if we were just lucky. The engineering group outranks us because all managers and executives are promoted from there. Those guys never live up to commitments. Last month I sent an inflammatory memo to a line manager because of his poor response to my requests. Now I get no support at all from him. This doesn't happen all of the time, but when it does, it's frustrating."

Project manager: "On large projects, how do we, the project managers, know when the project is in trouble? How do we decide when the project will fail? Some of our large projects are total disasters and should fail, but management comes to the rescue and pulls the best resources off of the good projects to cure the ailing projects. We then end up with six marginal projects and one partial catastrophe as opposed to six excellent projects and one failure. Why don't we just let the bad projects fail?"

Vice president: "We have to keep up our image for our customers. In most other companies, performance is sacrificed in order to meet time and cost. Here at QCI, with our professional integrity at stake, our engineers are willing to sacrifice time and cost in order to meet specifications. Several of our customers come to us because of this. Last year we had a project where, at the scheduled project termination date, engineering was able to satisfy only 75 percent of the customer's performance specifications. The project manager showed the results to the customer, and the customer decided to change his specification requirements to agree with the product that we designed. Our engineering people thought that this was a slap in the face and refused to sign off the engineering drawings. The problem went all the way up to the president for resolution. The final result was that the customer would give us an additional few months if we would spend our own money to try to meet the original specification. It cost us a bundle, but we did it because our integrity and professional reputation were at stake."

CAPITAL EQUIPMENT PROJECT MANAGEMENT

Project manager: "My biggest complaint is with this new priority scheduling computer package we're supposedly considering to install. The way I understand it, the computer program will establish priorities for *all* of the projects in-house, based on the feasibility study, cost-benefit analysis, and return on investment. Somehow I feel as though my projects will always be the lowest priority, and I'll never be able to get sufficient functional resources."

Project manager: "Every time I lay out a reasonable schedule for one of our capital equipment projects, a problem occurs in the manufacturing area and the functional employees are always pulled off of my project to assist manufacturing.And now I have to explain to everyone why I'm behind schedule. Why am I always the one to suffer?"

The vice president carefully weighed the remarks of his project managers. Now came the difficult part. What, if anything, could the vice president do to amend the situation given the current organizational environment?

QUESTIONS

1. Can 13 project managers be controlled and supervised effectively by one vice president?
2. Can the 13 managers under this vice president work effectively with the four product managers under the vice president of marketing/sales?

3. Why does the R&D project manager have built-in conflicts?
4. Should marketing have R&D project managers reporting to them?
5. What are the major problems with small-customer project management?
6. Should the project manager on large projects be permitted to perform marketing activities?
7. Should a company be willing to let some large projects fail?
8. Is it possible for a company to have such a strong technical community that technical integrity is more important than the project itself?
9. Is it possible that capital equipment projects almost always take a backseat to other projects?
10. What specific problems appear in the management of large projects?
11. What specific problems appear in the management of R&D projects?
12. Are there any strengths in the current QCI organization?
13. What type of project management structure is QCI using?
14. What possible recommendation could you make?

Fargo Foods

Fargo Foods is a $2 billion–a–year international food manufacturer with canning facilities in 22 countries. Fargo products include meats, poultry, fish, vegetables, vitamins, and cat and dog foods. Fargo Foods has enjoyed a 12.5 percent growth rate each of the past eight years primarily due to the low overhead rates in the foreign companies.

During the past five years, Fargo had spent a large portion of retained earnings on capital equipment projects in order to increase productivity without increasing labor costs. An average of three new production plants have been constructed in each of the last five years. In addition, almost every plant has undergone major modifications each year in order to increase productivity.

In 2010, the president of Fargo Foods implemented formal project management for all construction projects using a matrix. By 2014, it became obvious that the matrix was not operating effectively or efficiently. In December 2014, the author consulted for Fargo Foods by interviewing several of the key managers and many functional personnel. What follows are the several key questions and responses addressed to Fargo Foods.

Author: "Give me an example of one of your projects."

Manager: "The project begins with an idea. The idea can originate anywhere in the company. The planning group picks up the idea and determines the feasibility. The planning group then works informally with the various line organizations to determine rough estimates for time and cost. The results are then fed back to the planning group and to the top management planning and steering committees. If top management decides to undertake the project, then top management selects the project manager and we're off and running."

Author: "Do you have any problems with this arrangement?"

Manager: "You bet! Our executives have the tendency of considering rough estimates as detailed budgets and rough schedules as detailed schedules. Then they want to know why the line managers won't commit their best resources. We almost always end up with cost overruns and schedule slippages. To make matters even worse, the project managers do not appear to be dedicated to the projects. I really can't blame them. After all, they're not involved in planning the project, laying out the schedule, and establishing the budget. I don't see how any project manager can become dedicated to a plan in which the project manager has no input and may not even know the assumptions or considerations that were included. Recently some of our more experienced project managers have taken a stand on this and are virtually refusing to accept a project assignment unless they can do their own detailed planning at the beginning of the project in order to verify the constraints established by the planning group. If the project managers come up with different costs and schedules (and you know that they will), the planning group feels that they have just gotten slapped in the face. If the costs and schedules are the same, then the planning group runs upstairs to top management asserting that the project managers are wasting money by continuously wanting to replan."

Author: "Do you feel that replanning is necessary?"

Manager: "Definitely! The planning group begins their planning with a very crude statement of work, expecting our line managers, the true experts, to read in between the lines and fill in the details. The project managers develop a detailed statement of work and a work breakdown structure, thus minimizing the chance that anything would fall through the cracks. Another reason for replanning is that the ground rules have changed between the time that the project was originally adopted by the planning group and the time that the project begins implementation. Another possibility, of course, is that technology may have changed or people can be smarter now and can perform at a higher position on the learning curve."

Author: "Do you have any problems with executive meddling?"

Manager: "Not during the project, but initially. Sometimes executives want to keep the end date fixed but take their time in approving the project. As a result, the project manager may find himself a month or two behind scheduling before he even begins the project. The second problem is when the executive decides to arbitrarily change the end-date milestone but keep the front-end milestone fixed. On one of our projects, it was necessary to complete the project in half the time. Our line managers worked like dogs to get the job done. On the next project, the same thing happened, and, once again, the line managers came to the rescue. Now management feels that line managers cannot make good estimates and that they—the executives—can arbitrarily change the milestones on any project. I wish that they would realize what they're doing to us. When we put forth all of our efforts on one project, then all of the other projects suffer. I don't think our executives realize this."

Author: "Do you have any problems selecting good project managers and project engineers?"

Manager: "We made a terrible mistake for several years by selecting our best technical experts as the project managers. Today our project managers are doers, not managers. The project managers do not appear to have any confidence in our line people and often try to do all of the work themselves. Functional employees are taking technical direction from project managers and project engineers instead of line managers. I've heard one functional employee say, 'Here come those project managers again to beat me up. Why can't they leave me alone and let me do my job?' Our line employees now feel that this is the way that project management is supposed to work. Somehow, I don't think so."

Author: "Do you have any problems with the line manager–project manager interface?"

Manager: "Our project managers are technical experts and therefore feel qualified to do all of the engineering estimates without consulting with line managers. Sometimes this occurs because not enough time or money is allocated for proper estimating. This is understandable. But when project managers have enough time and money and refuse to get off their ivory towers and talk to the line managers, then the line managers will always find fault with project managers' estimates even if they are correct. Sometimes I just can't feel any sympathy for the project managers. There is one special case that I should mention. Many of our project managers do the estimating themselves but have courtesy enough to ask the line manager for his blessing. I've seen line managers who were so loaded with work that they look the estimate over for two seconds and say, 'It looks fine to me. Let's do it.' Then when the cost overrun appears, the project manager gets blamed."

Author: "Where are your project engineers located in the organization?"

Manager: "We're having trouble deciding that. Our project engineers are primarily responsible for coordinating the design efforts—that is, electrical, civil, HVAC, and so on. The design manager wants these people reporting to him if they are responsible for coordinating efforts in his shop. The design manager wants control of these people even if they have their title changed to assistant project managers. The project managers, on the other hand, want the project engineers to report to them with the argument that they must be dedicated to the project and must be willing to complete the effort within time, cost, and performance. Furthermore, the project managers argue that project engineers will be more likely to get the job done within the constraints if they are not under the pressure of being evaluated by the design manager. If I were the design manager, I would be a little reluctant to let someone from outside of my shop integrate activities that utilize the resources under my control. But I guess this gets back to interpersonal skills and the attitudes of the people. I do not want to see a brick wall set up between project management and design."

Author: "I understand that you've created a new estimating group. Why was that done?"

Manager: "In the past we have had several different types of estimates such as first guess, detailed, 10 percent complete, and so on. Our project managers are usually the first people at the job site and give a shoot-from-the-hip estimate. Our line managers do estimating as do some of our executives and functional employees. Because we're in a relatively slowly changing environment, we should have well-established standards, and the estimating department can maintain uniformity in our estimating policies. Since most of our work is approved based on first-guess estimates, the question is 'Who should give the first-guess estimate?' Should it be the estimator, who does not understand the processes but knows the estimating criteria, or the project engineer, who understands the processes but does not know the estimates, or the project manager, who is an expert in project management? Right now, we are not sure where to place the estimating group. The vice president of engineering has three operating groups beneath him—project management, design, and procurement. We're contemplating putting estimating under procurement, but I'm not sure how this will work."

Author: "How can we resolve these problems that you've mentioned?"

Manager: "I wish I knew!"

Government Project Management

A major government agency is organized to monitor government subcontractors as shown in Figure I. The vital characteristics of certain project office team members are:

- *Project manager:* Directs all project activities and acts as the information focal point for the subcontractor.
- *Assistant project manager:* Acts as chairman of the steering committee and interfaces with both in-house functional groups and contractor.
- *Department managers:* Act as members of the steering committee for any projects that utilize their resources. These slots on the steering committee must be filled by the department managers themselves, not by functional employees.
- *Contracts officer:* Authorizes all work directed by the project office to in- house functional groups and to the customer, and ensures that all work requested is authorized by the contract. The contracts officer acts as the focal point for all contractor cost and contractual information.

QUESTIONS

1. Explain how this structure should work.
2. Explain how this structure actually works.

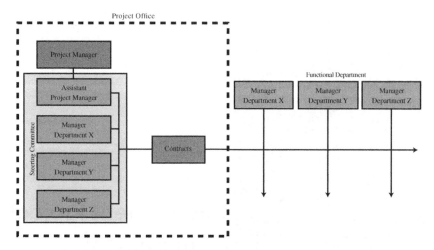

FIGURE I Project team organizational structure

3. Can the project manager be a military type who is reassigned after a given tour of duty?
4. What are the advantages and disadvantages of this structure?
5. Could this be used in industry?

Falls Engineering

Located in New York, Falls Engineering is a $850 million chemical and materials operation employing 950 people. The plant has two distinct manufacturing product lines: industrial chemicals and computer materials. Both divisions are controlled by one plant manager, but direction, strategic planning, and priorities are established by corporate vice presidents in Chicago. Each division has its own corporate vice president, list of projects, list of priorities, and manpower control. The chemical division has been at this location for the past 20 years. The materials division is, you might say, the tenant in the landlord–tenant relationship, with the materials division manager reporting dotted to the plant manager and solid to the corporate vice president. (See Figure I.)

The chemical division employed 3,000 people in 1998. By 2003, there were only 600 employees. In 2004, the materials division was formed and located on the chemical division site with a landlord–tenant relationship. The materials division has grown from $50 million in 2000 to $120 million in 2004. Today, the materials division employs 350 people.

All projects originate in construction or engineering but usually are designed to support production. The engineering and construction departments have projects that span the entire organization directed by a project coordinator. The project coordinator is a line employee who is temporarily assigned to coordinate a project in his line organization in addition to performing his line responsibilities. Assignments are made by the division managers (who report to the plant manager) and are based on technical expertise. The coordinators have monitoring

119

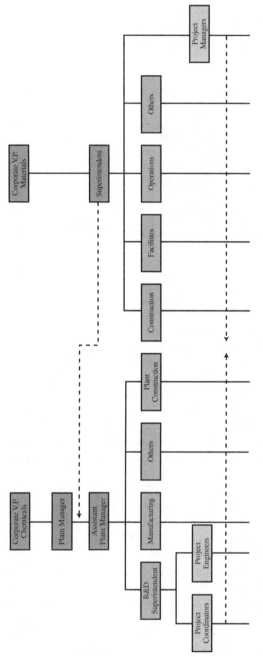

FIGURE I Falls Engineering organizational chart

authority only and are not noted for being good planners or negotiators. The coordinators report to their respective line managers.

Basically, a project can start in either division with the project coordinators. The coordinators draw up a large scope of work and submit it to the project engineering group, which arranges for design contractors, depending on the size of the project. Project engineering places it on their design schedule according to priority and produces prints and specifications, and receives quotes. A construction cost estimate is then produced following 60 to 75 percent design completion. The estimate and project papers are prepared, and the project is circulated through the plant and in Chicago for approval and authorization. Following authorization, the design is completed, and materials are ordered. Following design, the project is transferred to either of two plant construction groups for construction. The project coordinators than arrange for the work to be accomplished in their areas with minimum interference from manufacturing forces. In all cases, the coordinators act as project managers and must take the usual constraints of time, money, and performance into account.

Falls Engineering has 300 projects listed for completion between 2006 and 2008. In the last two years, fewer than 10 percent of the projects were completed within time, cost, and performance constraints. Line managers find it increasingly difficult to make resource commitments because crises always seem to develop, including a number of fires.

Profits are made in manufacturing, and everyone knows it. Whenever a manufacturing crisis occurs, line managers pull resources off the projects, and, of course, the projects suffer. Project coordinators are trying, but with very little success, to put some slack onto the schedules to allow for contingencies.

The breakdown of the 300 plant projects is shown next.

Number of Projects	$ Range
120	less than $50,000
80	$50,000–$200,000
70	$250,000–$750,000
20	$1–$3 million
10	$4–$8 million

Corporate realized the necessity for changing the organizational structure. A meeting was set up among the plant manager, plant executives, and corporate executives to resolve these problems once and for all. The plant manager decided to survey his employees concerning their feelings about the present organizational structure. Their comments are listed next.

● "The projects we have the most trouble with are the small ones under $200,000. Can we use informal project management for the small ones and formal project management on the large ones?"

- "Why do we persist in using computer programming to control our resources? These sophisticated packages are useless because they do not account for firefighting."
- "Project coordinators need access to various levels of management, in both divisions."
- "Our line managers do not realize the necessity for effective planning of resources. Resources are assigned based on emotions and not need."
- "Sometimes a line manager gives a commitment, but the project coordinator cannot force him to keep it."
- "Line managers always find fault with project coordinators who try to develop detailed schedules themselves."
- "If we continuously have to crash project time, doesn't that indicate poor planning?"
- "We need a career path in project coordination so that we can develop a body of good planners, communicators, and integrators."
- "I've seen project coordinators who have no interest in the job, cannot work with diverse functional disciplines, and cannot communicate. Yet someone assigned them as a project coordinator."
- "Any organizational system we come up with has to be better than the one we have now."
- "Somebody has to have total accountability. Our people are working on projects and, at the same time, do not know the project status, the current cost, the risks, and the end date."
- "One of these days I'm going to kill an executive while he's meddling in my project."
- "Recently management made changes requiring more paperwork for the project coordinators. How many hours a week do they expect me to work?"
- "I've yet to see any documentation detailing the job description of the project coordinator."
- "I have absolutely no knowledge about who is assigned as the project coordinator until work has to be coordinated in my group. Somehow, I'm not sure that this is the way the system should work."
- "I know that we line managers are supposed to be flexible, but changing the priorities every week isn't exactly my idea of fun."
- "If the projects start out with poor planning, then management does not have the right to expect the line managers always to come to the rescue."
- "Why is it that line managers always get blamed for schedule delays, even if it's the result of poor planning up front?"
- "If management doesn't want to hire additional resources, then why should the line managers be made to suffer? Perhaps we should cut out some of these useless projects. Sometimes I think management dreams up some of these projects simply to spend the allocated funds."
- "I have yet to see a project I felt had a realistic deadline."

After preparing alternatives and recommendations as plant manager, try to do some role-playing by putting yourself in the shoes of the corporate executives. Would you, as a corporate executive, approve the recommendation? Where does profitability, sales, return on investment, and so on enter into your decision?

QUESTIONS

1. Explain how the landlord–tenant relationship works.
2. How are priorities established for each division?
3. What is the working relationship between the plant manager and the superintendent for materials?
4. Can this dotted line be eliminated? If so, under what circumstances?
5. Can we establish one uniform set of priorities? If so, who will have to make concessions?
6. How are project coordinators selected?
7. Should the company have a career path in project coordination?
8. What is the difference between a project coordinator and a project manager?
9. On what kinds of projects are project coordinators being used?
10. Can these projects be run informally?
11. Of what value is computerized project management on capital equipment projects?
12. How should the company solve its project coordinator problems?
13. Who gets blamed if the project schedule is not met? If unrealistic estimates are given?
14. What is your recommendation to the plant manager? To the corporate vice presidents?
15. Will your answer to the last question have political ramifications?

White Manufacturing

In 2004, White Manufacturing realized the necessity for project management in the manufacturing group. A three-person project management staff was formed. Although the staff was shown on the organizational chart as reporting to the manufacturing operations manager, they actually worked for the vice president and had sufficient authority to integrate work across all departments and divisions. As in the past, the vice president's position was filled by the manufacturing operations manager. Manufacturing operations was directed by the former manufacturing manager, who came from manufacturing engineering. (See Figure I.)

In 2007, the manufacturing manager created a matrix in the manufacturing department with the manufacturing engineers acting as departmental project managers. This benefited both the manufacturing manager and the group project managers since all information could be obtained from one source. Work was flowing very smoothly.

In January 2008, the manufacturing manager resigned his position effective March, and the manufacturing engineering manager began packing his bags to move up to the vacated position. In February, the vice president announced that the position would be filled from outside. He said also that there would be an organizational restructuring and that the three project managers would now report to the manufacturing manager. When the three project managers confronted the manufacturing operations manager, he said, "We've hired the new man in at a very high salary. In order to justify this salary, we have to give him more responsibility."

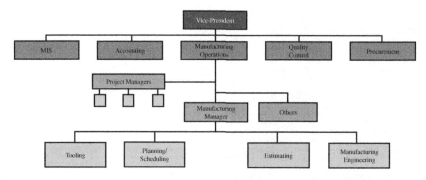

FIGURE I   White Manufacturing organizational structure

In March 2008, the new manager took over and immediately made two declarations:

1. The project managers will never go "upstairs" without first going through him.
2. The departmental matrix will be dissolved, and the department manager will handle all of the integration.

QUESTIONS

1. How do you account for the actions of the new department manager?
2. What would you do if you were one of the project managers?

Martig Construction Company

Martig Construction was a family-owned mechanical subcontractor business that had grown from $5 million in 2006 to $25 million in 2008. Although the gross profit had increased sharply, the profit as a percentage of sales declined drastically. The question was "Why the decline?" The following observations were made:

1. Since Martig senior died in July of 2008, Martig junior has tried unsuccessfully to convince the family to let him sell the business. Martig junior, as company president, has taken an average of eight days of vacation per month for the past year. Although the project managers are supposed to report to Martig, they appear to be calling their own shots and are in a continuous struggle for power.

2. The estimating department consists of one man, John, who estimates all jobs. Martig wins one job in seven. Once a job is won, a project manager is selected and is told that he must perform the job within the proposal estimates. Project managers are not involved in proposal estimates. They are required, however, to provide feedback to the estimator so that standards can be updated. This very seldom happens because of the struggle for power. The project managers are afraid that the estimator might be next in line for executive promotion since he is a good friend of Martig's.

3. The procurement function reports to Martig. Once the items are ordered, the project manager assumes procurement responsibility. Several times

in the past, the project manager has been forced to spend hours trying to overcome shortages or simply to track down raw materials. Most project managers estimate that approximately 35 percent of their time involves procurement.

4. Site superintendents believe they are the true project managers or at least are at the same level. The superintendents are very unhappy about not being involved in the procurement function and, therefore, look for ways to annoy the project managers. It appears that the more time project managers spend at the site, the longer the work takes; the feedback of information to the home office is also distorted.

QUESTIONS

1. What is the major problem with the case study?
2. How would you resolve the problem?
3. What should the working relationship be between project managers and site superintendants?
4. Does Martig have a good procurement function?
5. Does the estimating function appear to be performing correctly?

Part 5

NEGOTIATING FOR RESOURCES

In most organizations, project management is viewed as multiple-boss reporting. It is possible for the employees to report to one line manager and several project managers at the same time. This multiple boss reporting problem can greatly influence the way that the project manager negotiates for resources. Project managers must understand the skill level needed to perform the work, whether the resource would be needed on a part-time or full-time basis, and the duration of the effort for this worker.

Some people argue that today's project managers no longer have command of technology but possess more of an understanding of technology. If this is, in fact, the case, then the project managers might be better off negotiating for deliverables than for people. The argument is whether a project manager should manage people or manage deliverables.

Ducor Chemical

Ducor Chemical received a research and development (R&D) contract from one of its most important clients. The client had awarded Ducor a 12-month, sole-source contract for the R&D effort to create a new chemical that the client required for one of its future products. If Ducor could develop the product, the long-term production contract that would follow could generate significant profits over the next several years.

In addition to various lab personnel who would be used as needed, the contract mandated that a senior chemist be assigned for the duration of the project. In the past, senior chemists had been used mainly for internal rather than external customer projects. This would be the first time a senior chemist had been assigned to this client. With only four senior chemists on staff, the project manager expected the resource negotiation process with the lab manager to be an easy undertaking.

Project manager: "I understand you've already looked over the technical requirements, so you should understand the necessity for assigning your best senior scientist."

Lab manager: "All of my senior scientists are good. Any one of them can do the job. Based on the timing of your project, I have decided to assign John Thornton."

Project manager: "Just my luck! You assigned the only one I cannot work with effectively. I have had the misfortune of working with him before. He's extremely arrogant and unpleasant to work with."

Lab manager: "Perhaps so, but he got the job done, didn't he?"

Project manager: "Yes, he did. Technically, he is capable. However, his arrogant attitude and sarcasm produced a demoralizing atmosphere for my team. That project was about three months in length. This project is at least a year. Also, if follow-on work is generated, as I expect it to be, I'll be stuck with him for a long time. That's unacceptable to me."

Lab manager: "I'll talk to John and see if I can put a gag in his mouth. Anyway, you're a good project manager and you should know how to work with these technical and scientific prima donnas."

Project manager: "I'll never be able to maintain my sanity having to work with him full time for at least one year. Surely you can assign one of the other three senior chemists instead."

Lab manager: "Because of the nature of the other projects I have, John is the only senior chemist I can release for one full year. If your project were two or three months, then I might be able to give you one of the other senior chemists."

Project manager: "I feel like you are dumping Thornton on me without considering what is in the best interests of the project. Perhaps we should have the sponsor resolve this conflict."

Lab manager: "First of all, this is not a conflict. Second, threatening me with sponsor intervention will not help your case. Do you plan on asking for my resources or support ever again in the future? I'm like an elephant. I have a long memory. Third, my responsibility is to meet your deliverable in a manner that is in the best interest of the company.

"Try to look at resource assignments through my eyes. You're worried about the best interests of your project. I have to support some 20 projects and must make decisions in the best interests of the entire company. Benefiting one project at the expense of several other projects is not a good company decision. And I am paid to make sound *company* decisions, whereas you are paid to make a *project* decision."

Project manager: "My salary, promotion, and future opportunities rest solely on the success of this one project, not 20."

Lab manager: "Our relationship must be a partnership based on trust if project management is to succeed. You must trust me when I tell you that your deliverables will be accomplished within time, cost, and quality. It's my job to make that promise and to see that it is kept.

Project manager: "But what about morale? That should also be a factor. There is also another important consideration. The customer wants monthly team meetings, at our location, to assess progress."

Lab manager: "I know that. I read the requirements document. Why are the monthly meetings a problem?"

Project manager: "I have worked with this customer before. At the team meetings, they want to hear the technical status from the people doing the work rather than from the project manager. That means that John Thornton would be directly interfacing with the customer at least once a month. Thornton is a loose cannon, and there's no telling what words will come out of his mouth. If it were not for the interface meetings, I might be agreeable to accept Thornton. But based on previous experience, he simply does not know when to shut up! He could cause irrevocable damage to our project."

Lab manager: "I will take care of John Thornton. And to appease you, I will also attend each one of the customer interface meetings to keep Thornton in line. As far as I'm concerned, Thornton will be assigned and the subject is officially closed!"

THE PROJECT CONTINUED

John Thornton was assigned to the project team. During the second interface meeting, Thornton stood up and complained to the customer that some of the tests that the customer had requested were worthless, serving no viable purpose. Furthermore, Thornton asserted that if he were left alone, he could develop a product far superior to what the customer had requested.

The customer was furious over Thornton's remarks and asserted that they would now evaluate the project performance to date as well as Ducor's commitment to the project. After the evaluation, they would consider whether the project should be terminated or perhaps assigned to one of Ducor's competitors. The lab manager had not been present during either of the first two customer interface meetings.

QUESTIONS

1. How do we create a partnership between the project manager and line managers when the project manager focuses only on the best interest of his or her project and the line manager is expected to make impartial company decisions?
2. Who should have more of a say during negotiations for resources: the project manager or the line manager?
3. How should irresolvable conflicts over staffing between project and line managers be handled?
4. Should an external customer have a say in project staffing?
5. How do we remove an employee who is not performing as expected?
6. Should project managers negotiate for people or deliverables?

American Electronics International

On February 13, 2004, American Electronics International (AEI) was awarded a $30 million contract for R&D and production qualification for an advanced type of guidance system. During an experimental program that preceded this award and was funded by the same agency, AEI identified new materials with advanced capabilities that could easily replace existing field units. The program, entitled the Mask Project, would be 30 months in length, requiring the testing of 15 units. The Mask Project was longer than any other project that AEI had ever worked on. AEI personnel were now concerned about what kind of staffing problems there would be.

BACKGROUND

In June 2002, AEI won a one-year research project for new material development. Blen Carty was chosen as project manager. He had 25 years of experience with the company in both project management and project engineering positions. During the past five years, Blen had successfully performed as the project manager on R&D projects.

AEI used the matrix approach to structuring project management. Blen was well aware of the problems that can be encountered with this organizational form.

When it became apparent that a follow-on contract would be available, Blen felt that functional managers would be reluctant to assign key personnel full time to his project and lose their services for 30 months. Likewise, difficulties could be expected in staffing the project office.

During the proposal stage of the Mask Project, a meeting was held with Blen Carty; John Wallace, the director of project management; and Dr. Albert Runnels, the director of engineering. The purpose of the meeting was to satisfy a customer requirement that all key project members be identified in the management volume of the proposal.

John Wallace: "I'm a little reluctant to make any firm commitment. By the time your program gets off the ground, four of our other projects are terminating and several new projects are starting up. I think it's a little early to make firm selections."

Blen Carty: "But we have a proposal requirement. Thirty months is a long time to assign personnel for. We should consider this problem now."

Dr. Runnels: "Let's put the names of our top people into the proposal. We'll add several Ph.D.'s from our engineering community. That should beef up our management volume. As soon as we're notified of contract go-ahead, we'll see who's available and make the necessary assignments. This is a common practice in the industry."

COMPLETION OF THE MATERIAL DEVELOPMENT PROJECT

The material development program was a total success. From its inception, everything went smoothly. Blen staffed the project office with Richard Flag, a Ph.D. in engineering, to serve as project engineer. This was a risky move at first, because Richard had been a research scientist during his previous four years with the company. During the development project, however, Richard demonstrated that he could divorce himself from R&D and perform the necessary functions of a project engineer assigned to the project office. Blen was pleased with the way that Richard controlled project costs and directed activities.

Richard had developed excellent working relations with development lab personnel and managers. Richard permitted lab personnel to work at their own rate of speed provided that schedule dates were kept. Richard spent 10 minutes each week with each of the department managers informing them of the status of the project. The department managers liked this approach because they received firsthand (nonfiltered) information concerning the total picture, not necessarily on their own activities, and because they did not have to spend "wasted hours" in team meetings.

When it became evident that a follow-up contract might be available, Blen spent a large percentage of his time traveling to the customer, working out the details for future business. Richard then served as both project manager and project engineer.

The customer's project office was quite pleased with Richard's work. Information, both good and bad, was transmitted as soon as it became available. Nothing was hidden or disguised. Richard became familiar with all of the customer's project office personnel through the monthly technical interchange meetings.

At completion of the material development project, Blen and John decided to search for project office personnel and make recommendations to upper-level management. Blen wanted to keep Richard on board as chief project engineer. He would be assigned six engineers and would have to control all engineering activities within time, cost, and performance specifications. Although this would be a new experience for Richard, Blen felt that he could easily handle it.

Unfortunately, the grapevine was saying that Larry Gilbert was going to be assigned as chief project engineer for the Mask Project.

SELECTION PROBLEMS

On November 15, Dr. Runnels and Blen Carty had a meeting to select the key members of the project team.

Dr. Runnels: "Well, Blen, the time has come to decide on your staff. I want to assign Larry Gilbert as chief engineer. He's a good man and has 15 years' experience. What are your feelings on that?

Blen Carty: "I was hoping to keep Richard Flag on. He has performed well, and the customer likes working with him."

Dr. Runnels: "Richard does not have the experience necessary for that position. We can still assign him to Larry Gilbert and keep him in the project office."

Blen Carty: "I'd like to have Larry Gilbert working for Richard Flag, but I don't suppose that we'd ever get approval to have a grade-9 engineer working for a grade-7 engineer. Personally, I'm worried about Gilbert's ability to work with people. He has been so regimented in his ways that our people in the functional units have refused to work with him. He treats them as kids, always walking around with a big stick. One department manager said that if Gilbert becomes the boss, then it will probably result in cutting the umbilical cord between the project office and his department. His people refuse to work for a dictator. I have heard the same from other managers."

Dr. Runnels: "Gilbert gets the job done. You'll have to teach him how to be a Theory Y manager. You know, Blen, we don't have very many grade-9 engineering positions in this company. I think we should have a responsibility to our employees. I can't demote Gilbert into a lower slot. If I were to promote Flag and the project gets canceled, where would I reassign him? He can't go back to functional engineering. That would be a step down."

Blen Carty: "But Gilbert is so set in his ways. He's just totally inflexible. In addition, 30 months is a long time to maintain a project office. If he screws up, we'll never be able to replace positions in time without totally upsetting the customer. There seem to be an awful lot of people volunteering to work on the Mask Project. Is there anyone else available?"

Dr. Runnels: "People always volunteer for long-duration projects because it gives them a feeling of security. This even occurs among our dedicated personnel. Unfortunately, we have no other grade-9 engineers available. We could reassign one from another program, but I hate to do it. Our engineers like to carry a project through from start to finish. I think you had better spend some time with the functional managers making sure that you get good people."

Blen Carty: "I've tried that and am having trouble. The functional managers will not surrender their key people full time for 30 months. One manager wants to assign two employees to our project so that they can get on-the-job training. I told him that this project is considered strategic by our management and that we must have good people. The manager just laughed at me and walked away."

Dr. Runnels: "You know, Blen, you cannot have all top people. Our other projects must be manned. Also, if you were to use all seasoned veterans, the cost would exceed what we put into the proposal. You're just going to have to make do with what you can get. Prepare a list of the people you want and I'll see what I can do."

As Blen left the office, he wondered if Dr. Runnels would help him in obtaining key personnel.

QUESTIONS

1. Whose responsibility is it to staff the office?
2. What should be Blen Carty's role and Dr. Runnels's role?
3. Should Larry Gilbert be assigned?
4. How would you negotiate with the functional managers?

The Carlson Project

"I sympathize with your problems, Frank," stated Joe McGee, manager of project managers. "You know as well as I do that I'm supposed to resolve conflicts and coordinate efforts among all projects. Staffing problems are your responsibility."

Frank: Royce Williams has a resume that would choke a horse. I don't understand why he performs with a lazy, I-don't-care attitude. He has 15 years of experience in a project organizational structure, with 10 of those years being in project offices. He knows the work that has to be done."

McGee: "I don't think that it has anything to do with you personally. This happens to some of our best workers sooner or later. You can't expect guys to give 120 percent all of the time. Royce is at the top of his pay grade and, being an exempt employee, he doesn't get paid for overtime. He'll snap out of it sooner or later."

Frank: "I have deadlines to meet on the Carlson Project. Fortunately, the Carlson Project is big enough that I can maintain a full-time project office staff of eight employees, not counting myself.

"I like to have all project office employees assigned full time and qualified in two or three project office areas. It's a good thing that I have someone else checked out in Royce's area. But I just can't keep asking this other guy to do his own work and Royce's. This poor guy has been working 60 to 70 hours a week, and Royce has been doing only 40. That seems unfair to me."

McGee: "Look, Frank, I have the authority to fire him, but I'm not going to. It doesn't look good if we fire somebody because they won't work free overtime. Last year we had a case similar to this, where an employee refused to work on Monday and Wednesday evenings because it interfered with his MBA classes. Everyone knew he was going to resign the instant he finished his degree, and yet there was nothing that I could do."

Frank: "There must be other alternatives for Royce Williams. I've talked to him as well as to other project office members. Royce's attitude doesn't appear to be demoralizing the other members, but it easily could be in a short time."

McGee: "We can reassign him to another project, as soon as one comes along. I'm not going to put him on my overhead budget. Your project can support him for the time being. You know, Frank, the grapevine will know the reason for his transfer. This might affect your ability to get qualified people to volunteer to work with you on future projects. Give Royce a little time and see if you can work it out with him. What about this guy, Harlan Green, from one of the functional groups?"

Frank: "Two months ago, we hired Gus Johnson, a man with 10 years of experience. For the first two weeks that he was assigned to my project, he worked like hell and got the work done ahead of schedule. His work was flawless. That was the main reason why I wanted him. I know him personally, and he's one great worker.

"During weeks three and four, his work slowed down considerably. I chatted with him and he said that Harlan Green refused to work with him if he kept up that pace."

McGee: "Did you ask him why?"

Frank: "Yes. First of all, you should know that, for safety reasons, all men in that department must work in two- or three-men crews. Therefore, Gus was not allowed to work alone. Harlan did not want to change the standards of performance for fear that some of the other employees would be laid off.

"By the end of the first week, nobody in the department would talk to Gus. As a matter of fact, they wouldn't even sit with him in the cafeteria. So, Gus had to either conform to the group or remain an outcast. I feel partially responsible for what has happened, since I'm the one who brought him here.

"I know that has happened before, in the same department. I haven't had a chance to talk to the department manager yet. I have an appointment to see him next week."

McGee: "There are solutions to the problem, simple ones. But, again, it's not my responsibility. You can work it out with the department manager."

"Yeah," thought Frank. "But what if we can't agree?"

QUESTION

1. How do you think this situation was resolved?

Communication Failures

BACKGROUND

Herb had been with the company for more than eight years and had worked on various R&D and product enhancement projects for external clients. He had a Ph.D. in engineering and had developed a reputation as a subject matter expert. Because of his specialized skills, he worked by himself most of the time and interfaced with the various project teams only during project team meetings. All of that was about to change.

Herb's company had just won a two-year contract from one of its best customers. The first year of the contract would be R&D, and the second year would be manufacturing. The company made the decision that the person best qualified to be the project manager was Herb because of his knowledge of R&D and manufacturing. Unfortunately, Herb had never taken any courses in project management, and because of his limited involvement with previous project teams, there were risks in assigning him as the project manager. But management believed he could do the job.

THE TEAM IS FORMED

Herb's team consisted of 14 people, most of whom would be full time for at least the first year of the project. The four people Herb would be interfacing with on a daily basis were Alice, Bob, Betty, and Frank.

- Alice was a seasoned veteran who worked with Herb in R&D. She had been with the company longer than Herb and would coordinate the efforts of the R&D personnel.

- Bob also had been with the company longer that Herb and had spent his career in engineering. Bob would coordinate the engineering efforts and drafting.
- Betty was relatively new to the company. She would be responsible for all reports, records management, and procurements.
- Frank, a five-year employee with the company, was a manufacturing engineer. Unlike Alice, Bob, and Betty, Frank would be part time on the project until it was time to prepare the manufacturing plans.

For the first two months of the program, work seemed to be progressing as planned. Everyone understood their role on the project, and there were no critical issues.

FRIDAY THE 13TH

Herb held weekly teams meetings every Friday from 2:00 to 3:00 p.m. Unfortunately, the next team meeting would fall on Friday the 13th, and that bothered Herb because he was somewhat superstitious. He was considering canceling the team meeting just for that week but decided against it.

At 9:00 a.m. on Friday the 13th, Herb met with his project sponsor, as he always did. Two days before, Herb casually talked to his sponsor in the hallway and the sponsor told Herb that on Friday, he would like to discuss the cash flow projections for the next six months and ways to reduce some of the expenditures. The sponsor had seen some expenditures that bothered him. As soon as Herb entered the sponsor's office, the sponsor said: "It looks like you have no report with you. I specifically recall asking you for a report on the cash flow projections."

Herb was somewhat displeased over this. He specifically recalled that this was to be a discussion only, and no report was requested. But Herb knew that "rank has its privileges" and questioning the sponsor's communication skills would be wrong. Obviously, this was not a good start to Friday the 13th.

At 10:00 a.m., Alice came into Herb's office, and he could see from the expression on her face that she was somewhat distraught. Alice then said: "Herb, last Monday I told you that the company was considering me for promotion and the announcements would be made this morning. Well, I did not get promoted. How come you never wrote a letter of recommendation for me?"

Herb remembered the conversation vividly. Alice did say that she was being considered for promotion but never asked him to write a letter of recommendation. Did Alice expect Herb to read between the lines and try to figure out what she really meant?

Herb expressed his sincere apologies for what happened. Unfortunately, this did not make Alice feel any better as she stormed out of Herb's office. Obviously, Herb's day was getting worse.

No sooner had Alice exited the doorway to Herb's office than Bob entered. Herb could tell that Bob had a problem. Bob then stated:

In one of our team meetings last month, you stated that you had personally contacted some of my engineering technicians and told them to perform this week's tests at 70°F, 90°F, and 110°F. You and I know that the specifications called for testing at 60°F, 80°F, and 100°F. That's the way it was always done, and you were asking them to perform the tests at different intervals than the specifications called for.

Well, it seems that the engineering technicians forgot the conversation you had with them and did the tests according to the specification criteria. I assumed that you had followed up your conversation with them with a memo, but that was not the case. It seems that they forgot. When dealing with my engineering technicians, the standard rule is "If it's not in writing, then it hasn't been said." From now on, I would recommend that you let me provide the direction to my engineering technicians. My responsibility is engineering, and all requests of my engineering personnel should go through me.

Yes, Friday the 13th had become a very bad day for Herb. What else could go wrong? he wondered. It was now 11:30 a.m. and almost time for lunch. Herb was considering locking his office door so that nobody could find him and disconnecting his phone. But in walked Betty and Frank, and once again he could tell by their expressions that they had problems. Frank spoke first.

I just received confirmation from procurement that they purchased certain materials which we will need when we begin manufacturing. We are a year away from beginning manufacturing, and, if the final design changes in the slightest, we will be stuck with costly raw materials that cannot be used. Also, my manufacturing budget did not have the cash flow for early procurement. I should be involved in all procurement decisions involving manufacturing. I might have been able to get it cheaper than Betty did. So, how was this decision made without me?

Before Herb could say anything, Betty spoke up.

Last month, Herb, you asked me to look into the cost of procuring these materials. I found a great price at one of the vendors and made the decision to purchase them. I thought that this was what you wanted me to do. This is how we did it in the last company I worked for.

Herb then remarked: "I just wanted you to determine what the cost would be, not to make the final procurement decision, which is not your responsibility."

Friday the 13th was becoming possibly the worst day in Herb's life. He decided not to take any further chances. As soon as Betty and Frank left, Herb immediately sent out e-mails to all team members canceling the team meeting scheduled for 2:00 to 3:00 p.m. that day.

QUESTIONS

1. How important are communication skills in project management?
2. Was Herb the right person to be assigned as the project manager?
3. There were communications issues with Alice, Bob, Betty, and Frank. For each communication issue, where was the breakdown in communications: encoding, decoding, feedback, and so on?

Part 6

PROJECT ESTIMATING

Some people believe the primary critical factor for project success is the quality of the estimate. Unfortunately, not all companies have estimating databases, nor do all companies have good estimates. Some companies are successful estimating at the top levels of the work breakdown structure, while others are willing to spend the time and money estimating at the lower levels of the work breakdown structure.

In organizations that are project driven and survive on competitive bidding, good estimates are often massaged and then changed based on the belief by management that the job cannot be won without a lower bid. This built-in process can and does severely impact the project manager's ability to get people to be dedicated to the project's financial baseline.

Capital Industries

In the summer of 2006, Capital Industries undertook a material development program to see if a hard plastic bumper could be developed for medium-size cars. By January 2007, Project Bumper (as it was called by management) had developed a material that endured all preliminary laboratory testing.

One more step was required before full-scale laboratory testing: a three-dimensional stress analysis on bumper impact collisions. The decision to perform the stress analysis was the result of a concern on the part of the technical community that the bumper might not perform correctly under certain conditions. The cost of the analysis would require corporate funding over and above the original estimates. Since the current costs were identical to what was budgeted, the additional funding was a necessity.

Frank Allen, the project engineer in the Bumper Project Office, was assigned control of the stress analysis. Frank met with the functional manager of the engineering analysis section to discuss the assignment of personnel to the task.

Functional manager: "I'm going to assign Paul Troy to this project. He's a new man with a Ph.D. in structural analysis. I'm sure he'll do well."

Frank Allen: "This is a priority project. We need seasoned veterans, not new people, regardless of whether or not they have Ph.D.'s. Why not use some other project as a testing ground for your new employee?"

Functional manager: "You project people must accept part of the responsibility for on-the-job training. I might agree with you if we were talking about

147

blue-collar workers on an assembly line. But this is a college graduate, coming to us with a good technical background."

Frank Allen: "He may have a good background, but he has no experience. He needs supervision. This is a one-man task. The responsibility will be yours if he fouls up."

Functional manager: "I've already given him our book for cost estimates. I'm sure he'll do fine. I'll keep in close communication with him during the project."

Frank Allen met with Paul Troy to get an estimate for the job.

Paul Troy: "I estimate that 800 hours will be required."

Frank Allen: "Your estimate seems low. Most three-dimensional analyses require at least 1,000 hours. Why is your number so low?"

Paul Troy: "Three-dimensional analysis? I thought that it would be a two-dimensional analysis. But no difference; the procedures are the same. I can handle it."

Frank Allen: "Okay, I'll give you 1,100 hours. But if you overrun it, we'll both be sorry."

Frank Allen followed the project closely. By the time the costs were 50 percent completed, performance was only 40 percent. A cost overrun seemed inevitable. The functional manager still asserted that he was tracking the job and that the difficulties were a result of the new material properties. His section had never worked with materials like these before.

Six months later, Troy announced that the work would be completed in one week, two months later than planned. The two-month delay caused major problems in facility and equipment utilization. Project Bumper was still paying for employees who were waiting to begin full-scale testing.

On Monday mornings, the project office would receive the weekly labor monitor report for the previous week. This week the report indicated that the publications and graphics art department had spent over 200 man-hours last week in preparation of the final report. Frank Allen was furious. He called a meeting with Paul Troy and the functional manager.

Frank Allen: "Who told you to prepare a formal report? All we wanted was a go or no-go decision as to structural failure."

Paul Troy: "I don't turn in any work unless it's professional. This report will be documented as a masterpiece."

Frank Allen: "Your 50 percent cost overrun will also be a masterpiece. I guess your estimating was a little off!"

Paul Troy: "Well, this was the first time that I had performed a three-dimensional stress analysis. And what's the big deal? I got the job done, didn't I?"

QUESTIONS

1. Who is the best person qualified to make functional estimates?
2. Can this task be delegated?
3. Should the estimating of this task have been delegated?
4. Should Paul Troy have been delegated the responsibility for the estimate?
5. Should Frank Allen have sensed a communication problem between Paul Troy and the functional manager?
6. Should Frank Allen have tracked the project more closely because of the above-mentioned problems?
7. Does Paul Troy appear to realize that there are also time and cost constraints on a project?
8. Is it beneficial for project office personnel to know the manhour estimates?

Small-Project Cost Estimating at Percy Company

Paul graduated from college in June 2000 with a degree in industrial engineering. He accepted a job as a manufacturing engineer in the Manufacturing Division of Percy Company. His prime responsibility was performing estimates for the Manufacturing Division. Each estimate was then given to the appropriate project office for consideration. The estimation procedure history had shown the estimates to be valid.

In 2005, Paul was promoted to project engineer. His prime responsibility was the coordination of all estimates for work to be completed by all of the divisions. For one full year, Paul went by the book and did not do any estimating except for project office personnel manager. After all, he was now in the Project Management Division, which contained job descriptions including such words as "coordinating and integrating."

In 2006, Paul was transferred to small-program project management. This was a new organization designed to perform low-cost projects. The problem was that these projects could not withstand the expenses needed for formal divisional cost estimates. For five projects, Paul's estimates were right on the money. But the sixth project incurred a cost overrun of $20,000 in the Manufacturing Division.

In November 2007, a meeting was called to resolve the question of "Why did the overrun occur?" The attendees included the general manager, all division managers and directors, the project manager, and Paul. Paul now began to worry about what he should say in his defense.

QUESTIONS

1. Should Paul have been permitted to price out the jobs?
2. Should Paul have shown his estimates to the functional managers for their blessings?
3. Can this situation be corrected for small projects?
4. Should line managers be willing to price out small jobs using their line organization overhead cost account?
5. What are the long-term problems in the case study?
6. How can the lack of project planning and controlling throughout the life of a project affect the end cost of the project?

Cory Electric

"Frankly speaking, Jeff, I didn't think that we would stand a chance in winning this $20 million program. I was really surprised when they said that they'd like to accept our bid and begin contract negotiations. As chief contract administrator, you'll head up the negotiating team," remarked Gus Bell, vice president and general manager of Cory Electric. "You have two weeks to prepare your data and line up your team. I want to see you when you're ready to go."

Jeff Stokes was chief contract negotiator for Cory Electric, a $750-million-a-year electrical components manufacturer serving virtually every major U.S. industry. Cory Electric had a well-established matrix structure that had withstood 15 years of testing. Job casting standards were well established but did include some fat on the discretion of the functional manager.

Two weeks later, Jeff met with Gus Bell to discuss the negotiation process.

Gus Bell: "Have you selected an appropriate team? You had better make sure that you're covered on all sides."

Jeff: "There will be four, plus myself, at the negotiating table: the program manager; the chief project engineer who developed the engineering labor package; the chief manufacturing engineer who developed the production labor package; and a pricing specialist who has been on the proposal since the kickoff meeting. We have a strong team and should be able to handle any questions."

Gus Bell: "Okay, I'll take your word for it. I have my own checklist for contract negotiations. I want you to come back with a guaranteed fee of $1.6 million for our stockholders. Have you worked out the possible situations based on the negotiated costs?"

Jeff: "Yes! Our minimum position is $20 million plus an 8 percent profit. Of course, this profit percentage will vary depending on the negotiated cost. We can bid the program at a $15 million cost—that's $5 million below our target—and still book a $1.6 million profit by overrunning the cost-plus-incentive-fee contract. Here is a list of the possible cases." (See Table I.)

Gus Bell: "If we negotiate a cost overrun fee, make sure that cost accounting knows about it. I don't want the total fee to be booked as profit if we're going to need it later to cover the overrun. Can we justify our overhead rates, general and administrative costs, and our salary structure?"

Jeff: "That's a problem. You know that 20 percent of our business comes from Mitre Corporation. If they fail to renew our contract for another two-year follow-on effort, then our overhead rates will jump drastically. Which overhead rates should I use?"

TABLE I   COST POSITIONS

| Negotiated Cost | Negotiated Fee | | | | |
	%	Target Fee	Overrun Fee	Total Fee	Total Package
15,000,000	14.00	1,600,000	500,000	2,100,000	17,100,000
16,000,000	12.50	1,600,000	400,000	2,000,000	18,000,000
17,000,000	11.18	1,600,000	300,000	1,900,000	18,900,000
18,000,000	10.00	1,600,000	200,000	1,800,000	19,800,000
19,000,000	8.95	1,600,000	100,000	1,700,000	20,700,000
20,000,000	8.00	1,600,000	0	1,600,000	21,600,000
21,000,000	7.14	1,600,000	−100,000	1,500,000	*22,500,000
22,000,000	6.36	1,600,000	−200,000	1,400,000	23,400,000
23,000,000	5.65	1,600,000	−300,000	1,300,000	24,300,000
24,000,000	5.00	1,600,000	−400,000	1,200,000	

Assume total cost will be spent:

21,000,000	7.61	
22,000,000	7.27	Minimum position = $20,000,000
23,000,000	6.96	Minimum fee = 1,600,000 = 8% of minimum position
24,000,000	6.67	Sharing ratio = 90%/10%

* Buy-in

Gus Bell: "Let's put in a renegotiation clause to protect us against a drastic change in our business base. Make sure that the customer understands that as part of the terms and conditions. Are there any unusual terms and conditions?"

Jeff: "I've read over all terms and conditions, and so have all of the project office personnel as well as the key functional managers. The only major item is that the customer wants us to qualify some new vendors as sources for raw material procurement. We have included in the package the cost of qualifying two new raw material suppliers."

Gus Bell: "Where are the weak points in our proposal? I'm sure we have some."

Jeff: "Last month, the customer sent in a fact-finding team to go over all of our labor justifications. The impression that I get from our people is that we're covered all the way around. The only major problem might be where we'll be performing on our learning curve. We put into the proposal a 45 percent learning curve efficiency. The customer has indicated that we should be up around 50 to 55 percent efficiency, based on our previous contracts with him. Unfortunately, the contracts the customer referred to were four years old. Several of the employees who worked on those programs have left the company. Others are assigned to ongoing projects here at Cory. I estimate that we could put together about 10 percent of the people we used previously. That learning curve percentage will be a big point for disagreements. We finished off the previous programs with the customer at a 35 percent learning curve position. I don't see how they can expect us to be smarter, given these circumstances."

Gus Bell: "If that's the only weakness, then we're in good shape. It sounds like we have a foolproof audit trail. That's good! What's your negotiation sequence going to be?"

Jeff: "I'd like to negotiate the bottom line only, but that's a dream. We'll probably negotiate the raw materials, the man-hours and the learning curve, the overhead rate, and, finally, the profit percentage. Hopefully, we can do it in that order."

Gus Bell: "Do you think that we'll be able to negotiate a cost above our minimum position?"

Jeff: "Our proposal was for $22.2 million. I don't foresee any problem that will prevent us from coming out ahead of the minimum position. The 5 percent change in learning curve efficiency amounts to approximately $1 million. We should be well covered.

"The first move will be up to them. I expect that they'll come in with an offer of $18 to $19 million. Using the binary chop procedure, that'll give us our guaranteed minimum position."

Gus Bell: "Do you know the guys who you'll be negotiating with?"

Jeff: "Yes, I've dealt with them before. The last time, the negotiations took three days. I think we both got what we wanted. I expect this one to go just as smoothly."

Gus Bell: "Okay, Jeff. I'm convinced we're prepared for negotiations. Have a good trip."

The negotiations began at 9:00 a.m. on Monday morning. The customer countered the original proposal of $22.2 million with an offer of $15 million.

After six solid hours of arguments, Jeff and his team adjourned. Jeff immediately called Gus Bell at Cory Electric.

Jeff: "Their counteroffer to our bid is absurd. They've asked us to make a counteroffer to their offer. We can't do that. The instant we give them a counteroffer, we are in fact giving credibility to their absurd bid. Now they're claiming that, if we don't give them a counteroffer, then we're not bargaining in good faith. I think we're in trouble."

Gus Bell: "Has the customer done their homework to justify their bid?"

Jeff: "Yes. Very well. Tomorrow we're going to discuss every element of the proposal, task by task. Unless something drastically changes in their position within the next day or two, contract negotiations will probably take up to a month."

Gus Bell: "Perhaps this is one program that should be negotiated at the top levels of management. Find out if the person that you're negotiating with reports to a vice president and general manager, as you do. If not, break off contract negotiations until the customer gives us someone at your level. We'll negotiate this at my level, if necessary."

QUESTIONS

1. How many people should make up the negotiations team?
2. What is the role of the project manager?
3. What items are negotiated during contract negotiations?
4. What is the order of these items?
5. Should the negotiating parties be headed up by individuals who are the same rank?
6. Can a company bid on a contract that is well below its minimum cost estimate? If so, under what circumstances?

Camden Construction Corporation

"For five years, I've heard nothing but flimsy excuses from you people as to why the competition was beating us out in the downtown industrial building construction business," remarked Joseph Camden, president. "Excuses, excuses, excuses; that's all I ever hear! Only 15 percent of our business over the past five years has been in this area, and virtually all of that was with our established customers. Our growth rate is terrible. Everyone seems to just barely outbid us. Maybe our bidding process leaves something to be desired. If you three vice presidents don't come up with the answers, then we'll have three positions to fill by midyear

"We have a proposal request coming in next week, and I want to win it. Do you guys understand that?"

BACKGROUND

Camden Construction Corporation matured from a $1 million to a $26 million construction company between 1989 and 1999. Camden's strength was in its ability to work well with the customer. Its reputation for quality work far exceeded the local competitor's reputation. Camden's business grew rapidly for the next decade.

Most of Camden's contracts were with long-time customers who were willing to go sole-source procurement and pay the extra price for quality and service. With the economic downturn in 2008, Camden found that, unless it penetrated the competitive bidding market, its business base would decline.

In 2010, Camden was "forced" to go union in order to bid government projects. Unionization drastically reduced Camden's profit margin but offered a greater promise for increased business. Camden had avoided the major downtown industrial construction market. But with the availability of multimillion-dollar skyscraper projects, Camden wanted its share of the pot of gold at the base of the rainbow.

MEETING OF THE MINDS

On January 17, 2011, the three vice presidents met to consider ways of improving Camden's bidding technique.

V.P. finance: "You know, fellas, I hate to say it, but we haven't done a good job in developing a bid. I don't think that we've been paying enough attention to the competition. Now's the time to begin."

V.P. operations: "What we really need is a list of who our competitors have been on each project over the last five years. Perhaps we can find some bidding trends."

V.P. engineering: "I think the big number we need is to find out the overhead rates of each of the companies. After all, union contracts specify the rate at which the employees will work. Therefore, except for the engineering design packages, all of the companies should be almost identical in direct labor man-hours and union labor wages for similar jobs."

V.P. finance: "I think I can hunt down past bids by our competitors. Many of them are in public records. That'll get us started."

V.P. operations: "What good will it do? The past is past. Why not just look toward the future?"

V.P. finance: "What we want to do is to maximize our chances for success and maximize profits at the same time. Unfortunately, we can't do both of these things at the same time. We must find a compromise."

V.P. engineering: "Do you think that the competition looks at our past bids?"

V.P. finance: "They're stupid if they don't. What we have to do is to determine their target profit and target cost. I know many of the competitors personally and have a good feel for what their target profits are. We'll have to assume that their target direct costs equals ours; otherwise we will have a difficult time making a comparison."

V.P. engineering: "What can we do to help you?"

Finance: "You'll have to tell me how long it takes to develop the engineering design packages, and how our personnel in engineering design stack up against the competition's salary structure. See if you can make some contacts and find out how much money the competition put into some of their proposals for engineering design activities. That'll be a big help.

"We'll also need good estimates from engineering and operations for this new project we're suppose to bid. Let me pull my data together, and we'll meet again in two days, if that's all right with you two."

REVIEWING THE DATA

The executives met two days later to review the data. The vice president for finance presented the data on the three most likely competitors. (See Table I.) These companies were Ajax, Acme, and Pioneer. The vice president for finance said:

> In 2003, Acme was contract-rich and had a difficult time staffing all of its projects. In 2000, Pioneer was in danger of bankruptcy. It was estimated that it needed to win one or two in order to hold its organization together.
>
> Two of the 2002 companies were probably buy-ins based on the potential for follow-on work. The 2004 contract was for an advanced state-of-the-art project. It is estimated that Ajax bought in so that it could break into a new field.

TABLE I   PROPOSAL DATA SUMMARY (COST IN 10s OF 1,000s, US$)

Year	Acme	Ajax	Pioneer	Camden Bid	Camden Cost
2000	270	244	260	283	260
2000	260	250	233	243	220
2000	355	340	280	355	300
2001	836	830	838	866	800
2001	300	288	286	281	240
2001	570	560	540	547	500
2002	240*	375	378	362	322
2002	100*	190	180	188	160
2002	880	874	883	866	800
2003	410	318	320	312	280
2003	220	170	182	175	151
2004	400	300	307	316	283
2004	408	300*	433	449	400
2005	338	330	342	333	300
2005	817	808	800	811	700
2005	886	884	880	904	800
2006	384	385	380	376	325
2006	140	148	158	153	130
2007	197	193	188	200	165
2007	750	763	760	744	640

*Buy-in contracts

The vice presidents for engineering and operations presented data indicating that the total project cost (fully burdened) was approximately $5 million. "Well," thought the vice president of finance, "I wonder what we should bid so it we will have at least a reasonable chance of winning the contract."

QUESTIONS

1. How hungry are the competitors?
2. How hungry are we?
3. What has been the bidding history of the competitors?
4. What has been our bidding history?
5. How do we compare to the other competitors cost-wise?
6. How do we compare to the other competitors technology-wise?

The Estimating Problem

Barbara just received the good news: She was assigned as the project manager for a project that her company won as part of competitive bidding. Whenever a request for proposal (RFP) comes into Barbara's company, a committee composed mainly of senior managers reviews the RFP. If the decision is made to bid on the job, the RFP is turned over to the Proposal Department. Part of the Proposal Department is an estimating group that is responsible for estimating all work. If the estimating group has no previous history with some of the deliverables or work packages and is unsure about the time and cost for the work, the estimating team will then ask the functional managers for assistance with estimating.

Project managers like Barbara do not often participate in the bidding process. Usually their first knowledge about a project comes after the contract is awarded to their company and they are assigned as the project manager. Some project managers are highly optimistic and trust the estimates that were submitted in the bid implicitly unless, of course, a significant span of time has elapsed between the date of submittal of the proposal and the final contract award date. Barbara, however, is somewhat pessimistic. She believes that accepting the estimates as they were submitted in the proposal is like playing Russian roulette. As such, Barbara prefers to review the estimates. One of the most critical work packages in the project was estimated at 12 weeks using one grade 7 employee full time. Barbara had performed this task on previous projects, and it required one person full time for 14 weeks. Barbara asked the estimating group how they arrived at this estimate. The estimating group responded that they used the three-point estimate where the optimistic time was four weeks, the most likely time was 13 weeks, and the pessimistic time was 16 weeks.

Barbara believed that the three-point estimate was way off of the mark. The only way that this work package could ever be completed in four weeks would be for a very small project nowhere near the complexity of Barbara's project. Therefore, the estimating group was not considering any complexity factors when using the three-point estimate. Had the estimating group used the triangular distribution where each of the three estimates had an equal likelihood of occurrence, the final estimate would have been 13 weeks. This was closer to the 14 weeks that Barbara thought the work package would take. While a difference of one week seems small, it could have a serious impact on Barbara's project and incur penalties for late delivery.

Barbara was still confused and decided to talk to Peter, the employee assigned to do this task. Barbara had worked with Peter on previous projects. Peter was a grade 9 employee and considered to be an expert in this work package. As part of the discussions with Barbara, Peter said: "I have seen estimating data bases that include this type of work package, and they all estimate the work package at about 14 weeks. I do not understand why our estimating group prefers to use the three-point estimate."

"Does the typical data base account for project complexity when considering the estimates?" asked Barbara.

Peter responded:

> Some databases have techniques for considering complexity, but mostly they just assume an average complexity level. When complexity is important, as it is in our project, analogy estimating would be better. Using analogy estimating and comparing the complexity of the work package on this project to the similar works packages I have completed, I would say that 16–17 weeks is closer to reality, and let's hope I do not get removed from the project to put out a fire somewhere else in the company. That would be terrible. It is impossible for me to get it done in 12 weeks. And adding more people to this work package will not shorten the schedule. It may even make it worse.

Barbara then asked Peter one more question: "Peter, you are a grade 9 and considered as the subject matter expert. If a grade 7 had been assigned, as the estimating group had said, how long would it have taken the grade 7 to do the job?"

"Probably about 20 weeks or so," responded Peter.

QUESTIONS

1. How many different estimating techniques were discussed in the case?
2. If each estimate is different, how does a project manager decide that one estimate is better than another?
3. If you were the project manager, which estimate would you use?

The Singapore
Software Group (A)

BACKGROUND

The Singapore Software Group (SSG) was a medium-size company that had undergone significant growth over the past 20 years. Initially, the company provided software services just to the Pacific Rim countries. Now it serviced all of Asia and had contracts and partnerships with companies in Europe, South America, and North America.

SSG's strengths were in software development, database management, and management information systems. SSG created an excellent niche for itself and maintained a low-risk strategy whereby growth was funded out of cash flow rather than through bank borrowing. The low-risk strategy forced SSG to focus on providing the same type of high-quality deliverables to its existing and new clients rather than expanding into other software development markets with other products and services. Although SSG had a reputation for quality products and services, excellence in customer support, and competitive pricing, the software landscape was changing.

SSG had seen a 400 percent increase in the number of small companies entering the marketplace over the past several years and competing with it in its core lines of business. Its client base was under attack by other Southeast Asian countries that had lower salary structures and lower costs of living, thus allowing these new companies to put pressure on SSG's profit margins.

Under the direction of senior management, SSG performed a SWOT (strengths, weaknesses, opportunities, and threats) analysis. SSG's strengths were quite clear: highly talented and dedicated personnel, a fairly young labor force, and significant knowledge related to its existing products and services. Its internal weaknesses were also quite apparent; although SSG provided training and educational opportunities to its workforce, the opportunities were limited to educational programs that would directly support existing products and services only; for the business to expand, significant expenses would be incurred for retraining its labor force or workers with new talents would need to be hired; and last, the time necessary to expand the business may be too long to take advantage of an opportunity.

The threats to SSG were real. The company did not know if it could maintain its current growth rate. SSG had already seen signs that its growth rate might be deteriorating since many of its critical clients were seeking out competitors that possessed skills that SSG did not have. There was no question that the opportunities for growth existed if SSG could expand its skills in a timely manner and penetrate new fields. The questions, of course, were What new skills do we need? and What new products can we offer? There were opportunities for joint ventures and partnerships, but SSG preferred to remain independent and often had a go-it-alone mentality.

NEED FOR GROWTH

In a bold move, SSG began training its workforce in the type of software development necessary to support the iPhone, iPad, and other screen-interfacing software packages. Although SSG had some limited experience in this area, software to support touch screens was seen as the future. This included software to support games, telecommunications, photography, and videotaping.

SSG realized that training alone would not be sufficient. Time was a real serious constraint for the plans for continued growth. It would be necessary to hire additional staff with expertise in computer engineering rather than just computer programming. SSG was fortunate to be able to attract highly talented people with the expertise needed to compete in this area. Some of the new hires even came from SSG's competitors who were looking at opportunities to enter this marketplace.

It took almost a year and significant expense to build the in-house skills that SSG desperately needed for future growth. With some in-house experimental work, SSG was able to eliminate many of the software bugs that plagued the first-generation touch-screen products and even exceeded performance in some cases. But all this was just part of educational development and limited R&D. What was needed now were contracts.

REQUEST FOR PROPOSAL FROM TAIWAN TECHNOLOGIES

Taiwan Technologies (TT) was one of SSG's most important clients. SSG was on TT's preferred supplier list and was awarded more contracts from TT than from any other client. Many of the contracts were awarded on points, including past performance, rather than simply being the lowest bidder.

TT was in the process of designing new products to enter the smartphone marketplace and compete with other smartphone suppliers. TT's major strengths were in redesigning someone else's products and improving the performance and quality. Without having to recover vast R&D costs that others were incurring, TT was able to become a low-cost supplier. TT's strengths were manufacturing-driven, and it possessed very limited capability in software development.

SSG was one of five companies invited to bid on creating the software. The problem was that TT's design efforts were still in progress and SSG's efforts would be done in parallel with TT's work. TT's specifications were only partially complete. The final designs would not be known until perhaps six months into the project.

To complicate matters further, TT was requiring that the contract be a firm-fixed-price effort. Usually, parallel development work is done with cost-reimbursable contracts. With a firm-fixed-price contract, SSG could be exposed to significant risks, especially if downstream scope changes resulted in rework. The risks could be partially mitigated through a formalized change control process. Since the number and magnitude of the downstream scope changes were unknown, having the project manager on board the project on a full-time basis was critical.

PROPOSAL KICKOFF MEETING

SSG's senior management made an immediate decision that SSG would be bidding on this contract. In attendance at the bidding kickoff meeting were senior representatives from all of the groups that would be supplying support for the project. Also in attendance was the chief executive officer (CEO), who announced that Jim Kirby would be the project manager and would be assigned full time for the duration of the project. The project manager would be expected to work 2,000 hours of direct labor.

Frank Ling (Business Analyst): "As the business analyst, it is my responsibility to make sure that we have the right business case for this project. The following information in the business case is critical for project planning:

- The market demand for the TT products could be millions of units a year.
- Downstream product upgrades will provide SSG with a long-term cash flow stream.
- SSG views this project as an essential component of our strategic plan.
- The project may require some technological breakthroughs and, as yet, we are unsure what these breakthroughs might be. However, we feel confident that we can do it in a timely manner.

- Even though we have people trained in this technology, this is a completely new type of project for SSG. We know there are risks.
- Corporate legal says that TT's product requirements thus far pose no legal headaches for SSG.

"We expect that the business case may change until such time as TT finalizes its product description and requirements. In this regard, I will be working closely with Jim Kirby on the impact that the changes will have on the business case and the project. Parallel development projects are always difficult.

"SSG's interest in winning this bid is of the utmost importance, but we do not want to win it at a price that is so low that we are losing money. We need to know as soon as possible what the realistic cost is to meet their requirements. The successful completion of this contract will 'open doors' for us elsewhere."

Kathryn James (VP, Human Resources): "According to the RFP, the go-ahead date is July 1, 2011, and a completion date of June 30, 2012. At SSG, we typically evaluate people for promotion in the first two weeks of December, and the promotions go into effect the beginning of January. We also provide cost-of-living adjustments and salary increases the first week of January as well. We expect that the average cost-of-living and salary adjustment to be 6 percent beginning January 1. For those who receive promotions, they will receive an additional 10 percent salary increase on average.

"There are other facts that must be considered since the program is a year in length. Last year, the average person in the company had:

- Three weeks of paid vacation
- 10 paid holidays
- 4 days of paid sick leave
- 5 days of training paid by corporate
- 2 days of jury duty

"These costs are paid out of corporate overhead and are not part of direct labor. I am also providing you with the salary structure for the departments that are expected to provide support for the project. (See Table I.) If overtime is required, and I assume it will be, all workers will be paid at time and a half, regardless of pay grade.

"The overhead rates for each of the departments are also included in Table I. However, on overtime, the overhead rates are only 50 percent of the overhead rates on regular time."

Paul Creighton (Chief Financial Officer): "Kathryn has provided you with the overhead rates for our departments. We expect these overhead rates to remain the same for the duration of the project. Also, our corporate general and administrative costs (G&A) will remain fixed at 8 percent. The G&A cost are included in all of our contracts and are a necessity to support corporate headquarters. This

TABLE I   SALARY INFORMATION FOR 2011

Department	Pay Grade	2011 Median Hourly Salary, $	Overhead, %
Project Manager	9	56	100
Systems Programmer	5	38	150
Systems Programmer	6	41	150
Systems Programmer	7	45	150
Systems Programmer	8	49	150
Software Programmer	5	38	150
Software Programmer	6	41	150
Software Programmer	7	45	150
Software Programmer	8	49	150
Software Engineer	6	41	150
Software Engineer	7	45	150
Software Engineer	8	49	150
Software Engineer	9	55	150
Manufacturing Engineer	6	41	250
Manufacturing Engineer	7	45	250
Manufacturing Engineer	8	49	250
Manufacturing Engineer	9	55	250

contract is a firm-fixed-price effort, which you all know provides significant risk to the seller. To mitigate some of our possible risk, I want a 15 percent profit margin included in the contract. In the past, the contract profit margins on TT contracts ranged from 10 percent to 15 percent. The higher end of the range was always on the firm-fixed-price contracts.

"I know this is the first time we have worked on a contract like this and that there are risks. I am not opposed to adding in a management reserve as protection. But before you go overboard in adding in a large management reserve, just remember that we want to win the contract."

Ellen Pang (VP, Computer Technologies): "I consider this project to be at the top of the priority list for SSG. Therefore, we will assign the appropriate resources with the necessary skill levels to get the job done. I will assign five people full time for the duration of the project; one from systems programming, three from software programming, and one from software engineering. I believe that each employee will be required to work at least 2,000 hours of direct labor on this project, with the hours broken down equally each month.

"I am not sure right now which employees I will assign because the go-ahead date is a few months away. But I will keep my promise that there will be five workers and that they will be committed full time with no responsibilities on other projects. In addition, I am hiring a consultant with expertise in this type of project. The cost for the consultant for the duration of the project will be $75,000."

Eric Tong (VP, Manufacturing): "I'm sure you all know from newspapers and TV news broadcasts about the problems smartphone manufacturers were having with the casing and cover. To avoid having the same problems and alienating TT, I will be assigning one of my manufacturing engineers who is an expert in value engineering and quality. I expect him to be assigned for roughly 600 hours beginning some time after January, 2012.

"The RFP requires that we experiment with various size touch screens to see if the software is affected by the screen size and screen thickness. This could be part of the problem that other suppliers were having. This experimental work is also included as part of the manufacturing engineer's job."

"I estimate that we will need about $6,000 in material costs. We should probably include a scrap factor as well, but I am unsure right now how much of a scrap factor is reasonable."

Bruce Clay (Proposal Manager): "TT wants the proposal in their hands within 30 days. I think that's enough time to make our estimates. Here is a copy of a small proposal we did for TT a couple of years ago. (See Table II.) It should give you an idea how we price out our projects for competitive bidding.

"Together with your individual estimates, I also need a listing of all of the assumptions you made in arriving at your estimate(s). This is critical information for risk management and decisions on scope changes."

TABLE II   TYPICAL PROJECT PRICING SUMMARY

Dept.	Direct Labor			Overhead		
	Hours	**Rate**	**Dollars**	**%**	**Dollars**	**Total**
Eng.	1000	$42.00	$42,000	110	$46,200	$88,200
Manu.	500	$35.00	$17,500	200	$35,000	$52,500
					Total Labor	$140,700
				Other: Subcontracts	$10,000	
				Consultants	$ 2,000	$12,000
				Total Labor and material:		$152,700
				Corporate G & A: 10%		$ 15,270
						$167,970
				Profit: 15%		$ 25,196
						$193,166

The Singapore Software Group (B)

DETERMINING WHAT TO BID

You and your team have carefully reviewed the risks and the associated costs. It is pretty obvious that there is quite a bit of risk exposure on this project. Had the RFP stated that the contract would be cost reimbursable, your exposure to risks might be less.

The decision made by your team is to recommend to senior management that a 15 percent management reserve be added into the contract and to submit a bid price of $2,279,762. You present your recommendations to senior management for review and approval. Your company has a policy that any bids over $500,000 have to be reviewed and approved by an executive committee. You tell senior management about your concerns over the risks and request that they contact TT to see if the contract type could be changed to a cost-reimbursable contract type. Your CEO informs you and your team that he has good working relations with TT because of all of the previous contracts SSG did for them. The CEO then states: "Taiwan Technologies has no intention of changing the contract type. I already asked them about this, and they will not make any changes to the solicitation package. Furthermore, I have been informed from reliable sources that they have committed only $1.5 million for this contract and are pretty sure that they will get bidders at this price."

The CEO then asks you to go back to the drawing board, sharpen your pencil, and see what financial risks SSG would be exposed to if SSG submitted a bid of $1.5 million.

The Singapore
Software Group (C)

MANAGEMENT'S DECISION

You and your team have carefully reviewed costs and present your findings to management. You are surprised by the fact that management actually seems pleased that the loss would be only $106,780. Obviously, management has been thinking about this for some time, but you did not know about it.

Management tells you that they are willing to submit a bid of $1.5 million to TT. The CEO looks over your pricing sheet and says:

> We need a pricing sheet that gets us to exactly $1.5 million on price. Work backward to generate the numbers. I want the pricing sheet to show a profit of $170,000. Leave in the scrap factor on materials, but eliminate the consultant. We will pay for the consultant's time using another source of internal funds, but do not identify him on this proposal. Also, eliminate the overtime hours and overtime costs. And for simplicity sake, just give us a total burdened labor cost rather than breaking it down by department.

The Singapore Software Group (D)

ANOTHER CRITICAL DECISION

As always, management was correct. You submitted a bid of $1.5 million and were awarded the contract. Work has been taking place as planned. There have been some scope changes, but they had only a minor impact on the cost and schedule thus far.

By the end of the eighth month of the project, your team makes a hardware and software breakthrough that may revolutionize touch-screen technology. You are pretty sure that none of TT's competitors has this technology and that TT will certainly capture a large portion of the marketplace when its products are introduced. The new technology could be applied to laptops and PCs as well.

SSG believes that this new technology can generate a significant cash flow stream for years to come. However, there is a serious problem. Because the contract is a firm-fixed-price effort, the intellectual property rights are owned by TT. SSG has limited usage of the technology and cannot license it to other companies.

You present your concerns to management. A week later, you and your team are asked to appear before the senior management committee. The CEO says:

> I have explained our position to Taiwan Technologies. They have agreed to allow us to change the contract type from firm-fixed-price to cost sharing. In the cost-sharing contract, the profit is removed from consideration and SSG will pay

40 percent of all costs and TT will pay the remaining 60 percent of the costs up to a maximum of 60 percent of the total cost line in the proposal. This is a win-win situation for both parties. Furthermore, TT will allow SSG to have shared intellectual property rights, but only after TT's products have been in the marketplace for 90 days.

The CEO then asks you to recalculate your numbers and see how much money SSG had to pay out of pocket to develop this technology, including the cost of the consultant.

To Bid or Not to Bid

BACKGROUND

Marvin was the president and CEO of his company. The decision of whether to bid on a job above a certain dollar value rested entirely on his shoulders. In the past, his company would bid on all jobs that were a good fit with the company's strategic objectives, and the company's win-to-loss ratio was excellent. But bidding on this job would be difficult. The client was requesting certain information in the RFP that Marvin did not want to release. If Marvin did not comply with the requirements of the RFP, his company's bid would be considered as nonresponsive.

BIDDING PROCESS

Marvin's company was highly successful at winning contracts through competitive bidding. The company was project-driven, and all revenue that came into the company came through winning contracts. Almost all of the clients provided the company with long-term contracts as well as follow-on contracts.

Almost all of the contracts were firm-fixed-price contracts. Business was certainly good, at least up until now.

Marvin established a policy whereby 5 percent of sales would be used for responding to RFPs. This was referred to as a bid-and-proposal (B&P) budget. The cost for bidding on contracts was quite high, and clients knew that requiring the company to spend a great deal of money bidding on a job might force a no-bid on the job. That could eventually hurt the industry by reducing the number of bidders in the marketplace.

Marvin's company used parametric and analogy estimating on all contracts. This allowed Marvin's people to estimate the work at level 1 or level 2 of the work breakdown structure (WBS). From a financial perspective, this was the most cost-effective way to bid on a project, knowing full well that there were risks with the accuracy of the estimates at these levels of the WBS. But over the years, continuous improvements to the company's estimating process reduced much of the uncertainty in the estimates.

NEW RFP

One of Marvin's most important clients announced it would be going out for bids for a potential 10-year contract. This contract was larger than any other contract that Marvin's company had ever received and could provide an excellent cash flow stream for 10 years or even longer. Winning the contract was essential.

Because most of the previous contracts were firm-fixed-price, only summary-level pricing at the top two levels of the WBS was provided in the proposal. That was usually sufficient for the company's clients to evaluate the cost portion of the bid.

The RFP was finally released. For this project, the contract type would be cost reimbursable. A WBS created by the client was included in the RFP, and the WBS was broken down into five levels. Each bidder had to provide pricing information for each work package in the WBS. By doing this, the client could compare the cost of each work package from each bidder. The client would then be comparing apples and apples from each bidder rather than apples and oranges. To make matters better for the customer, each bidder had to agree to use the WBS created by the client during project execution and to report costs according to the WBS.

Marvin saw the risks right away. If he decided to bid on the job, the company would be releasing its detailed cost structure to the client. All costs would then be clearly exposed to the client. If Marvin were to bid on this project, releasing the detailed cost information could have a serious impact on future bids, even if contracts in the future were firm-fixed-price.

Marvin convened a team composed of his senior officers. During the discussions that followed, the team identified the pros and cons of bidding on the job:

Pros

- A lucrative 10-year (or longer) contract
- The ability to have the client treat Marvin's company as a strategic partner rather than just a supplier
- Possibly lower profit margins on this and other future contracts but greater overall profits and earnings per share because of the larger business base
- Establishment of a workable standard for winning more large contracts

Cons

- Release of the company's cost structure
- Risk that competitors will see the cost structure and hire away some of the company's talented people by offering them more pay
- Inability to compete on price and having entire cost structure exposed could be a limiting factor on future bids
- If the company does not bid on this job, the company could be removed from the client's bidder list
- Clients must force Marvin's company to accept lower profit margins

Marvin then asked the team, "Should we bid on the job?"

QUESTIONS

1. What other factors should Marvin and his team consider?
2. Should they bid on the job?

Part 7

PROJECT PLANNING

Perhaps the most important phase of any project is planning. If the planning is performed effectively and the workers participate in the development of the plan, the chances of success are greatly enhanced. Yet even with the best-prepared plan, changes will occur.

Good project planning begins with a definition of the requirements, such as the statement of work, work breakdown structure, specifications, timing, and spending curve. Effective planning also assumes that the project manager understands the business case and the accompanying assumptions and constraints.

Greyson Corporation

Greyson Corporation was formed in 1970 by three scientists from the University of California. The major purpose of the company was research and development for advanced military weaponry. Following World War II, Greyson became a leader in the field of research and development. By the mid-1980s, Greyson employed over 200 scientists and engineers.

The fact that Greyson handled only research and development (R&D) contracts was advantageous. First of all, all of the scientists and engineers were dedicated to R&D activities; they did not have to share their loyalties with production programs. Second, a strong functional organization was established. The project management function was the responsibility of the functional manager whose department would perform the majority of the work. Working relationships between departments were excellent.

By the late 1980s, Greyson was under new management. Almost all R&D programs called for establishment of qualification and production planning. As a result, Greyson decided to enter into the production of military weapons as well and capture some of the windfall profits of the production market. This required a major reorganization from a functional to a matrix structure. Personnel problems occurred, but none that proved a major catastrophe.

In 1994, Greyson entered into the aerospace market with the acquisition of a subcontract for the propulsion unit of the Hercules missile. The contract was projected at $200 million over a five-year period, with excellent possibilities for follow-on work. Between 1994 and 1998, Greyson developed a competent technical staff composed mainly of young, untested college graduates. The majority of the original employees who were still there were in managerial positions. Greyson never had any layoffs. In addition, Greyson had excellent career development programs for almost all employees.

Between 1997 and 2001, the Department of Defense procurement for new weapons systems was on the decline. Greyson relied heavily on their two major production programs, Hercules and Condor II, both of which gave great promise for continued procurement. Greyson also had some 30hirty smaller R&D contracts as well as two smaller production contracts for hand weapons.

Because R&D money was becoming scarce, Greyson's management decided to phase out many of the R&D activities and replace them with lucrative production contracts. Greyson believed that it could compete with anyone in regard to low-cost production. Under this philosophy, the R&D community was reduced to minimum levels necessary to support in-house activities. The director of engineering froze all hiring except for job shoppers with special talents. All nonessential engineering personnel were transferred to production units.

In 2002, Greyson entered into competition with Cameron Aerospace Corporation for development, qualification, and testing of the Navy's new Neptune missile. The competition was an eight-motor shoot-off during the last 10 months of 2003. Cameron Corporation won the contract owing to technical merit. Greyson Corporation, however, had gained valuable technical information in rocket motor development and testing. The loss of the Neptune Program made it clear to Greyson's management that aerospace technology was changing too fast for Greyson to maintain a passive position. Even though funding was limited, Greyson increased its technical staff and soon found great success in winning R&D contracts.

By 2005, Greyson had developed a solid aerospace business base. Profits had increased by 30 percent. Greyson Corporation expanded from a company with 200 employees in 1994 to 1,800 employees in 2005. The Hercules Program, which began in 1994, was providing yearly follow-on contracts. All indications projected a continuation of the Hercules Program through 2002.

Cameron Corporation, in contrast, had found 2005 a difficult year. The Neptune Program was the only major contract that it maintained. The current production buy for the Neptune missile was scheduled for completion in August 2005 with no follow-on work earlier than January 2006. Cameron Corporation anticipated that overhead rates would increase sharply prior to next buy. The cost per motor would increase from $55,000 to $75,000 for a January procurement, $85,000 for a March procurement, and $125,000 for an August procurement. In February 2005, the Navy asked Greyson Corporation if it would be interested in submitting a sole-source bid for production and qualification of the Neptune missile. The Navy considered Cameron's position uncertain and wanted to maintain a qualified vendor should Cameron Corporation decide to get out of the aerospace business.

Greyson submitted a bid of $30 million for qualification and testing of 30 Neptune motors over a 30-month period beginning in January 2006. Current testing of the Neptune missile indicated that the minimum motor age life would extend through January 2009. This meant that production funds over the next 30 months could be diverted toward requalification of a new vendor, and production requirements for 2009 still could be met.

In August 2005, on delivery of the last Neptune rocket to the Navy, Cameron Corporation announced that without an immediate production contract for Neptune follow-on work, it would close its doors and get out of the aerospace business. Cameron invited Greyson Corporation to interview all of its key employees for possible work on the Neptune Requalification Program. Greyson hired 35 of Cameron's key people to begin work in October 2005. The key people would be assigned to ongoing Greyson programs to become familiar with Greyson methods. Greyson's lower-level management was very unhappy about bringing in these employees for fear that they would be placed in slots that could have resulted in promotions for some of Greyson's people. Management then decreed that these 35 people would work solely on the Neptune Program, and other vacancies would be filled, as required, from the Hercules and Condor II programs. Greyson estimated that the cost of employing these 35 people was approximately $150,000 per month, almost all of which was being absorbed through overhead. Without these 35 people, Greyson did not believe that it would have won the contract as sole-source procurement. Other competitors could have grabbed these key people and forced an open-bidding situation.

Because of the increased overhead rate, Greyson maintained a minimum staff to prepare for contract negotiations and document preparation. To minimize costs, the directors of engineering and program management gave the Neptune program office the authority to make decisions for departments and divisions that were without representation in the program office. Top management had complete confidence in the program office personnel because of their past performance on other programs and years of experience.

In December 2005, the Department of Defense announced that spending was being curtailed sharply and that funding limitations made it impossible to begin the qualification program before July 2006. To make matters worse, consideration was being made for a compression of the requalification program to 25 motors in a 20-month period. However, long-lead funding for raw materials would be available.

After lengthy consideration, Greyson decided to maintain its current position and retain the 35 Cameron employees by assigning them to in-house programs. The Neptune program office was still maintained for preparations to support contract negotiations, rescheduling of activities for a shorter program, and long-lead procurement.

In May 2006, contract negotiations began between the Navy and Greyson. At the beginning of contract negotiations, the Navy stated the three key elements for negotiations:

1. Maximum funding was limited to the 2005 quote for a 30-motor/30-month program.
2. The amount of money available for the last six months of 2006 was limited to $3.7 million.
3. The contract would be cost plus incentive fee.

After three weeks of negotiations there appeared a stalemate. The Navy contended that the production man-hours in the proposal were at the wrong level on the learning curves. It was further argued that Greyson should be a lot "smarter" now because of the 35 Cameron employees and because of experience learned during the 2001 shoot-off with Cameron Corporation during the initial stages of the Neptune Program.

Since the negotiation teams could not agree, top-level management of the Navy and Greyson Corporation met to iron out the differences. An agreement was finally reached on a figure of $28.5 million. This was $1.5 million below Greyson's original estimate to do the work. Management, however, felt that, by tightening their belts, the work could be accomplished within budget.

The program began on July 1, 2006, with the distribution of the department budgets by the program office. Almost all of the department managers were furious. Not only were the budgets below their original estimates, but the 35 Cameron employees were earning salaries above the department mean salary, thus reducing total man-hours even further. Almost all department managers asserted that cost overruns would be the responsibility of the program office and not the individual departments.

By November 2006, Greyson was in trouble. The Neptune Program was on target for cost but 35 percent behind for work completion. Department managers refused to take responsibility for certain tasks that were usually considered to be joint department responsibilities. Poor communication between program office and department managers provided additional discouragement. Department managers refused to have their employees work on Sunday.

Even with all this, program management felt that catch-up was still possible. The 35 former Cameron employees were performing commendable work equal to their counterparts on other programs. Management considered that the potential cost overrun situation was not in the critical stage and that more time should be permitted before considering corporate funding.

In December 2006, the Department of Defense announced that there would be no further buys of the Hercules missile. This announcement was a severe blow to Greyson's management. Not only was the company in danger of having to lay off 500 employees, but overhead rates would rise considerably. There was an indication last year that there would be no further buys, but management did not consider the indications positive enough to require corporate strategy changes.

Although Greyson was not unionized, there was a possibility of a massive strike if Greyson career employees were not given seniority over the 35 former Cameron employees in the case of layoffs.

By February 2007, the cost situation was clear:

1. The higher overhead rates threatened to increase total program costs by $1 million on the Neptune Program.
2. Because the activities were behind schedule, the catch-up phases would have to be made in a higher salary and overhead rate quarter, thus increasing total costs further.
3. Inventory costs were increasing. Items purchased during long-lead funding were approaching shelf-life limits. Cost impact might be as high as $1 million.

The vice president and general manager considered the Neptune Program critical to the success and survival of Greyson Corporation. The directors and division heads were ordered to take charge of the program. The following options were considered:

1. Perform overtime work to get back on schedule.
2. Delay program activities in hopes that the Navy can come up with additional funding.
3. Review current material specifications in order to increase material shelf life, thus lowering inventory and procurement costs.
4. Begin laying off noncritical employees.
5. Purchase additional tooling and equipment (at corporate expense) so that schedule requirements could be met on target.

On March 1, 2007, Greyson gave merit salary increases to the key employees on all in-house programs. At the same time, Greyson laid off 700 employees, some of whom were seasoned veterans. By March 15, Greyson employees formed a union and went out on strike.

QUESTIONS

1. What are the critical issues in the case?
2. How would you resolve each issue?

Teloxy
Engineering (A)

Teloxy Engineering has received a onetime contract to design and build 10,000 units of a new product. During the proposal process, management felt that the new product could be designed and manufactured at a low cost. One of the ingredients necessary to build the product was a small component that could be purchased for $60 in the marketplace, including quantity discounts. Accordingly, management budgeted $650,000 for the purchasing and handling of 10,000 components plus scrap.

During the design stage, your engineering team informs you that the final design will require a somewhat higher-grade component that sells for $72 with quantity discounts. The new price is substantially higher than you had budgeted for. This will create a cost overrun.

You meet with your manufacturing team to see if it can manufacture the component at a cheaper price than buying it from the outside. Your manufacturing team informs you that it can produce a maximum of 10,000 units, just enough to fulfill your contract. The setup cost will be $100,000 and the raw material cost is $40 per component. Since Teloxy has never manufactured this product before, manufacturing expects the following defects:

% defective	0	10	20	30	40
probability of occurrence (%)	10	20	30	25	15

All defective parts must be removed and repaired at a cost of $120 per part.

QUESTIONS

1. Using expected value, is it economically better to make or buy the component?
2. Strategically thinking, why might management opt for other than the most economical choice?

Teloxy
Engineering (B)

Your manufacturing team informs you that it has found a way to increase the size of the manufacturing run from 10,000 to 18,000 units, in increments of 2,000 units. However, the setup cost will be $150,000, and defects will cost the same $120 for removal and repair.

QUESTIONS

1. Calculate the economic feasibility of make or buy.
2. Should the probability of defects change if we produce 18,000 units as opposed to 10,000 units?
3. Would your answer to question 1 change if Teloxy management believes that follow-on contracts will be forthcoming? What would happen if the probability of defects changes to 15 percent, 25 percent, 40 percent, 15 percent, and 5 percent due to learning-curve efficiencies?

Payton Corporation

Payton Corporation had decided to respond to a government request for proposal for the R&D phase on a new project. The statement of work specified that the project must be completed within 90 days after go-ahead and that the contract would be at a fixed cost and fee.

The majority of the work would be accomplished by the development lab. According to government regulations, the estimated cost must be based on the *average* cost of the entire department, which was $19.00 per hour (unburdened).

Payton won the contract for a total package (cost plus fee) of $305,000. After the first weekly labor report was analyzed, it became evident that the development lab was spending $28.50 per hour. The project manager decided to discuss the problem with the manager of the development lab.

Project manager: "Obviously you know why I'm here. At the rate that you're spending money, we'll overrun our budget by 50 percent."

Lab manager: "That's your problem, not mine. When I estimate the cost to do a job, I submit only the hours necessary based on historical standards. The pricing department converts the hours to dollars based on department averages."

Project manager: "Well, why are we using the most expensive people? Obviously there must be lower-salaried people capable of performing the work."

Lab manager: "Yes, I do have lower-salaried people, but none who can complete the job within the two months required by the contract. I have to use people high on the learning curve, and they're not cheap. You should have told the pricing department to increase the average cost for the department."

Project manager: "I wish I could, but government regulations forbid this. If we were ever audited, or if this proposal were compared to other salary structures in other proposals, we would be in deep trouble. The only legal way to accomplish this would be to set up a new department for those higher-paid employees working on this project. Then the average department salary would be correct.

"Unfortunately the administrative costs of setting up a temporary unit for only two months is prohibitive. For long-duration projects, this technique is often employed.

"Why couldn't you have increased the hours to compensate for the increased dollars required?"

Lab manager: "I have to submit labor justifications for all hours I estimate. If I were to get audited, my job would be on the line. Remember, we had to submit labor justification for all work as part of the proposal.

"Perhaps next time management might think twice before bidding on a short-duration project. You might try talking to the customer to get his opinion."

Project manager: "His response would probably be the same regardless of whether I explained the situation to him before we submitted the proposal or now, after we have negotiated it. There's a good chance that I've just lost my Christmas bonus."

QUESTIONS

1. What is the basis for the problem?
2. Who is at fault?
3. How can the present situation be corrected?
4. Is there any way this situation can be prevented from recurring?
5. How would you handle this situation on a longer-duration project, say one year, assuming that multiple departments are involved and that no new departments were established other than possibly the project office?
6. Should a customer be willing to accept monetary responsibility for this type of situation, possibly by permitting established standards to be deviated from? If so, then how many months should be considered as a short-duration project?

Kemko Manufacturing

BACKGROUND

Kemko Manufacturing was a 50-year-old company that had a reputation for manufacturing high-quality household appliances. Kemko's growth was rapid during the 1990s. It grew by acquiring other companies. Kemko now had more than 25 manufacturing plants throughout the United States, Europe, and Asia.

Originally, each manufacturing plant that was acquired wanted to maintain its own culture, and quite often each was allowed to remain autonomous from corporate at Kemko provided that work was progressing as planned. But as Kemko began acquiring more companies, growing pains made it almost impossible to allow each plant to remain autonomous.

Each company had its own way of handling raw material procurement and inventory control. All purchase requests above a certain dollar value had to be approved by corporate. At corporate, there was often confusion over the information in all of the forms since each plant had its own documentation for procurement. Corporate was afraid that, unless it established a standardized procurement and inventory control system across all plants, cash flow problems and loss of corporate control over inventory could take its toll in the near future.

PROJECT IS INITIATED

Because of the importance of the project, senior management asked Janet Adams, director of information technology (IT), to take control of the project personally.

Janet had more than 30 years of experience in IT and fully understood how scope creep can create havoc on a large project.

Janet selected her team from IT and set up an initial kickoff date for the project. In addition to the mandatory presence of all of her team members, she also demanded that each manufacturing plant assign at least one representative and that all plant representatives be in attendance at the kickoff meeting. At the meeting, Janet said:

> I asked all of you here because I want you to have a clear understanding of how I intend to manage this project. Our executives have given us a timetable for this project and my greatest fear is scope creep. "Scope creep" is the growth of or enhancements to the project's scope as the project is being developed. On many of our other projects, scope creep has lengthened the project and driven up the cost. I know that scope creep isn't always evil and that it can happen in any life cycle phase.
>
> The reason why I have asked all of the plant representatives to attend this meeting is because of the dangers of scope creep. Scope creep has many causes, but it is generally the failure of effective up-front planning. When scope creep exists, people generally argue that it is a natural occurrence and we must accept the fact that it will happen. That's unacceptable to me!
>
> There will be no scope changes on this project, and I really mean it when I say this. The plant representatives must meet on their own and provide us with a detailed requirements package. I will not allow the project to officially begin until we have a detailed listing of the requirements. My team will provide you with some guidance, as needed, in preparing the requirements.
>
> No scope changes will be allowed once the project begins. I know that there may be some requests for scope changes, but all requests will be bundled together and worked on later as an enhancement project. This project will be implemented according to the original set of requirements. If I were to allow scope changes to occur, this project would run forever. I know some of you do not like this, but this is the way it will be on this project.

There was dead silence in the room. Janet could tell from the expressions on the faces of the plant representatives that they were displeased with her comments. Some of the plants were under the impression that the IT group was supposed to prepare the requirements package. Now Janet had transferred the responsibility to them, the user group, and they were not happy. Janet made it clear that user involvement would be essential for the preparation of the requirements.

After a few minutes of silence, the plant representatives said that they were willing to do this and it would be done correctly. Many of the representatives understood user requirements documentation. They would work together and come to an agreement on the requirements. Janet again stated that her team would support the plant representatives but that the burden of responsibility would rest solely on the plants. The plants would get what they ask for and nothing more. Therefore, they must be quite clear up front in their requirements.

While Janet was lecturing to the plant representatives, the IT team members were just sitting back smiling. Their job was about to become easier, or at least they thought so. Janet then addressed the IT team members:

Now I want to address the IT personnel. The reason why we are all in attendance at this meeting is because I want the plant representatives to hear what I have to say to the IT team. In the past, the IT teams have not been without some blame for scope creep and schedule elongation. So, here are my comments for the IT personnel:

- It is the IT team's responsibility to make sure that they understand the requirements as prepared by the plant representatives. Do not come back to me later telling me that you did not understand the requirements because they were poorly defined. I am going to ask every IT team member to sign a document stating that they have read over the requirements and fully understand them.
- Perfectionism is not necessary. All I want you to do is to get the job done.
- In the past we have been plagued with "featuritis," where many of you have added in your own bells and whistles unnecessarily. If that happens on this project, I will personally view this as a failure by you, and it will reflect in your next performance review.
- Sometimes people believe that a project like this will advance their career, especially if they look for perfectionism and bells and whistles. Trust me when I tell you this can have the opposite effect.
- Back-door politics will not be allowed. If any of the plant representatives come to you looking for ways to sneak in scope changes, I want to know about it. And if you make the changes without my permission, you may not be working for me much longer.
- I, and only I, have signature authority for scope changes.
- This project will be executed using detailed planning rather than rolling wave or progressive planning. We should be able to do this once we have clearly defined requirements.

Now, are there any questions from anyone?

The battle lines were now drawn. Some believed that it was Janet against the team, but most understood her need to do this. However, whether the project could work this way was still questionable.

QUESTIONS

1. Was Janet correct in the comments she made to the plant representatives?
2. Was Janet correct in the comments she made to the IT team members?
3. Is it always better on IT projects to make changes using enhancement projects or should we allow changes to be made as we go along?
4. What is your best guess on what happened?

Chance of a Lifetime

BACKGROUND

Sometimes in life opportunities come up for project managers, and they must evaluate the risks and the rewards. This case involves an experienced project manager with a well-secured position in a large company who was given the opportunity to join a start-up company. Unfortunately, although some of the critical decisions were his hands, they had a serious effect on his future and career.

SIGNS OF A POOR ECONOMY

Jason was a high school science teacher who liked to dabble in the small laboratory in the basement of his home. For two years, Jason had been experimenting with the design of long-lasting batteries that could be used in battery-powered cars. Jason was successful in designing two different types of batteries that had much longer lives than existing batteries under development by larger companies. Jason took out patents on his designs and tried to sell them to the larger companies. Unfortunately, with the low cost of gasoline at the pumps, the larger companies were not interested in Jason's ideas or his patents.

Economists, however, were predicting that within the next year, the cost of gasoline at the pumps could increase by 50 percent or more from $2 per gallon to $3 or more per gallon. If that happened, Jason believed that there would be a significant interest in electric-powered vehicles.

Jason believed that the timing was right to go out on his own. He was earning a little over $40,000 a year as a high school science teacher. He was married with two children, and there were significant financial risks in going out on his own. Despite the risks, his family was supportive of his decision to start up his own company.

Jason needed start-up funding. His family was willing to provide him with $50,000, but Jason knew that this was certainly not enough. This money could be gone in two months or less. He had a friend who had contacts with investment bankers and personal investors. Jason's friend was also an accountant. Originally Jason thought that the best approach would be to go with investment bankers who were willing to lend him $2 million. Financially, that sounded good. But the investment bankers wanted 75 percent of the company and complete decision-making authority. Jason was reluctant to give up control.

His friend was able to convince a group of investors to provide start-up funds of $500,000. These investors were willing to agree to a 49 percent share in the company. And if Jason were able to repay them their initial $500,000 plus a $100,000 profit at the end of the first year, the investors would return to Jason 44 percent of the company. That way the investors would have a 5 percent ownership in the company and have recovered all of their costs plus a 20 percent profit in just one year. Jason found this deal found attractive. His friend agreed to work as Jason's accountant on a part-time basis for $10,000 per year plus 10 percent ownership in the company. The ownership, however, did not include any voting rights and would not be in effect until the beginning of the third year of operations.

HIRING A PROJECT MANAGER

Jason believed that eventually he could sell his patents at a reasonable price. But the real big money would be in obtaining contracts to install his batteries in cars, and the market was somewhat limited. He could work with the automotive manufacturers as a contractor performing the installation work. He could also work with government agencies creating fleets of electric-powered vehicles for them. In any event, Jason needed a project manager.

After a lengthy search, Jason hired Craig, a 20-year project manager with an automobile supplier and with extensive knowledge of batteries. Craig's salary would be $50,000 initially and also included 35 percent ownership in the company after two years. Once again, the ownership did not come with voting rights.

Craig would have to write proposals and prepare project plans, schedules, budgets, and stakeholder reports. He would be the prime interface between the clients and the company after contract award.

It was clear at this point that a large portion of the start-up costs would be spent just on writing unsolicited proposals for work on electric-powered vehicles and batteries. Contracts had to be won before the start-up funds were expended.

The moment of truth was now at hand; Jason quit his job as a high school science teacher and started up his company.

AWARDING CONTRACTS

The unsolicited proposals sent to automotive manufacturers fell on deaf ears. Even though the cost of gasoline was increasing, the automotive manufacturers could not see any future in electric-powered vehicles. If necessary, they could also spend billions of dollars to compete with Jason's company.

Government agencies, however, were very interested in Jason's ideas. Within three months, Jason's company had received government contracts to convert some existing gas-powered government vehicles to electric-powered vehicles.

Jason and Craig were now drawing salaries in excess of $125,000 thanks to the government contracts. They rented a large warehouse and converted it into a facility where mechanics could work on cars. They also hired four licensed auto mechanics.

Life was good. Jason's dream was coming true. His salary was three times his salary as a high school science teacher. He was rapidly paying off his $500,000 start-up debt. The price of gasoline at the pumps was still rising and approaching $3 a gallon. The news media was discussing the need for electric-powered vehicles.

SEVERAL MONTHS LATER

As the contracts with the government agencies began approaching the completion phase, Jason and Craig began writing unsolicited proposals for follow-on work. Gas prices appeared to have leveled off, but the news media was still selling the need for electric vehicles.

A large automobile manufacturer in the United States as well a battery manufacturer approached Jason about buying out his company. If Jason agreed to sell, then Jason, Craig, and possibly even the accountant could become millionaires overnight.

Craig and the accountant wanted to sell and take their winnings. But Jason was enamored of his title of president and drawing a salary three times what he was earning as a teacher. Craig tried to explain to Jason that the company had no real business plan and that living day to day is not good and, if the government failed to renew the contracts, the company would go under.

Jason refused to listen. Both the automotive manufacturer and the battery company told Jason that their offer was good for only one week. This was certainly not enough time to wait and see if the contracts would be renewed with the government. A decision had to be made.

Once again, Jason refused to consider selling the company. Craig and the accountant told Jason that they might consider legal action to try to force him to sell, but without voting rights, that would be a difficult case to win.

Jason informed the automotive manufacturer and the battery company that he had no intention of selling. Two weeks later, as Jason and his team were finishing up the government contracts, the government agencies announced that the contracts would not be renewed. Within a week, Jason's company was out of business. Jason returned to teaching high school science and Craig fortunately was rehired by his previous employer.

QUESTIONS

1. Was Jason right in wanting to start up his own company?
2. Did Craig make a good decision in giving up a potential $150,000 salary as a project manager to work with Jason?
3. How does a project manager convince executives that they (the executives) are making bad business decisions? How many clients did Jason's company have? Who were the company's competitors, and what was the financial strength of the competition?
4. If you were in Jason's position, would you have sold the company? If so, what would you then do with your life?
5. Is it true that some project managers put their careers at stake each time they take on a new project? Can we call this career risk management?

Part 8

PROJECT SCHEDULING

Once project planning is completed, the next step is to schedule the project according to some timeline. Doing this requires knowledge of the activities, the necessary depth of the activities, the dependencies between the activities, and the duration of the activities.

Effective scheduling allows us to perform what-if exercises, develop contingency plans, determine the risks in the schedule, perform trade-offs, and minimize paperwork during customer review meetings. Although there are four basic scheduling techniques, they all utilize the same basic principles and common terminology.

Crosby
Manufacturing
Corporation

"I've called this meeting to resolve a major problem with our management cost and control system [MCCS]," remarked Wilfred Livingston, president of Crosby Manufacturing Corporation. "We're having one hell of a time trying to meet competition with our antiquated MCCS reporting procedures. Last year we were considered nonresponsive to three large government contracts because we could not adhere to the customer's financial reporting requirements. The government has recently shown a renewed interest in Crosby Manufacturing. If we can computerize our project financial reporting procedure, we'll be in great shape to meet the competition head on. The customer might even waive the financial reporting requirements if we show our immediate intent to convert."

Crosby Manufacturing was a \$250-million-a-year electronics component manufacturing firm in 2005, at which time Wilfred "Willy" Livingston became president. His first major act was to reorganize the 700 employees into a modified matrix structure. This reorganization was the first step in Livingston's long-range plan to obtain large government contracts. The matrix provided the customer focal point policy that government agencies prefer. After three years, the matrix seemed to be working. Now the company could begin the second phase, an improved MCCS policy.

On October 20, 2007, Livingston called a meeting with department managers from project management, cost accounting, management information systems (MIS), data processing, and planning.

Livingston: "We have to replace our current computer with a more advanced model so as to update our MCCS reporting procedures. In order for us to grow, we'll have to develop capabilities for keeping two or even three different sets of books for our customers. Our current computer does not have this capability. We're talking about a sizable cash outlay, not necessarily to impress our customers, but to increase our business base and grow. We need weekly, or even daily, cost data so as to better control our projects."

MIS manager: "I guess the first step in the design, development, and implementation process would be the feasibility study. I have prepared a list of the major topics which are normally included in a feasibility study of this sort." [See Exhibit I.]

EXHIBIT I. FEASIBILITY STUDY

- Objectives of the study
- Costs
- Benefits
- Manual or computer-based solution?
- Objectives of the system
- Input requirements
- Output requirements
- Processing requirements
- Preliminary system description
- Evaluation of bids from vendors
- Financial analysis
- Conclusions

Livingston: "What kind of costs are you considering in the feasibility study?"

MIS manager: "The major cost items include input–output demands; processing; storage capacity; rental, purchase, or lease of a system; nonrecurring expenditures; recurring expenditures; cost of supplies; facility requirements; and training requirements. We'll have to get a lot of this information from the electronic data processing (EDP) department."

EDP manager: "You must remember that, for a short period of time, we'll end up with two computer systems in operation at the same time. This cannot be helped. However, I have prepared a typical (abbreviated) schedule of my own. [See Table I.] You'll notice from the right-hand column that I'm somewhat optimistic as to how long it should take us."

Livingston: "Have we prepared a checklist on how to evaluate a vendor?"

EDP manager: "Besides the benchmark test, I have prepared a list of topics that we must include in evaluating any vendor. (See Exhibit II.). We should plan to

TABLE.I   TYPICAL SCHEDULE (IN MONTHS)

Activity	Normal Time to Complete	Crash Time to Complete
Management go-ahead	0	0
Release of preliminary system specs.	6	2
Receipt of bids on specs.	2	1
Order hardware and systems software	2	1
Flow charts completed	2	2
Applications programs completed	3	6
Receipt of hardware and systems software	3	3
Testing and debugging done	2	2
Documentation, if required	2	2
Changeover completed	22	15*

*This assumes that some of the activities can be run in parallel, instead of in series.

EXHIBIT II. VENDOR SUPPORT
EVALUATION FACTORS

- Availability of hardware and software packages
- Hardware performance, delivery, and past track record
- Vendor proximity and service-and-support record
- Emergency backup procedure
- Availability of applications programs and their compatibility with our other systems
- Capacity for expansion
- Documentation
- Availability of consultants for systems programming and general training
- Who burdens training cost?
- Risk of obsolescence
- Ease of use

---

call on or visit other installations that have purchased the same equipment and see the system in action. Unfortunately, we may have to commit real early and begin developing software packages. As a matter of fact, using the principle of concurrency, we should begin developing our software packages right now."

Livingston: "Because of the importance of this project, I'm going to violate our normal structure and appoint Tim Emary from our planning group as project leader. He's not as knowledgeable as your people are in regard to computers, but he does know how to lay out a schedule and get the job done. I'm sure your people will give him all the necessary support he needs. Remember, I'll be behind this project all the way. We're going to convene again one week from today, at which time I expect to see a detailed schedule with all major milestones, team meetings, design review meetings, et cetera, shown and identified. I'd like the project to be complete in 18 months, if possible. If there are risks in the schedule, identify them. Any questions?"

The Scheduling Dilemma

BACKGROUND

Sarah's project had now become more complex than she had anticipated. Sarah's company had a philosophy that the project manager would be assigned during proposal preparation, assist in the preparation of the proposal, and take on the role of the project manager after contract award, assuming the company would be awarded the contract.

Usually, contract go-ahead would take place within a week or two after contract award. That made project staffing relatively easy for most of the project managers. It also allowed the company to include in the proposal a detailed schedule based on resources that would be assigned upon contract award and go-ahead. During proposal preparation, the functional managers would anticipate who would be available for assignment to this project over the next few weeks. The functional managers could then estimate with reasonable accuracy the duration and effort required based on the grade level of the resources to be assigned. Since the go-ahead date was usually within two weeks of contract award and the contract award was usually within a week or so after proposal submittal, the schedule that appeared in the proposal was usually the same schedule for the actual project with very few changes. This entire process was based on the actual availability of resources rather than the functional managers assuming unlimited resources and using various estimating techniques.

Although this approach worked well on most projects, Sarah's new project had a go-ahead date of three months after contract award. For the functional managers, this created a problem estimating the effort and duration. Estimating now had to be made based on the assumption of unlimited availability rather than the availability of limited resources. Functional managers were unsure as to who would be available three or four months from now, yet some type of schedule had to appear in the proposal.

Sarah knew the risks. When the estimates were being prepared for her proposal, the functional managers assumed that the average worker in the department would be available and assigned to the project after go-ahead. The effort and duration estimates were then made based on the average employee. If, after go-ahead, above-average employees would be assigned to Sarah's project, she could possibly see the schedule accelerated but had to make sure that cost overruns did not happen because the fully loaded salary of the workers might be higher that what was estimated in the proposal. If below-average workers are assigned, a schedule slippage might occur, and Sarah would have to look at possible schedule compression techniques, hopefully without incurring added costs.

AWARD OF CONTRACT

Sarah's company was awarded the contract. Sarah had silently hoped that the company would not get the contract, but it did. As expected, the go-ahead date was three months from now. This created a problem for Sarah because she was unsure as to when to begin the preparation of the detailed schedule. The functional managers told her that they could not commit to an effort and duration based on actual limited resource availability until somewhere around two to three weeks prior to the actual go-ahead date. The resources were already spread thin across several projects, and many of the projects were having trouble. Sarah was afraid that the worst-case scenario would come true and that the actual completion date would be longer than what was in the proposal. Sarah was certainly not happy about explaining this to the client should it be necessary to do so.

APPROACHING GO-AHEAD DATE

As the go-ahead date neared, Sarah negotiated with the functional managers for resources. Unfortunately, her worst fears came true when, for the most part, she was provided with only average or above-average resources. The best resources were in demand elsewhere, and it was obvious that they would not be available for her project.

Using the efforts and durations provided by the functional managers, Sarah prepared the new schedule. Much to her chagrin, she would be at least two weeks late on the four-month project. The client would have to be told about this. But before telling the client, Sarah decided to look at ways to compress the schedule.

Working overtime was a possibility, but Sarah knew that overtime could lead to burned-out workers and increased chances of making mistakes. Also, Sarah knew that the workers really did not want to work overtime. Crashing the project by adding more resources was impossible because no other resources were available. Outsourcing some of the work was not possible as well because the statement of work identified proprietary information provided by the client and that the contract would not allow any outsourcing of the work to a third party. Because of the nature of the work, doing some of the work in parallel rather than series was not possible. There was always a chance that the assigned resources could get the job done ahead of schedule, but Sarah believed that a schedule delay was inevitable.

TIME FOR A DECISION

Sarah had to make a decision about when and how to inform the client of the impending schedule delay. If she told the truth to the client right now, the client might understand but might also believe that her company had lied in the proposal. That would be an embarrassment for her company. If she delayed informing the client, there might a chance, however slim, that the original schedule in the proposal would be adhered to. If the client was informed at the last minute about the delay, it could be costly for the client and equally embarrassing for her company.

QUESTIONS

1. Is this a common situation for most companies or an exception to the rule?
2. Can policies be established as part of competitive bidding to alleviate the pain of this occurring on other possible contracts where contract go-ahead date is several months after contract award?
3. Is it possible to convince a client that the schedule (and possibly the budget) is just a rough guess during competitive bidding and that finalization of the schedule (and budget) can be made only after go-ahead?
4. What schedule compression techniques were considered in the case? Were there any techniques she did not consider?
5. Was Sarah correct in her analysis that these techniques probably would not work on her project?
6. If one of these techniques were to be used, which one has the greatest likelihood for possible schedule compression?

Part 9

PROJECT EXECUTION

The best-prepared plans can result in a project failure because of poor execution. Project execution involves the working relationships among the participants and whether or not they support project management. There are two critical working relationships: the project–line manager interface and the project–executive management interface.

Other factors can affect the execution of a project. These include open communications, honesty, and integrity in dealing with customers, truth in negotiations, and factual status reporting. Execution can also be influenced by the quality of the original project plan. A project plan based on faulty or erroneous assumptions can destroy morale and impact execution.

The Blue Spider Project

"This is impossible! Just totally impossible! Ten months ago I was sitting on top of the world. Upper-level management considered me one of the best, if not the best, engineer in the plant. Now look at me! I have bags under my eyes, I haven't slept soundly in the last six months, and here I am, cleaning out my desk. I'm sure glad they gave me back my old job in engineering. I guess I could have saved myself a lot of grief and aggravation had I not accepted the promotion to project manager."

HISTORY

Gary Anderson had accepted a position with Parks Corporation right out of college. With a Ph.D. in mechanical engineering, Gary was ready to solve the world's most traumatic problems. At first, Parks Corporation offered Gary little opportunity to do the pure research that he eagerly wanted to undertake. However, things soon changed. Parks grew into a major electronics and structural design corporation during the big boom of the late 1950s and early 1960s when Department of Defense (DoD) contracts were plentiful.

Parks Corporation grew from a handful of engineers to a major DoD contractor, employing some 6,500 people. During the recession of the late 1960s, money became scarce and major layoffs resulted in lowering the employment level to 2,200 employees. At that time, Parks decided to get out of the research and development (R&D) business and compete as a low-cost production facility while maintaining an engineering organization solely to support production requirements.

After attempts at virtually every project management organizational structure, Parks Corporation selected the matrix form. Each project had a program manager who reported to the director of program management. Each project also maintained an assistant project manager—normally a project engineer—who reported directly to the project manager and indirectly to the director of engineering. The program manager spent most of his time worrying about cost and time, whereas the assistant program manager worried more about technical performance.

With the poor job market for engineers, Gary and his colleagues began taking coursework toward MBA degrees in case the job market deteriorated further. In 1995, with the upturn in DoD spending, Parks had to change its corporate strategy. Parks had spent the last seven years bidding on the production phase of large programs. Now, however, with the new evaluation criteria set forth for contract awards, those companies winning the R&D and qualification phases had a definite edge on being awarded the production contract. The production contract was where the big profits could be found. In keeping with this new strategy, Parks began to beef up its R&D engineering staff. By 1998, Parks had increased in size to 2,700 employees. The increase was mostly in engineering. Experienced R&D personnel were difficult to find for the salaries that Parks was offering. Parks was, however, able to lure some employees away from competitors, but it relied mostly on the younger, inexperienced engineers fresh out of college.

With the adoption of this corporate strategy, Parks Corporation administered a new wage and salary program that included job upgrading. Gary was promoted to senior scientist, responsible for all R&D activities performed in the mechanical engineering department. Gary had distinguished himself as an outstanding production engineer during the past several years, and management felt that his contribution could be extended to R&D as well.

In January 1998, Parks Corporation decided to compete for Phase I of the Blue Spider Project, an R&D effort that, if successful, could lead into a $500 million program spread out over 20 years. The Blue Spider Project was an attempt to improve the structural capabilities of the Spartan missile, a short-range tactical missile used by the Army. The Spartan missile was exhibiting fatigue failure after six years in the field. This was three years less than what the original design specifications called for. The Army wanted new materials that could result in a longer life for the Spartan missile.

Lord Industries was the prime contractor for the Army's Spartan Program. Parks Corporation would be a subcontractor to Lord if it could successfully bid and win the project. The criteria for subcontractor selection were based not only on low bid but also on technical expertise as well as management performance on other projects. Parks's management felt that it had a distinct advantage over most of the other competitors because the company had successfully worked on other projects for Lord Industries.

THE BLUE SPIDER PROJECT KICKOFF

On November 3, 1997, Henry Gable, Parks's director of engineering, called Gary Anderson into his office and said:

> Gary, I've just been notified through the grapevine that Lord will be issuing the request for proposal [RFP] for the Blue Spider Project by the end of this month, with a 30-day response period. I've been waiting a long time for a project like this to come along so that I can experiment with some new ideas that I have. This project is going to be my baby all the way! I want you to head up the proposal team. I think it must be an engineer. I'll make sure that you get a good proposal manager to help you. If we start working now, we can get close to two months of research in before proposal submittal. That will give us a one-month's edge on our competitors.

Gary was pleased to be involved in such an effort. He had absolutely no trouble in getting functional support for the R&D effort necessary to put together a technical proposal. All of the functional managers continually remarked to Gary, "This must be a biggy. The director of engineering has thrown all of his support behind you."

On December 2, the RFP was received. The only trouble area that Gary could see was that the technical specifications stated that all components must be able to operate normally and successfully through a temperature range of –65°F to 145°F. Current testing indicated the Parks Corporation's design would not function above 130°F. An intensive R&D effort was conducted over the next three weeks. Everywhere Gary looked, it appeared that the entire organization was working on his technical proposal.

A week before the final proposal was to be submitted, Gary and Henry Gable met to develop a company position concerning the inability of the preliminary design material to be operated above 130°F.

Gary Anderson: "Henry, I don't think it is going to be possible to meet specification requirements unless we change our design material or incorporate new materials. Everything I've tried indicates we're in trouble."

Gable: "We're in trouble only if the customer knows about it. Let the proposal state that we expect our design to be operative up to 155°F. That'll please the customer."

Anderson: "That seems unethical to me. Why don't we just tell them the truth?"

Gable: "The truth doesn't always win proposals. I picked you to head up this effort because I thought that you'd understand. I could have just as easily selected one of our many moral project managers. I'm considering you for program manager after we win the program. If you're going to pull this conscientious crap on me like the other project managers do, I'll find someone else. Look at it this way;

later we can convince the customer to change the specifications. After all, we'll be so far downstream that he'll have no choice."

After two solid months of 16-hour days for Gary, the proposal was submitted. On February 10, 1998, Lord Industries announced that Parks Corporation would be awarded the Blue Spider Project. The contract called for a 10-month effort, negotiated at $2.2 million at a firm-fixed price.

SELECTING THE PROJECT MANAGER

Following contract award, Henry Gable called Gary in for a conference.

Gable: "Congratulations, Gary! You did a fine job. The Blue Spider Project has great potential for ongoing business over the next 10 years, provided that we perform well during the R&D phase. Obviously you're the most qualified person in the plant to head up the project. How would you feel about a transfer to program management?"

Gary Anderson: "I think it would be a real challenge. I could make maximum use of the MBA degree I earned last year. I've always wanted to be in program management."

Gable: "Having several masters' degrees, or even doctorates for that matter, does not guarantee that you'll be a successful project manager. There are three requirements for effective program management: You must be able to communicate both in writing and orally; you must know how to motivate people; and you must be willing to give up your car pool. The last one is extremely important in that program managers must be totally committed and dedicated to the program, regardless of how much time is involved.

"But this is not the reason why I asked you to come here. Going from project engineer to program management is a big step. There are only two places you can go from program management—up the organization or out the door. I know of very, very few engineers who failed in program management and were permitted to return."

Anderson: "Why is that? If I'm considered to be the best engineer in the plant, why can't I return to engineering?"

Gable: "Program management is a world of its own. It has its own formal and informal organizational ties. Program managers are outsiders. You'll find out. You might not be able to keep the strong personal ties you now have with your fellow employees. You'll have to force even your best friends to comply with your standards. Program managers can go from program to program, but functional departments remain intact.

"I'm telling you all this for a reason. We've worked well together the past several years. But if I sign the release so that you can work for Elliot Grey in program management, you'll be on your own, like hiring into a new company. I've already signed the release. You still have some time to think about it."

Anderson: "One thing I don't understand. With all of the good program managers we have here, why am I being given this opportunity?"

Gable: "Almost all of our program managers are over 45 years old. This resulted from our massive layoffs several years ago when we were forced to lay off the younger, inexperienced program managers. You were selected because of your age and because all of our other program managers have worked only on production-type programs. We need someone at the reins who knows R&D. Your counterpart at Lord Industries will be an R&D type. You have to fight fire with fire.

"I have an ulterior reason for wanting you to accept this position. Because of the division of authority between program management and project engineering, I need someone in program management whom I can communicate with concerning R&D work. The program managers we have now are interested only in time and cost. We need a manager who will bend over backward to get performance also. I think you're that man. You know the commitment we made to Lord when we submitted that proposal. You have to try to achieve that. Remember, this program is my baby. You'll get all the support you need. I'm tied up on another project now. But when it's over, I'll be following your work like a hawk. We'll have to get together occasionally and discuss new techniques.

"Take a day or two to think it over. If you want the position, make an appointment to see Elliot Grey, the director of program management. He'll give you the same speech I did. I'll assign Paul Evans to you as chief project engineer. He's a seasoned veteran and you should have no trouble working with him. He'll give you good advice. He's a good man."

THE WORK BEGINS

Gary Anderson accepted the new challenge. His first major hurdle occurred in staffing the project. The top priority given to him to bid the program did not follow through for staffing. The survival of Parks Corporation depended on the profits received from its production programs. Gary found that, in keeping with this philosophy, engineering managers (even his former boss) were reluctant to give up their key people to the Blue Spider Program. However, with a little support from Henry Gable, Gary formed an adequate staff for the program.

Right from the start, Gary was worried that the test matrix called out in the technical volume of the proposal would not produce results that could satisfy specifications. Gary had 90 days after go-ahead during which to identify the raw materials that could satisfy specification requirements. He and Paul Evans held a meeting to map out their strategy for the first few months.

Gary Anderson: "Well, Paul, we're starting out with our backs against the wall on this one. Any recommendations?"

Paul Evans: "I also have my doubts about the validity of this test matrix. Fortunately, I've been through this before. Gable thinks this is his project, and he'll sure as hell try to manipulate us. I have to report to him every morning at 7:30 a.m. with the raw data results of the previous day's testing. He wants to see it before you do. He also stated that he wants to meet with me alone.

"Lord will be the big problem. If the test matrix proves to be a failure, we're going to have to change the scope of effort. Remember, this is a firm-fixed-price contract. If we change the scope of work and do additional work in the earlier phases of the program, then we should prepare a trade-off analysis to see what we can delete downstream so as to not overrun the budget."

Anderson: "I'm going to let the other project office personnel handle the administrating work. You and I are going to live in the research labs until we get some results. We'll let the other project office personnel run the weekly team meetings."

For the next three weeks Gary and Paul spent virtually 12 hours per day, seven days a week, in the R&D lab. None of the results showed any promise. Gary kept trying to set up a meeting with Henry Gable but always found him unavailable.

During the fourth week, Gary, Paul, and the key functional department managers met to develop an alternate test matrix. The new test matrix looked good. Gary and his team worked frantically to develop a new workable schedule that would not have impact on the second milestone, which was to occur at the end of 180 days. The second milestone was the final acceptance of the raw materials and preparation of production runs of the raw materials to verify that there would be no scale-up differences between lab development and full-scale production.

Gary personally prepared all of the technical handouts for the interchange meeting. After all, he would be the one presenting all of the data. The technical interchange meeting was scheduled for two days. On the first day, Gary presented all of the data, including test results, and the new test matrix. Lord Industries, the customer, appeared displeased with the progress to date and decided to have its own in-house caucus that evening to go over the material that was presented.

The following morning, a spokesman for Lord Industries stated its position:

> First of all, Gary, we're quite pleased to have a project manager who has such a command of technology. That's good. But every time we've tried to contact you last month, you were unavailable or had to be paged in the research laboratories. You did an acceptable job presenting the technical data, but the administrative data was presented by your project office personnel. We at Lord do not think that you're maintaining the proper balance between your technical and administrative responsibilities. We prefer that you personally give the administrative data and your chief project engineer present the technical data.
>
> We did not receive any agenda. Our people like to know what will be discussed, and when. We also want a copy of all handouts to be presented at least

three days in advance. We need time to scrutinize the data. You can't expect us to walk in here blind and make decisions after seeing the data for ten minutes.

To be frank, we feel that the data to date is totally unacceptable. If the data does not improve, we will have no choice but to issue a work stoppage order and look for a new contractor. The new test matrix looks good, especially since this is a firm-fixed-price contract. Your company will bear the burden of all costs for the additional work. A trade-off with later work may be possible, but this will depend on the results presented at the second design review meeting, 90 days from now.

We have decided to establish a customer office at Parks to follow your work more closely. Our people feel that monthly meetings are insufficient during R&D activities. We would like our customer representative to have daily verbal meetings with you or your staff. He will then keep us posted. Obviously, we had expected to review much more experimental data than you have given us.

Many of our top-quality engineers would like to talk directly to your engineering community, without having to continually waste time by having to go through the project office. We must insist on this last point. Remember, your effort may be only $2.2 million, but our total package is $100 million. We have a lot more at stake than you people do. Our engineers do not like to get information that has been filtered by the project office. They want to help you.

And last, don't forget that you people have a contractual requirement to prepare complete minutes for all interchange meetings. Send us the original for signature before going to publication."

Although Gary was unhappy with the first team meeting, especially with the requests made by Lord Industries, he felt that the client had sufficient justification for its comments. Following the team meeting, Gary personally prepared the complete minutes. "This is absurd," thought Gary. "I've wasted almost one entire week doing nothing more than administrative paperwork. Why do we need such detailed minutes? Can't a rough summary suffice? Why is it that customers want everything documented? That's like an indication of fear. We've been completely cooperative with them. There has been no hostility between us. If we've gotten this much paperwork to do now, I hate to imagine what it will be like if we get into trouble."

A NEW ROLE

Gary completed and distributed the minutes to the customer and to all key team members.

For the next five weeks testing went according to plan, or at least Gary thought that it had. The results were still poor. Gary was so caught up in administrative paperwork that he hadn't found time to visit the research labs in over a month. On a Wednesday morning, Gary entered the lab to observe the morning testing. Upon arriving in the lab, Gary found Paul Evans, Henry Gable, and two technicians testing a new material, JXB-3.

Gable: "Gary, your problems will soon be over. This new material, JXB-3, will permit you to satisfy specification requirements. Paul and I have been testing it for two weeks. We wanted to let you know but were afraid that if the word leaked out to the customer that we were spending their money for testing materials that were not called out in the program plan, they would probably go crazy and might cancel the contract. Look at these results. They're super!"

Gary Anderson: "Am I supposed to be the one to tell the customer now? This could cause a big wave."

Gable: "There won't be any wave. Just tell them that we did it with our own internal R&D funds. That'll please them because they'll think we're spending our own money to support their program."

Before presenting the information to Lord, Gary called a team meeting to present the new data to the project personnel. At the team meeting, one functional manager spoke out, saying: "This is a hell of a way to run a program. I like to be kept informed about everything that's happening here at Parks. How can the project office expect to get support out of the functional departments if we're kept in the dark until the very last minute? My people have been working with the existing materials for the last two months and you're telling us that it was all for nothing. Now you're giving us a material that's so new that we have no information on it whatsoever. We're now going to have to play catch-up, and that's going to cost you plenty."

One week before the 180-day milestone meeting, Gary Anderson submitted the handout package to Lord Industries for preliminary review. An hour later the phone rang.

Lord Industries: We've just read your handout. Where did this new material come from? How come we were not informed that this work was going on? You know, of course, that our customer, the Army, will be at this meeting. How can we explain this to them? We're postponing the review meeting until all of our people have analyzed the data and are prepared to make a decision.

"The purpose of a review or interchange meeting is to exchange information when *both* parties have familiarity with the topic. Normally, we require almost weekly interchange meetings with our other customers because we don't trust them. We disregard this policy with Parks Corporation based on past working relationships. But with the new state of developments, you have forced us to revert to our previous position, since we now question Parks Corporation's integrity in communicating with us. At first we believed this was due to an inexperienced program manager. Now we're not sure."

Gary Anderson: "I wonder if the real reason we have these interchange meetings isn't to show our people that Lord Industries doesn't trust us. You're creating a hell of a lot of work for us, you know."

Lord Industries: "You people put yourself in this position. Now you have to live with it."

Two weeks later, Lord reluctantly agreed that the new material offered the greatest promise. Three weeks later, the design review meeting was held. The Army was definitely not pleased with the prime contractor's recommendation to put a new, untested material into a multimillion-dollar effort.

THE COMMUNICATIONS BREAKDOWN

During the week following the design review meeting, Gary planned to make the first verification mix in order to establish final specifications for selection of the raw materials. Unfortunately, the manufacturing plans were a week behind schedule, primarily because of Gary, since he had decided to reduce costs by accepting the responsibility for developing the bill of materials himself.

Gary Anderson called a meeting to consider rescheduling of the mix.

Gary Anderson: "As you know, we're about a week to 10 days behind schedule. We'll have to reschedule the verification mix for late next week."

Production manager: "Our resources are committed until a month from now. You can't expect to simply call a meeting and have everything reshuffled for the Blue Spider Program. We should have been notified earlier. Engineering has the responsibility for preparing the bill of materials. Why aren't they ready?"

Engineering integration: "We were never asked to prepare the bill of materials. But I'm sure that we could get it out if we work our people overtime for the next two days."

Anderson: "When can we remake the mix?"

Production manager: "We have to redo at least 500 sheets of paper every time we reschedule mixes. Not only that, we have to reschedule people on all three shifts. If we are to reschedule your mix, it will have to be performed on overtime. That's going to increase your costs. If that's agreeable with you, we'll try it. But this will be the first and last time that production will bail you out. There are procedures that have to be followed."

Testing engineer: "I've been coming to these meetings since we kicked off this program. I think I speak for the entire engineering division when I say that the role that the director of engineering is playing in this program is suppressing individuality among our highly competent personnel. In new projects, especially those involving R&D, our people are not apt to stick their necks out. Now our people are becoming ostriches. If they're impeded from contributing, even in their own slight way, then you'll probably lose them before the project gets completed. Right now I feel that I'm wasting my time here. All I need are minutes of the team meetings and I'll be happy. Then I won't have to come to these pretend meetings anymore."

The purpose of the verification mix was to make a full-scale production run of the material to verify that there would be no material property changes in scale-up from the small mixes made in the R&D laboratories. After testing, it became obvious that the wrong lots of raw materials were used in the production verification mix.

Lord Industries called a meeting for an explanation of why the mistake had occurred and what the alternatives were.

Lord Industries: "Why did the problem occur?"

Gary Anderson: "Well, we had a problem with the bill of materials. The result was that the mix had to be made on overtime. And when you work people on overtime, you have to be willing to accept mistakes as being a way of life. The energy cycles of our people are slow during the overtime hours."

Lord Industries: "The ultimate responsibility has to be with you, the program manager. We at Lord think that you're spending too much time doing and not enough time managing. As the prime contractor, we have a hell of a lot more at stake than you do. From now on we want documented weekly technical interchange meetings and closer interaction by our quality control section with yours."

Anderson: "These additional team meetings are going to tie up our key people. I can't spare people to prepare handouts for weekly meetings with your people."

Lord Industries: "Team meetings are a management responsibility. If Parks does not want the Blue Spider Program, I'm sure we can find another subcontractor. All you, Gary, have to do is give up taking the material vendors to lunch, and you'll have plenty of time for handout preparation."

Gary left the meeting feeling as if he had gotten raked over the coals. For the next two months, Gary worked 16 hours a day, almost every day. Gary did not want to burden his staff with the responsibility of the handouts, so he began preparing them himself. He could have hired additional staff, but with such a tight budget, and having to remake verification mix, cost overruns appeared inevitable.

As the end of the seventh month approached, Gary was feeling pressure from within Parks Corporation. The decision-making process appeared to be slowing down, and Gary found it more and more difficult to motivate his people. In fact, the grapevine was referring to the Blue Spider Project as a loser, and some of his key people acted as if they were on a sinking ship.

By the time the eighth month rolled around, the budget had nearly been expended. Gary was tired of doing everything himself. "Perhaps I should have stayed an engineer," he thought. He and Elliot Grey had a meeting to see what could be salvaged. Grey agreed to get Gary additional corporate funding to complete the project. "But performance must be met, since there is a lot riding on the Blue Spider Project," asserted Grey. He called a team meeting to identify the program status.

Gary Anderson: "It's time to map out our strategy for the remainder of the program. Can engineering and production adhere to the schedule that I have laid out before you?"

Team member, engineering: "This is the first time that I've seen this schedule. You can't expect me to make a decision in the next 10 minutes and commit the resources of my department. We're getting a little unhappy being kept in the dark until the last minute. What happened to effective planning?

Anderson: "We still have effective planning. We must adhere to the original schedule, or at least try to adhere to it. This revised schedule will do that."

Team member, engineering: "Look, Gary! When a project gets in trouble, it is usually the functional departments that come to the rescue. But if we're kept in the dark, then how can you expect us to come to your rescue? My boss wants to know, well in advance, every decision that you're contemplating with regard to our departmental resources. Right now, we—"

Anderson: "Granted, we may have had a communications problem. But now we're in trouble and have to unite forces. What is your impression as to whether your department can meet the new schedule?"

Team member, engineering: "When the Blue Spider Program first got in trouble, my boss exercised his authority to make all departmental decisions regarding the program himself. I'm just a puppet. I have to check with him on everything."

Team member, production: "I'm in the same boat, Gary. You know we're not happy having to reschedule our facilities and people. We went through this once before. I also have to check with my boss before giving you an answer about the new schedule."

The following week, the verification mix was made. Testing proceeded according to the revised schedule, and it looked as if the total schedule milestones could be met, provided that specifications could be adhered to.

Because of the revised schedule, some of the testing had to be performed on holidays. Gary wasn't pleased with asking people to work on Sundays and holidays, but he had no choice, since the test matrix called for testing to be accomplished at specific times after end of mix.

A team meeting was called on Wednesday to resolve the problem of who would work on the holiday, which would occur on Friday, as well as staffing Saturday and Sunday. During the team meeting, Gary became quite disappointed. Phil Rodgers, who had been Gary's test engineer since the project started, was assigned to a new project that the grapevine called Gable's new adventure. His replacement was a relatively new man, with the company only eight months. For an hour and a half, the team members argued about the little problems and continually avoided the major question, stating that they would first have to coordinate commitments with their bosses. It was obvious to Gary that his team members were afraid to make major decisions and therefore ate up a lot of time on trivial problems.

On the following day, Thursday, Gary went to see the department manager responsible for testing, in hopes that he could use Phil Rodgers that weekend.

Department manager: "I have specific instructions from the boss [director of engineering] to use Phil Rodgers on the new project. You'll have to see the boss if you want him back."

Gary Anderson: "But we have testing that must be accomplished this weekend. Where's the new man you assigned yesterday?"

Department manager: "Nobody told me you had testing scheduled for this weekend. Half of my department is already on an extended weekend vacation, including Phil Rodgers and the new man. How come I'm always the last to know when we have a problem?"

Anderson: "The customer is flying down his best people to observe this weekend's tests. It's too late to change anything. You and I can do the testing."

Department manager: "Not on your life. I'm staying as far away as possible from the Blue Spider Project. I'll get you someone, but it won't be me. That's for sure!"

The weekend's testing went according to schedule. The raw data was made available to the customer under the stipulation that the final company position would be announced at the end of the next month, after the functional departments had a chance to analyze it.

Final testing was completed during the second week of the ninth month. The initial results looked excellent. The materials were within contract specifications, and although they were new, both Gary and Lord's management felt that there would be little difficulty in convincing the Army that this was the way to go. Henry Gable visited Gary and congratulated him on a job well done.

All that now remained was the making of four additional full-scale verification mixes in order to determine how much deviation there would be in material properties between full-size production-run mixes. Gary tried to get the customer to concur (as part of the original trade-off analysis) that two of the four production runs could be deleted. Lord Industries' management refused, insisting that contractual requirements must be met at the expense of the contractor.

The following week, Elliot Grey called Gary in for an emergency meeting concerning expenditures to date.

Elliot Grey: "Gary, I just received a copy of the financial planning report for last quarter in which you stated that both the cost and performance of the Blue Spider Project were 75 percent complete. I don't think you realize what you've done. The target profit on the program was $200,000. Your memo authorized the vice president and general manager to book 75 percent of that, or $150,000, for corporate profit spending for stockholders. I was planning on using all $200,000 together with the additional $300,000 I personally requested from corporate headquarters to bail you out. Now I have to go back to the vice president and general manager and tell them that we've made a mistake and that we'll need an additional $150,000."

Gary Anderson: "Perhaps I should go with you and explain my error. Obviously, I take all responsibility."

Grey: "No, Gary. It's our error, not yours. I really don't think you want to be around the general manager when he sees red at the bottom of the page. It takes an act of God to get money back once corporate books it as profit. Perhaps you should reconsider project engineering as a career instead of program management. Your performance hasn't exactly been sparkling, you know."

Gary returned to his office quite disappointed. No matter how hard he worked, the bureaucratic red tape of project management seemed always to do him in. But late that afternoon, Gary's disposition improved. Lord Industries called to say that, after consultation with the Army, Parks Corporation would be awarded a sole-source contract for qualification and production of Spartan missile components using the new longer-life raw materials. Both Lord and the Army felt that the sole-source contract was justified, provided that continued testing showed the same results, since Parks Corporation had all of the technical experience with the new materials.

Gary received a letter of congratulations from corporate headquarters but no additional pay increase. The grapevine said that a substantial bonus was given to the director of engineering.

During the 10th month, results were coming back from the accelerated aging tests performed on the new materials. The results indicated that although the new materials would meet specifications, the age life would probably be less than five years. These numbers came as a shock to Gary. He and Paul Evans had a conference to determine the best strategy to follow.

Gary Anderson: "Well, I guess we're now in the fire instead of the frying pan. Obviously, we can't tell Lord Industries about these tests. We ran them on our own. Could the results be wrong?"

Evans: "Sure, but I doubt it. There's always margin for error when you perform accelerated aging tests on new materials. There can be reactions taking place that we know nothing about. Furthermore, the accelerated aging tests may not even correlate well with actual aging. We must form a company position on this as soon as possible."

Anderson: "I'm not going to tell anyone about this, especially Henry Gable. You and I will handle this. It will be my throat if word of this leaks out. Let's wait until we have the production contract in hand."

Evans: "That's dangerous. This has to be a company position, not a project office position. We had better let them know upstairs."

Anderson: "I can't do that. I'll take all responsibility. Are you with me on this?"

Evans: "I'll go along. I'm sure I can find employment elsewhere when we open Pandora's box. You had better tell the department managers to be quiet also."

Two weeks later, as the program was winding down into the testing for the final verification mix and final report development, Gary received an urgent phone call asking him to report immediately to Henry Gable's office.

Gable: "When this project is over, you're through. You'll never hack it as a program manager or possibly even a good project engineer. We can't run projects around here without honesty and open communications. How the hell do you expect top management to support you when you start censoring bad news to the top? I don't like surprises. I like to get the bad news from the program manager and project engineers, not secondhand from the customer. And, of course, we cannot forget the cost overrun. Why didn't you take some precautionary measures?"

Gary Anderson: "How could I when you were asking our people to do work such as accelerated aging tests that would be charged to my project and was not part of program plan? I don't think I'm totally to blame for what's happened."

Gable: "Gary, I don't think it's necessary to argue the point any further. I'm willing to give you back your old job, in engineering. I hope you didn't lose too many friends while working in program management. Finish up final testing and the program report. Then I'll reassign you."

Gary returned to his office and put his feet up on the desk. "Well," he thought, "perhaps I'm better off in engineering. At least I can see my wife and kids once in a while." As he began writing the final report, the phone rang.

Functional manager: "Hello, Gary. I just thought I'd call to find out what charge number you want us to use for experimenting with this new procedure to determine accelerated age life."

Gary Anderson: "Don't call me! Call Gable. After all, the Blue Spider Project is his baby."

QUESTIONS

1. If you were Gary Anderson, would you have accepted this position after the director stated that this project would be his baby all the way?
2. Do engineers with MBA degrees aspire to high positions in management?
3. Was Gary qualified to be a project manager?
4. What are the moral and ethical issues facing Gary?
5. What authority does Gary have and to whom does he report?
6. Is it true that, when you enter project management, you either go up the organization or out the door?
7. Is it possible for an executive to take too much of an interest in an R&D project?

8. Should Paul Evans have been permitted to report information to Gable before reporting it to the project manager?
9. Is it customary for the project manager to prepare all of the handouts for a customer interchange meeting?
10. What happens when a situation of mistrust occurs between the customer and contractor?
11. Should functional employees of the customer and contractor be permitted to communicate with one another without going through the project office?
12. Did Gary demonstrate effective time management?
13. Did Gary understand production operations?
14. Are functional employees authorized to make project decisions?
15. On R&D projects, should profits be booked periodically or at project termination?
16. Should a project manager ever censor bad news?
17. Could the above-mentioned problems have been resolved if there had been a single methodology for project management in place?
18. Can a single methodology for project management specify morality and ethics in dealing with customers? If so, how do we handle situations where the project manager violates protocol?
19. Could the lessons learned on success and failure during project debriefings cause a major change in the project management methodology?

Corwin Corporation

By June 2003, Corwin Corporation had grown into a $950 million-per-year corporation with an international reputation for manufacturing low-cost, high-quality rubber components. Corwin maintained more than a dozen different product lines, all of which were sold as off-the-shelf items in department stores, hardware stores, and automotive parts distributors. The name "Corwin" was now synonymous with "quality." This provided management with the luxury of having products that had extremely long life cycles.

Organizationally, Corwin had maintained the same structure for more than 15 years. (See Figure I.) The top management of Corwin Corporation was highly conservative and believed in using a marketing approach to find new markets for existing product lines rather than exploring for new products. Under this philosophy, Corwin maintained a small R&D group whose mission was simply to evaluate state-of-the-art technology and its application to existing product lines.

Corwin's reputation was so good that it continually received inquiries about the manufacturing of specialty products. Unfortunately, the conservative nature of Corwin's management created a "don't rock the boat" atmosphere opposed to taking any type of risks. A management policy was established to evaluate all specialty-product requests. The policy required answering yes to the following questions:

- Will the specialty product provide the same profit margin (20 percent) as existing product lines?
- Is there a chance for follow-on contracts?

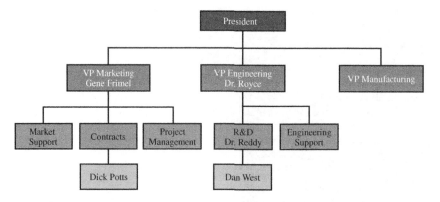

FIGURE I   Organizational chart for Corwin Corporation

- Can the specialty product be developed into a product line?
- Can the specialty product be produced with minimum disruption to existing product lines and manufacturing operations?

These stringent requirements forced Corwin not to bid on more than 90 percent of all specialty-product inquiries.

Corwin Corporation was a marketing-driven organization, although manufacturing often had different ideas. Almost all decisions were made by marketing with the exception of product pricing and estimating, which was a joint undertaking between manufacturing and marketing. Engineering was considered as merely a support group to marketing and manufacturing.

For specialty products, the project managers would always come out of marketing, even during the R&D phase of development. The company's approach was that if the specialty product should mature into a full product line, then there should be a product line manager assigned right at the onset.

THE PETERS COMPANY PROJECT

In 2000, Corwin accepted a specialty-product assignment from Peters Company because of the potential for follow-on work. In 2001, 2002, and again in 2003, profitable follow-on contracts were received, and a good working relationship developed, despite Peters' reputation for being a difficult customer to work with.

On December 7, 2002, Gene Frimel, the vice president of marketing at Corwin, received a rather unusual phone call from Dr. Frank Delia, the marketing vice president at Peters Company.

Frank Delia: "Gene, I have a rather strange problem on my hands. Our R&D group has $250,000 committed for research toward development of a new rubber product material, and we simply do not have the available personnel or talent to undertake the project. We have to go outside. We'd like your company to do the work. Our testing and R&D facilities are already overburdened."

Gene Frimel: "Well, as you know, Frank, we are not a research group, even though we've done this once before for you. And furthermore, I would never be able to sell our management on such an undertaking. Let some other company do the R&D work and then we'll take over on the production end."

Delia: "Let me explain our position on this. We've been burned several times in the past. Projects like this generate several patents, and the R&D company almost always requires that our contracts give it royalties or first refusal for manufacturing rights."

Frimel: "I understand your problem, but it's not within our capabilities. This project, if undertaken, could disrupt parts of our organization. We're already operating lean in engineering."

Delia: "Look, Gene! The bottom line is this: We have complete confidence in your manufacturing ability to such a point that we're willing to commit to a five-year production contract if the product can be developed. That makes it extremely profitable for you."

Frimel: "You've just gotten me interested. What additional details can you give me?"

Delia: "All I can give you is a rough set of performance specifications that we'd like to meet. Obviously, some trade-offs are possible."

Frimel: "When can you get the specification sheet to me?"

Delia: "You'll have it tomorrow morning. I'll ship it overnight express."

Frimel: "Good! I'll have my people look at it, but we won't be able to get you an answer until after the first of the year. As you know, our plant is closed down for the last two weeks in December, and most of our people have already left for extended vacations."

Delia: "That's not acceptable! My management wants a signed, sealed, and delivered contract by the end of this month. If this is not done, corporate will reduce our budget for 2003 by $250,000, thinking that we've bitten off more than we can chew. Actually, I need your answer within 48 hours so that I'll have some time to find another source."

Frimel: "You know, Frank, today is December 7, Pearl Harbor Day. Why do I feel as though the sky is about to fall in?"

Delia: "Don't worry, Gene! I'm not going to drop any bombs on you. Just remember, all that we have available is $250,000, and the contract must be a firm-fixed-price effort. We anticipate a six-month project with $125,000 paid on contract signing and the balance at project termination."

Frimel: "I still have that ominous feeling, but I'll talk to my people. You'll hear from us with a go or no-go decision within 48 hours. I'm scheduled to go on a Caribbean cruise, and my wife and I are leaving this evening. One of my people will get back to you on this matter."

Gene Frimel had a problem. All bid and no-bid decisions were made by a four-man committee composed of the president and the three vice presidents. The president and the vice president for manufacturing were on vacation. Frimel met with Dr. Royce, the vice president of engineering, and explained the situation.

Royce: "You know, Gene, I totally support projects like this because it would help our technical people grow intellectually. Unfortunately, my vote never appears to carry any weight."

Frimel: "The profitability potential as well as the development of good customer relations makes this attractive, but I'm not sure we want to accept such a risk. A failure could easily destroy our good working relationship with Peters Company."

Royce: "I'd have to look at the specification sheets before assessing the risks, but I would like to give it a shot."

Frimel: "I'll try to reach our president by phone."

By late afternoon, Frimel was fortunate enough to be able to contact the president and received a reluctant authorization to proceed. The problem now was how to prepare a proposal within the next two or three days and be ready to make an oral presentation to Peters Company.

Frimel: "The boss gave his blessing, Royce, and the ball is in your hands. I'm leaving for vacation, and you'll have total responsibility for the proposal and presentation. Delia wants the presentation this weekend. You should have his specification sheets tomorrow morning."

Royce: "Our R&D director, Dr. Reddy, left for vacation this morning. I wish he were here to help me price out the work and select the project manager. I assume that, in this case, the project manager will come out of engineering rather than marketing."

Frimel: "Yes, I agree. Marketing should not have any role in this effort. It's your baby all the way. And as for the pricing effort, you know our bid will be for $250,000. Just work backward to justify the numbers. I'll assign one of our contracting people to assist you in the pricing. I hope I can find someone who has experience in this type of effort. I'll call Delia and tell him we'll bid it with an unsolicited proposal."

Royce selected Dan West, one of the R&D scientists, to act as the project leader. Royce had severe reservations about doing this without the R&D director, Dr. Reddy, being actively involved. But with Reddy on vacation, Royce had to make an immediate decision.

TABLE I   PROPOSAL COST SUMMARIES

Direct labor and support	$ 30,000
Testing (30 tests at $2,000 each)	60,000
Overhead at 100%	90,000
Materials	30,000
General and administrative (G&A), 10%	21,000
Total	$ 231,000
Profit	19,000
Total	$ 250,000

On the following morning, the specification sheets arrived and Royce, West, and Dick Potts, a contracts man, began preparing the proposal. West prepared the direct labor man-hours, and Royce provided the costing data and pricing rates. Potts, being completely unfamiliar with this type of effort, simply acted as an observer and provided legal advice when necessary. Potts allowed Royce to make all decisions even though the contracts man was considered the president's official representative.

Finally completed two days later, the proposal was actually a 10-page letter that simply contained the cost summaries (see Table I) and the engineering intent. West estimated that 30 tests would be required. The test matrix described the test conditions only for the first five tests. The remaining 25 test conditions would be determined at a later date, jointly by Peters and Corwin personnel.

On Sunday morning, a meeting was held at Peters Company, and the proposal was accepted. Delia gave Royce a letter of intent authorizing Corwin Corporation to begin working on the project immediately. The final contract would not be available for signing until late January, and the letter of intent simply stated that Peters Company would assume all costs until such time that the contract was signed or the effort terminated.

West was truly excited about being selected as the project manager and being able to interface with the customer, a luxury that was usually given only to marketing personnel. Although Corwin Corporation was closed for two weeks over Christmas, West still went into the office to prepare the project schedules and to identify the support he would need in the other areas, thinking that if he presented this information to management on the first day back to work, they would be convinced that he had everything under control.

THE WORK BEGINS

On the first working day in January 2003, a meeting was held with the three vice presidents and Dr. Reddy to discuss the support needed for the project.

(West was not in attendance at this meeting, although all participants had a copy of his memo.)

Reddy: "I think we're heading for trouble in accepting this project. I've worked with Peters Company previously on R&D efforts, and they're tough to get along with. West is a good man, but I would never have assigned him as the project leader. His expertise is in managing internal rather than external projects. But, no matter what happens, I'll support West the best I can."

Royce: "You're too pessimistic. You have good people in your group and I'm sure you'll be able to give him the support he needs. I'll try to look in on the project every so often. West will still be reporting to you for this project. Try not to burden him too much with other work. This project is important to the company."

West spent the first few days after vacation soliciting the support that he needed from the other line groups. Many of the other groups were upset that they had not been informed earlier and were unsure as to what support they could provide. West met with Reddy to discuss the final schedules. Reddy said:

"Your schedules look pretty good, Dan. I think you have a good grasp on the problem. You won't need very much help from me. I have a lot of work to do on other activities, so I'm just going to be in the background on this project. Just drop me a note every once in a while telling me what's going on. I don't need anything formal. Just a paragraph or two will suffice."

By the end of the third week, all of the raw materials had been purchased, and initial formulations and testing were ready to begin. In addition, the contract was ready for signature. The contract contained a clause specifying that Peters Company had the right to send an in-house representative into Corwin Corporation for the duration of the project. Peters Company informed Corwin that Patrick Ray would be the in-house representative, reporting to Delia, and would assume his responsibilities on or about February 15.

By the time Pat Ray appeared at Corwin Corporation, West had completed the first three tests. The results were not what was expected but indicated that Corwin was heading in the right direction. Pat Ray's interpretation of the tests was completely opposite to that of West. Ray thought that Corwin was "way off base" and that redirection was needed.

Pat Ray: "Look, Dan! We have only six months to do this effort, and we shouldn't waste our time on marginally acceptable data. These are the next five tests I'd like to see performed."

Dan West: "Let me look over your request and review it with my people. That will take a couple of days, and, in the meanwhile, I'm going to run the other two tests as planned."

Ray's arrogant attitude bothered West. However, West decided that the project was too important to "knock heads" with Ray and simply decided to cater to Ray the best he could. This was not exactly the working relationship that West expected to have with the in-house representative.

West reviewed the test data and the new test matrix with engineering personnel, who felt that the test data was inconclusive as yet and preferred to withhold their opinion until the results of the fourth and fifth tests were made available. Although this displeased Ray, he agreed to wait a few more days if it meant getting Corwin Corporation on the right track.

The fourth and fifth tests appeared to be marginally acceptable, just as the first three had been. Corwin's engineering people analyzed the data and made their recommendations.

Dan West: "Pat, my people feel that we're going in the right direction and that our path has greater promise than your test matrix."

Pat Ray: "As long as we're paying the bills, we're going to have a say in what tests are conducted. Your proposal stated that we would work together in developing the other test conditions. Let's go with my test matrix. I've already reported back to. my boss that the first five tests were failures and that we're changing the direction of the project."

West: "I've already purchased $30,000 worth of raw materials. Your matrix uses other materials and will require additional expenditures of $12,000."

Ray: "That's your problem. Perhaps you shouldn't have purchased all of the raw materials until we agreed on the complete test matrix."

During the month of February, West conducted 15 tests, all under Ray's direction. The tests were scattered over such a wide range that no valid conclusions could be drawn. Ray continued sending reports back to Delia confirming that Corwin was not producing beneficial results and that there was no indication that the situation would reverse itself. Delia ordered Ray to take any steps necessary to ensure a successful completion of the project.

Ray and West met again as they had done for each of the past 45 days to discuss the status and direction of the project.

Pat Ray: "Dan, my boss is putting tremendous pressure on me for results, and thus far I've given him nothing. I'm up for promotion in a couple of months and I can't let this project stand in my way. It's time to completely redirect the project."

Dan West: "Your redirection of the activities is playing havoc with my scheduling. I have people in other departments who just cannot commit to this continual rescheduling. They blame me for not communicating with them when, in fact, I'm embarrassed to."

Ray: "Everybody has their problems. We'll get this problem solved. I spent this morning working with some of your lab people in designing the next 15 tests. Here are the test conditions."

West: "I certainly would have liked to be involved with this. After all, I thought I was the project manager. Shouldn't I have been at the meeting?"

Ray: "Look, Dan! I really like you, but I'm not sure that you can handle this project. We need some good results immediately, or my neck will be stuck out for the next four months. I don't want that. Just have your lab personnel start on these tests, and we'll get along fine. Also, I'm planning on spending a great deal of time in your lab area. I want to observe the testing personally and talk to your lab personnel."

West: "We've already conducted 20 tests, and you're scheduling another 15 tests. I priced out only 30 tests in the proposal. We're heading for a cost overrun condition."

Ray: "Our contract is a firm-fixed-price effort. Therefore, the cost overrun is your problem."

West met with Dr. Reddy to discuss the new direction of the project and potential cost overruns. West brought along a memo projecting the costs through the end of the third month of the project. (See Table II.)

Reddy told West: "I'm already overburdened on other projects and won't be able to help you out. Royce picked you to be the project manager because he felt that you could do the job. Now, don't let him down. Send me a brief memo next month explaining the situation, and I'll see what I can do. Perhaps the situation will correct itself."

During March, the third month of the project, West received almost daily phone calls from the people in the lab stating that Pat Ray was interfering with their job. In fact, one phone call stated that Ray had changed the test conditions from what was agreed on in the latest test matrix. When West confronted Ray on his meddling, Ray asserted that Corwin personnel were very unprofessional

TABLE II   PROJECTED COST SUMMARY AT THE END OF THE THIRD MONTH

	Original Proposal Cost Summary for Six-Month Project	Total Project Costs Projected at End of Third Month
Direct labor/support	$ 30,000	$ 15,000
Testing	60,000 (30 tests)	70,000 (35 tests)
Overhead	90,000 (100%)	92,000 (120%)*
Materials	30,000	50,000
G&A	21,000 (10%)	22,700 (10%)
Totals	$231,000	$249,700

*Total engineering overhead was estimated at 100 percent, whereas the R&D overhead was 120 percent.

in their attitude and that he thought this was being carried down to the testing as well. Furthermore, Ray demanded that one of the functional employees be removed immediately from the project because of incompetence. West stated that he would talk to the employee's department manager. Ray, however, felt that this would be useless and said, "Remove him or else!" The functional employee was removed from the project.

By the end of the third month, most Corwin employees were disenchanted with the project and were looking for other assignments. West attributed this to Ray's harassment of the employees. To aggravate the situation even further, Ray met with Royce and Reddy and demanded that West be removed and a new project manager be assigned.

Royce refused to remove West as project manager and ordered Reddy to take charge and help West get the project back on track. Reddy said: "You've kept me in the dark concerning this project, West. If you want me to help you, as Royce requested, I'll need all the information tomorrow, especially the cost data. I'll expect you in my office tomorrow morning at 8:00 a.m. I'll bail you out of this mess."

West prepared the projected cost data for the remainder of the work and presented the results to Dr. Reddy. (See Table III.) Both West and Reddy agreed that the project was now out of control and severe measures would be required to correct the situation, in addition to more than $250,000 in corporate funding.

Reddy: "Dan, I've called a meeting for 10:00 a.m. with several of our R&D people to completely construct a new test matrix. This is what we should have done right from the start."

West: "Shouldn't we invite Ray to attend this meeting? I'm sure he'd want to be involved in designing the new test matrix."

Reddy: "I'm running this show now, not Ray! Tell Ray that I'm instituting new policies and procedures for in-house representatives. He's no longer authorized to

TABLE III   ESTIMATE OF TOTAL PROJECT COMPLETION COSTS

Direct labor/support	$ 47,000*
Testing (60 tests)	120,000
Overhead (120%)	200,000
Materials	103,000
G&A	47,000
	$ 517,000
Peters contract	250,000
Overrun	$ 267,000

*Includes Dr. Reddy

visit the labs at his own discretion. He must be accompanied by either you or me. If he doesn't like these rules, he can get out. I'm not going to allow that guy to disrupt our organization. We're spending our money now, not his."

West met with Ray and informed him of the new test matrix as well as the new policies and procedures for in-house representatives. Ray was furious over the new turn of events and stated that he was returning to Peters Company for a meeting with Delia.

On the following Monday, Frimel received a letter from Delia stating that Peters Company was officially canceling the contract. The reasons given by Delia were as follows:

1. Corwin had produced absolutely no data that looked promising.
2. Corwin continually changed the direction of the project and did not appear to have a systematic plan of attack.
3. Corwin did not provide a project manager capable of handling such a project.
4. Corwin did not provide sufficient support for the in-house representative.
5. Corwin's top management did not appear to be sincerely interested in the project and did not provide sufficient executive-level support.

Royce and Frimel met to decide on a course of action in order to sustain good working relations with Peters Company. Frimel wrote a strong letter refuting all of the accusations in the Peters letter, but to no avail. Even the fact that Corwin was willing to spend $250,000 of its own funds had no bearing on Delia's decision. The damage was done. Frimel was now thoroughly convinced that a contract should not be signed on Pearl Harbor Day.

QUESTIONS

1. What were the major mistakes made by Corwin?
2. Should Corwin have accepted the assignment?
3. Should companies risk bidding on projects based on rough-draft specifications?
4. Should the shortness of the proposal preparation time have required more active top management involvement before the proposal went out of house?
5. Are there any risks in not having the vice president for manufacturing available during the go or no-go bidding decision?
6. Explain the attitude of Dick Potts during the proposal activities.
7. None of the executives expressed concern when Dr. Reddy said, "I would never have assigned him [West] as the project leader." How do you account for the executives' lack of concern?
8. How important is it to inform line managers of proposal activities even if the line managers are not required to provide proposal support?

9. Explain Dr. Reddy's attitude after go-ahead.
10. How should West have handled the situation where Pat Ray's opinion of the test data was contrary to that of Corwin's engineering personnel?
11. How should West have reacted to Ray's informing Delia that the first five tests were failures?
12. Is immediate procurement of all materials a mistake?
13. Should Pat Ray have been given the freedom to visit laboratory personnel at any time?
14. Should an in-house representative have the right to remove a functional employee from the project?
15. Financially, how should the extra tests have been handled?
16. Explain Dr. Reddy's attitude when told to assume control of the project.
17. Delia's letter, stating the five reasons for canceling the project, was refuted by Frimel, but with no success. Could Frimel's early involvement as a project sponsor have prevented this?
18. In retrospect, would it have been better to assign a marketing person as project manager?
19. Your company has a single methodology for project management. A powerful customer offers you a special project that does not fit into your methodology. Should a project be refused simply because it is not a good fit with your methodology?
20. Should a customer be informed that only projects that fit your methodology would be accepted?

Quantum Telecom

In June 2013, the executive committee of Quantum Telecom reluctantly approved two R&D projects that required technical breakthroughs. To make matters worse, the two products had to be developed by the summer of 2014 and introduced into the marketplace quickly. The life expectancy of both products was estimated to be less than one year because of the rate of change in technology. Yet, despite these risks, the two projects were fully funded. Two senior executives were assigned as the project sponsors, one for each project.

Quantum Telecom had a world-class project management methodology with five life-cycle phases and five gate-review meetings. The gate-review meetings were go/no-go decision points based on present performance and future risks. Each sponsor was authorized and empowered to make any and all decisions relative to projects, including termination.

Company politics always played an active role in decisions to terminate a project. Termination of a project often impacted the executive sponsor's advancement opportunities because the projects were promoted by sponsors and funded through the sponsors' organizations.

During the first two gate-review meetings, virtually everyone recommended the termination of both projects. Technical breakthroughs seemed unlikely, and the schedule appeared unduly optimistic. But terminating the projects this early would certainly not reflect favorably on the sponsors. Reluctantly, both sponsors agreed to continue the projects to the third gate review in hopes of a miracle.

During the third gate review, the projects were still in peril. Although the technical breakthrough opportunity now seemed plausible, the launch date would have to slip, thus giving Quantum Telecom a window of only six months to sell the products before obsolescence would occur.

By the fourth gate review, the technical breakthrough had not yet occurred but still seemed plausible. Both project managers were still advocating the cancellation of the projects, and the situation was getting worse. Yet, in order to save face within the corporation, both sponsors allowed the projects to continue to completion. They asserted: "If the new products could not be sold in sufficient quantity to recover the R&D costs, then the fault lies with marketing and sales, not with us." The sponsors were now off the hook, so to speak.

Both projects were completed six months late. The sales force could not sell as much as one unit, and obsolescence occurred quickly. Marketing and sales were blamed for the failures, not the project sponsors.

QUESTIONS

1. How do we eliminate politics from gate-review meetings?
2. How can we develop a methodology where termination of a project is not viewed as a failure?
3. Were the wrong people assigned as sponsors?
4. What options are available to a project manager when there exists a disagreement between the sponsor and the project manager?
5. Can your answer to question 4 be outlined as part of the project management methodology?

The Trophy Project

The ill-fated Trophy Project was in trouble right from the start. Reichart, who had been an assistant project manager, was involved with the project from its conception. When the Trophy Project was accepted by the company, Reichart was assigned as the project manager. The program schedules started to slip from day 1, and expenditures were excessive. Reichart found that the functional managers were charging direct labor time to his project but working on their own pet projects. When he complained of this, he was told not to meddle in the functional manager's allocation of resources and budgeted expenditures. After approximately six months, Reichart was requested to make a progress report directly to corporate and division staffs.

Reichart took this opportunity to bare his soul. The report substantiated that the project was forecasted to be one complete year behind schedule. Reichart's staff, as supplied by the line managers, was inadequate to maintain the current pace, let alone make up any time that had already been lost. The estimated cost at completion at this interval showed a cost overrun of at least 20 percent. This was Reichart's first opportunity to tell his story to people who were in a position to correct the situation. The result of Reichart's frank, candid evaluation of the Trophy Project was very predictable. Nonbelievers finally saw the light, and line managers realized that they had a role to play in the completion of the project. Most of the problems were now out in the open and could be corrected with adequate staffing and resources. Corporate staff ordered immediate remedial action and staff support to provide Reichart a chance to bail out his program.

The results were not at all what Reichart had expected. He no longer reported to the project office; he now reported directly to the operations manager. Corporate staff's interest in the project became very intense, requiring a 7:00 a.m. meeting every Monday for complete review of the project status and plans for recovery. Reichart found himself spending more time preparing paperwork, reports, and projections for his Monday morning meetings than he did administering the Trophy Project. The main concern of corporate was to get the project back on schedule. Reichart spent many hours preparing the recovery plan and establishing manpower requirements to bring the program back onto the original schedule.

Group staff, in order to closely track the progress of the Trophy Project, assigned an assistant program manager. The assistant program manager determined that a sure cure for the Trophy Project would be to computerize the various problems and track the progress through a very complex computer program. Corporate provided Reichart with 12 additional staff members to work on the computer program. In the meantime, nothing changed. The functional managers still did not provide adequate staff for recovery, as they assumed that the additional manpower Reichart had received from corporate would accomplish that task.

After approximately $50,000 was spent on the computer program to track the problems, it was found that the computer could not handle the program objectives. Reichart discussed this problem with a computer supplier and found that $15,000 more was required for programming and additional storage capacity. It would take two months for installation of the additional storage capacity and completion of the programming. At this point, the decision was made to abandon the computer program.

Reichart was now a year and a half into the program with no prototype units completed. The program was still nine months behind schedule with the overrun projected at 40 percent of budget. The customer had been receiving reports on a timely basis and was well aware that the Trophy Project was behind schedule. Reichart had spent a great deal of time with the customer explaining the problems and the plan for recovery. Another problem that Reichart had to contend with was that the vendors who were supplying components for the project were also running behind schedule.

One Sunday morning, while Reichart was in his office putting together a report for the client, a corporate vice president came in. "Reichart," he said, "in any project I look at the top sheet of paper, and the man whose name appears at the top of the sheet is the one I hold responsible. For this project, your name appears at the top of the sheet. If you cannot bail this thing out, you are in serious trouble in this corporation." Reichart did not know which way to turn or what to say. He had no control over the functional managers who were creating the problems, but he was the person who was being held responsible.

After another three months, the customer, becoming impatient, realized that the Trophy Project was in serious trouble and requested that the division general manager and his entire staff visit the customer's plant to give a progress and get-well report within a week. The division general manager called Reichart into his office and said, "Reichart, go visit our customer. Take three or four functional line people with you and try to placate him with whatever you feel is necessary." Reichart and four functional line people visited the customer and gave a four-and-a-half-hour presentation defining the problems and the progress to that point. The customer was very polite and even commented that it was an excellent presentation, but the content was totally unacceptable. The program was still six to eight months late, and the customer demanded progress reports on a weekly basis. The customer made arrangements to assign a representative in Reichart's department to be on-site at the project on a daily basis and to interface with Reichart and his staff as required. After this turn of events, the program became very hectic.

The customer representative demanded constant updates and problem identification and then became involved in attempting to solve these problems. This involvement created many changes in the program and the product in order to eliminate some of the problems. Reichart had trouble with the customer and did not agree with the changes in the program. He expressed his disagreement vocally when, in many cases, the customer felt the changes were at no cost. This caused a deterioration of the relationship between client and producer.

One morning Reichart was called into the division general manager's office and introduced to Mr. "Red" Baron. Reichart was told to turn over the reins of the Trophy Project to Red immediately. "Reichart, you will be temporarily reassigned to some other division within the corporation. I suggest you start looking outside the company for another job." Reichart looked at Red and asked, "Who did this? Who shot me down?"

Red was program manager on the Trophy Project for approximately six months, after which, by mutual agreement, he was replaced by a third project manager. The customer reassigned his local program manager to another project. With the new team, the Trophy Project was finally completed one year behind schedule and at a 40 percent cost overrun.

QUESTIONS

1. Did the project appear to be planned correctly?
2. Did functional management seem to be committed to the project?
3. Did senior management appear supportive and committed?
4. Can a single methodology for project management be designed to force cooperation to occur between groups?
5. Is it possible or even desirable for strategic planning for project management to include ways to improve cooperation and working relationships, or is this beyond the scope of strategic planning for project management?

Margo Company

"I've called this meeting, gentlemen, because that paper factory we call a computer organization is driving up our overhead rates," snorted Richard Margo, president, as he looked around the table at the vice presidents of project management, engineering, manufacturing, marketing, administration, and information systems. "We seem to be developing reports faster than we can update our computer facility. Just one year ago, we updated our computer, and now we're operating three shifts a day, seven days a week. Where do we go from here?"

V.P. information: "As you all know, Richard asked me, about two months ago, to investigate this gigantic increase in the flow of paperwork. There's no question that we're getting too many reports. The question is, are we paying too much money for the information that we get? I've surveyed all of our departments and their key personnel. Most of the survey questionnaires indicate that we're getting too much information. Only a small percentage of each report appears to be necessary. In addition, many of the reports arrive too late. I'm talking about scheduled reports, not planning, demand, or exception reports."

V.P. project management: "Every report people may receive is necessary for us to make decisions effectively with regard to planning, organizing, and controlling each project. My people are the biggest users, and we can't live with fewer reports."

V.P. information: "Can your people live with less information in each report? Can some of the reports be received less frequently?"

V.P. project management: "Some of our reports have too much information in them. But we need them at the frequency we have now."

V.P. engineering: "My people utilize about 20 percent of the information in most of our reports. Once our people find the information they want, the report is discarded. That's because we know that each project manager will retain a copy. Also, only the department managers and section supervisors read the reports."

V.P. information: "Can engineering and manufacturing get the information they need from other sources, such as the project office?"

V.P. project management: "Wait a minute! My people don't have time to act as paper pushers for each department manager. We all know that the departments can't function without these reports. Why should we assume the burden?"

V.P. information: "All I'm trying to say is that many of our reports can be combined into smaller ones and possibly made more concise. Most of our reports are flexible enough to meet changes in our operating business. We have two sets of reports: one for the customer and one for us. If the customer wants the report in a specific fashion, he pays for it. Why can't we act as our own customer and try to make a reporting system that we can all use?"

V.P. engineering: "Many of the reports obviously don't justify the cost. Can we generate the minimum number of reports and pass it on to someone higher or lower in the organization?"

V.P. project management: "We need weekly reports, and we need them on Monday mornings. I know our computer people don't like to work on Sunday evenings, but we have no choice. If we don't have those reports on Monday mornings, we can't control time, cost, and performance."

V.P. information: "There are no reports generated from the pertinent data in our original computer runs. This looks to me like every report is a one-shot deal. There has to be room for improvement.

"I have prepared a checklist for each of you with four major questions: Do you want summary or detailed information? How do you want the output to look? How many copies do you need? How often do you need these reports?"

Richard Margo: "In project organizational forms, the project exists as a separate entity except for administrative purposes. These reports are part of that administrative purpose. Combining this with the high cost of administration in our project structure, we'll never remain competitive unless we lower our overhead. I'm going to leave it up to you guys. Try to reduce the number of reports, but don't sacrifice the necessary information you need to control the projects and your resources."

Project Overrun

The Green Company production project was completed three months behind schedule and at a cost overrun of approximately 60 percent. Following submittal of the final report, Phil Graham, the director of project management, called a meeting to discuss the problems encountered on the project.

Phil Graham: "We're not here to point the finger at anyone. We're here to analyze what went wrong and to see if we can develop any policies and/or procedures that will prevent this from happening in the future. What went wrong?"

Project manager: "When we accepted the contract, Green did not have a fixed delivery schedule for us to go by because they weren't sure when their new production plant would be ready to begin production activities. So, we estimated 3,000 units per month for months 5 through 12 of the project. When Green found that the production plant would be available two months ahead of schedule, they asked us to accelerate our production activities. So, we put all of our production people on overtime in order to satisfy their schedule. This was our mistake, because we accepted a fixed delivery date and budget before we understood everything."

Functional manager: "Our problem was that the customer could not provide us with a fixed set of specifications, because the final set of specifications depended on OSHA and EPA requirements, which could not be confirmed until initial testing of the new plant. Our people, therefore, were asked to commit to man-hours before specifications could be reviewed.

"Six months after project go-ahead, Green Company issued the final specifications. We had to remake 6,000 production units because they did not live up to the new specifications."

Project manager: "The customer was willing to pay for the remake units. This was established in the contract. Unfortunately, our contract people didn't tell me that we were still liable for the penalty payments if we didn't adhere to the original schedule."

Phil Graham: "Don't you feel that misinterpretation of the terms and conditions is your responsibility?"

Project manager: "I guess I'll have to take some of the blame."

Functional manager: "We need specific documentation on what to do in case of specification changes. I don't think that our people realize that user approval of specification is not a contract agreed to in blood. Specifications can change, even in the middle of a project. Our people must understand that, as well as the necessary procedures for implementing change."

Phil Graham: "I've heard that the functional employees on the assembly line are grumbling about the Green Project. What's their gripe?"

Functional manager: "We were directed to cut out all overtime on all projects. But when the Green Project got into trouble, overtime became a way of life. For nine months, the functional employees on the Green Project had as much overtime as they wanted. This made the functional employees on other projects very unhappy.

"To make matters worse, the functional employees got used to a big take-home paycheck and started living beyond their means. When the project ended, so did their overtime. Now they claim that we should give them the opportunity for more overtime. Everybody hates us."

Phil Graham: "Well, now we know the causes of the problem. Any recommendations for cures and future prevention activities?"

QUESTIONS

1. What are the critical issues in the case?
2. How should they be resolved?

The Automated Evaluation Project

"No deal!" said the union. "The current method of evaluating government employees at this agency is terrible, and if a change doesn't occur, we'll be in court seeking damages."

In 1984, a government agency approved and initiated an ambitious project, part of which was to develop an updated, automated evaluation system for the 50,000 employees located throughout the United States. The existing evaluation system was antiquated. Although forms were used for employee evaluation, standardization was still lacking. Not all promotions were based on performance. Often they were based on time in grade, the personal whims of management, or friendships. Some divisions seemed to promote employees faster than others. The success or failure of a project could also seriously impact performance opportunities. Some type of standardization was essential.

In June 1985, a project manager was finally assigned and brought on board. The assignment of the project manager was based on rank and availability at that time rather than the requirements of the project. Team members often possessed a much better understanding of the project than did the project manager.

The project manager, together with his team, quickly developed an *action plan*. The action plan did *not* contain a work breakdown structure but did contain a statement of work which called out high-level deliverables that would be essential for structured analyses, design, and programming. The statement of work and deliverables were more in compliance with agency requirements for structured analyses, design, and programming than with the project's requirements. The entire action plan was prepared by the project office, which was composed of eight employees.

Bids from outside vendors were solicited for the software packages, with the constraint that all deliverables must be operational on existing agency hardware. In October 1985, the award was made by the project office to Primco Corporation with work scheduled to begin in December 1985.

In the spring of 1986, it became apparent that the project was running into trouble and disaster was imminent. Three major problems faced the project manager. As stated by the project manager:

1. The requirements for the project had to be changed because of new regulations for government worker employee evaluation.
2. Primco did not have highly skilled personnel assigned to the project.
3. The agency did not have highly skilled personnel from the functional areas assigned to the project.

The last item was controversial. The line managers at the agency contended that they had assigned some of their best people and that the real problem was that the project manager was trying to make all of the decisions himself without any input from the assigned personnel. The employees contended that proper project management practices were not being used. The project was being run like a dictatorship rather than a democracy. Several employees felt that they were not treated as part of the project team.

According to one team member:

> The project manager keeps making technical decisions without any solid foundation to support his views. Several of us in the line organization have significantly more knowledge than does the project manager, yet he keeps overriding our recommendations and decisions. Perhaps he has that right, but I dislike being treated as a second-class citizen. If the project manager has all of this technical knowledge, then why does he need us?

In June 1986, the decision was made by the project manager to ask one of the assistant agency directors to tell the union that the original commitment date of January 1987 would not be met. A stop-work order was issued to Primco, thus canceling the contract.

The original action plan called for the use of existing agency hardware. However, because of unfavorable publicity about hardware and software problems at the agency during the spring of 1986, the agency felt that the UNIVAC System would not support the additional requirements, and system overload might occur. Now hardware, as well as software, would be needed.

To help maintain morale, the project manager decided to perform as much of the work as possible in-house, even though the project lacked critical resources and was already more than one year late. The project office took what was developed thus far and tried to redefine the requirements.

With the support of senior management at the agency, the original statement of work was thrown away and a new statement of work was prepared. "It was like starting over right from the beginning," remarked one of the employees. "We never looked back at what was accomplished thus far. It was a whole new project!" With the support of the agency's personnel office, the new requirements were finally completed in February of 1987.

The union, furious over the schedule slippage, refused to communicate with the project office and senior management. The union's contention was that an "illegal" evaluation system was in place, and the current system could not properly validate performance review requirements. The union initiated a lawsuit against the agency seeking damages in excess of $21 million.

In November 1986, procurement went out for bids for both hardware and a database management system. The procurement process continued until June 1987, when it was canceled by another government agency responsible for procurement. No reason was ever provided for the cancellation.

Seeking alternatives, the following decisions were made:

1. Use rented equipment to perform the programming.
2. Purchase a database management system from ITEKO Corporation, provided that some customization could be accomplished. The new database management system was scheduled to be released to the general public in about two months.

The database management system was actually in the final stages of development. ITEKO Corporation promised the agency that a fully operational version, with the necessary customization, could be provided quickly. Difficulties arose with the use of the ITEKO package. After hiring a consultant from ITEKO, it was found that the ITEKO package was a beta rather than a production version. Despite these setbacks, personnel kept programming on the leased equipment with the hope of eventually purchasing a Micronet Hardware System. ITEKO convinced the agency that the Micronet hardware system was the best system available to support the database management system. The Micronet hardware was then added to the agency's equipment contract but later disallowed on September 29, 1987, because it was not standard agency equipment.

On October 10, 1987, the project office decided to outsource some of the work using a small/minority business procurement strategy for hardware to support the ITEKO package. The final award was made in November 1987, subject to software certification by the one of the agency's logistics centers. Installation in all of the centers was completed between November and December 1987.

QUESTIONS

1. Is there anything in the case that indicates the maturity level of project management at the agency around 1985–1986?
2. What are the major problems in the case?
3. Who was at fault?
4. How do you prevent this from occurring on other projects?

The Rise, Fall, and Resurrection of Iridium: A Project Management Perspective

The Iridium Project was designed to create a worldwide wireless handheld mobile phone system with the ability to communicate anywhere in the world at any time. Executives at Motorola regarded the project as the eighth wonder of the world. But more than a decade later and after investing billions of dollars, Iridium had solved a problem that very few customers needed solved. What went wrong? How did the Iridium Project transform from a leading-edge technical marvel to a multibillion-dollar blunder? Could the potential catastrophe have been prevented?

> What it looks like now is a multibillion-dollar science project. There are fundamental problems: The handset is big, the service is expensive, and the customers haven't really been identified.
> **—Chris Chaney, Analyst, A.G. Edwards, 1999**

> There was never a business case for Iridium. There was never market demand. The decision to build Iridium wasn't a rational business decision. It was more of a religious decision. The remarkable thing is that this happened at a big corporation, and that there was not a rational decision-making process in place to pull the plug. Technology for technology's sake may not be a good business case.
> **—Herschel Shosteck, Telecommunication Consultant**

Iridium is likely to be some of the most expensive space debris ever.
—William Kidd, Analyst, C.E. Unterberg, Towbin

In 1985, Bary Bertiger, chief engineer in Motorola's strategic electronics division, and his wife, Karen, were on a vacation in the Bahamas. Karen tried to make a cellular telephone call back to her home near the Motorola facility in Chandler, Arizona, to close a real-estate transaction. Unsuccessful, she asked her husband why it would not be possible to create a telephone system that would work anywhere in the world, even in remote locations.

At this time, cell technology was in its infancy but was expected to grow at an astounding rate. AT&T projected as many as 40 million subscribers by 2000.[1] Cell technology was based on tower-to-tower transmission, as shown in Exhibit I. Each tower, or "gateway" ground station. reached a limited geographic area or cell and had to be within the satellite's field of view. Cell phone users likewise had to be near a gateway that would uplink the transmission to a satellite. The satellite would then downlink the signal to another gateway that would connect the transmission to a ground telephone system. This type of communication is often referred to as bent pipe architecture. Physical barriers between the senders/receivers and the gateways, such as mountains, tunnels, and oceans, created interference problems and therefore limited service to high-density communities. Simply stated, cell phones couldn't leave home. And, if they did, there would be additional "roaming" charges. To make matters worse, every country had its own standards, and some cell phones were inoperable when traveling in other countries.

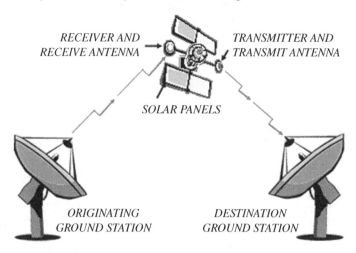

RECEIVER AND RECEIVE ANTENNA *TRANSMITTER AND TRANSMIT ANTENNA*

SOLAR PANELS

ORIGINATING GROUND STATION *DESTINATION GROUND STATION*

FIGURE I Typical satellite communication architecture

[1] Judith Bird, "Cellular Technology in Telephones," *data processing* 27, no. 8 (October 1985): 37.

Communications satellites, in use since the 1960s, were typically geostationary satellites that orbited at altitudes of more than 22,300 miles. At this altitude, three geosynchronous satellites and just a few gateways could cover most of Earth. But satellites at this altitude meant large phones and annoying quarter-second voice delays. Comsat's Planet 1 phone, for example, weighed in at a computer case–size 4.5 pounds. Geosynchronous satellites require signals with a great deal of power. Small mobile phones, with a 1-watt signal, could not work with satellites positioned at this altitude. If the power output of mobile phones was increased, human tissue would be damaged. The alternative was to move the satellites closer to Earth so that less power would be needed. This would require significantly more satellites closer to Earth as well as additional gateways. Geosynchronous satellites, which are 100 times farther away from Earth than low-Earth-orbiting (LEO) satellites, could require almost 10,000 times as much power as LEOs, if everything else were the same. [2]

When Bary Bertiger returned to Motorola, he teamed up with Dr. Raymond Leopold and Kenneth Peterson to see if such a worldwide system could be developed while overcoming all of the limitations of existing cell technology. There was also the problem that LEO satellites would be orbiting Earth rapidly and going through damaging temperature variations—from the heat of the Sun to the cold shadow of Earth.[3] The LEO satellites would most likely need to be replaced every 5 years. Numerous alternative terrestrial designs were discussed and abandoned. In 1987 research began on a constellation of LEO satellites moving in polar orbits that could communicate directly with telephone systems on the ground and with one another.

Iridium's innovation was to use a large constellation of LEO satellites approximately 400 to 450 miles in altitude. Because Iridium's satellites were closer to Earth, the phones could be much smaller and the voice delay would be imperceptible. But there were still major technical design problems. With the existing design, a large number of gateways would be required, thus substantially increasing the cost of the system. As they left work one day in 1988, Dr. Leopold proposed a critical design element. The entire system would be inverted whereby the transmission would go from satellite to satellite until the transmission reached the satellite directly above the person who would be receiving the message. With this approach, only one gateway Earth station would be required to connect mobile-to-landline calls to existing land-based telephone systems. This was considered to be the sought-after solution and was immediately written in outline format on a whiteboard in a security guard's office. Thus arose the idea behind a worldwide

---

[2] Ibid.

[3] Bruce Gerding, "Personal Communications via Satellite: An Overview," *Telecommunications* 30, no. 2 (February 1996): 35, 77.

wireless handheld mobile phone with the ability to communicate anywhere and anytime.

NAMING THE PROJECT

Motorola cellular telephone system engineer Jim Williams, from the Motorola facility near Chicago, suggested the name "Iridium." The proposed 77-satellite constellation reminded him of the electrons that encircle the nucleus in the classical Bohr model of the atom. When he consulted the periodic table of elements to discover which atom had 77 electrons, he found iridium—a creative name that had a nice ring. Fortunately, the system had not yet been scaled back to 66 satellites, or else he might have suggested the name "Dysprosium."

OBTAINING EXECUTIVE SUPPORT

Initially, Bertiger's colleagues and superiors at Motorola had rejected the Iridium concept because of its cost. Originally, the Iridium concept was considered perfect for the U.S. government. Unfortunately, the era of lucrative government-funded projects was coming to an end, and it was unlikely that the government would fund a project of this magnitude. However, the idea behind the Iridium concept intrigued Durrell Hillis, the general manager of Motorola's Space and Technology Group. Hillis believed that Iridium was workable if it could be developed as a commercial system. Hillis instructed Bertiger and his team to continue working on the Iridium concept but to keep it quiet. "I created a bootleg project with secrecy so no one in the company would know about it," Hillis recalls. He was worried that if word leaked out, the ferociously competitive business units at Motorola, all of which had to fight for R&D funds, would smother the project with nay saying. [4]

After 14 months of rewrites on the commercialized business plan, Hillis and the Iridium team leaders presented the idea to Robert Galvin, Motorola's chairman at the time, who gave approval to go ahead with the project. Galvin, and later his successor and son, Christopher Galvin, viewed Iridium as a potential symbol of Motorola's technological prowess and believed that this would become the eighth wonder of the world. In one of the initial meetings, Robert Galvin turned to John Mitchell, Motorola's president and chief operating officer, and said, "If you don't write out a check for this, John, I will, out of my own pocket." [5]

To the engineers at Motorola, the challenge of launching Iridium's constellation provided considerable motivation. They continued developing the project that resulted in initial service in November 1998 at a total cost of over $5 billion.

---

[4] David S. Bennahum, "The United Nations of Iridium," *Wired*, issue 6.10 (October 1998): 194.

[5] Quentin Hardy, "How a Wife's Question Led Motorola to Chase a Global Cell-Phone Plan," *Wall Street Journal*, December 16, 1996, p. A1.

LAUNCHING THE VENTURE

On June 26, 1990, Hillis and his team formally announced the launch of the Iridium Project to the general public. The response was not very pleasing to Motorola: There was skepticism that this would be a new technology, the target markets were too small, the revenue model was questionable, obtaining licenses to operate in 170 countries could be a problem, and the cost of a phone call might be overpriced. Local phone companies that Motorola assumed would buy into the project viewed Iridium as a potential competitor since the Iridium system bypassed traditional landlines. In many countries, postal, telephone, and telegraph operators are state owned and a major source of revenue because of the high profit margins. Another issue was that the Iridium Project was announced before permission was granted by the Federal Communications Commission (FCC) to operate at the desired frequencies.

Both Mitchell and Galvin made it clear that Motorola would not go it alone and absorb the initial financial risk for a hefty price tag of about $3.5 billion. Funds would need to be obtained from public markets and private investors. In order to minimize Motorola's exposure to financial risk, Iridium would need to be set up as a project-financed company. Project financing involves the establishment of a legally independent project company where the providers of funds are repaid out of cash flow and earnings and where the assets of the unit (and only the unit) are used as collateral for the loans. Debt repayment would come from the project company only rather than from any other entity. A risk with project financing is that the capital assets may have a limited life. The potential limited life constraint often makes it difficult to get lenders to agree to long-term financial arrangements.

Another critical issue with project financing especially for high-tech projects is that the projects are generally long term. It would be nearly eight years before service would begin, and in terms of technology, eight years is an eternity. The Iridium Project was certainly a bet on the future. And if the project were to fail, Motorola could be worth nothing after liquidation.

In 1991, Motorola established Iridium Limited Liability Corporation (Iridium LLC) as a separate company. In December of 1991, Iridium promoted Leo Mondale to vice president of Iridium International. Financing the project was still a critical issue. Mondale decided that, instead of having just one gateway, there should be as many as 12 regional gateways that plugged into local, ground-based telephone lines. This would make Iridium a truly global project rather than appearing to be an American-based project designed to seize market share from state-run telephone companies. This would also make it easier to get regulatory approval to operate in 170 countries. Investors would pay $40 million for the right to own their own regional gateway. As stated by Flower: "The motive of the investors is clear: They are taking a chance on owning a slice of a de-facto world monopoly. Each of them will not only have a piece of the company, they will own the Iridium gateways and act as the local distributors in their respective home markets. For them it's a game worth playing." [6]

There were political ramifications with selling regional gateways. What if in the future the U.S. government forbids shipment of replacement parts to certain gateways? What if sanctions are imposed? What if Iridium were to become a political tool during international diplomacy because of the number of jobs it creates?

In addition to financial incentives, gateway owners were granted seats on the board of directors of Iridium LLC. As described by David Bennahum, reporter for *Wired*:

> Four times a year, 28 Iridium board members from 17 countries gather to coordinate overall business decisions. They met around the world, shuttling between Moscow, London, Kyoto, Rio de Janeiro, and Rome, surrounded by an entourage of assistants and translators. Resembling a United Nations in miniature, board meetings were conducted with simultaneous translation in Russian, Japanese, Chinese, and English. [7]

The partner with the largest equity share was Motorola. For its contribution of $400 million, Motorola originally received an equity stake of 25 percent, and 6 of the 28 seats on Iridium's board. Additionally, Motorola made loan guarantees to Iridium of $750 million, with Iridium holding an option for an additional $350 million loan.

For its part, Iridium agreed to $6.6 billion in long-term contracts with Motorola that included $3.4 billion for satellite design and launch and $2.9 billion for operations and maintenance. Iridium also exposed Motorola to developing satellite technology that would provide the latter with significant expertise in building satellite communications systems, as well as vast intellectual property.

THE IRIDIUM SYSTEM

The operational version of the Iridium system is a satellite-based, wireless personal communications network providing a robust suite of voice features to virtually any destination anywhere on Earth.

The Iridium system comprises three principal components: the satellite network, the ground network, and the Iridium subscriber products including phones and pagers. The design of the Iridium network allows voice and data to be routed virtually anywhere in the world. Voice and data calls are relayed from one satellite

---

[6] Joe Flower, "Iridium," *Wired*, issue 1.05, November, 1993.

[7] Bennahum, "The United Nations of Iridium," p. 136.

Information for "The Iridium System" and "The Terrestrial and Space-Based Network" is adapted from material found on the Iridium website, www.Iridium.com.

to another until they reach the satellite above the Iridium subscriber unit (handset) and the signal is relayed back to Earth.

THE TERRESTRIAL AND SPACE-BASED NETWORK

The Iridium constellation consists of 66 operational satellites and 11 spares orbiting in a constellation of six polar planes. Each plane has 11 mission satellites performing as nodes in the telephony network. The remaining 11 satellites orbit as spares ready to replace any unserviceable satellite. This constellation ensures that every region on the globe is covered by at least one satellite at all times.

The satellites are in a near-polar orbit at an altitude of 485 miles. They circle Earth once every 100 minutes traveling at a rate of 16,832 miles per hour. The satellite weight is 1,500 pounds. Each satellite is approximately 40 feet in length and 12 feet in width. In addition, each satellite has 48 spot beams, 30 miles in diameter per beam.

Each satellite is cross-linked to four other satellites: two satellites in the same orbital plane and two in an adjacent plane. The ground network is comprised of the system control segment and telephony gateways used to connect into the terrestrial telephone system. The system control segment is the central management component for the Iridium system. It provides global operational support and control services for the satellite constellation, delivers satellite-tracking data to the gateways, and performs the termination control function of messaging services. The system control segment consists of three main components: four telemetry tracking and control sites, the operational support network, and the satellite network operation center. The primary linkage between the System Control Segment, the satellites, and the gateways is via K-band feeder links and cross-links throughout the satellite constellation.

Gateways are the terrestrial infrastructure that provides telephony services, messaging, and support to the network operations. The key features of gateways are their support and management of mobile subscribers and the interconnection of the Iridium network to the terrestrial phone system. Gateways also provide network management functions for their own network elements and links.

PROJECT INITIATION: DEVELOPING THE BUSINESS CASE

For the Iridium Project to be a business success rather than just a technical success, there had to exist an established customer base. Independent studies conducted by A.T. Kearney, Booz, Allen & Hamilton, and Gallup indicated that 34 million people had a demonstrated need for mobile satellite services, with that number expected to grow to 42 million by 2002. Of these 42 million, Iridium anticipated 4.2 million to be satellite-only subscribers, 15.5 million satellite and world terrestrial roaming subscribers, and 22.3 million terrestrial roaming-only subscribers.

A universal necessity in conducting business is ensuring that you are never out of touch. Iridium would provide this unique solution to business with the essential communications tool. This proposition of one phone, one number with the capability to be accessed anywhere, anytime was a message that target markets—global travelers; mining, rural, and maritime industries; governments; disaster relief; and community aid groups—would readily embrace.

Also at the same time of Iridium's conception, another potentially lucrative opportunity in the telecommunications marketplace. When users of mobile or cellular phones crossed international borders, they soon discovered that there existed a lack of common standards, thus making some phones inoperable. Motorola viewed this as an opportunity to create a worldwide standard allowing phones to be used anywhere in the world.

The expected breakeven market for Iridium was estimated between 400,000 and 600,000 customers globally, assuming a reasonable usage rate per customer per month. With a launch date for Iridium service established for 1998, Iridium hoped to recover all of its investment within one year. By 2002, Iridium anticipated a customer base of 5 million users. The initial Iridium target market had been the vertical market, those of the industry, government, and world agencies that have defended needs and far-reaching communication requirements. Also important would be both industrial and public sector customers. Often isolated in remote locations outside of cellular coverage, industrial users were expected to use handheld Iridium satellite services to complement or replace their existing radio or satellite communications terminals. The vertical markets for Iridium would include:

- Aviation
- Construction
- Disaster relief/emergency
- Forestry
- Government
- Leisure travel
- Maritime
- Media and entertainment
- Military
- Mining
- Oil and gas
- Utilities

Using its own marketing resources, Iridium appeared to have identified an attractive market segment after having screened over 200,000 people, interviewed 23,000 people from 42 countries, and surveyed over 3,000 corporations.

Iridium would also need regional strategic partners, not only for investment purposes and to share the risks but to provide services throughout their territories. The strategic regional partners or gateway operating companies would have exclusive rights to their territories and were obligated to market and sell Iridium services. The gateways would also be responsible for end user sales, activation and deactivation of Iridium services, account maintenance, and billing.

Iridium would need each country to grant full licenses for access to the Iridium system. Iridium would need to identify the "priority" countries that would account for the majority of the business plan.

Because of the number of countries involved in the Iridium network, Iridium would need to establish global Customer Care Centers for support services in all languages. No matter where an Iridium user was located, he or she would have access to a customer service representative in the user's native language. The Customer Care Centers would be strategically located to offer support 24 hours a day, seven days a week, and 365 days a year.

THE "HIDDEN" BUSINESS CASE

The decision by Motorola to invest heavily into the Iridium Project may have been driven by a secondary or hidden business case. Over the years, Motorola achieved a reputation of being a first mover (i.e., first to market). With the Iridium Project, Motorola was poised to capture first-mover advantage in providing global telephone service via LEO satellites. In addition, even if the Iridium Project never resulted in providing service, Motorola would still have amassed valuable intellectual property that could make it the major player for years to come in satellite communications. Robert and Christopher Galvin also may have desired to have their names etched in history as pioneers in satellite communication.

RISK MANAGEMENT

Good business cases identify the risks that the project must consider. For simplicity's sake, the initial risks associated with the Iridium Project could be classified as technology, development, financial, or marketing risks.

Technology Risks

Although Motorola had some technology available for the Iridium Project, there was still the need to develop additional technology, specifically satellite communications technology. The development process was expected to take years and would eventually result in numerous patents.

Mark Gercenstein, Iridium's vice president of operations, explained the system's technological complexity: "More than 26 completely impossible things had to happen first, and in the right sequence (before we could begin operations)—like getting capital, access to the marketplace, global spectrum, the same frequency band in every country of operations."[8]

While there was still some risk in the development of new technology, Motorola had the reputation of being a high-tech, can-do company. The engineers at Motorola believed that they could bring forth miracles in technology. Motorola also had a reputation for being a first mover with new ideas and products, and there was no reason to believe that this would not happen on the Iridium Project. There was no competition for Iridium at the inception of the project.

Because the project schedule was more than a decade in duration, there was the risk of technology obsolescence. This required that certain assumptions be made concerning technology a decade downstream. Developing a new product is relatively easy if the environment is stable. But in a high-tech environment that is both turbulent and dynamic, it is extremely difficult to determine how customers will perceive and evaluate the product 10 years later.

Development Risks

The satellite communication technology, once developed, had to be manufactured, tested, and installed in the satellites and ground equipment. Even though the technology existed or would exist, there were still transitional or development risks from engineering, to manufacturing, to implementation that would bring with them additional problems that were not contemplated or foreseen.

Financial Risks

The cost of the Iridium Project would most certainly be measured in the billions of dollars. This would include the costs for technology development and implementation, the manufacture and launch of satellites, the construction of ground support facilities, and marketing and supervision. Raising money from Wall Street's credit and equity markets was years away. Investors were unlikely to put up the necessary hundreds of millions of dollars on merely an idea or a vision. The technology needed to be developed and possibly accompanied by the launch of a few satellites before the credit and equity markets would come on board.

Private investors were a possibility, but the greatest source of initial funding would have to come from the members of the Iridium consortium. Although sharing the financial risks among the membership seemed appropriate, there was no question that bank loans and lines of credit would be necessary. Since the Iridium Project was basically an idea, the banks would require some form of collateral or

---

[8] Peter Grams and Patrick Zerbib, "Caring for Customers in a Global Marketplace," *Satellite Communications* (October 1998): 25.

guarantee for the loans. Motorola, being the largest stakeholder (and also with the deepest pockets), would need to guarantee the initial loans.

Marketing Risks

The marketing risks were certainly the greatest risks facing the members of the Iridium consortium. Once again, the risks were shared among the membership where each member was expected to sign up customers in its geographic area.

Each consortium member had to aggressively sign up customers for a product that did not exist yet, where no prototypes existed to be shown to the customers, where limitations on the equipment were unknown as yet, and where significant changes in technology could occur between the time the customer signed up and the time the system was ready for use. Companies that saw the need for Iridium today may not see the same need 10 years later.

Motivating the consortium partners to begin marketing immediately would be extremely difficult since marketing material was nonexistent. There was also the very real fear that the consortium membership would be motivated more by the technology than by the necessary size of the customer base required.

The risks were interrelated. The financial risks were highly dependent on the marketing risks. If a sufficient customer base could not be signed up, there could be significant difficulty in raising capital.

THE COLLECTIVE BELIEF

Although the literature does not clearly identify it, there was most likely a collective belief among the workers assigned to the Iridium Project. A collective belief is a fervent, and perhaps blind, desire to achieve that can permeate an entire team, the project sponsor, and even the most senior levels of management. A collective belief can make a rational organization act in an irrational manner.

When a collective belief exists, people are selected based on their support for that belief. Nonbelievers are pressured into supporting the collective belief, and team members are not allowed to challenge the results. As the collective belief grows, both advocates and nonbelievers are trampled. The pressure of the collective belief can outweigh the reality of the results.

There are several characteristics of the collective belief, which is why some large, high-tech projects are often difficult to kill:

- Inability or refusal to recognize failure
- Refusing to see the warning signs
- Seeing only what you want to see
- Fear of exposing mistakes
- Viewing bad news as a personal failure
- Viewing failure as a sign of weakness

- Viewing failure as damage to one's career
- Viewing failure as damage to one's reputation

THE EXIT CHAMPION

Project champions do everything possible to make their projects successful. But what if the project champions and the project team have blind faith in the success of the project? What happens if the strongly held convictions and the collective belief disregard early warning signs of imminent danger? What happens if the collective belief drowns out dissent?

In such cases, an exit champion must be assigned. The exit champion sometimes needs to have some direct involvement in the project in order to have credibility. Exit champions must be willing to put their reputations on the line and possibly face the likelihood of being cast out from the project team. According to Isabelle Royer:

> Sometimes it takes an individual, rather than growing evidence, to shake the collective belief of a project team. If the problem with unbridled enthusiasm starts as an unintended consequence of the legitimate work of a project champion, then what may be needed is a countervailing force—an exit champion. These people are more than devil's advocates. Instead of simply raising questions about a project, they seek objective evidence showing that problems in fact exist. This allows them to challenge—or, given the ambiguity of existing data, conceivably even to confirm—the viability of a project. They then take action based on the data.[9]

The larger the project and the greater the financial risk to the firm, the higher up the exit champion should reside. On the Iridium Project, the collective belief originated with Galvin, Motorola's CEO. Therefore, who could possibly function as the exit champion on the Iridium Project? Since it most likely should be someone higher up than Galvin, the exit champion should have been someone on the board of directors or even the entire Iridium board of directors.

Unfortunately, the entire Iridium board of directors was also part of the collective belief and shirked its responsibility for oversight on the Iridium Project. In the end, Iridium had no exit champion. Large projects incur large cost overruns and schedule slippages. Once a project has started, making the decision to cancel is very difficult, according to David Davis.

> The difficulty of abandoning a project after several million dollars have been committed to it tends to prevent objective review and recosting. For this reason, ideally an independent management team—one not involved in the projects development—should do the recosting and, if possible, the entire review. . . . If

---

[9] Isabelle Royer, "Why Bad Projects Are So Hard to Kill," *Harvard Business Review* (February 2003): 11. Copyright © 2003 by the Harvard Business School Publishing Corporation. All rights reserved.

the numbers do not hold up in the review and recosting, the company should abandon the project. The number of bad projects that make it to the operational stage serves as proof that their supporters often balk at this decision.

. . . Senior managers need to create an environment that rewards honesty and courage and provides for more decision making on the part of project managers. Companies must have an atmosphere that encourages projects to succeed, but executives must allow them to fail. [10]

The longer the project, the greater the necessity for the exit champions and project sponsors to make sure that the business plan has exit ramps such that the project can be terminated before massive resources are committed and consumed. Unfortunately, when a collective belief exists, exit ramps are purposefully omitted from the project and business plans.

IRIDIUM'S INFANCY YEARS

By 1992, the Iridium Project attracted such stalwart companies as General Electric, Lockheed, and Raytheon. Some companies wanted to be involved to be part of the satellite technology revolution while others were afraid of falling behind the technology curve. In any event, Iridium was lining up strategic partners, but slowly.

The Iridium Plan, submitted to the FCC in August 1992, called for a constellation of 66 satellites, expected to be in operation by 1998, more powerful than originally proposed, thus keeping the project's cost at the previously estimated $3.37 billion. But the Iridium Project, while based on lofty forecasts of available customers, was now attracting other companies competing for FCC approval on similar satellite systems. These companies included Loral Corp., TRW Inc., and Hughes Aircraft Co., a unit of General Motors Corp. At least nine companies were competing for the potential billions of dollars in untapped revenue possible from satellite communications.

Even with the increased competition, Motorola was signing up partners. Motorola had set an internal deadline of December 15, 1992, to find the necessary funding for Iridium. Signed letters of intent were received from the Brazilian government and United Communications Co., of Bangkok, Thailand, to buy 5 percent stakes in the project, each now valued at about $80 million. The terms of the agreement implied that the Iridium consortium would finance the project with roughly 50 percent equity and 50 percent debt.

When the December 15 deadline arrived, Motorola was relatively silent on the signing of funding partners, fueling speculation that it was having trouble. Motorola did admit that the process was time-consuming because some investors required government approval before proceeding. Motorola was expected to announce at some point, perhaps in the first half of 1993, whether it was ready

---

[10] David Davis, "New Projects: Beware of False Economics," *Harvard Business Review* (March–April 1985): 100–101.

to proceed with the next step, namely receiving enough cash from its investors, securing loans, and ordering satellite and group equipment.

As the competition increased, so did the optimism about the potential size of the customer base.

> "We're talking about a business generating billions of dollars in revenue," says John Mitchell, Vice Chairman at Motorola. "Do a simple income extrapolation," adds Edward J. Nowacki, a general manager at TRW's Space & Electronics Group, Redondo Beach, Calif., which plans a $1.3 billion, 12-satellite system called Odyssey. "You conclude that even a tiny fraction of the people around the world who can afford our services will make them successful." Mr. Mitchell says that if just 1% to 1.5% of the expected 100 million cellular users in the year 2000 become regular users at $3 a minute, Iridium will breakeven. How does he know this? "Marketing studies," which he won't share. TRW's Mr. Nowacki says Odyssey will blanket the Earth with two-way voice communication service priced at "only a slight premium" to cellular. "With two million subscribers we can get a substantial return on our investment," he says. "Loral Qualcomm Satellite Services, Inc. aims to be the 'friendly' satellite by letting phone-company partners use and run its system's ground stations," says Executive Vice President Anthony Navarra. "By the year 2000 there will be 15 million unserved cellular customers in the world," he says.[11]

But while Motorola and other competitors were trying to justify their investment with "inflated market projections" and a desire from the public for faster and clearer reception, financial market analysts were not so benevolent. First, market analysts questioned the size of the customer base that would be willing to pay $3,000 or more for a satellite phone in addition to $3 to $7 per minute for a call. Second, the system required a line-of-sight transmission, which meant that the system would not work in buildings or in cars. If a businessman were attending a meeting in Bangkok and needed to call his company, he must exit the building, raise the antenna on his $3,000 handset, point the antenna toward the heavens, and then make the call. Third, the low-flying satellites would eventually crash into Earth's atmosphere every five to seven years because of atmospheric drag and would need to be replaced. That would most likely result in high capital costs. And fourth, some industry analysts believed that the start-up costs would be closer to $6 to $10 billion rather than the $3.37 billion estimated by Iridium. In addition, the land-based cellular phone business was expanding in more countries, thus creating another competitive threat for Iridium.

The original business case needed to be reevaluated periodically. But with strong collective beliefs and no exit champions, the fear of a missed opportunity, irrespective of the cost, took center stage.

---

[11] John J. Keller, "Telecommunications: Phone Space Race Has Fortune at Stake," *Wall Street Journal*, January 18, 1993, p. B1.

Reasonably sure that 18 out of 21 investors were on board, Motorola hoped to start launching test satellites in 1996 and begin commercial service by 1998. But critics argued that Iridium might be obsolete by the time it actually started working.

Eventually, Iridium was able to attract financial support from these strategic partners:

- AIG Affiliated Companies
- China Great Wall Industry Corporation (CGWIC)
- Iridium Africa Corporation (based in Cape Town)
- Iridium Canada, Inc.
- Iridium India Telecom Private Ltd, (ITIL)
- Iridium Italia S.p.A.
- Iridium Middle East Corporation
- Iridium SudAmerica Corporation
- Khrunichev State Research and Production Space Center
- Korea Mobile TELECOM
- Lockheed Martin
- Motorola
- Nippon Iridium Corporation
- Pacific Electric Wire & Cable Co. Ltd (PEWC)
- Raytheon
- STET
- Sprint
- Thai Satellite Telecommunications Co., Ltd.
- Verbacom

Seventeen of the strategic partners also participated in gateway operations with the creation of operating companies.

The Iridium board of directors consisted of 28 telecommunications executives. All but one board member was a member of the consortium as well. This made it very difficult for the board to fulfill its oversight obligation, effectively giving the members vested/financial interests in the Iridium Project.

In August 1993, Lockheed announced that it would receive $700 million in revenue for satellite construction. Lockheed would build the satellite structure, solar panels, attitude and propulsion systems, along with other parts and would supply engineering support. Motorola and Raytheon Corp. would build the satellite's communications gear and antenna.

In April 1994, McDonnell Douglas Corp. received from Iridium a $400 million contract to launch 40 satellites for Iridium. Other contracts for launch services would be awarded to Russia's Khrunichev Space Center and China's Great Wall Industry Corporation, both members of the consortium. The lower-cost contracts with Russia and China were putting extraordinary pressure on U.S. providers to lower their costs.

Also at the same time, one of Iridium's competitors, the Globalstar system, which was a 48-satellite mobile telephone system led by Loral Corporation, announced that it intended to charge 65 cents per minute in the areas it served. Iridium's critics were arguing that Iridium would be too pricey to attract a high volume of callers.[12]

DEBT FINANCING

In September 1994, Iridium said that it had completed its equity financing by raising an additional $733.5 million. This brought the total capital committed to Iridium through equity financing to $1.57 billion. The completion of equity financing permitted Iridium to enter into debt financing to build the global wireless satellite network.

In September 1995, Iridium announced that it would be issuing $300 million 10-year senior-subordinated discounted notes rated Caa by Moody's and CCC+ by Standard & Poor's, via the investment banker Goldman Sachs Inc. The bonds were considered to be high-risk, high-yield "junk" bonds after investors concluded that the rewards were not worth the risk.

The rating agencies cited the reasons for the low rating to be yet-unproven sophisticated technology and the fact that a significant portion of the system's hardware would be located in space. But there were other serious concerns:

- The ultimate cost of the Iridium Project would be more like $6 billion or higher rather than $3.5 billion, and it was unlikely that Iridium would recover that cost.
- Iridium would be hemorrhaging cash for several more years before service would begin.
- The optimistic number of potential customers for satellite phones might not choose the Iridium system.
- The number of competitors had increased since the Iridium concept was first developed.
- If Iridium defaulted on its debt, the investors could lay claim to Iridium's assets. But what would investors do with more than 66 satellites in space, waiting to disintegrate upon reentering the atmosphere?

Iridium was set up as "project financing" in which case, if a default occurred, only the assets of Iridium could be attached. With project financing,

---

[12] Jeff Cole, "McDonnell Douglas Said to Get Contract to Launch 40 Satellites for Iridium Plan," *Wall Street Journal*, April 12, 1994, p. A4.

the consortium's investors would be held harmless for any debt incurred from the stock and bond markets and could simply walk away from Iridium. Those who invested in the equity and credit markets well understood the risks associated with project financing.

Goldman Sachs Inc., the lead underwriter for the securities offering, determined that for the bond issue to be completed successfully, there would need to exist a completion guarantee from investors with deep pockets, such as Motorola. Goldman Sachs cited a recent $400 million offering by one of Iridium's competitors, Globalstar, which had a guarantee from the managing general partner, Loral Corp.[13]

Because of investor concern, Iridium withdrew its planned $300 million debt offering. Also, Globalstar, even with its loan guarantee, eventually withdrew its $400 million offering. Investors wanted both an equity position in Iridium and a 20 percent return. Additionally, Iridium would need to go back to its original 17-member consortium and arrange for internal financing.

In February 1996, Iridium had raised an additional $315 million from the 17-member consortium and private investors. In August 1996, Iridium had secured a $750 million credit line with 62 banks coarranged by Chase Securities Inc., a unit of Chase Manhattan Corp. and the investment banking division of Barclays Bank PLC. The credit line was oversubscribed by more than double its original goal because the line of credit was backed by a financial guarantee by Motorola and its AAA credit rating. Because of the guarantee by Motorola, the lending rate was slightly more than the 5.5 percent baseline international commercial lending rate and significant lower than the rate in the $300 million bond offering that was eventually recalled.

Despite this initial success, Iridium still faced financial hurdles. By the end of 1996, Iridium planned on raising more than $2.65 billion from investors. It was estimated that more than 300 banks around the globe would be involved and that this would be the largest private debt placement ever. Iridium believed that this debt placement campaign might not be that difficult since the launch date for its services was getting closer.

THE M-STAR PROJECT

In October 1996, Motorola announced that it was working on a new project dubbed M-Star, which would be a $6.1 billion network of 72 low-orbit satellites capable of worldwide voice, video, and high-speed data links targeted at the international community. The project was separate from the Iridium venture and was expected to take four years to complete after FCC approval. According to

---

[13] Quentin Hardy, "Iridium Pulls $300 Million Bond Offer; Analysts Cite Concerns about Projects," *Wall Street Journal*, September 22, 1995, p. A5.

Bary Bertiger, now corporate vice president and general manager of Motorola's satellite communications group, "Unlike Iridium, Motorola has no plans to detach M-Star as a separate entity. We won't fund it ourselves, but we will have fewer partners than in Iridium."[14]

The M-Star Project raised some eyebrows in the investment community. Iridium employed 2,000 people but M-Star had only 80. The Iridium Project generated almost 1,100 patents for Motorola and that intellectual property would most likely be transferred to M-Star. Also, Motorola had three contracts with Iridium for construction and operation of the global communication system providing for approximately $6.5 billion in payments to Motorola over a 10-year period that began in 1993. Was M-Star being developed at the expense of Iridium? Could M-Star replace Iridium? What would happen to the existing 17-member consortium at Iridium if Motorola were to withdraw its support in lieu of its own internal competitive system?

A NEW CEO

In 1996, Iridium began forming a very strong top management team with the hiring of Dr. Edward Staiano as CEO and vice chairman. Prior to joining Iridium in 1996, Staiano had worked for Motorola for 23 years, during which time he developed a reputation for being hard-nosed and unforgiving. During his final 11 years with Motorola, Staiano led the company's General Systems Sector to record growth levels. In 1995, the division accounted for approximately 40 percent of Motorola's total sales of $27 billion. In leaving Motorola's payroll for Iridium's, Staiano gave up a $1.3-million-per-year contract with Motorola for a $500,000 base salary plus 750,000 Iridium stock options that vested over a five-year period. Staiano commented, "I was spending 40 percent to 50 percent of my time [at Motorola] on Iridium anyway. . . If I can make Iridium's dream come true, I'll make a significant amount of money."[15]

SATELLITE LAUNCHES

At 11:28 a.m. on a Friday morning the second week of January 1997, a Delta 2 rocket carrying a Global Positioning System (GPS) exploded upon launch, scattering debris above its Cape Canaveral launch pad. The launch, which was originally scheduled for the third quarter of 1996, would certainly have an impact on Iridium's schedule, while an industry board composed of representatives from

---

[14] Quentin Hardy, "Motorola Is Plotting New Satellite Project—M-Star Would Be Faster than the Iridium System, Pitched to Global Firms," Wall Street Journal, October 14, 1996, p. B4.

[15] Quentin Hardy, "Staiano Is Leaving Motorola to Lead Firm's Iridium Global Satellite Project," *Wall Street Journal*, December 10, 1996, p. B8.

McDonnell-Douglas and the Air Force determined the cause of the explosion. Other launches had already been delayed for a variety of technical reasons.

In May of 1997, after six failed tries, the first five Iridium satellites were launched. Iridium still believed that the target date for launch of service, September 1998, was still achievable but that all slack in the schedule had been eliminated due to the earlier failures.

By this time, Motorola had amassed tremendous knowledge on how to mass-produce satellites. As described by Bennahum:

> The Iridium constellation was built on an assembly line, with all the attendant reduction in risk and cost that comes from doing something over and over until it is no longer an art but a process. At the peak of this undertaking, instead of taking 18 to 36 months to build one satellite, the production lines disgorged a finished bird every four and a half days, sealed it in a container, and placed it on the flatbed of an idling truck that drove it to California or Arizona, where a waiting Boeing 747 carried it to a launch pad in the mountains of Taiyuan, China, or on the steppes of Baikonur in Kazakhstan.[16]

AN INITIAL PUBLIC OFFERING

Iridium was burning cash at the rate of $100 million per month. Iridium filed a preliminary document with the Securities and Exchange Commission for an initial public offering (IPO) of 10 million shares to be offered at $19 to $21 a share. Because of the launch delays, the IPO was delayed.

In June of 1997, after the first five satellites were placed in orbit, Iridium filed for an IPO of 12 million shares priced at $20 per share. This would cover about three months of operating expenses including satellite purchases and launch costs. The majority of the money would go to Motorola.

SIGNING UP CUSTOMERS

The reality of the Iridium concept was now at hand. All that was left to do was to sign up 500,000 to 600,000 customers, as predicted, to use the service. Iridium set aside $180 million for a marketing campaign including advertising, public relations, and a worldwide, direct mail effort. Part of the advertising campaign included direct mail translated into 13 languages, ads on television and on airlines, airport booths, and Internet web pages.

How to market Iridium was a challenge. People would certainly hate the phone. According to John Windolph, executive director of marketing communications at Iridium, "It's huge! It will scare people. It is like a brick-size device with an antenna like a stout bread stick. If we had a campaign that featured our product,

---

[16] Bennahum, "The United Nations of Iridium."

we'd lose." The decision was to focus on the fears of being out of touch. Thus, the marketing campaign began. But Iridium still did not have a clear picture of who would subscribe to the system. An executive earning $700,000 would probably purchase the bulky phone, have his or her assistant carry the phone in his or her briefcase, be reimbursed by the company for the use of the phone, and pay $3 to $7 per minute for calls, also a business expense. But were there 600,000 executives worldwide who needed the service?

Several other critical questions needed to be addressed. How do we hide or downplay the $3,400 purchase price of the handset and the usage cost of $7 per minute? How do we avoid discussions about competitors that are offering similar services at a lower cost? With operating licenses in about 180 countries, do we advertise in all of them? Do we take out ads in *Oil and Gas Daily*? Do we advertise in girlie magazines? Do we use full-page or double-page spreads?

Iridium had to rely heavily on its "gateway" partners for marketing and sales support. Iridium itself would not be able to reach the entire potential audience. Would the gateway partners provide the required marketing and sales support? Do the gateway partners know how to sell the Iridium system and the associated products?

The answer to these questions appeared quickly.

> Over a matter of weeks, more than one million sales inquiries poured into Iridium's sales offices. They were forwarded to Iridium's partners—and many of them promptly disappeared, say several Iridium insiders. With no marketing channels and precious few sales people in place, most global partners were unable to follow up on the inquiries. A mountain of hot sales tips soon went cold.[17]

IRIDIUM'S RAPID ASCENT

On November 1, 1998, the Iridium system was officially launched. It was truly a remarkable feat that the 11-year project was finally launched, just a little more than a month late.

> After 11 years of hard work, we are proud to announce that we are open for business. Iridium will open up the world of business, commerce, disaster relief, and humanitarian assistance with our first-of-its-kind global communications service. . . . The potential use of Iridium products is boundless. Business people who travel the globe and want to stay in touch with home and office, industries that operate in remote areas—all will find Iridium to be the answer to their communications needs.[18]

---

[17] Leslie Cauley, "Losses in Space—Iridium's Downfall: The Marketing Took a Back Seat to Science—Motorola and Partners Spent Billions on Satellite Links for a Phone Few Wanted," *Wall Street Journal*, August 18, 1999, p. A1.

[18] Excerpts from the Iridium press release, November 1, 1998.

On November 2, 1998, Iridium began providing service. With the Iridium system finally up and running, most financial analysts issued "buy" recommendations for Iridium stock with expected yearly revenues of $6 to $7 billion within five years. On January 25, 1999, Iridium held a news conference to discuss its earnings for the fourth quarter of 1998. Ed Staiano, CEO of Iridium stated:

> In the fourth quarter of 1998, Iridium made history as we became the first truly global mobile telephone company. Today, a single wireless network, the Iridium Network, covers the planet. And we have moved into 1999 with an aggressive strategy to put a large number of customers on our system, and quickly transform Iridium from a technological event to a revenue generator. We think the prospects for doing this are excellent. Our system is performing at a level beyond expectations.
>
> Financing is now in place through projected cash flow positives. Customer interest remains very high and a number of potentially large customers have now evaluated our service and have given it very high ratings. With all of this going for us, we are in position to sell the service and that is precisely where we are focusing the bulk of our efforts.[19]

Roy Grant, chief financial officer of Iridium, stated:

> Last week Iridium raised approximately $250 million through a very successful 7.5 million-share public offering. This offering had three major benefits. It provided $250 million of cash to our balance sheet. It increased our public float to approximately 20 million shares. And it freed up restrictions placed on $300 million of the $350 million of Motorola guarantees. These restrictions were placed on that particular level of guarantees by our bankers in our $800 million secured credit facility.
>
> With this $250 million, combined with the $350 million of additional guarantees from Motorola, this means we have approximately $600 million of funds in excess of what we need to break cash flow breakeven. This provides a significant contingency for the company.[20]

DECEMBER 1998

In order to make its products and services known to travelers, Iridium agreed to acquire Claircom Corporation from AT&T and Rogers Cantel Mobile Communications for about $65 million. Claircom provided in-flight telephone systems for U.S. planes as well as equipment for international carriers. The purchase of Claircom would be a marketing boost for Iridium.

---

[19] Excerpts from the Iridium conference call, January 25, 1999.

[20] Ibid.

The problems with large, long-term technology projects were now appearing in the literature. As described by Bennahum:

> "This system does not let you do what a lot of wired people want to do," cautions Professor Heather Hudson, who runs the telecommunications program at the University of San Francisco and studies the business of wireless communications. "Nineteen-nineties technologies are changing so fast that it is hard to keep up. Iridium is designed from a 1980s perspective of a global cellular system. Since then, the Internet has grown and cellular telephony is much more pervasive. There are many more opportunities for roaming than were assumed in 1989. So there are fewer businesspeople who need to look for an alternative to a cell phone while they are on the road."[21]

Additionally, toward the late 1990s, some industry observers felt that Motorola had an additional incentive to ensure that Iridium succeeded, irrespective of the costs—namely, protecting its reputation. Between 1994 and 1997, Motorola had suffered slowing sales growth, a decline in net income, and declining margins. Moreover, the company had experienced several previous business mishaps, including a failure to anticipate the cellular industry's switch to digital cell phones, which played a major role in Motorola's more than 50 percent share-price decline in 1998.

IRIDIUM'S RAPID DESCENT

It took more than a decade for the Iridium Project to ascend and only a few months to descend. In the first week of March, almost five weeks after the January teleconference, Iridium's financial woes began to surface. Iridium had expected 200,000 subscribers by the end of 1998 and additional subscribers at a rate of 40,000 per month. Iridium's bond covenants stated a target of 27,000 subscribers by the end of March. Failure to meet such a small target could send investor confidence spiraling downward. Iridium had only 10,000 subscribers. The market that was out there 10 years ago was not the market that was there today. Also, 10 years ago Iridium faced little competition.

Iridium cited the main causes of the shortfall in subscriptions as shortages of phones, glitches in some of the technology, software problems, and, most important, a lack of trained sales channels. Iridium found out that it had to train a sales staff and that Iridium itself would have to sell the product, not its distributors. The investor community did not appear pleased with the sales problem, which should have been addressed years ago, not four months into commercial service.

Iridium's advertising campaign was dubbed "Calling Planet Earth" and promised that you had the freedom to communicate anytime and anywhere. This was

---

[21] Bennahum, "The United Nations of Iridium."

TABLE I COVENANTS ON THE CREDIT AGREEMENT

Date	Cumulative Cash Revenue ($ Millions)	Cumulative Accrued Revenue ($ Millions)	Number of Satellite Phone Subscribers	Number of System Subscribers
March 31, 1999	$ 4	$ 30	27,000	52,000
June 30, 1999	50	150	88,000	213,000
Sept. 30, 1999	220	470	173,000	454,000

*Total system subscribers include users of Iridium's phone, fax, and paging services.

Source: Iridium World Communications Ltd., 1998 Annual Report.

not exactly true; the system could not work in buildings or even cars. Furthermore, Iridium underestimated the amount of time subscribers would require to examine and test the system before signing on. In some cases, this would be six months.

Many people blamed marketing and sales for Iridium's rapid descent:

> True, Iridium committed so many marketing and sales mistakes that its experiences could form the basis of a textbook on how not to sell a product. Its phones started out costing $3,000, were the size of a brick, and didn't work as promised. They weren't available in stores when Iridium ran a $180 million advertising campaign. And Iridium's prices, which ranged from $3.00 to $7.50 a call, were out of this world.[22]

Iridium's business plan was flawed. With service beginning on November 2, 1998, it was unlikely that 27,000 subscribers would be on board by March of 1999, given the time required to test the product. The original business plan required that the consortium market and sell the product prior to the onset of service. But selling the service from just a brochure was almost impossible. Subscribers want to touch the phone, use it, and test it prior to committing to a subscription.

Iridium announced that it was entering into negotiations with its lenders to alter the terms of an $800 million secured credit agreement due to the weaker-than-expected subscriber and revenue numbers (Table I).

The stock, which had traded as high as almost $73 per share, was now at approximately $20 per share. And in yet another setback, the chief financial officer, Roy Grant, resigned.

April 1999

Iridium's CEO, Ed Staiano, resigned at the April 22 board meeting. Sources believed that Staiano resigned when the board nixed his request for additional funds to develop Iridium's own marketing and distribution team rather than

---

[22] James Surowiecki, "The Latest Satellite Startup Lifts Off. Will It Too Explode?" *Fortune Magazine*, October 25, 1999, pp. 237–254.

relying on its strategic partners. Sources also stated another issue: Staiano had cut costs to the bare bones at Iridium but could not get Motorola to reduce its lucrative $500 million service contract with Iridium. Some people believed that Staiano wanted to reduce the Motorola service contract by up to 50 percent. John Richardson, the CEO of Iridium Africa Corp., was assigned as interim CEO. Richardson's expertise was in corporate restructuring. For the quarter ending March, Iridium said it had a net loss of $505.4 million, or $3.45 a share. The stock fell to $15.62 per share. Iridium managed to attract just 10,294 subscribers five months after commercial rollout.

One of Richardson's first tasks was to revamp Iridium's marketing strategy. Iridium was unsure as to what business it was in. According to Richardson,

> The message about what this product was and where it was supposed to go changed from meeting to meeting. . . . One day, we'd talk about cellular applications, the next day it was a satellite product. When we launch in November, I'm not sure we had a clear idea of what we wanted to be.[23]

May 1999

Iridium officially announced that it did not expect to meet its targets specified under the $800 million loan agreement. Lenders granted Iridium a two-month extension. The stock dropped to $10.44 per share, partly due to a comment by Motorola that it might withdraw from the ailing venture.

Wall Street began talking about the possibility of bankruptcy. But Iridium stated that it was revamping its business plan and by the end of May hoped to have chartered a new course for its financing. Iridium also stated in a regulatory filing that it was uncertain whether it would have enough cash to complete the agreement to purchase Claircom Communications Group Inc., an in-flight telephone service provider, for the promised $65 million in cash and debt.

Iridium had received extensions on debt payments because the lending community knew that moving from a project plan to an operating business was no small feat. Another reason why the banks and creditors were willing to grant extensions was because bankruptcy was not a viable alternative. The equity partners owned all of the Earth stations, all distribution, and all regulatory licenses. If the banks and creditors forced Iridium into bankruptcy, they could end up owning a satellite constellation that could not talk to the ground or gateways.

June 1999

Iridium received an additional 30-day extension beyond the two-month extension it had already received. Iridium was given until June 30 to make a $90 million

---

bond payment. Iridium began laying off 15 percent of its 550-employee work-force including two senior officers. The stock had now sunk to $6 per share, and the bonds were selling at 19 cents on the dollar. John Richardson, CEO of Iridium, said: "We did all of the difficult stuff well, like building the network, and did all of the no-brainer stuff at the end poorly."[24]

In a later interview, John Richardson stated:

> Iridium's major mistake was a premature launch for a product that wasn't ready. People became so obsessed with the technical grandeur of the project that they missed fatal marketing traps. . . . Iridium's international structure has proven almost impossible to manage: the 28 members of the board speak multiple languages, turning meetings into mini-U.N. conferences complete with headsets translating the proceedings . . . into five languages.
>
> We're a classic MBA case study in how not to introduce a product. First we created a marvelous technological achievement. Then we asked how to make money on it.[25]

Iridium was doing everything possible to avoid bankruptcy. Time was what Iridium needed. Some industrial customers would take six to nine months to try out a new product but would be reluctant to subscribe if it appeared that Iridium would be out of business in six months. In addition, Iridium's competitors were lowering their prices significantly, putting further pressure on Iridium. Richardson then began providing price reductions of up to 65 percent off the original price for some of Iridium's products and services.

July 1999

The banks and investors agreed to give Iridium yet a third extension to August 11 to meet its financial covenants. Everyone seemed to understand that the restructuring effort was much broader than originally contemplated.

Motorola, Iridium's largest investor and general contractor, admitted that the project might have to be shut down and liquidated as part of bankruptcy proceedings unless a restructuring agreement could be reached. Motorola also stated that if bankruptcy occurred, Motorola would continue to maintain the satellite network, but for a designated time period only.

Iridium had asked its consortium investors and contractors to come up with more money. But to many consortium members, it looked like they would be throwing good money after bad. Several partners made it clear that they would simply walk away from Iridium rather than providing additional funding. That could have a far-reaching effect on the service at some locations.

---

[24] Hawn, "High Wireless Act."
[25] Cauley, "Losses in Space—Iridium's Downfall."

Therefore, all partners had to be involved in the restructuring. Wall Street analysts expected Iridium to be allowed to repay its cash payments on its debt over several years or offer debt holders an equity position in Iridium. It was highly unlikely that Iridium's satellites orbiting Earth would be auctioned off in bankruptcy court.

August 1999

On August 12, Iridium filed for bankruptcy protection. This was like having "a dagger stuck in their heart" for a company that a few years earlier had predicted financial breakeven in just the first year of operations. This was one of the 20 largest bankruptcy filings up to this time. The stock, which had been trading as little as $3 per share, was suspended from the NASDAQ on August 13, 1999.

The cost of Iridium's phone calls had been reduced to around $1.40 to $3 per minute, and the handsets were reduced to $1,500 per unit.

There was little hope for Iridium. Both the business plan and the technical plan were flawed. The business plan for Iridium seemed like it came out of the film *Field of Dreams*, where an Iowa corn farmer was compelled to build a baseball field in the middle of a corn crop. A mysterious voice in his head said, "Build it and they will come." In the film, he did, and they came. This made for a good plot for a Hollywood movie, but it made a horrible business plan. In 1992, Herschel Shosteck, a telecommunications consultant, said: "If you build Iridium, people may come. But what is more likely is, if you build something cheaper, people will come to that first."

The technical plan was designed to build the holy grail of telecommunications. Unfortunately, after billions were spent, the need for the technology changed over time. The engineers who designed the system, many of whom had worked previously on military projects, lacked an understanding of the word "affordability" and the need for marketing a system to more than just one customer, namely the DoD. "Satellite systems are always far behind the technology curve. Iridium was completely lacking the ability to keep up with Internet time," stated Bruce Egan, senior fellow at Columbia University's Institute for Tele-Information.[26]

September 1999

Leo Mondale resigned as Iridium's chief financial officer. Analysts believed that his resignation was because a successful restructuring was no longer possible. According to one analyst, "If they [Iridium] were close [to a restructuring plan], they wouldn't be bringing in a whole new team."

---

[26] Paterik, "Iridium Alive and Well."

THE IRIDIUM "FLU"

The bankruptcy of Iridium was having a flulike effect on the entire industry. ICO Global Communications, one of Iridium's major competitors, filed for bankruptcy protection just two weeks after the Iridium filing. ICO failed to raise $500 million it sought from public rights offerings that had already been extended twice. Another competitor, the Globalstar Satellite Communications System, was still financially sound. Anthony Navarro, Globalstar's chief operating officer, stated, "They [Iridium] set everybody's expectations way too high."[27]

SEARCHING FOR A WHITE KNIGHT

Iridium desperately needed a qualified bidder who would function as a white knight. It was up to the federal bankruptcy court to determine whether someone was a qualified bidder. A qualified bidder was required to submit a refundable cash deposit or letter of credit issued by a respected bank that would equal the greater of $10 million or 10 percent of the value of the amount bid to take control of Iridium.

According to bankruptcy court filing, Iridium was generating revenue of $1.5 million per month. On December 9, 1999, Motorola agreed to a $20 million cash infusion for Iridium. Iridium needed a white knight quickly or it could run out of cash by February 15, 2000. With a monthly operating cost of $10 million and a staggering cost of $300 million every few years for satellite replenishment, it was questionable if anyone could make a successful business from Iridium's assets because of asset specificity.

Cellular phone entrepreneur Craig McCaw planned on a short-term cash infusion while he considered a much larger investment to rescue Iridium. He was also leading a group of investors who pledged $1.2 billion to rescue the ICO satellite system, which had filed for bankruptcy protection shortly after the Iridium filing.[28]

Several supposed white knights came forth, but Craig McCaw's group was regarded as the only credible candidate. Although McCaw's proposed restructuring plan was not fully disclosed, it was expected that Motorola's involvement would be that of a minority stakeholder. Also, under the restructuring plan, Motorola would reduce its monthly fee for operating and maintaining the Iridium system from $45 million to $8.8 million.[29]

---

[27] Quentin Hardy, "Surviving Iridium," *Forbes*, September 6, 1999, pp. 216–217.

[28] "Craig McCaw Plans Cash Infusion to Support Cash-Hungry Iridium," *Wall Street Journal*, February 7, 2000, p.1

[29] "Iridium Set to Get $75 Million from Investors Led by McCaw," *Wall Street Journal*, February 10, 2000, p.1.

DEFINITION OF FAILURE: OCTOBER 1999

The Iridium network was an engineering marvel. Motorola's never-say-die attitude created technical miracles and overcame NASA-level technical problems. Iridium overcame global political issues, international regulatory snafus, and a range of other geopolitical issues on seven continents. The Iridium system was, in fact, what Motorola's Galvin called the eighth wonder of the world.

But did the bankruptcy indicate a failure for Motorola? Absolutely not! Motorola collected $3.65 billion in Iridium contracts. Assuming $750 million in profit from these contracts, Motorola's net loss on Iridium was about $1.25 billion. Simply stated, Motorola spent $1.25 billion for a project that would have cost it perhaps as much as $5 billion out of its own pocket had it wished to develop the technology itself. Iridium provided Motorola with more than 1,000 patents in building satellite communication systems. Iridium allowed Motorola to amass a leadership position in the global satellite industry. Motorola was also signed up as the prime contractor to build the 288-satellite "Internet in the Sky" dubbed the Teledesic Project. Backers of the Teledesic Project, which had a price tag of $15 billion to transmit data, video, and voice, included Boeing, Microsoft's chairman Bill Gates, and cellular magnate Craig McCaw. Iridium had enhanced Motorola's reputation for decades to come.

Motorola stated that it had no intention of providing additional funding to ailing Iridium, unless of course other consortium members provided funding. Several members of the consortium stated that they would not provide any additional investment and were considering liquidating their involvement in Iridium.[30]

In March 2000, McCaw withdrew his offer to bail out Iridium even at a deep discount, stating that he would spend his efforts on salvaging the ICO satellite system instead. This, in effect, signed Iridium's death certificate. One reason for McCaw's reluctance to rescue Iridium may have been the discontent of some investors who would have been completely left out as part of the restructuring effort, thus losing perhaps their entire investment.

THE SATELLITE DEORBITING PLAN

With the withdrawal of McCaw's financing, Iridium notified the U.S. bankruptcy court that Iridium had not been able to attract a qualified buyer by the deadline assigned by the court. Iridium would terminate its commercial service after 11:59 p.m. on March 17, 2000, and it would begin the process of liquidating its assets.

---

[30] Scott Thurm, "Motorola Inc., McCaw Shift Iridium Tactics." *Wall Street Journal*, February 18, 2000, p. 1.

Immediately following the Iridium announcement, Motorola issued this press release:

> Motorola will maintain the Iridium satellite system for a limited period of time while the deorbiting plan is being finalized. During this period, we also will continue to work with the subscribers in remote locations to obtain alternative communications. However, the continuation of limited Iridium service during this time will depend on whether the individual gateway companies, which are separate operating companies, remain open. . . .
>
> In order to support those customers who purchased Iridium service directly from Motorola, Customer Support Call Centers and a website that are available 24 hours a day, seven days a week have been established by Motorola. Included in the information for customers is a list of alternative satellite communications services.

The deorbiting plan would likely take two years to complete at a cost of $50 to $70 million. This would include all 66 satellites and the other 22 satellites in space serving as spare or decommissioned failures. Iridium would most likely deorbit the satellites four at a time by firing their thrusters to drop them into the atmosphere where they would burn up.

IRIDIUM IS RESCUED FOR $25 MILLION

In November 2000, a group of investors led by an airline executive won bankruptcy court approval to form Iridium Satellite Corporation and purchase all remaining assets of failed Iridium Corporation. The purchase was at a fire-sale price of $25 million, which was less than a penny on the dollar. As part of the proposed sale, Motorola would turn over responsibility for operating the system to Boeing. Although Motorola would retain a 2 percent stake in the new system, it would have no further obligations to operate, maintain, or decommission the constellation.

Almost immediately after the announcement, Iridium Satellite was awarded a $72 million contract from the Defense Information Systems Agency, which is part of the DoD. Dave Oliver, principal deputy undersecretary of Defense for Acquisition, stated: "Iridium will not only add to our existing capability, it will provide a commercial alternative to our purely military systems. This may enable real civil/military dual use, keep us closer to leading edge technologically, and provide a real alternative for the future."[31]

Iridium had been rescued from the brink of extinction. As part of the agreement, the newly formed company acquired all of the assets of the original Iridium and its subsidiaries. This included the satellite constellation, the terrestrial

---

[31] "DoD Awards $72 Million to Revamp Iridium," Satellite Today 3, no. 227, December 7, 2000, p. 1.

network, Iridium real estate, and the intellectual property originally developed by Iridium. Because of the new company's significantly reduced cost structure, it was able to develop a workable business model based on a targeted market for Iridium's products and services. Weldon Knape, CEO, of World Comminication Center Inc. stated: "Everyone thinks the Iridium satellites crashed and burned, but they're all still up there."[32]

A new Iridium phone costs $1,495 and is the size of a cordless home phone. Older, larger models start at $699 or one can be rented for about $75 per week. Service costs $1 to $1.60 a minute.[33]

EPILOGUE

In a press release on February 6, 2006, Iridium Satellite declared that 2005 was the best year ever. The company had 142,000 subscribers, which was a 24 percent increase from 2004, and the 2005 revenue was 55 percent greater than in 2004. According to Carmen Lloyd, Iridium's CEO, "Iridium is on an exceptionally strong financial foundation with a business model that is self-funding."

For the year ending 2006, Iridium had $212 million in sales and $54 million in profit. Iridium had 180,000 subscribers and a forecasted growth rate of 14 to 20 percent per year. Iridium had changed its business model, focusing on sales and marketing first and hype second. This allowed it to reach out to new customers and new markets.[34]

SHAREHOLDER LAWSUITS

The benefit to Motorola, potentially at the expense of Iridium and its investors, did not go unnoticed. At least 20 investor groups filed suit against Motorola and Iridium, citing:

- Motorola milked Iridium and used the partners' money to finance its own foray into satellite communication technology.
- By using Iridium, Motorola ensured that its reputation would not be tarnished if the project failed.
- Most of the money raised through the IPOs went to Motorola for designing most of the satellite and ground-station hardware and software.
- Iridium used the proceeds of its $1.45 billion in bonds, with interest rates from 10.875 to 14 percent, mainly to pay Motorola for satellites.

---

[32] Paterik, "Iridium Alive and Well."

[33] Ibid.

[34] Adapted from Reena Jana, "Companies Known for Inventive Tech Were Dubbed the Next Big Thing and Then Disappeared. Now They're Back and Growing," *Business Week*, Innovation, April 10, 2007.

- Defendants falsely reported achievable subscriber numbers and revenue figures.
- Defendants failed to disclose the seriousness of technical issues.
- Defendants failed to disclose delays in handset deliveries.
- Defendants violated covenants between itself and its lenders.
- Defendants delayed disclosure of information, provided misleading information, and artificially inflated Iridium's stock price.
- Defendants took advantage of the artificially inflated price to sell significant amounts of their own holdings for millions of dollars in personal profit.

THE BANKRUPTCY COURT RULING

On September 4, 2007, after almost 10 months, the bankruptcy court in Manhattan ruled in favor of Motorola and irritated the burned creditors that had hoped to get a $3.7 billion judgment against Motorola. The judge ruled that even though the capital markets were "terribly wrong" about Iridium's hopes for huge profits, Iridium was "solvent" during the critical period when it successfully raised rather impressive amounts of debt and equity in the capital markets.

The court said that even though financial experts now know that Iridium was a hopeless one-way cash flow, flawed technology project, and doomed business model, Iridium was solvent at the critical period of fund raising. Even when the bad news began to appear, Iridium's investors and underwriters still believed that Iridium had the potential to become a viable enterprise.

The day after the court ruling, newspapers reported that Iridium LLC, the now–privately held company, was preparing to raise about $500 million in a private equity offering to be followed by an initial public offering within the next year or two.

QUESTIONS

1. What should be the role of a project sponsor on the Iridium Project compared to the traditional project sponsor? Are there differences, and do your answers depend on the project's length, technical complexity, and/or size?
2. Who, if anyone, was most likely the project sponsor for the Iridium Project prior to the creation of the Iridium Limited Liability Partnership (LLP) in 1991?
3. Who, if anyone, was most likely the project sponsor for the Iridium Project at the end of 1996 when Staiano was brought on board as CEO?
4. At what level of management should the project sponsor reside on a project such as the Iridium Project, considering the project's length, dollar value, and complexity?
5. Was sponsorship on the Iridium Project performed by one person or was it committee sponsorship? If it was committee sponsorship, then who sat on the committee?

6. What are the differences between a project sponsor and a project champion? Can the project sponsor and project champion be the same person? If so, what are the risks? Did the Iridium Project have a project sponsor, project champion, or both? If you believe that both existed, then who filled each position?
7. What should be the role description of an exit champion, and on what type of projects should an exit champion exist?
8. Should an exit champion be an active member of the project team? Can the champion be an outsider to Motorola or Iridium?
9. Are the length, dollar value, and complexity of the project factors to consider in whether an exit champion should be assigned? If so, are there other factors should be considered as well?
10. Was there an exit champion on the Iridium Project? If so, who was the exit champion?
11. If an exit champion did not exist, explain why not. Who should have been the exit champion, if one did not exist?
12. What is the relationship, if any, between an exit champion and a project champion or project sponsor?
13. What are the advantages and disadvantages of a collective belief?
14. Did the Iridium Project have a business case? If so, what was it?
15. Was there the possibility of a hidden business case?
16. Assuming that a business case existed, who developed the business case?
17. How long is a business case valid before it must be reexamined?
18. Can a project be a technical success and a commercial failure?

Health Care
Partners, Inc.

BACKGROUND

Health Care Partners (HCP) was a 40-year-old company providing health care benefits for large corporations. In order to keep health care costs down for its clients, HCP needed to get a large group of physicians and health care providers in the group who were willing to accept the cost reimbursement rates established by HCP. HCP provided reimbursement rates that were in line with its competitors. Although HCP had some success in getting service providers into its network, several hospitals and physicians resisted joining the network because HCP had a reputation for reimbursing service providers slowly.

In order to pay service providers more quickly, HCP had to modernize its operations and eliminate a lot of the paperwork that generated delays in payments. HCP upgraded its computers quickly. But the real problem was software. No software was readily available in the marketplace to satisfy the needs of HCP. HCP hired a software development company, SoftSmart, to assist its information technology (IT) personnel in the development of the package. SoftSmart was provided with offices for on-site personnel on the same floor as HCP's IT personnel. HCP had budgeted $15 million for the entire project, entitled QuickPay, and had promised all physicians and hospitals in its network that the system would be up and running in a year or less.

FIRST QUARTERLY REVIEW MEETING

In the first quarterly review meeting, which was attended only by HCP personnel, Paul Harris, the chief information officer, stated how furious he was:

> Why can't I get a straight answer from anyone on the status of the QuickPay Project? We're spending $15 million and nobody seems to know what's happening. Whenever I ask a question, it appears that all I get in response is bad news. Why aren't there any metrics for me to look at each week or each month?
>
> Since the project began about three months ago, I have seen requests for more than 200 scope changes. Now I'm told that we will probably be missing deadlines and the schedule slippages cannot be corrected. We have escalating costs because of the scope creep and it looks like we'll have some deterioration in value expected for our clients.

Evelyn Williams, the project manager for HCP, spoke up.

> When we hired SoftSmart, we gave them a fixed price contract. We had no idea how many scope changes they wanted, but we assumed that there would be a small number. We were a little naïve. Last week, when we asked them for their position on the status of their work, they said that they cannot provide us with detailed status information because they say that it depends on the number of scope changes that we approve. Their schedules keep changing.

Paul Harris was furious. It appeared that the company would be spending significantly more than $15 million, and he could not get metrics, schedules, or effective status reporting. This project had the potential to be a colossal disaster.

Paul demanded that the company now have monthly rather than quarterly review meetings with him and possibly other senior management personnel. He was convinced that everyone understood what had to be done, but he was equally unsure as to whether they would do it.

REVIEW MEETING AT END OF MONTH 4

Paul Harris was still quite unhappy after seeing the data in the review meeting. There were schedules and metrics. During the briefing, Paul was told that work was progressing but not as fast as originally hoped for. However, the metrics provided no useful information and the schedules had a series of footnotes at various locations stating dependencies on the approval of various scope changes. Once again, Harris found it difficult to determine the true status of the project.

Evelyn Williams, the HCP project manager, spoke up again.

> We're making progress in status reporting but not as fast as I would have liked. Some of the team members from SoftSmart are reluctant to provide us with good

metrics. They tell us that they simply do not believe in the use of metrics, probably because they are afraid of what the metrics might reveal. That's why they often select the easiest metrics to report or those that provide the least amount of information. Some of our own personnel are infatuated with metrics, and we simply cannot afford to create all of the metrics that these people desire. I'm not sure right now which way we should go or what would be a reasonable compromise. To make matters worse, we have lost some of our key personnel to other projects.

CRITICAL DECISIONS

There was now no question in Paul Harris's mind that things were not going as planned. The morale of the team was poor; key personnel had left the project, probably by choice; and status reporting was unacceptable. SoftSmart was probably taking advantage of HCP by pushing through questionable but profitable scope changes, and the end date would most likely slip. The decision was clear; there was a definite need for a health check on the project.

Several questions had to be answered before officially conducting the health check. First, should the health check be performed using internal personnel such as representatives from the project management office? The project manager would most certainly not be the person allowed to perform the health check. Using project management office personnel is an option, but they may have friendships and loyalties to some people on the project team and may not be honest in their conclusions as to the real status and health of the project. External facilitators may be the best choice, provided they can operate free of politics and create an environment such that people will feel free to vent their personal feelings. They also bring to the table experience in conducting health checks in other companies.

The second question is whether the interviewees will be honest in their responses to the health check facilitators. Paul believed that the personnel at HCP would be honest. However, the real issue may be with SoftSmart. HCP may not be able to get SoftSmart to agree to the interviews, and, even if the interviews were conducted, SoftSmart personnel would most likely not provide honest responses. Therefore, it would have to be a health check at HCP only.

Third, Paul was unsure as to how other executives at HCP would respond when hearing the truth about the QuickPay Project. The results of the health check could cause other issues previously hidden to surface. This could make the situation worse than it is already. People could lose their jobs or be demoted. However, there could also be good news and the early detection of problems that could have led to disasters later on.

Paul concluded that there was really no choice: A health check must be conducted. HCP must know the true status. HCP must identify issues early such that sufficient time exists for corrective action.

Paul called an emergency meeting of the QuickPay Project team and asked SoftSmart representatives to be present as well. When informed that he was authorizing an outside company to come in and conduct a health check, several team members expressed their dissatisfaction. One team member argued that outside resources do not understand the HCP culture or the project and that this would be a waste of time. Another argued that the project was already in financial distress and the cost of the health check would make matters worse. A third team member asserted that critical resources would be tied up in interviews. SoftSmart argued that, by the time the results of the health check are known, it may be too late to make changes because of the ongoing scope changes occurring on the project. Harris held his ground and stated emphatically that the health check would be performed and that everyone would be expected to support the company conducting the health check.

THE HEALTH CHECK

HCP hired Pegasus Consulting, which had experience in health checks on IT projects and also some experience with hospitals and the health care profession. An agreement was reached that the health check would be completed within three weeks, just prior to the project review meeting scheduled for the end of the fifth month of the QuickPay Project.

Pegasus spent part of the first week reviewing the business case for the project and the project's history over the past four months. During the remainder of the first week and all of the second week, Pegasus interviewed project personnel from HCP to discover the facts. The interview sessions went well, and the interviewees were quite honest in their opinion on the status of the project and what needed to be done to correct the deteriorating situation. A few of the on-site representatives from SoftSmart were also interviewed, but Pegasus believed that their contributions to the health check were meaningless.

By the end of the third week, Pegasus had prepared its report and was ready to brief Harris on the findings. At the briefing meeting, a spokesperson for Pegasus made the following statements:

> There are several issues, but the most critical one is the schedule. It is our opinion that the project will be at least three months late. We did a root cause analysis and discovered the following:
>
> ● Performance is not following the baseline because the baseline is continuously changing due to the number of scope changes. While we believe . . . some of the scope changes are necessary, many could have been delayed and performed later as an enhancement project.
> ● HCP will not be able to meet its forecasts. The culprit appears to be the requirements package which was ill-defined. Had the requirements package been properly prepared, the original forecast may have been achievable.

- The benefits and value expected, as identified in the business case, seem realistic and it is our opinion that both HCP and its clientele will receive these benefits.
- The governance for the project is poorly structured. For a project of this magnitude and with the associated risks, a governance committee should have been established. The chief information officer should not be the only person responsible for governance on this project.
- Risk mitigation has not been performed by the project team. We cannot find any plan on risk management and this is unheard of in a project of this magnitude. This reflects poorly upon the project manager and the assigned team.
- It is our opinion that HCP may have assigned the wrong project manager. Several of the interviewees commented that they had little faith in the ability of Evelyn Williams to manage this project. Several of the interviewees had worked for her on previous projects. We recommend that she be replaced.
- We were not able to find any contingency plans. Everyone told us that they wanted to develop contingency plans but the number and frequency of the scope changes made this impossible.

There are opportunities for some corrective action. Included in our report is a fix-it plan that we believe will work. However, even with the implementation of the fix-it plan, the project will still be about three months late.

Paul Harris did not seem surprised with the findings of the health check. He read the final report and believed that the recommended fix-it plan could work. But now Paul had to prepare for two more meetings in which he was expected to report on the findings of the health check: an executive staff meeting and the five-month QuickPay Project review meeting.

QUESTIONS

1. Is it customary for a company like HCP to hold quarterly review meetings without having representation from its contractor, SoftSmart?
2. Can project status be determined without the use of metrics?
3. Should the project manager have taken the lead for the establishment of project metrics for the QuickPay Project?
4. If you were Paul Harris, how would you have reacted after being briefed at the first quarterly review meeting?
5. Do 200 scope changes in the first three months of the project indicate that there was a poor definition of the requirements?
6. Once the bad news appeared in the first quarterly review meeting, should HCP have gone to weekly rather than monthly review meetings?

7. Why did it appear that HCP's team members did not want to establish metrics?
8. Why did it appear that SoftSmart's personnel also did not want to establish metrics?
9. Would the establishment of metrics have had any impact on the project up to that point?
10. Is the turnover of key personnel in the first three months of the project an indication of looming disaster?
11. Is it advisable to perform a health check after just the first three months of the project?
12. What were the three life-cycle phases that Pegasus used in performing the health check?
13. Did Pegasus do its job correctly?
14. Was three weeks sufficient time for Pegasus to do its job?
15. Was it appropriate for Pegasus to recommend that Evelyn Williams, the project manager, be dismissed?
16. What should Paul Harris report to the executive staff?
17. What should Paul Harris say to the QuickPay team concerning the health check?
18. What is the proper way for Paul Harris to remove Evelyn Williams from the project?

McRoy Aerospace

McRoy Aerospace was a highly profitable company that built cargo planes and refueling tankers for the armed forces. It had been doing this for more than 50 years and was highly successful. But because of a downturn in the government's spending on these types of planes, McRoy decided to enter the commercial aviation aircraft business, specifically wide-body planes that would seat up to 400 passengers, and compete head on with Boeing and Airbus Industries.

During the design phase, McRoy found that the majority of the commercial airlines would consider purchasing its plane provided that the costs were lower than those of the other aircraft manufacturers. Although the actual purchase price of the plane was a consideration for buyers, greater interest was in the life-cycle cost of maintaining the operational readiness of the aircraft, specifically the maintenance costs.

Operations and support costs were a considerable expense, and maintenance requirements were regulated by the government for safety reasons. The airlines make money when planes are in the air rather than sitting in maintenance hangars. Each maintenance depot maintained an inventory of spare parts so that, if a part did not function properly, it could be removed from a plane and replaced with a new part. The damaged part would be sent to the manufacturer for repairs or replacement. Inventory costs could be significant but were considered a necessary expense to keep the planes flying.

One of the issues facing McRoy was the mechanisms for the eight doors on the aircraft. Each pair of doors had its own mechanisms, which appeared to be restricted by their location in the plane. If McRoy could come up with a single design for all four pairs of doors, it would significantly lower the inventory costs for the airlines and the necessity to train mechanics on four sets of mechanisms. On the cargo planes and refueling tankers, each pair of doors had a unique mechanism. For commercial aircraft, finding one design for all doors would be challenging.

Mark Wilson, one of the department managers at McRoy's design center, assigned Jack, the best person he could think of, to work on this extremely challenging project. If anyone could accomplish it, it was Jack. If Jack could not do it, Mark sincerely believed it could not be done.

The successful completion of this project would be seen as a value-added opportunity for McRoy's customers and could make a tremendous difference from a cost and efficiency standpoint. McRoy would be seen as an industry leader in life-cycle costing, and this could make the difference in getting buyers to purchase commercial planes from McRoy Aerospace.

The project was to design an opening/closing mechanism that was the same for all of the doors. Until now, each door on the plane had a different set of open/close mechanisms, which made the design, manufacturing, maintenance, and installation processes more complex, cumbersome, and costly.

Without a doubt, Jack was the best—and probably the only—person to make this happen, even though the equipment engineers and designers all agreed that it could not be done. Mark put all of his cards on the table when he presented the challenge to Jack. He told him that his only hope was for Jack to take on this project and explore it from every possible, out-of-the-box angle he could think of. But Jack said right off the bat that this might not be possible. Mark was not happy hearing Jack say this right away, but he knew Jack would do his best.

Jack spent two months looking at the problem and simply could not come up with the solution needed. Jack decided to inform Mark that a solution was not possible. Both Jack and Mark were disappointed that a solution could not be found.

"I know you're the best, Jack," stated Mark. "I can't imagine anyone else even coming close to solving this critical problem. I know you put forth your best effort and the problem was just too much of a challenge. Thanks for trying. But if I had to choose one of your coworkers to take another look at this project, who might have even half a chance of making it happen? Who would you suggest? I just want to make sure that we have left no stone unturned," he said glumly. Mark's words caught Jack by surprise. Jack thought for a moment, and you could practically see the wheels turning in his mind. Was he thinking about who could take this project on and waste more time trying to find a solution? No, Jack's wheels were turning on the subject of the challenging problem itself. A glimmer of an idea whisked through his brain and he said, "Can you give me a few days to think about some things, Mark?" he asked pensively.

Mark had to keep the little glimmer of a smile from erupting on his face. "Sure, Jack," he said. "Like I said before, if anyone can do it, it's you. Take all the time you need."

A few weeks later, the problem was solved, and Jack's reputation rose to even greater heights than before.

QUESTIONS

1. Was Mark correct in what he said to get Jack to continue investigating the problem?
2. Should Mark just have given up on the idea rather than what he said to Jack?
3. Should Mark have assigned this to someone else rather than giving Jack a second chance? If so, how might Jack have reacted?
4. What should Mark have done if Jack still was not able to resolve the problem?
5. Would it make sense for Mark to assign this problem to someone else now, after Jack could not solve the problem the second time around?
6. What other options, if any, were now available to Mark?

The Poor Worker

Paula, the project manager, was reasonably happy the way that work was progressing on the project. The only issue was the work being done by Frank. Paula knew from the start of the project that Frank was a mediocre employee who often was regarded as a troublemaker. The tasks that Frank was expected to perform were not overly complex, and the line manager assured Paula during the staffing function that Frank could do the job. The line manager also informed Paula that Frank demonstrated behavioral issues on other projects and sometimes had to be removed from projects. Frank was a chronic complainer and found fault with everything and everybody. But the line manager also assured Paula that Frank's attitude was changing and that the line manager would get actively involved if any of these issues began to surface on Paula's project. Reluctantly, Paula agreed to allow Frank to be assigned to her project.

Unfortunately, Frank was not performing his work on the project to Paula's standards. Paula had told Frank on more than one occasion what she expected from him, but he persisted in doing things his own way. Paula was now convinced that the situation was getting worse. Frank's work packages were coming in late and sometimes over budget. Frank continuously criticized Paula's performance as a project manager, and his attitude was beginning to affect the performance of some other team members. Frank was lowering the morale of the team. It was obvious that Paula had to take some action.

QUESTIONS

1. What options are available to Paula?
2. If Paula decides to try to handle the situation first by herself rather than approach the line manager, what should she do and in what order?
3. If all of Paula's attempts fail to change Frank's attitude and the line manager refuses to remove him, what options are available to Paula?
4. What rights, if any, does Paula have with regard to wage and salary administration regarding this employee?

The Prima Donna

Ben was placed in charge of a one-year project. Several of the work packages had to be accomplished by the Mechanical Engineering Department and required three people to be assigned full time for the duration of the project. When the project was originally proposed, the Mechanical Engineering Department manager estimated that he would assign three of his grade 7 employees to do the job. Unfortunately, the start date of the project was delayed by three months, and the department manager was forced to assign the resources he planned to use to another project. The resources available for Ben's project at the new starting date were two grade 6s and a grade 9.

The department manager assured Ben that these three employees could adequately perform the required work and that Ben would have these three employees full time for the duration of the project. Furthermore, if any problems occurred, the department manager made it clear to Ben that he personally would get involved to make sure that the work packages and deliverables were completed correctly.

Ben did not know any of the three employees personally. But since a grade 9 was considered a senior subject matter expert pay grade, Ben made the grade 9 the lead engineer representing his department on the project. It was common practice for the seniormost person assigned from each department to act as the lead and even as an assistant project manager. The lead was often allowed to interface with customers at information exchange meetings.

By the end of the first month of the project, work was progressing as planned. Although most of the team seemed happy to be assigned to the project and team morale was high, the two grade 6 team members in the Mechanical Engineering Department were disenchanted with the project. Ben interviewed those employees to see why they were somewhat unhappy. One employee stated:

> The grade 9 wants to do everything himself. He simply does not trust us. Every time we use certain equations to come up with a solution, he must review everything we did in microscopic detail. He has to approve everything. The only time he does not micromanage us is when we have to make copies of reports. We do not feel that we are part of the team.

Ben was unsure how to handle the situation. Resources are assigned by the department managers and usually cannot be removed from a project without the permission of the department managers. Ben met with the Mechanical Engineering Department manager, who stated:

> The grade 9 that I assigned is probably the best worker in my department. Unfortunately, he's a prima donna. He trusts nobody else's numbers or equations other than his own. Whenever coworkers perform work, he feels obligated to review everything that they have done. Whenever possible, I try to assign him to one-person activities so that he will not have to interface with anyone. But I have no other one-person assignments right now, which is why I assigned him to your project. I was hoping he would change his ways and work as a real team member with the two grade 6 workers, but I guess not. Don't worry about it. The work will get done, and get done right. We'll just have to allow the two grade 6 employees to be unhappy for a little while.

Ben understood what the department manager said but was not happy about the situation. Forcing the grade 9 to be removed could result in the assignment of someone with lesser capabilities, and this could impact the quality of the deliverables from the Mechanical Engineering Department. Leaving the grade 9 in place for the duration of the project would alienate the two grade 6 employees, and their frustration and morale issues could infect other team members.

QUESTIONS

1. What options are available to Ben?
2. Is there a risk in leaving the situation as is?
3. Is there a risk in removing the grade 9 employee?

The Team Meeting

BACKGROUND

Every project team has team meetings. The hard part is deciding when during the day to have the meeting.

KNOW YOUR ENERGY CYCLE

Vince had been a morning person ever since graduating from college. He enjoyed getting up early. He knew his own energy cycle and that he was more productive in the morning than in the afternoon.

Vince would come into work at 6:00 a.m., two hours before the normal work-force would show up. Between 6:00 a.m. and noon, Vince would keep his office door closed and often would not answer the phone. Vince did this to prevent others from interrupting his most productive time. Vince considered time robbers such as unnecessary phone calls lethal to project success. This gave Vince six hours of productive time each day to do the necessary project work. After lunch, Vince would open his office door, and anyone could talk with him.

A TOUGH DECISION

Vince's energy cycle worked well, at least for him. But Vince had just become the project manager on a large project. Vince knew that he might have to sacrifice some of his precious morning time for team meetings. It was customary for each project team to have a weekly team meeting, and most project team meetings seemed to be held in the morning.

Initially, Vince decided to go against tradition and hold team meetings between 2:00 and 3:00 p.m. This would allow him to keep his precious morning time for his own productive work. During the afternoon team meeting, Vince was somewhat disturbed when there was very little discussion on some critical issues and people appeared to be looking at their watches. Finally, Vince understood the problem. A large portion of his team members were manufacturing personnel who started work as early as 5:00 a.m. The manufacturing personnel were ready to go home at 2:00 p.m. and were tired.

The following week Vince changed the team meeting time to 11:00 to 12:00 noon. It was evident to him that he had to sacrifice some of his morning time. But once again, during the team meetings, there really wasn't very much discussion about some critical issues on the project, and the manufacturing personnel were looking at their watches. Vince was disappointed. As he exited the conference room, one of the manufacturing personnel commented to him, "Don't you know that the manufacturing people usually go to lunch around 11:00 a..m.?"

Vince came up with a plan for the next team meeting. He sent out e-mails to all team members stating that the meeting would be at 11:00 to 12:00 noon, as before, but the project would pick up the cost for providing lunch in the form of pizzas and salads. Much to Vince's surprise, this worked well. The atmosphere in the team meeting improved significantly. There were meaningful discussions and decisions were being made instead of creating action items for future team meetings. It suddenly became an informal rather than a formal team meeting. Although Vince's project could certainly incur the cost of pizzas, salads, and soft drinks for team meetings, this might set a bad precedent if this would happen at each team meeting. At the next team meeting, the team decided that it would be nice if the pizza lunches could happen once or twice a month. It was decided to leave the time for the team meetings the same at 11:00 to 12:00 noon, but team members would bring their own lunches, and the project would provide soft drinks and perhaps some cookies or brownies.

QUESTIONS

1. How should a project manager determine when (i.e., time of day) to hold a team meeting? What factors should be considered?
2. What mistakes did Vince make initially?
3. If you were an executive in this company, would you allow Vince to continue having pizza at team meetings?

The Management Control Freak

BACKGROUND

The company hired a new vice president for the Engineering Department, Richard Cramer. Unlike his predecessor, Richard ruled with an iron fist and was a true micromanager. This played havoc with the project managers in the department because Richard wanted to be involved in all decisions, regardless of how small.

WHAT TO DO

Anne was an experienced project manager who had been with the company for more than 20 years. She had a reputation for being an excellent project manager, and people wanted to work on her projects. She knew how to get the most out of her team and delegated as much decision making as possible to team members. Her people skills were second to none.

A few months before Richard was hired, Anne was assigned to a two-year project for one of the company's most important clients. Anne had worked on projects for this client previously, and the results had been well received. The client actually requested that Anne be assigned to this project.

Almost all of Anne's team members had worked for her before. Some had even asked to work for her on this project. Anne knew some of the people personally and trusted their decision-making skills. Having people assigned who have worked with you previously is certainly considered a plus.

Work progressed smoothly until about the third week after Richard Cramer came on board. In a meeting with Anne, Richard commented:

> I have established a policy that I will be the project sponsor for all projects where the project managers report to someone in Engineering. I know that the vice president for marketing had been your sponsor for previous projects with this client, but all of that will now change. I have talked with the vice president for marketing, and he understands that I will now be your sponsor. I just cannot allow anyone from outside of Engineering to be a sponsor of a project that involves critical engineering decisions and where the project managers come from Engineering. So, Anne, I will be your sponsor from now on, and I want you to talk to my secretary and set up weekly briefings for me on the status of your project. This is how I did it in my previous company, and it worked quite well.

These comments didn't please Anne. The vice president for marketing was quite friendly with the client, and now things were changing. Anne understood Richard's reasons for wanting to do this but certainly was not happy about it.

Over the next month, Anne found that her working relationship with Richard was getting progressively worse, and it was taking its toll on the project. Richard was usurping Anne's authority and decision making. On previous projects, Anne would meet with the sponsor about every two weeks for about 15 minutes. Her meetings with Richard were now weekly and were lasting for more than one hour. Richard wanted to see all of the detailed schedules and wanted a signature block for himself on all documents that involved engineering decisions. There was no question in Anne's mind that Richard was a true micromanager.

At the next full team meeting, some workers complained that Richard was calling them directly, without going through Anne, and making some decisions that Anne did not know about. The workers were receiving directions from Richard that conflicted with Anne's directions. Anne could tell that morale was low and heard people mumbling about wanting to get off of this project.

At Anne's next meeting with Richard, she made it quite clear about how upset she was with his micromanagement of the project. If this continued, she would have a very unhappy client. Richard again asserted how he had to be involved in all technical decisions and that this was his way of managing. He also stated that, if Anne was unhappy, he could find someone else to take over her job as the project manager.

Something had to be done. This situation could not be allowed to continue without damaging the project further. Anne thought about taking her concerns directly to the president but realized that nothing would probably change. And if that happened, Anne could be worse off.

Anne then came up with a plan. She would allow Richard to micromanage and even help him do so. There was a risk in doing this, and Anne could very well lose her job. But she decided to go ahead with her plan. For the next several weeks, Anne and all team members refused to make even the smallest decisions themselves. Instead, they brought all of the decisions directly to Richard. Richard was even getting phone calls at home from the team members on weekends, during the dinner hour, late at night, and early Sunday mornings.

Richard was now being swamped with information overload and was spending a large portion of his time making mundane decisions on Anne's project. In the next team sponsor briefing meeting with Anne, Richard stated: "I guess that you've taught me a lesson. If it's not broken, then there isn't any reason to fix it. I guess that I came across too strong and made things worse. What can we do to repair the damage I may have done?"

Anne could not believe that Richard was saying these words. She was speechless. She thought for a moment and then went over to the whiteboard in Richard's office. She took a marker and drew a vertical line down the center of the board. She put her name to the left of the line and Richard's name to the right of the line. She then said: "I'm putting my responsibilities as a project manager under my name, and I'd like you to put your responsibilities as a sponsor under your name. However, the same responsibility cannot appear under both names."

An hour later, Anne and Richard came to an agreement on what each other's responsibilities should be. Anne walked out of Richard's office somewhat relieved that she was still employed.

QUESTIONS

1. When someone hires into a company, is there any way of telling whether he or she is a control freak?
2. If someone higher in rank than you turns out to be a control freak, how long should you wait before confronting them?
3. Do you believe that Anne handled the situation correctly?
4. Could Anne's decision on how to handle the situation result in her getting removed as a project manager or even fired?
5. What other ways were available to Anne for handling the situation?

The Skills Inventory
Project

The Riverside Software Group (RSG) was a small software company that specialized in software to support the human resources departments of both large and small corporations. RSG had been in business for more than 30 years and had an excellent reputation and an abundance of repeat business.

In 2011, RSG was awarded a contract from a Fortune 100 company to develop an inventory skills software package. The Fortune 100 company maintained a staff of more than 10,000 project managers worldwide and a total employment of more than 150,000. Although the company sold products and services across the world, it was also marketed as a global business solutions provider. Since most of the work was global, RSG utilized virtual teams on almost all projects. The difficulty was in creating the virtual teams. Quite often, the project managers had limited knowledge of the capabilities of employees around the world, and this made it difficult to establish a project team with the best available resources. What was needed was an inventory skills matrix for all employees.

The contract with RSG was not that complex. Whenever the Fortune 100 company would complete a project, either for an external client or one of its worldwide clients, the entire project team would use the software to update their resumes, including the new skills they developed, the chemical or specialized processes they were now familiar with, and whatever additional information would be valuable to their company in determining the best available personnel for the next project. The project team also had to identify in the software program the lessons that were learned on that project, the best practices that were captured, the metrics and key performance indicators that were used, and other such factors that could benefit the company in the future.

RSG saw this as an excellent opportunity. The client had done its homework well and created a detailed requirements package. Neither RSG nor the client expected any significant scope changes since the requirements were reasonably well established. The contract was a firm-fixed-price effort of $1.2 million for labor and materials, an additional $150,000 in profit, and with a scheduled completion date of 12 months.

Within the first two months of the project, RSG realized that this software package had tremendous potential and could be sold to many of its clients around the world. RSG estimated that clients would pay at least $75,000 for such a package and also pay additional costs for possible customization. The problem was that the contract with the client was firm-fixed-price and all of the intellectual property rights stayed with the client.

If RSG agreed to allow the client to sell the package to other customers, RSG would probably have to spend about $10,000 in preliminary customization for each client. Detailed customization would be billed separately to each client. Additional costs, including documentation, packaging, and shipping/handling, would be about $5,000. Therefore, even adding in a small financial reserve of $5,000 as a risk factor for other design contingencies, RSG's cost per package would be about $20,000, and it could sell for $75,000. Marketing and sales personnel believed that at least 100 of these packages could be sold worldwide.

Given the potential of this effort, the company had to come up with a plan on how they would approach the Fortune 100 client and request a change in the contract. The simplest solution would be to make the client a 50–50 partner, but that could create problems with enhancements and upgrades to the package downstream. The second approach would be to see if the client would allow the contract to change to a cost-sharing effort. The profit of $150,000 would be removed from the $1,350,000 contract that now existed, and the remaining question would be the cost-sharing split. Originally, RSG considered proposing a 70–30 or 60–40 spilt with the Fortune 100 client paying the greater percentage of the cost. However, to make new plan attractive to the client, RSG decided to offer the client a 40–60 split with the 60 percent paid for by RSG.

QUESTIONS

1. If you were the client, would you accept this offer?
2. If the client accepts the offer, is it a win–win situation?
3. If the client accepts the change to the contract, how much profit will RSG make if it can sell 100 units?

Part 10

CONTROLLING PROJECTS

Controlling projects is a necessity so that meaningful and timely information can be obtained to satisfy the needs of the project's stakeholders. This control process includes measuring resources consumed, measuring status and accomplishments, comparing measurements to projections and standards, and providing effective diagnosis and replanning.

For cost control to be effective, both the scheduling and the estimating systems must be somewhat disciplined in order to prevent arbitrary and inadvertent budget or schedule changes. Changes must be disciplined and result only from a deliberate management action. This includes distribution of allocated funds and redistribution of funds held in reserve.

The Two-Boss Problem

On May 15, 2011, Brian Richards was assigned full time to Project Turnbolt by Fred Taylor, manager of the thermodynamics department. All work went smoothly for four and one-half of the five months necessary to complete this effort. During this period of successful performance, Brian had good working relations with Edward Compton (the Project Turnbolt engineer) and Fred.

Fred treated Brian as a Theory Y employee. Once a week Fred and Brian would chat about the status of Brian's work. Fred would always conclude their brief meeting with "You're doing a fine job, Brian. Keep it up. Do anything you have to do to finish the project."

During the last month of the project, Brian began receiving conflicting requests from the project office and the department manager regarding the preparation of the final report. Edward Compton told him that the final report was to be assembled in viewgraph format (i.e., "bullet" charts) for presentation to the customer at the next technical interchange meeting. The project did not have the funding necessary for a comprehensive engineering report.

The thermodynamics department, however, had a policy that all engineering work done on new projects would be documented in a full and comprehensive report. This new policy had been implemented about one year earlier when Fred Taylor became department manager. Rumor had it that Fred wanted formal reports so that he could put his name on them and either publish them or present them at technical meetings. All work performed in the thermodynamics department required Fred's signature before it could be released to the project office as an official company position. Upper-level management did not want its people to publish and therefore did not maintain a large editorial or graphic arts department. Personnel desiring to publish had to get the department manager's approval and, on approval, had to prepare the entire report themselves, without any "overhead" help. Since Fred had taken over the reins as department head, he had presented three papers at technical meetings.

A meeting was held among Brian Richards, Fred Taylor, and Edward Compton.

Edward: "I don't understand why we have a problem. All the project office wants is a simple summary of the results. Why should we have to pay for a report that we don't want or need?"

Fred: "We have professional standards in this department. All work that goes out must be fully documented for future use. I purposely require that my signature be attached to all communications leaving this department. This way we obtain uniformity and standardization. You project people must understand that, although you can institute or own project policies and procedures within the constraints and limitations of company policies and procedures, we department personnel also have standards. Your work must be prepared within our standards and specifications."

Edward: "The project office controls the purse strings. The project office specified that only a survey report was necessary. Furthermore, if you want a more comprehensive report, then you had best do it on your own overhead account. The project office isn't going to foot the bill for your publications."

Fred: "The customary procedure is to specify in the program plan the type of report requested from the departments. Inasmuch as your program plan does not specify this, I used my own discretion as to what I thought you meant."

Edward: "But I told Brian Richards what type of report I wanted. Didn't he tell you?"

Fred: "I guess I interpreted the request a little differently from what you had intended. Perhaps we should establish a new policy that all program plans must specify reporting requirements. This would alleviate some of the misunderstandings, especially since my department has several projects going on at one time. In addition, I am going to establish a policy for my department that all requests for interim, status, or final reports be given to me directly. I'll take personal charge of all reports."

Edward: "That's fine with me! And for your first request, I'm giving you an order that I want a survey report, not a detailed effort."

Brian: "Well, since the meeting is over, I guess I'll return to my office (and begin updating my résumé just in case)."

QUESTIONS

1. What are the major issues in the case?
2. How should these issues be resolved?

The Bathtub Period

The award of the Scott contract on January 3, 1987, left Park Industries elated. If managed correctly, the Scott Project offered tremendous opportunities for follow-on work over the next several years. Park's management considered the Scott Project as strategic in nature.

The Scott Project was a 10-month endeavor to develop a new product for Scott Corporation. Scott informed Park Industries that sole-source production contracts would follow, for at least five years, assuming that the initial research and development (R&D) effort proved satisfactory. All follow-on contracts were to be negotiated on a year-to-year basis.

Jerry Dunlap was selected as project manager. Although he was young and eager, he understood the importance of the effort for future growth of the company. Jerry was given some of the best employees to fill out his project office as part of Park's matrix organization. The Scott Project maintained a project office of seven full-time people, including Jerry, throughout the duration of the project. In addition, eight people from the functional department were selected to work as functional project team members, four full time and four half time.

Although the workload fluctuated, the manpower level for the project office and team members was constant for the duration of the project at 2,080 hours per month. The company assumed that each hour worked incurred a cost of $60.00 per person, fully burdened.

At the end of June, with four months remaining on the project, Scott Corporation informed Park Industries that, owing to a projected cash flow problem, follow-on work would not be awarded until the first week in March 1988. This posed a tremendous problem for Jerry Dunlap because he did not wish to break up the project office. If he permitted his key people to be assigned to other projects, there would be no guarantee that he could get them back at the beginning of the follow-on work. Good project office personnel are always in demand.

Jerry estimated that he needed $40,000 per month during the "bathtub" period to support and maintain his key people. Fortunately, the bathtub period fell over Christmas and New Year's, a time when the plant would be shut down for 17 days. Between the vacation days that his key employees would be taking and the small special projects that this people could be temporarily assigned to on other programs, Jerry revised his estimate to $125,000 for the entire bathtub period.

At the weekly team meeting, Jerry told the program team members that they would have to "tighten their belts" in order to establish a management reserve of $125,000. The project team understood the necessity for this action and began rescheduling and replanning until a management reserve of this size could be realized. Because the contract was firm-fixed-price, all schedules for administrative support (i.e., project office and project team members) were extended through February 28 on the supposition that this additional time was needed for final cost data accountability and program report documentation.

Jerry informed his boss, Frank Howard, the division head for project management, of the problems with the bathtub period. Frank was the intermediary between Jerry and the general manager. Frank agreed with Jerry's approach to the problem and asked to be kept informed.

On September 15, Frank told Jerry that he wanted to "book" the management reserve of $125,000 as excess profit since it would influence his (Frank's) Christmas bonus. Frank and Jerry argued for a while, with Frank constantly saying "Don't worry! You'll get your key people back. I'll see to that. But I want those uncommitted funds recorded as profit and the program closed out by November 1."

Jerry was furious with Frank's lack of interest in maintaining the current organizational membership.

QUESTIONS

1. Should Jerry go to the general manager?
2. Should the key people be supported on overhead?
3. If this were a cost-plus program, would you consider approaching the customer with your problem in hopes of relief?
4. If you were the customer of this cost-plus program, what would your response be for additional funds for the bathtub period, assuming cost overrun?
5. Would your previous answer change if the program had the money available as a result of an underrun?
6. How do you prevent this situation from recurring on all yearly follow-on contracts?

Irresponsible Sponsors

BACKGROUND

Two executives in this company each funded a "pet" project that had little chance of success. Despite repeated requests by the project managers to cancel the projects, the sponsors decided to throw away good money after bad money. The sponsors then had to find a way to prevent their embarrassment from such blunders from becoming apparent to all.

STORY LINE

Two vice presidents came up with ideas for pet projects and funded the projects internally using money from their functional areas. Both projects had budgets close to $2 million and schedules of approximately one year. These were somewhat high-risk projects because they both required that a similar technical breakthrough be made. There was no guarantee that the technical breakthrough could be made at all. And even if the technical breakthrough could be made, both executives estimated that the shelf life of both products would be about one year before becoming obsolete but that they could easily recover their R&D costs.

These two projects were considered pet projects because they were established at the personal request of two senior managers and without any real business case. Had these projects been required to go through the formal process of portfolio selection of projects, neither would have been approved. The budgets for these projects were way out of line for the value that the company would receive, and the return on investment would be below minimum levels even if the technical break-through could be made. Personnel from the project management office (PMO) who were actively involved in the portfolio selection of projects also stated that they would never recommend approval of a project where the end result would have a shelf life of one year or less. Simply stated, these projects existed merely for the satisfaction of the two executives and to get them prestige with their colleagues.

Nevertheless, both executives found money for their projects and were willing to let them go forward without the standard approval process. Each executive was able to get an experienced project manager from his group to manage the pet project.

GATE-REVIEW MEETINGS

At the first gate-review meeting, both project managers stood up and recom-mended that their projects be canceled and the resources assigned to other, more promising projects. They both stated that the technical breakthrough needed could not be made in a timely manner. Under normal conditions, both of these project managers should have received medals for bravery in recommending that their projects be canceled. These recommendations certainly appeared to be in the best interests of the company.

But both executives were not willing to give up that easily. Canceling both projects would be humiliating for the executives who were sponsoring these pro-jects. Instead, both executives stated that the projects were to continue until the next gate-review meeting, at which time a decision would be made for possible cancellation of both projects.

At the second gate-review meeting, both project managers once again recom-mended that their projects be canceled. And, as before, both executives asserted that the projects should continue to the next gate-review meeting before a decision would be made.

As luck would have it, the necessary technical breakthrough was finally made, but six months late. That meant that the window of opportunity to sell the products and recover the R&D costs would be six months rather than one year. Unfortunately, the thinking in the marketplace was that these products would be obsolete in six months, and no sales occurred of either product.

Both executives had to find a way to save face and avoid the humiliation of having to admit that they squandered a few million dollars on two useless R&D projects.

This could very well impact their year-end bonuses.

QUESTIONS

1. Is it customary for companies to allow executives to have pet or secret projects that do not follow the normal project approval process?
2. Who got promoted and who got fired? In other words, how did the executives save face?

The Need for Project Management Metrics (A)

Everybody knew that this would be a very unpleasant meeting. The selling price of the company's stock was near a five-year low. The company had just been downgraded by one of the rating agencies. Several Wall Street analysts wanted more information on what new projects the company was working on and the potential strength of the R&D projects in the company's pipeline. And to make matters worse, the company was forced to slash the company's dividend payment to conserve cash.

Unlike other companies that could produce new products quickly and deliver them to the marketplace with minimal cost, this company struggled. Historically, it was more of a follower than a leader. The company had problems when it came to creating value through innovation processes. In the past, whenever new products were created, the company was good at adding value to existing products using process reengineering, product modifications and enhancements, quality initiatives, and business process improvements. But this alone would not get the company through the current turbulent economic times. Innovation was a necessity and needed to happen quickly. The creation of customer-recognized value was needed, and the company was struggling.

As Al Grey, president and chief executive officer, entered the room, everyone could see that he was not happy. The company was in trouble and nobody really had a plan for how to rectify the situation. Finger-pointing and the laying of blame elsewhere had become the norm. The company had a multitude of talented people, but their achievements were less than par. This was particularly true of the R&D Group and the Engineering Departments responsible for the development of new products.

Al Grey stood up and addressed the senior staff:

I believe that I have identified the root cause of our innovation problems and we should be able to come up with a solution. I'm passing out to each of you two sheets of paper. In the first sheet (see Figure I), I have identified seven R&D projects that we considered to be total failures. You'll notice that five of the seven projects that failed consumed some of our most talented people and yet there were no new products developed from these five projects or from the other two projects. You'll also notice in which life cycle phase we made the decision to pull the plug on these projects. Several of these projects had gone through to completion before we discovered, or should I say were willing to admit, that the projects would produce no fruitful results.

Over the past year, we worked on a total of 12 high-priority R&D projects where we were convinced that success would be forthcoming. Only five of these projects generated any revenue stream, and none of the five was considered a "home run" producing the desired cash flow. In the second sheet of paper (see Table I), I have shown how much money we have squandered on the seven projects that failed. We threw away millions of dollars.

Several of you have pointed the finger at the project management office and tried to put the entire blame on their shoulders, claiming that the process we use for our portfolio selection of projects was at fault. While it is true that perhaps our process, like any other process, could be improved, all of us in this room still agreed with their selection recommendations. Not all projects that we recommend will be successful and any executive in this room who believes that they will all be a success is a fool.

Some projects will fail, but with a failure rate of seven projects out of 12, the company's growth could be limited.

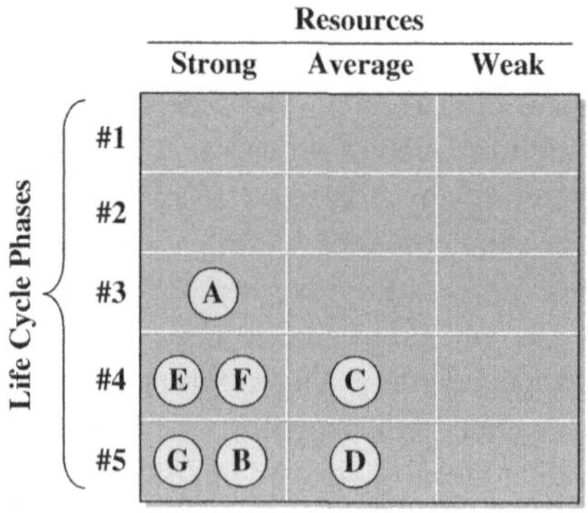

FIGURE I   Failure identification per life cycle phase

TABLE I   R&D TERMINATION COSTS AND REASONS FOR FAILURE

Project	Original Budget	Expenditure at Termination	Reason for Failure
A	$2,200,000	$1,150,000	Objective too optimistic
B	$3,125,000	$3,000,000	Could not make breakthrough
C	$2,200,000	$1,735,000	Structural integrity test failure
D	$5,680,000	$5,600,000	Vendors could not perform
E	$4,900,000	$3,530,000	Product safety test failures
F	$4,100,000	$3,200,000	Specification limits unreachable
G	$6,326,700	$6,200,000	Could not make breakthrough
Total	**$28,531,700**	**$24,415,000**	

Everyone in the room reviewed the two sheets. There was dead silence. No one wanted to speak. Using the PMO as the scapegoat would no longer work. There was a feeling that Al Grey was about to blame someone for the calamity, but no one knew who it might be.

Al Grey continued:

> Given the reality that some projects will fail, why must we squander so much money by waiting until we get to the last two life-cycle phases of our five-phase methodology before we are willing to admit that the project might fail? Why can't one of the project managers stand up in any of the earlier phases and state that the project should be cancelled?

Everyone then looked at Doug Wilson, vice president for engineering and R&D, for his response to the question. Even though a PMO existed, most of the project managers were engineers who reported directly to Doug. The project managers were "solid" to engineering but "dotted" to the PMO just for project status reporting. Doug said:

> I'm going to defend my people. They work hard and have a history of producing results, profitable results at that! I know that some of our more recent projects have come in late, have been over budget, and the results have not been there. Our projects were challenging and sometimes this happens. It is uncalled for to blame my people for these seven failures.

Ann Hawthorne, vice president for marketing, decided to intervene:

> I have worked with engineers for decades. They are highly optimistic and believe that whatever plan they develop will work correctly the first time. They refuse to admit that projects have budgets and schedules. The goal of every engineer I have ever worked with is to exceed the specifications rather than just meet them, and they

want to do it on someone else's budget. When you perform R&D on government contracts, you can always request more money and a schedule extension, and you'll probably get it from the government. But our engineers are spending *our* money, not funds provided from an external source. It's hard for me to believe that they couldn't identify early on that some of these projects should have been cancelled.

Everyone looked at each other wondering who would be next to be blamed for the problems. Al Grey then spoke out again.

Whenever I review one of our project status reports, all I see is information on budgets and schedules. The rest of the information is obscure or hidden and sometimes makes no sense to me. Sometimes there are comments about risks. Why isn't it possible to establish some metrics, other than time or cost metrics, that can provide us with meaningful information such that we can make informed decisions in some of the earlier life-cycle phases? I see this as being the critical issue that must be resolved quickly.

After a brief discussion, everyone seemed to agree that better metrics could alleviate some of the problems. But getting agreement on the identification of the problem was a lot easier than finding a solution. New issues on how to perform metrics management would now be surfacing, and most of the people in the room had limited experience with metrics management.

The company had a PMO that reported to Carol Daniels, chief information officer. The PMO was created for several reasons, including the development of an enterprise project management methodology, support for the senior staff in the project portfolio selection process, and creation of executive-level dashboards that would provide information on the performance of the strategic plan. Carol then commented:

Our PMO has expertise with metrics, but business-based metrics rather than project-based metrics. Our dashboards contain information on financial metrics such as profitability, market share, number of new customers, percentage of our business that is repeat business, customer satisfaction, quality survey results, and so forth. I will ask the PMO to take the lead in this, but I honestly have no clue how long it will take or the complexities with designing project-based metrics.

Everyone seemed relieved and pleased that Carol Daniels would take the lead role in establishing a project-based metric system. But there were still concerns and issues that needed to be addressed, and it was certainly possible that this solution could not be achieved even after the expenditure of significant time and effort on metric management.

Al Grey then stated that he wanted another meeting with the executive staff scheduled in a few days where the only item up for discussion would be the plan for developing the metrics. Everyone in the room was given the same action item in preparation for the next meeting: "Prepare a list of what metrics you feel are necessary for early-on informed project decision making and what potential problems we must address in order to accomplish this."

QUESTIONS

In answering these questions, do not look for the perfect answer. The problem of determining the correct metrics to use is quite common today and plagues executives, stakeholders, and other decision makers involved in projects. There may be several answers to each question based on your interpretation of the situation.

1. Can the failure of R&D actually be this devastating to a company?
2. When project work goes bad and failures occur, is it common practice for finger-pointing and the laying of blame to occur, even at executive levels?
3. What information is found in Figure I?
4. What information is lacking in Figure I?
5. Is it possible that highly talented resources can overthink an R&D project to the point where they look for the most complicated solution rather than the simplest solution?
6. Is it good or bad to have five R&D projects out of 12 completed successfully?
7. Can a PMO prevent failure?
8. Is it inherently dangerous to encourage a project manager to recommend that his or her project be terminated during early life-cycle phases?
9. Who, if anyone, should be blamed for the failure of the projects in this case study?
10. Is Ann Hawthorne's description of engineers realistic?
11. Would you agree with Al Grey that the real cause of the failures appears to be a lack of good metrics? If this is the cause, then how do you justify that other projects were successful?
12. Should the PMO take the lead in the establishment of the metrics?
13. Are a few days enough time for the follow-on meeting, and should the executives attend?

The Need for Project Management Metrics (B)

WHERE DO WE BEGIN?

The weekly executive staff meeting concluded, and everyone felt confident that the company was now heading in the right direction. Al Grey sent out a companywide e-mail letting everyone know what was about to happen and that the company needed everyone's cooperation to make this metrics management initiative succeed. Al stated:

> As you all know, today's business environment is changing rapidly. We can no longer rely solely upon our existing product lines for continuous growth. In the past, we have captured best practices and lesson learned, and this has improved the efficiency and effectiveness of our operations which then added to profits. Unfortunately, the best practices and lessons learned that we captured did not directly provide benefits to our innovation processes.
>
> Because we are now in a dynamic rather than stable business environment, we must rely heavily upon the creation of new products to achieve sustained growth. Our customers are demanding new products with higher quality and at a lower cost. Customers are now looking at how our products provided value to them and sometimes the importance of perceived value takes precedence over cost and quality considerations.
>
> We must now redefine our innovations processes to meet rapidly changing consumer demands. Our business development managers are being challenged to identify the value of business opportunities for new products that do not yet

exist. Our R&D staff must develop these products and we must have an innovation process in place that allows us to achieve our strategic objectives.

Because of the turbulent business environment, time is no longer a luxury but a critical constraint in our innovation process. With limited resources to work with, we must be absolutely sure that we are working on the right mix of projects. We are in the process of developing a metrics management system to allow us to make better decisions with regard to the selection and development of new products with exceptional value. The metrics we create will help us ensure that we are creating products that have value. Metrics management is essential. We must know if we are heading in the right direction and if the light at the end of the tunnel is reachable. If the metrics indicate that we cannot achieve our goals on a particular project, then we must pull the plug and assign the resources to those other projects that provide business value opportunities.

We are establishing a metric management team to develop this capability. The metrics management team will report to the PMO. I expect all of you to assist the team in carrying out their mission if they ask for your input and assistance.

MEETING OF THE MINDS

Al Grey was convinced that he was on the right track in his quest for a metrics management system. Rather than leave team member assignments to chance, Al personally selected the members of the task force. He knew each team member personally and was convinced that they could live up to the challenge. The six team members were:

1. John, representing the PMO and the team leader
2. Patsy, representing marketing
3. Carol, representing new business development
4. Allen, representing engineering
5. Barry, representing R&D
6. Paul, representing manufacturing

The team met and began discussing the challenge. The first step was to get a good understanding of what metrics are and how the company can benefit from their use. Everyone seemed to understand that a company cannot manage innovation projects without having good metrics and reasonably accurate measurement that can provide complete or nearly complete information for decision makers. Furthermore, since most of the company's projects were becoming more complex, it would become harder to determine true progress without effective metrics.

The team prepared a list of the benefits of using metrics. The list included:

● To improve performance for the future
● To improve future estimating
● To validate baselines

- To validate if we are hitting our targets or getting better or worse
- To catch mistakes before they lead to other perhaps more serious mistakes
- To improve client satisfaction
- A means of capturing best practices and lessons learned

Although everyone agreed on the benefits of metrics, John expressed his concern that the team must remain focused. He stated:

> It takes companies years to achieve all of those benefits. We simply do not have the luxury of doing that. We must focus on our primary mission, which is the establishment of metrics for our innovation process. We need objective, measurable attributes of project performance in order to make informed decisions. We must be able to use the metrics to predict project success and failure. Therefore, we must establish some type of priority for what type of metrics we will develop first.

The team decided that the primary focus should be to establish metrics that can be used as a means of continuous health checks on innovation projects. The metrics have to serve as early warning signs or risk triggers.

But deciding what to do and being able to do it were two separate activities. The business side of the company had been using metrics for some time. These were metrics related to market share, profitability, cash flow, and other such high-level measurements. The innovation metrics to be developed would be more detailed, and this could alienate the culture to the point where there would be more resistance than support. Allen commented: "Engineers do not like constant supervision. They like the freedom to create, and I am sure the same holds true in R&D. If we develop metrics that are too detailed and our people believe that the metrics are being used to spy on them, I feel that we will get a lot of resistance."

Patsy then recounted her observations with the metrics used in marketing and sales:

> I agree with Allen's comments. In marketing, we had some resistance as well, but for different reasons. Some people felt that the metrics were a waste of time and tied up valuable resources doing measurements. Some of the metrics we needed were not readily accessible from our information systems. Every executive wanted a different set of business metrics, and it was impossible for us to get agreement on what metrics were actually needed. We had to make software changes to some of our information systems, and that took time and money. Also, people felt that, since these were high-level metrics and updated monthly or quarterly, it was sometimes too late to make changes that were necessary to improve our business.

Everyone at the meeting had smiles on their faces ready to accept the new challenge. Patsy then showed the team Figure I, which compares financial metrics with the metrics that would have to be developed for projects. The smiles soon

Variable	Business/Financial	Project
Focus	Financial measurement	Project performance
Intent	Meeting strategic goals	Meeting project objectives, milestones and deliverables
Reporting	Monthly or quarterly	Real time data
Items to be looked at	Profitability, market share, repeat business, number of new customers, etc.	Adherence to competing constraints, validation and verification of performance
Length of use	Decades of even longer	Life of the project
Use of the data	Information flow and changes to the strategy	Corrective action to maintain baselines
Target audience	Executive management	Stakeholders and working levels

FIGURE I Differences between financial and project-based metrics

disappeared because the team now realized that they might not be able to draw on marketing expertise in creating project-based metrics. There were significant differences.

Now the team was coming to the realization that this problem with metrics was more complicated than they originally thought. They began questioning whether they could get everything done in a timely manner.

Realizing that the team was getting a little nervous, John stepped into the conversation.

I know you are all a little nervous now, but let's solve the problem with small rather than large steps. As I see it, there are four questions we should concern ourselves with:

1. What metrics should we select?
2. How will the metrics be measured?
3. How will the metrics be reported?
4. How will management react to the information?

The first two questions are probably the most important, and this is where we should start. I'm convinced we can do this, and in a reasonable time frame.

The team established an action item for the next meeting whereby all of the team members would interview their people and come up with a possible list of metrics. The meeting adjourned.

QUESTIONS

1. Is there a relationship between the capturing of best practices and the development of new metrics?
2. Is the makeup of the team correct? Should someone from senior management also have been part of the team?
3. Does it seem reasonable that some people might feel that metrics can be a spying machine?
4. Are the four questions posed by John correct?
5. What metrics would you include in the list that may be appropriate for innovation projects?

The Need for Project Management Metrics (C)

SELECTING THE RIGHT METRICS

The team reconvened with each member bringing a list of possible metrics. Each team member had interviewed people in his or her own group and at all levels of management. All members knew that their lists were highly subjective, and they now had to combine their thought processes with the entire team and see if they could come up with a more objective list. The first step was to combine the metrics, as shown in Exhibit I.

EXHIBIT I. COMBINED LISTING OF METRICS

- Percentage of work packages adhering to the schedule
- Percentage of work packages adhering to the budget
- Number of assigned resources versus planned resources
- Quality of the assigned resources versus planned resources
- Percentage of actual versus planned baselines completed to date
- Percentage of actual versus planned best practices used
- Project complexity factor
- Customer satisfaction ratings
- Number of critical assumptions made

(Continued)

(*Continued*)

- Percentage of critical assumptions that have changed
- Number of cost revisions
- Number of schedule revisions
- Number of scope change control meetings
- Number of critical constraints
- Percentage of work packages with a critical risk designation
- Net operating margins
- Number of unstaffed hours
- Turnover of key personnel, in number or percentage
- Percentage of labor hours on overtime
- Schedule variance (SV)
- Cost variance (CV)
- Schedule performance index (SPI)
- Cost performance index (CPI)

Although many of the metrics seemed worthy of consideration, there was a consensus that the list might be too long. John spoke first.

If we accept all of these as workable metrics, we may do more harm than good. All metrics need to be measured, and too many metrics will force team members to steal time from other work to do the measurement and reporting. Some of these metrics have little or no value for innovation projects. If we were to provide all of this information to the executives, they may not be able to determine what information is critical.

Carol spoke next.

I agree with John's remarks that the list is too long. But playing the devil's advocate, providing too few metrics can be equally as bad. Providing too few metrics can be disastrous if executives overreact to bad news on just a couple of metrics. They may not see the true story. We may need to educate the executives on how to understand the metrics. If we do not provide the right information, then executives may not be able to make informed decisions in a timely manner.

Patsy then added her comments.

I was part of the committee that established the financial and business metrics a few years ago. After several meetings, we established a business metric selection process that stated that whatever metrics we selected had to be worth collecting; we had to be sure that we would use what we collected; we had to make sure that the metrics were informative; and we eventually had to train our people in the use and value of these metrics.

It was now apparent that metric selection would be critical. The team knew that no matter how large or how small the final list would be, there would be naysayers who would argue that the benefits do not justify the cost and that metric measurement is a waste of time and useless.

Barry stated that friends in other companies maintain a metrics library the same way that other companies maintain a best practices library. Although his contacts were not willing to provide a list of the exact metrics in their library, they were willing to provide the categories of metrics in the library. The categories are shown in Exhibit II.

EXHIBIT II. CATEGORIES OF METRICS

- Quantitative metrics (planning dollars or hours as a percentage of total labor)
- Practical metrics (improved efficiencies)
- Directional metrics (risk ratings getting better or worse)
- Actionable metrics (affect change such as the number of unstaffed hours)
- Financial metrics (profit margins, return on investment, etc.)
- Milestone metrics (number of work packages on time)
- End result or success metrics (customer satisfaction)

QUESTIONS

1. What are the risks of reporting on too many metrics?
2. What are the risks of reporting on too few metrics?
3. Using Table I, categorize the metrics in Exhibit I.

TABLE I   CATEGORIZING THE METRICS

Metric	Quantitative	Practical	Directional	Actionable	Financial	Milestone	End Result
Percentage of work packages on schedule							
Percentage of work packages on budget							
Number of assigned versus planned resources							
Quality of assigned versus planned resources							
Percentage of actual versus planned baselines completed							
Percentage of actual versus planned best practices used							
Complexity factor							
Customer satisfaction rating							
Number of critical assumptions							
Percentage of critical assumptions that have changed							
Number of cost revisions							
Number of schedule revisions							
Number of scope change control meetings							
Number of critical constraints							
Percentage of work packages with a critical designation							
Net operating margin							

The Need for Project Management Metrics (D)

CONVERTING METRICS TO KEY PERFORMANCE INDICATORS

Although the list of metrics was good, it was apparent that the list was too long for senior management. The metrics had to be converted to key performance indicators (KPIs). Although most companies use metrics alone for measurement purposes, they seem to have a poor understanding of what constitutes a KPI, especially for projects. The ultimate purpose of a KPI is to measure items relevant to performance and to provide information on controllable factors appropriate for informed decision making such that it leads to positive outcomes. On innovation projects, KPIs drive change but do not prescribe a specific course of action. Not all metrics are KPIs. A KPI is a metric specifically related to decision making. All KPIs have targets. If we are meeting or exceeding the target, then that's good. If we are not, then we must decide whether a correction is possible or whether we should cancel the project. On innovation projects, KPIs serve as early indicators of success or failure.

Several team members appeared confused over the difference between metrics and KPIs. John then said:

> All KPIs are metrics, but not all metrics are KPIs. As an example, executives should not be concerned with the number of unstaffed man-hours or the quality of the assigned resources. However, if it looks like the project may be in trouble, then the executives should have the right to "drill down" to more detailed

levels of information. KPIs should be viewed as high-level metrics whereas more detailed information may appear as just metrics. I know the difference may not be clear to some of you, but the difference is there. On one project, a measurement can be treated as a KPI and on another project it may appear as just a pure metric.

Allen then made the following comments: "Given what you just said, each innovation project can have a different set of KPIs. Therefore, I recommend that, for each project, we look at the entire list of metrics and decide which should be treated as KPIs."

Patsy then interjected: "In marketing, we have just eight metrics, and we report on these same eight metrics every quarter. We have been doing this for the past several years with our financial scorecards. Now you're saying that project-based metrics and KPIs can change from project to project."

John responded:

Not only can they change from project to project, they can also change during each life-cycle phase. If a project gets into trouble, I expect some executives may want to see other metrics or KPIs reported such that they can make better decisions. We must also be prepared for the situation where each executive may want to see a different set of KPIs. They have that right, and we must live with it.

The team realized that converting metrics to KPIs would not be an easy task. It would be important to limit the number of KPIs so that everyone could focus on the same KPIs and understand them. Having too many KPIs might distract the project team and the executives from what is really important. This, in turn, could slow down projects because of excessive measurements and blur sight of actual performance.

John then stated: "It is my experience that companies often create too many KPIs rather than too few. Hopefully, we can overcome that temptation. I did some research and found a list of attributes that KPIs should have. The list is shown in Exhibit I. Perhaps this list can help us differentiate KPIs from pure metrics."

EXHIBIT I. KPI SELECTION CRITERIA

- *Predictive:* Able to predict the future of this trend
- *Measurable:* Can be expressed quantitatively
- *Actionable:* Triggers changes that may be necessary
- *Relevant:* Directly related to the success or failure of the project
- *Automated:* Reporting minimizes the chance of human error
- *Few in number:* Only what is necessary

QUESTIONS

1. Is there a simple way to differentiate between a metric and a KPI?
2. What factors determine how many KPIs should be reported?
3. Using the KPI selection criteria in Exhibit I, complete Table I.
4. How many of the metrics in Table I are now considered as KPIs?

TABLE I   CATEGORIZING THE METRICS

Metric	Predictive	Measurable	Actionable	Relevant	Automated
Percentage of work packages on schedule					
Percentage of work packages on budget					
Number of assigned versus planned resources					
Quality of assigned versus planned resources					
Percentage of actual versus planned baselines completed					
Percentage of actual versus planned best practices used					
Complexity factor					
Customer satisfaction rating					
Number of critical assumptions					
Percentage of critical assumptions that have changed					
Number of cost revisions					
Number of schedule revisions					
Number of scope change control meetings					
Number of critical constraints					
Percentage of work packages with a critical designation					
Net operating margin					
Number of unstaffed hours					
Turnover of key personnel, In # or %					
Percentage of labor hours on overtime					
Schedule variance (SV)					
Cost variance (CV)					
Schedule performance index (SPI)					
Cost performance index (CPI)					

The Need for Project Management Metrics (E)

FROM KPI SELECTION TO KPI MEASUREMENT

The team felt reasonably comfortable that they understood the differences between metrics and KPIs. Now came perhaps the biggest challenge: the need for KPI measurements. For decades, the only metrics the company looked at were time and cost metrics. The measurements came from time cards and were reported through the company's project management information system. Time and cost metrics were considered to be objective measurements, even though management often questioned how valuable they were in predicting the success or failure of a project.

The team knew the questions that now needed to be addressed:

Measurements

- What should be measured?
- When should it be measured?
- How should it be measured?
- Who will perform the measurement?

Collecting Information and Reporting

- Who will collect the information?
- When will the information be collected?
- When and how will the information be reported?

Patsy commented:

In marketing, the eight metrics we have on our scorecard are direct measurements and absolutely objective. But for project-based metrics, I believe that most of the metrics will come from highly subjective calculations. Perhaps after we use all of these project-based metrics for a while the way we calculate them will become more objective rather than subjective, but that may be years from now.

Also, I'm not sure we will be able to come up with measurements for all of the KPIs we selected. This may be a challenge beyond our team's capability. And even if we can come up with a measurement approach, how will we know if each project team can perform the measurement?

John had some experience with measurements and added:

Anything can be measured as long as we do not insist on perfect measurements. The alternative to perfect measurements is no measurements at all, and that's really bad. Work that gets measured gets done! If it cannot be measured, then it cannot be managed, and that would defeat the purpose of having KPIs. My experience is that you never really understand anything until you try to measure it.

In the past, we looked only at those metrics that were easy to measure, such as time and cost. Everything else was difficult to measure and therefore ignored. Now we are realizing that all metrics must somehow be measured and reported. Perhaps in the future more sophisticated measurement techniques will be available to us. But for now, we must use what we have and what we understand.

Allen knew that John's comments were correct and said:

There are numerous ways that measurements can be made. We use a variety of techniques in engineering. We can measure things in numbers, dollars, headcount, and ratings such as good, neutral or bad. Some measurements will be quantitative whereas others will be qualitative.

The team spent several hours looking at various measurement techniques for each of the KPIs. It was obvious that no single measurement method would be appropriate for all of them. The team was now somewhat perplexed as to what measurement techniques to look at. Carol then said:

We cannot separate measurement from reporting. We must look at them together. For example, if we are $15,000 over budget, is that really bad and should we give consideration to canceling this project? Perhaps $15,000 over budget is acceptable to management. Perhaps it might even be looked at as being good.

My concern is that all measurements should be made from a target or reference point so that we can determine if this is a good or bad situation. I believe that we must establish targets for each of the KPIs.

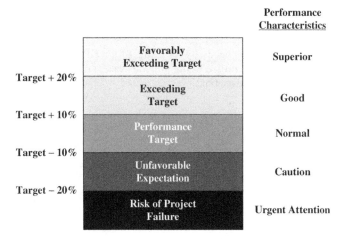

FIGURE I Generic boundary box

The team felt that Carol's comments provided them with some direction as to where to proceed. After a few more hours of deliberation, the team came up with a generic boundary box approach for establishing targets for each of the KPIs. The generic boundary box is shown in Figure I.

Each boundary box would have five levels. At the beginning of the project, the project manager would work with the sponsor or stakeholders to establish a reasonable performance target for this KPI. For example, if we consider cost as a KPI, then the actual cost ±10 percent might be considered an acceptable performance target and something management can live with. If we are under budget by 10 to 20 percent, then this could be considered as exceeding the target by a small amount. If we are under budget by more than 20 percent, then this could be regarded as superior performance.

The same scenario holds true if we are over budget. If we are over budget by 10 to 20 percent, this could be viewed as a caution, especially if this number becomes more unfavorable each reporting period. If the costs are over budget by more than 20 percent, then management may wish to consider canceling the project.

The team felt that this approach was workable. However, it would be highly subjective because on one project the normal range boundaries may be at ±10 percent but on another project they may be at ±5 percent. But everyone knew that, with experience, the ranges would be tightened up.

Most likely no single metric or KPI would dictate that cancellation would be necessary. However, looking at all of the KPIs together and possibly in combination with some other metrics, termination may be the only viable alternative.

QUESTIONS

1. Can any KPI be measured?
2. Does Exhibit I satisfy the necessity for a target for a KPI?
3. For each of the KPIs identified in the previous case, establish what you consider to be a reasonable boundary box.

The Need for Project Management Metrics (F)

NEED FOR KPIS THAT MEASURES VALUE

For almost a year, the company had some degree of success with the metrics for innovation projects, even though the measurements were highly subjective. The metrics management initiative team met occasionally to reassess performance and look for ways to improve the process. Decisions to cancel some projects that appeared unable to meet the objectives were being made much earlier than before. Resources were then assigned to other projects. Projects that went through to completion were generating sales. But management was still unhappy. The products that were being developed were not meeting sales goals. Perhaps some of the metrics had to be changed.

Al Grey met with the metrics management initiative team to express his concerns.

> The metrics that were developed seemed rather attractive at first. But as often happens, some of the more important metrics that are needed are not discovered until some time later. This is the case now.
>
> Many of the projects that went through to completion were done within the time and cost constraints. Unfortunately, being within time and cost does not mean that there will be value in the end result such that the customers will be pleased.
>
> We need not just products but products that possess the necessary value so that clients faced with a purchase decision will select our products. While we do not always know the customer's definition of value, we must still try to develop value-based metrics to allow us to work on the right projects. Value-based

metrics are also a necessity to help us make the right tradeoffs on some of these innovation projects.

The team was not sure where to begin. It took quite a bit of time to establish targets and measurements for the metrics and KPIs that were currently being used. But value-based metrics for innovation projects were something entirely new to most of the team. After a lengthy discussion, Patsy stated:

> We've done a few surveys in marketing, and the following five points were found to be important value contributors to our clients:
>
> 1. Product quality
> 2. Product cost
> 3. Product safety
> 4. Product features
> 5. Delivery date
>
>   I believe that we should be able to take advantage of this research for establishing some value metrics for projects. But we must remember that this list was generated from surveys of our customers based on their definition of quality. Measuring a project's value requires a rigorous value measurement approach. It is not just measuring value; it is measuring customer value. This list that I just gave you may have to change for use on our internal projects.

John then reinforced Patsy's comments by stating that the PMO is looking at a new way of doing the portfolio selection of projects. Each project must provide value in one or more of the four value quadrants: internal value, financial value, future value, and customer-oriented value. Many of the value characteristics are the same for more than one quadrant. As an example, the value contributors Patsy identified were applicable to future value and customer-oriented value.

The team believed that Patsy's and John's comments had tremendous merit. But using all five of these contributors as separate metrics would simply add more metrics to the list. The team came up with another idea. It would use the same five criteria mentioned, but product cost would be replaced by innovation project cost and delivery date would be replaced by innovation project completion date. Using this concept, weighting factors could be assigned to each component of the new value metric. That would show what percentage of the value metric came from each of these factors. As an example:

Product quality	10%
Project cost	20%
Product safety	20%
Product features	30%
Project completion date	20%

Now there was just one value metric, but it had five components. Using the generic boundary box in the previous case study, the team established a quantitative value measurement system where points could be assigned, such as:

	Points
Superior performance	4
Good performance	3
Standard performance	2
Below-par performance	1
Risk of failure	0

As an example, the team considered the five value components measured and reported in Table I:

TABLE I   FIVE VALUE COMPONENTS

Value Component	Weighting Factor (%)	Value Measurement	Value Contribution
Quality	10	3	0.3
Cost	20	2	0.4
Safety	20	4	0.8
Features	30	2	0.6
Timing	20	3	0.6
Totals	**100**		**2.70**

If all of the value components simply met their targets or were expected to meet their targets or standard performance, all value measurements would be 2.0 and the value contribution would be 2.0 as well, which is the standard target value. But since the total value contribution is 2.70, this project is producing added value and should be considered for continuation. If the value contribution were 1.75, then the project may not be producing the desired value (i.e., targeted value) and should be reevaluated for continuation.

This approach was also highly subjective in assigning the weighting factors and the boundary box measurement. But over time, it could become more of an objective rather than a subjective technique.

For this technique to work well, each project manager would have to work with his or her sponsor to determine the weighting factors. The weighting factors also could change during the project. For example, Table II shows how weighting factors can change if a project is currently having a significant schedule slippage or cost overruns.

If weighting was adjusted to a higher percentage in one area, there would have to be significant value contributions elsewhere to compensate for the potential problem. As the company gets more experienced with this approach, more

TABLE II WEIGHTING FACTORS

Value Component	Normal Weighting Factor (%)	Weighting Factors If There is Significant Schedule Slippage (%)	Weighting Factors If There is Significant Cost Overrun (%)
Quality	10	10	10
Cost	20	20	40
Safety	20	10	10
Features	30	20	20
Timing	20	40	20

than five value components can be used, and the decision on what value components to use will occur at the onset of each project.

The subjectivity of the approach still bothered some team members. The team then decided that, to reduce some of the subjectivity, there should be ranges in the weighting factors, as shown in Table III and where the nominal values need not be the middle value of the range:

TABLE III RANGE OF WEIGHTING VALUES

Value Component	Minimal Weighting Value (%)	Maximum Weighting Value (%)	Nominal Weighting Value (%)
Quality	10	40	20
Cost	10	50	20
Safety	10	4	20
Features	20	40	30
Schedule	10	50	20

Realizing that management would like to be briefed on their recommendations, team members prepared a list of topics to be discussed in the briefing:

- Every project will have just one value metric or value KPI.
- There will be a maximum of five components for each value metric.
- The weighting factors and measurement techniques will be established by the project manager and the stakeholders at the onset of the project.
- The target boundary boxes will be established by the project manager and the PMO.

Even though the team felt comfortable with this approach, there were still questions that needed to be addressed, but perhaps not immediately:

- What if only three of the five components of value could be measured at a point in time, such as in early life-cycle phases?

- Does the project need be a certain percentage complete before the value metric has any real meaning and should be considered?
- In cases where only some components can be measured, should the weighting factors be changed, normalized to 100 percent, or left alone?
- Who will make decisions regarding changes in the weighting factors as the project progresses through its life-cycle phases?
- Can the measurement technique for a given component change over each life-cycle phase, or must it be the same throughout the project?
- Can we reduce the subjectivity of the process?

Even though questions persisted, the team realized that some form of template had to be developed specifically for value metric reporting. After some deliberation, team members came up with the template shown in Figure I. Now they had to wait until this subjective approach was tried in several projects.

QUESTIONS

1. What factors would make this process more subjective than objective?
2. If these innovation projects were for external rather than internal clients, who should have more of an influence in the selection of the value components: the customers or the contractor doing the work?
3. Can the value component change over the life-cycle phases? If so, under what circumstances?
4. At what value contribution level would a project definitely be canceled?
5. Under what conditions would a project still be allowed to continue even if it falls below the acceptable value threshold limit?

Project Title:	Smart Phone Redesign
Project Manager:	Carol Grady
Planning Date:	November 12, 2010
Plan Revision Date:	January 15, 2011
Revision Number:	3

Comments:

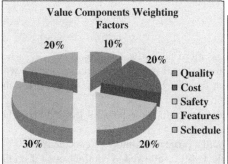

Value Components Weighting Factors

20%   10%   20%
- Quality
- Cost
- Safety
- Features
- Schedule
30%   20%

Boundary Box/Target Value

	Performance Characteristics	Value Points
Very Favorably Exceeding Target	Superior	4
Exceeding Target	Good	3
Performance Target	Normal	2
Unfavorable Expectation	Caution	1
Risk of Project Failure	Urgent Attention	0

Value Component	Weighting Factor	Measurement Technique	Value Measurement	Value Contribution
Quality	10%	Statistical Measurement	3	0.3
Cost	20%	Direct Counting	2	0.4
Safety	20%	Direct Measurement	4	0.8
Features	30%	Observation	2	0.6
Schedule	20%	Direct Counting	3	0.6

Total = 2.7

FIGURE I   Value metric report

The Need for Project Management Metrics (G)

NEED TO CANCEL PROJECTS

Even with the use of value metrics, the company realized that not many projects were being canceled. Everyone knew that creating products within time and cost constraints would be difficult. Although the company was good at linking its innovation projects to a business strategy, it was poor at cost estimating. Even when a project was selected and properly linked to a business strategy, there was still a fuzzy front end where detailed requirements were almost impossible to develop. On innovation projects, it was common practice to use "rolling-wave planning," where more detail was added to the requirements as the work progressed. Simply stated, if you can lay out a detailed plan for innovation, then you do not have an innovation project.

Estimating the time and especially the cost of an innovation breakthrough was almost impossible. Effective innovation leaders are those who have a fervent belief in the project, refuse to let the project die, and often use faulty rationalizations as to why the project should continue regardless of what the value metric measurement shows. Some people believe that effective innovation leaders are those who see a future that does not exist yet for what they are developing. This belief generates a reluctance to terminate projects.

The company needed to do a much better job of pulling the plug on projects. Every project seemed to develop a life of its own, and no one had the heart to cancel them, regardless of the value metric measurement results. No one seemed to have the authority to cancel the projects. Once a project was terminated, the company would ask, "Why didn't we do this earlier?" or "Why did we approve this project in the first place?" No best practices or lesson learned were ever captured related to mechanisms for canceling projects.

Al Grey met with the metric management team again and asked them for their assistance.

> We need to do a better job on canceling projects. I know this is not the reason why your team was created, but I value your input. Perhaps metrics management in another form is the solution, but I am not sure. I have looked at three mechanisms for canceling projects and perhaps you can give me your opinion of the advantages and disadvantages of each method.
>
> First, our senior people seem to get involved in these projects at a point where they can be the least helpful. They seem to avoid identification with any project that might damage their career. Their involvement appears only after they have someone to blame other than themselves if the project is terminated. So, in the first method, we could assign a project sponsor from the senior levels of management to each of these innovation projects, and the sponsor must then be involved all the way through.
>
> The second method involves lower and middle management. Right now, lower and middle management are backed into a corner because they may be involved in some of the projects yet have no decision-making authority for canceling them. To make matters worse, the people on project teams often are not honest with lower and middle management regarding the real status of the projects. Now senior management begins to wonder if there is frank disclosure coming up to their levels. Perhaps lower and middle management should serve as project sponsors and be actively involved in innovation projects from cradle to grave.
>
> Although project sponsorship seems like the right idea, I have read about some of the risks in assigning sponsors. The risks include:
>
> - Seeing what they want to see
> - Refusing to accept or admit defeat or failure
> - Viewing bad news as a personal failure
> - Fear of exposing mistakes to others
> - Viewing failure as a sign of weakness
> - Viewing failure as damage to one's reputation
> - Viewing failure as damage to one's career
>
> As a result, sponsors may not want to cancel projects. Therefore, perhaps we should assign an exit champion. The exit champion would be someone from the executive levels of management and a person who has no vested interest in the workings of the project. The exit champion will determine periodically if the project should continue. If the exit champion determines that cancellation is

the best option, then he or she will present the findings to the executive steering committee. The executive steering committee will have the authority to override the findings of the project sponsor in favor of the findings of the exit champion.

QUESTIONS

1. What are the advantages and disadvantages of each approach?
2. Which approach would you pick?
3. Can the exit champion use a different criterion, such as looking only at return on investment?

The Need for Project Management Metrics (H)

NEED FOR CAPACITY METRICS

Prosperity never comes without headaches. The new metrics measurement system was working quite well. Some projects were canceled, as expected, and most of the projects that went through to completion were considered successes. Actually, the company was becoming more successful than it had originally hoped, and this was placing a strain on manufacturing capacity.

Al Grey convened the metrics management team once again to see if any metrics could be created to assist with capacity planning. He stated:

> As you all know, the company has been relatively conservative in the past when it came to adding more manufacturing capacity. Sometimes we were too conservative and ended up with a large backlog of orders, which, in turn, alienated some of our customers. The success of our innovation processes has created more new products than we can currently manage with our existing capacity. Manufacturing personnel are working overtime and weekends to try to keep up with demand. We are managing somehow at present, but we have many new innovation projects in the queue. Are there any metrics we can develop and use during our innovation processes that will give us some insight on future capacity needs?

Al Grey handed out Figure I and explained that, with the existing products in manufacturing, capacity would be lagging demand by almost 10,000 units per year beginning in 2012. The conservative nature of the company was based on the

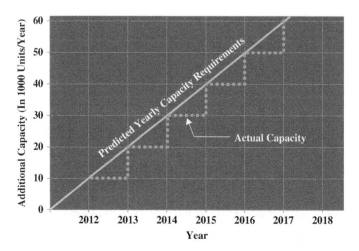

FIGURE I Capacity lagging demand

belief that unused capacity was a cost not worth considering. That thinking has since disappeared.

Al Grey then handed out Figure II, which showed that the company was considering yearly capacity increases beginning in 2012 to alleviate this pressure on

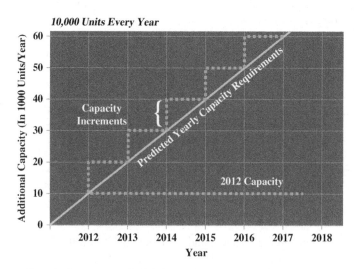

FIGURE II Capacity increments of 10,000 yearly to meet demand

manufacturing. Yearly increases of 10,000 units would satisfy current and projected demands for the company's existing products. Al Grey then handed out Figure III, which showed that adding capacity at 20,000 units every two years could be a better approach.

But there was still another capacity issue. With the success in the company's innovation processes, new products were being developed that could make the capacity problem worse. The company wanted metrics to show what capacity requirements would be needed for the projects that were now part of the innovation processes. The company knew the difficulties with capacity planning projections and knew that alternative sources of capacity could be used, as shown in Figure IV, but this would be just a temporary solution.

Al Grey then continued:

> The problem is more complex than just adding capacity. In the past, it has taken us between three months and six months from the time the innovation project is completed to the time when manufacturing begins and we can start delivering the products. During that window, we prepare our manufacturing plans and conduct our procurements. Sometimes procurement alone can last for three months or longer. So, what I am saying is that any metric you can provide on the overall probability of success and how much capacity we will need will be helpful. The company understands the risks in what we are asking you to do and is willing to accept those risks.

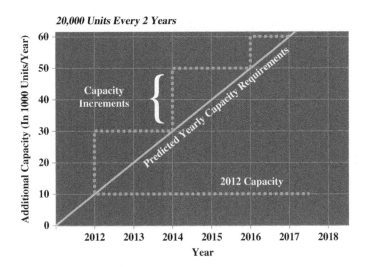

FIGURE III   Capacity increments of 20,000 units every two years

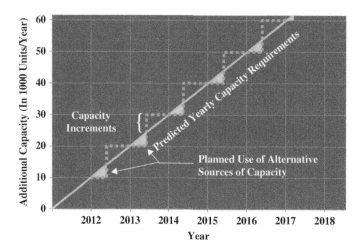

FIGURE IV Using alternative sources of capacity increments yearly

QUESTIONS

1. Can a metric for predicting success be developed?
2. Can a metric for predicting capacity needs be developed?
3. What are the risks assuming these two metrics can be created?

Part 11

PROJECT RISK MANAGEMENT

In today's world of project management, perhaps the single most important skill that a project manager can possess is risk management. Risk management includes identifying the risks, assessing the risks either quantitatively or qualitatively, choosing the appropriate method for handling the risks, and then monitoring and documenting the risks.

Effective risk management requires that the project manager be proactive and demonstrate a willingness to develop contingency plans, actively monitor the project, and be willing to respond quickly when a serious risk event occurs. Time and money are required for effective risk management to take place.

The Space Shuttle
Challenger Disaster

On January 28, 1986, the space shuttle *Challenger* lifted off the launch pad at 11:38 a.m., beginning the flight of Mission 51-L. Approximately 75 seconds into the flight, the *Challenger* was engulfed in an explosive burn, and all communication and telemetry ceased. Seven brave crewmembers lost their lives. On board the *Challenger* were Francis R. (Dick) Scobee (commander), Michael John Smith (pilot), Ellison S. Onizuka (mission specialist 1), Judith Arlene Resnik (mission specialist 2), Ronald Erwin McNair (mission specialist 3), S. Christa McAuliffe (payload specialist 1), and Gregory Bruce Jarvis (payload specialist 2). A faulty seal, or O-ring, on one of the two solid rocket boosters (SRBs) caused the accident.

Following the accident, significant energy was expended trying to ascertain whether the accident had been predictable. Controversy arose from the desire to assign, or to avoid, blame. Some publications called it a management failure, specifically in risk management, while others called it a technical failure.

Whenever accidents had occurred in the past at the National Aeronautics and Space Administration (NASA), an internal investigation team had been formed.

But in this case, perhaps because of the visibility, the White House took the initiative in appointing an independent commission. There did exist significant justification for the commission. NASA was in a state of disarray, especially in the management ranks. The agency had been without a permanent administrator for almost four months. The turnover rate at the upper echelons of management was high, and there seemed to be a lack of direction from the top down.

This mission had been known as the Teacher in Space mission, and Christa McAuliffe, a Concord, New Hampshire, schoolteacher, had been selected from a list of over 10,000 applicants. The nation knew the names of all of the crewmembers on board *Challenger.* The mission had been highly publicized for months, stating that Christa McAuliffe would be teaching students from aboard the *Challenger* on day 4 of the mission.

The Presidential Commission consisted of the following members:

- William P. Rogers, chairman: Former secretary of state under President Nixon and attorney general under President Eisenhower.
- Neil A. Armstrong, vice chairman: Former astronaut and spacecraft commander for Apollo 11.
- David C. Acheson: Former senior vice president and general counsel, Communications Satellite Corporation (1967–1974), and a partner in the law firm of Drinker Biddle & Reath.
- Dr. Eugene E. Covert: Professor and head, Department of Aeronautics and Astronautics at Massachusetts Institute of Technology.
- Dr. Richard P. Feynman: Physicist and professor of theoretical physics at California Institute of Technology; Nobel Prize winner in Physics, 1965
- Robert B. Hotz: Editor-in-chief of *Aviation Week & Space Technology* magazine (1953–1980).
- Major General Donald J. Kutyna, USAF: Director of Space Systems and Command, Control, Communications.
- Dr. Sally K. Ride: Astronaut and mission specialist on STS-7, launched on June 18, 1983, making her the first American woman in space. She also flew on mission 41-G, launched October 5, 1984. She held a doctorate in physics from Stanford University (1978) and was still an active astronaut.
- Robert W. Rummel: Vice president of Trans World Airlines and president of Robert W. Rummel Associates, Inc., of Mesa, Arizona.
- Joseph F. Sutter: Executive vice president of the Boeing Commercial Airplane Company.
- Dr. Arthur B. C. Walker, Jr.: Astronomer and professor of applied physics; formerly associate dean of the Graduate Division at Stanford University, and consultant to Aerospace Corporation, Rand Corporation, and the National Science Foundation.
- Dr. Albert D. Wheelon: Executive vice president, Hughes Aircraft Company.
- Brigadier General Charles Yeager, USAF (retired): Former experimental test pilot. He was the first person to break the sound barrier and the first to fly at a speed of more than 1,600 miles an hour.
- Dr. Alton G. Keel, Jr., Executive Director: Detailed to the Commission from his position in the Executive Office of the President, Office

of Management and Budget, as associate director for National Security and International Affairs; formerly assistant secretary of the Air Force for Research, Development and Logistics, and Senate Staff.

The Commission interviewed more than 160 individuals, and more than 35 formal panel investigative sessions were held generating almost 12,000 pages of transcript. Almost 6,300 documents totaling more than 122,000 pages, along with hundreds of photographs, were examined and made a part of the Commission's permanent database and archives. These sessions and all the data gathered added to the 2,800 pages of hearing transcript generated by the Commission in both closed and open sessions. Unless otherwise stated, all of the quotations and memos in this case study are from direct testimony cited in the *Report by the Presidential Commission* (*RPC*) (http://history.nasa.gov/rogersrep/genindex.htm).

BACKGROUND TO THE SPACE TRANSPORTATION SYSTEM

During the early 1960s, NASA's strategic plans for post-*Apollo* manned space exploration rested on a three-legged stool. The first leg was a reusable space transportation system, the space shuttle, which could transport people and equipment to low-Earth orbits and then return to Earth in preparation for the next mission. The second leg was a manned space station that would be resupplied by the space shuttle and serve as a launch platform for space research and planetary exploration. The third leg would be planetary exploration to Mars. But by the late 1960s, the United States was involved in the Vietnam War, which was becoming costly. In addition, confidence in the government was eroding because of civil unrest and assassinations. With limited funding due to budgetary cuts and with the lunar landing missions coming to an end, prioritization of projects was necessary. With a Democratic Congress continuously attacking the cost of space exploration and minimal support from President Nixon, the space program was left standing on one leg only, the space shuttle.

President Nixon made it clear that funding all the programs NASA envisioned would be impossible and that funding for even one program on the order of the *Apollo* program was likewise not possible. President Nixon seemed to favor the space station concept, but this required the development of a reusable space shuttle. Thus NASA's Space Shuttle Program became the near-term priority.

One of the reasons for the high priority given to the Space Shuttle Program was a 1972 study completed by Dr. Oskar Morgenstern and Dr. Klaus Heiss of the Princeton-based Mathematica organization. The study showed that the space shuttle would be able to orbit payloads for as little as $100 per pound based on 60 launches per year with payloads of 65,000 pounds. This provided tremendous promise for military applications such as reconnaissance and weather satellites as well as for scientific research.

Unfortunately, the pricing data were somewhat tainted. Much of the cost data were provided by companies that hoped to become NASA contractors and that therefore provided unrealistically low cost estimates in hopes of winning future bids. The actual cost per pound would prove to be more than 20 times the original estimate. Furthermore, the main engines never achieved the 109 percent of thrust that NASA desired, thus limiting the payloads to 47,000 pounds instead of the predicted 65,000 pounds. In addition, the European Space Agency began developing the capability to place satellites into orbit and began competing with NASA for the commercial satellite business.

NASA SUCCUMBS TO POLITICS AND PRESSURE

To retain shuttle funding, NASA was forced to make a series of major concessions. First, facing a highly constrained budget, NASA sacrificed the research and development (R&D) necessary to produce a truly reusable shuttle and instead accepted a design that was only partially reusable, eliminating one of the features that had made the shuttle attractive in the first place. SRBs were used instead of safer liquid-fueled boosters because they required a much smaller R&D effort. Numerous other design changes were made to reduce the level of R&D required.

Second, to increase its political clout and to guarantee a steady customer base, NASA enlisted the support of the United States Air Force. The Air Force could provide the considerable political clout of the Department of Defense and it used many satellites, which required launching. However, Air Force support did not come without a price. The shuttle payload bay was required to meet Air Force size and shape requirements, which placed key constraints on the ultimate design. Even more important was the Air Force requirement that the shuttle be able to launch from Vandenberg Air Force Base in California. This constraint required a larger cross range than the Florida site, which, in turn, decreased the total allowable vehicle weight. The weight reduction required the elimination of the design's air breathing engines, resulting in a single-pass unpowered landing. This greatly limited the safety and landing versatility of the vehicle.[1]

As the year 1986 began, there was extreme pressure on NASA to "Fly out the Manifest." From its inception, the Space Shuttle Program had been plagued by exaggerated expectations, funding inconsistencies, and political pressure. The ultimate vehicle and mission design were shaped almost as much by politics as by physics. President Kennedy's declaration that the United States would land a man on the moon before the end of the decade (the 1960s) had provided NASA's *Apollo* program with high visibility, a clear direction, and powerful political

---

[1] Kurt Hoover and Wallace T. Fowler, "Studies in Ethics, Safety and Liability for Engineers." www .tsgc.utexas. edu/archive/general/ethics/shuttle.html page 2.

backing. The Space Shuttle Program was not as fortunate; it had neither a clear direction nor consistent political backing.

Cost containment became a critical issue for NASA. In order to minimize cost, NASA designed a space shuttle system that utilized both liquid and solid propellants. Liquid-propellant engines are more easily controllable than solid-propellant engines. Flow of liquid propellant from the storage tanks to the engine can be throttled and even shut down in case of an emergency. Unfortunately, an all-liquid-fuel design was prohibitive because a liquid-fuel system is significantly more expensive to maintain than a solid-fuel system.

Solid-fuel systems are less costly to maintain. However, once a solid-propellant system is ignited, it cannot be easily throttled or shut down. Solid-propellant rocket motors burn until all of the propellant is consumed. This could have a significant impact on safety, especially during launch, at which time the SRBs are ignited and have maximum propellant loads. Also, SRBs can be designed for reusability, whereas liquid engines are generally used only once.

The final design that NASA selected was a compromise of both solid- and liquid-fuel engines. The space shuttle would be a three-element system composed of the orbiter vehicle, an expendable external liquid-fuel tank carrying[2] The orbiter's engines were liquid fuel because of the necessity for throttle capability. The two SRBs would provide the added thrust necessary to launch the space shuttle into its orbiting altitude.

In 1972, NASA selected Rockwell as the prime contractor for building the orbiter. Many industry leaders believed that other competitors that had actively participated in the *Apollo* program had a competitive advantage. Rockwell, however, was awarded the contract. Rockwell's proposal did not include an escape system. NASA officials decided against the launch escape system since it would have added too much weight to the shuttle at launch and was very expensive. There was also some concern on how effective an escape system would be if an accident occurred during launch when all of the engines were ignited. Thus, the Space Shuttle Program became the first U.S. manned spacecraft without a launch escape system for the crew.

In 1973, NASA went out for competitive bidding for the SRBs. The competitors were Morton-Thiokol, Inc. (MTI) (henceforth called Thiokol), Aerojet General, Lockheed, and United Technologies. The contract was eventually awarded to Thiokol because of its low cost, $100 million lower than the nearest competitor. Some believed that other competitors that ranked higher in technical design and safety should have been given the contract. NASA believed that Thiokol-built solid rocket motors would provide the lowest cost per flight.

---

[2] The terms "solid rocket booster" (SRB) and "solid rocket motor" (SRM) will be used interchangeably.

SOLID ROCKET BOOSTERS

Thiokol's SRBs had a height of approximately 150 feet and a diameter of 12 feet. The empty weight of each booster was 192,000 pounds, and the full weight was 1,300,000 pounds. Once ignited, each booster provided 2.65 million pounds of thrust, which is more than 70 percent of the thrust needed to lift off the launch pad.

Thiokol's design for the boosters was criticized by some competitors and even by some NASA personnel. The boosters were to be manufactured in four segments and then shipped from Utah to the launch site, where the segments would be assembled into a single unit. The Thiokol design was largely based on the segmented design of the Titan III solid rocket motor produced by United Technologies in the 1950s for Air Force satellite programs. Satellite programs were unmanned efforts.

The four solid rocket sections made up the case of the booster, which essentially encased the rocket fuel and directed the flow of the exhaust gases. This is shown in Figure I. The cylindrical shell of the case is protected from the propellant by a layer of insulation. The mating sections of the field joint are called the tang and the clevis. One hundred seventy-seven pins spaced around the circumference of each joint hold the tang and the clevis together. The joint is sealed in three ways. First, zinc chromate putty is placed in the gap between the mating segments and their insulation. This putty protects the second and third seals, which are rubberlike rings, called O-rings. The first O-ring is called the primary O-ring and is lodged in the gap between the tang and the clevis. The last seal is called the secondary O-ring, which is identical to the primary O-ring except it is positioned farther downstream in the gap. Each O-ring is 0.280 inches in diameter. The placement of each O-ring can be seen in Figure II. Another component of the field joint is called the leak check port, which is shown in Figure III. The leak check port is designed to allow technicians to check the status of the two O-ring seals. Pressurized air is inserted through the leak check port into the gap between the two O-rings. If the O-rings maintain the pressure and do not let the pressurized air past the seal, the technicians know the seal is operating properly.[3] In the Titan III assembly process, the joints between the segmented sections contained one O-ring. Thiokol's design had two O-rings instead of one. The second O-ring was initially considered redundant but was included to improve safety.

The purpose of the O-rings was to seal the space in the joints such that the hot exhaust gases could not escape and damage the case of the boosters.

Both the Titan III and shuttle O-rings were made of Viton rubber, which is an elastomeric material. For comparison, rubber is also an elastomer. The elastomeric material used is a fluoroelastomer, which is an elastomer that contains fluorine. This material was chosen because of its resistance to high temperatures

---

[3] "The *Challenger* Accident: Mechanical Causes of the *Challenger* Accident," University of Texas, http://www.me.utexas.edu/~uer/challenger/chall2.html, pp. 1–2.

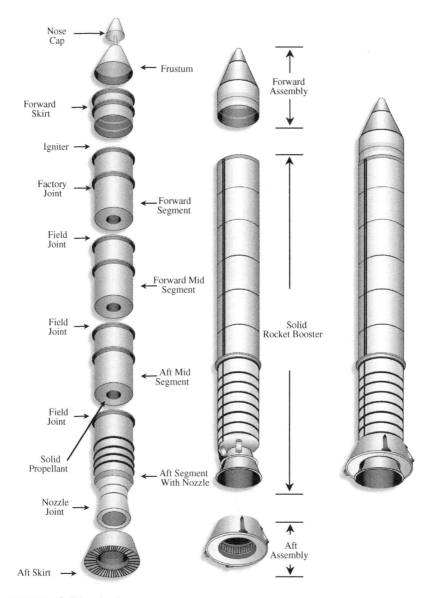

FIGURE I Solid rocket booster

and its compatibility with the surrounding materials. The Titan III O-rings were molded in one piece, whereas the shuttle's SRB O-rings would be manufactured in five sections and then glued together. Routinely, repairs would be necessary for inclusions and voids in the rubber received from the material suppliers.

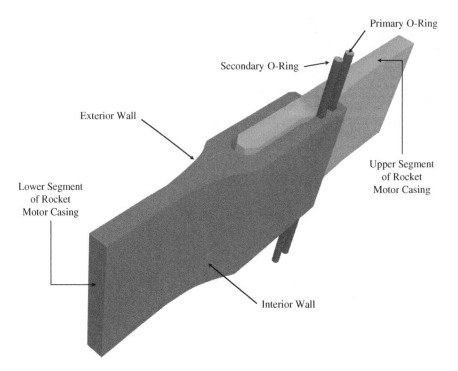

FIGURE II   Location of the O-rings

BLOWHOLES

The primary purpose of the zinc chromate putty was to act as a thermal barrier that protected the O-rings from the hot exhaust. As mentioned, the O-ring seals were tested using the leak check port to pressurize the gap between the seals. During the test, the secondary seal was pushed down into the same, seated position as it occupied during ignition pressurization. However, because the leak check port was between the two O-ring seals, the primary O-ring was pushed up and seated against the putty. The position of the O-rings during flight and their position during the leak check test is shown in Figure III. During early flights, engineers worried that, because the putty above the primary seal could withstand high pressures, the presence of the putty would prevent the leak test from identifying problems with the primary seal. They contended that the putty would seal the gap during testing regardless of the condition of the primary seal. Since the proper operation of the primary seal was essential, engineers decided to increase the pressure used during the test to above the pressure that the putty could withstand. This would ensure that the primary O-ring was properly sealing the gap without the aid of

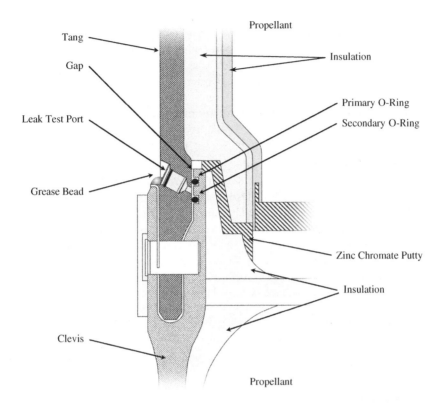

FIGURE III   Cross section showing the leak test port

the putty. Unfortunately, during this new procedure, the high-test pressures blew holes through the putty before the primary O-ring could seal the gap.

Since the putty was on the interior of the assembled SRB, technicians could not mend the blowholes in the putty. As a result, this procedure left small, tunneled holes in the putty. These holes would allow focused exhaust gases to contact a small segment of the primary O-ring during launch. Engineers realized that this was a problem but decided to test the seals at the high pressure despite the formation of blowholes rather than risking a launch with a faulty primary seal.

The purpose of the putty was to prevent the hot exhaust gases from reaching the O-rings. For the first nine successful shuttle launches, NASA and Thiokol used asbestos-bearing putty manufactured by the Fuller-O'Brien Company of San Francisco. However, because of the notoriety of products containing asbestos and the fear of potential lawsuits, Fuller-O'Brien stopped manufacturing the putty that had served the shuttle so well. This created a problem for NASA and Thiokol.

The new putty selected came from Randolph Products of Carlstadt, New Jersey. Unfortunately, with the new putty, blowholes and O-ring erosion were becoming more common to a point where the shuttle engineers became worried. Yet the new putty was still used on the boosters. Following the *Challenger* disaster, testing showed that, at low temperatures, the Randolph putty became much stiffer than the Fuller-O'Brien putty and lost much of its stickiness.[4]

O-RING EROSION

If the hot exhaust gases penetrated the putty and contacted the primary O-ring, the extreme temperatures would break down the O-ring material. Because engineers were aware of the possibility of O-ring erosion, the joints were checked after each flight for evidence of erosion. The amount of O-ring erosion found on flights before the new high-pressure leak check procedure was around 12 percent. After the new high-pressure leak test procedure, the percentage of O-ring erosion was found to increase by 88 percent. In some cases, high percentages of O-ring erosion allowed the exhaust gases to pass the primary O-ring and begin eroding the secondary O-ring. Some managers argued that some O-ring erosion was "acceptable" because the O-rings were found to seal the gap even if they were eroded by as much as one-third their original diameter.[5] The engineers believed that the design and operation of the joints were acceptable risks because a safety margin could be identified quantitatively. This numerical boundary would become an important precedent for future risk assessment.

JOINT ROTATION

During ignition, the internal pressure from the burning fuel applies approximately 1,000 pounds per square inch on the case wall, causing the walls to expand. Because the joints are generally stiffer than the case walls, each section tends to bulge out. The swelling of the solid rocket sections causes the tang and the clevis to become misaligned; this misalignment is called joint rotation. A diagram showing a field joint before and after joint rotation is seen in Figure IV. The problem with joint rotation is that it increases the gap size near the O-rings. This increase in size is extremely fast, which makes it difficult for the O-rings to follow the increasing gap and keep the seal.[6]

Prior to ignition, the gap between the tang and the clevis is approximately 0.004 inches. At ignition, the gap enlarges to between 0.042 and 0.060 inches *for a maximum of 0.60 seconds* and then returns to its original position.

---

[4] Ibid., p. 3.
[5] Ibid., p. 4.
[6] Ibid.

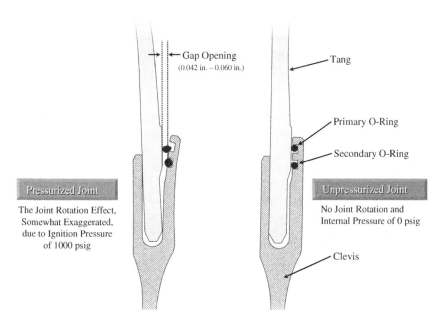

FIGURE IV   Field joint rotation

O-RING RESILIENCE

The term "O-ring resilience" refers to the ability of the O-ring to return to its original shape after it has been deformed. This property is analogous to the ability of a rubber band to return to its original shape after it has been stretched. As with a rubber band, the resiliency of an O-ring is related directly to its temperature. As the temperature of the O-ring gets lower, the O-ring material becomes stiffer. Tests have shown that an O-ring at 75°F is five times more responsive in returning to its original shape than an O-ring at 30°F. This decrease in O-ring resiliency during a cold-weather launch would make the O-ring much less likely to follow the increasing gap size during joint rotation. As a result of poor O-ring resiliency, the O-ring would not seal properly.[7]

THE EXTERNAL TANK

The solid rockets are each joined forward and aft to the external liquid fuel tank. They are not connected to the orbiter vehicle. The solid-rocket motors are mounted

---

[7] Ibid., pp. 4–5.

first, and the external liquid-fuel tank is put between them and connected. Then the orbiter is mounted to the external tank at two places in the back and one place forward. Those connections carry all of the structural loads for the entire system at liftoff and through the ascent phase of flight. Also connected to the orbiter, under the orbiter's wing, are two large propellant lines 17 inches in diameter. The one on the port side carries liquid hydrogen from the hydrogen tank in the back part of the external tank. The line on the right side carries liquid oxygen from the oxygen tank at the forward end, inside the external tank.[8]

The external tank contains about 1.6 million pounds of propellant, or about 526,000 gallons. The orbiter's three engines burn the liquid hydrogen and liquid oxygen at a ratio of 6:1 and at a rate equivalent to emptying out a family swimming pool every 10 seconds. Once ignited, the exhaust gases leave the orbiter's three engines at approximately 6,000 miles per hour. After the fuel is consumed, the external tank separates from the orbiter, falls to Earth, and disintegrates in the atmosphere on reentry.

THE SPARE PARTS PROBLEM

In March 1985, NASA's administrator, James Beggs, announced that there would be one shuttle flight per month for all of fiscal year 1985. In actuality, there were only six flights. Repairs became a problem. Continuous repairs were needed on the heat tiles required for reentry, the braking system, and the main engines' hydraulic pumps. Parts were routinely borrowed from other shuttles. The cost of spare parts was excessively high, and NASA was looking for cost containment.

RISK IDENTIFICATION PROCEDURES

The necessity for risk management was apparent right from the start. Prior to the launch of the first shuttle in April of 1981, hazards were analyzed and subjected to a formalized hazard reduction process as described in the NASA Handbook, NHB5300.4. The process required that the credibility and probability of the hazards be determined. A Senior Safety Review Board was established for overseeing the risk assessment process. For the most part, the risk assessment process was qualitative. The conclusion reached was that no single hazard or combination of hazards should prevent the launch of the first shuttle *as long as the aggregate risk remained acceptable.*

NASA used a rather simplistic safety (risk) classification system. A quantitative method for risk assessment was not in place at NASA because gathering the data needed to generate statistical models would be expensive and labor-intensive. If the risk identification procedures were overly complex, NASA would have been buried in paperwork due to the number of components on the space shuttle. The risk classification system selected by NASA is shown in Table I.

---

[8] *Report by the Presidential Commission (RPC),* p. 50.

TABLE I   RISK CLASSIFICATION SYSTEM

Level	Description
Criticality 1 (C1)	Loss of life and/or vehicle if the component fails.
Criticality 2 (C2)	Loss of mission if the component fails.
Criticality 3 (C3)	All others.
Criticality 1R (C1R)	Redundant components exist. The failure of both could cause loss of life and/or vehicle.
Criticality 2R (C2R)	Redundant components exist. The failure of both could cause loss of mission.

From 1982 on, the O-ring seal was labeled Criticality 1. By 1985, there were 700 components identified as Criticality 1.

TELECONFERENCING

The Space Shuttle Program involves a vast number of people at both NASA and the contractors. Because of the geographical separation between NASA and the contractors, it became impractical to have continuous meetings. Travel between Thiokol in Utah and the Cape in Florida took one day each way. Therefore, teleconferencing became the primary method of communication and a way of life. Interface meetings were still held, but the emphasis was on teleconferencing. All locations could be linked together in one teleconference and data could be faxed back and forth as needed.

PAPERWORK CONSTRAINTS

With the rather optimistic flight schedule provided to the news media, NASA was under scrutiny and pressure to deliver. For fiscal 1986, the mission manifest called for 16 flights. The pressure to meet schedule was about to take its toll. Safety problems had to be resolved quickly.

As the number of flights scheduled began to increase, so did the requirements for additional paperwork. The majority of the paperwork had to be completed prior to NASA's Flight Readiness Review (FRR) meetings. Approximately one week prior to every flight, flight operations and cargo managers were required to endorse the commitment of flight readiness to the NASA associate administrator for space flight at the FRR meeting. The responsible project/element managers would conduct pre-FRR meetings with their contractors, center managers, and the NASA Level II manager. The content of the FRR meetings included the following:

- Determine overall status as well as establish the baseline in terms of significant changes since the last mission.
- Review significant problems resolved since the last review and significant anomalies from the previous flight.

- Review all open items and constraints remaining to be resolved before the mission.
- Present all new waivers since the last flight.

NASA personnel were working excessive overtime, including weekends, to fulfill the paperwork requirements and prepare for the required meetings. As the number of space flights increased, so did the paperwork and overtime.

The paperwork constraints were affecting the contractors as well. Additional paperwork requirements existed for problem solving and investigations. On October 1, 1985, an interoffice memo was sent from Scott Stein, space booster project engineer at Thiokol, to Bob Lund, vice president for engineering at Thiokol, and other selected managers concerning the O-Ring Investigation Task Force:

> We are currently being hog-tied by paperwork every time we try to accomplish anything. I understand that for production programs, the paperwork is necessary. However, for a priority, short schedule investigation, it makes accomplishment of our goals in a timely manner extremely difficult, if not impossible. We need the authority to bypass some of the paperwork jungle. As a representative example of problems and time that could easily be eliminated, consider assembly or disassembly of test hardware by manufacturing personnel. . . . I know the established paperwork procedures can be violated if someone with enough authority dictates it. We did that with the DR system when the FWC hardware "Tiger Team" was established. If changes are not made to allow us to accomplish work in a reasonable amount of time, then the O-ring investigation task force will never have the potency necessary to resolve problems in a timely manner.[9]

Both NASA and the contractors were now feeling the pressure caused by the paperwork constraints.

ISSUING WAIVERS

One quick way of reducing paperwork and meetings was to issue a waiver. Historically, a waiver was a formalized process that allowed an exception to a rule, a specification, a technical criterion, or a risk. Waivers were ways to reduce excessive paperwork requirements. Project managers and contract administrators had the authority to issue waivers, often with the intent of bypassing standard protocols in order to maintain a schedule. The use of waivers had been in place well before the manned space program even began. What is important here was *not* NASA's use of the waiver but the *justification* for the waiver, given the risks.

NASA had issued waivers on both Criticality 1 status designations and launch constraints. In 1982, the SRBs were designated C1 by the Marshall Space Flight Center because failure of the O-rings could have caused loss of crew and the shuttle. This meant that the secondary O-rings were not considered redundant. The SRB project manager at Marshall, Larry Malloy, issued a waiver just in time

---

[9] Reproduced in *RPC*, p. 253

for the next shuttle launch to take place as planned. Later, the O-rings designation went from C1 to C1R (i.e., a redundant process), thus partially avoiding the need for a waiver. The waiver was a necessity to keep the shuttle flying according to the original manifest.

Having a risk identification of C1 was not regarded as a sufficient reason to cancel a launch. It simply meant that component failure could be disastrous. It implied that this might be a potential problem that needed attention. If the risks were acceptable, NASA could still launch. A more serious condition was the issuing of launch constraints. Launch constraints were official NASA designations for situations in which mission safety was a serious enough problem to justify a decision not to launch. But once again, a launch constraint did not imply that the launch should be delayed. It meant that this was an important problem and needed to be addressed.

Following the 1985 mission that showed O-ring erosion and exhaust gas blow-by, a launch constraint was imposed. Yet on each of the next five shuttle missions, NASA's Malloy issued a launch constraint waiver allowing the flights to take place on schedule without any changes to the O-rings.

Were the waivers a violation of serious safety rules just to keep the shuttle flying? The answer is *no*. NASA had protocols such as policies, procedures, and rules for adherence to safety. Waivers were also protocols but for the purpose of deviating from other existing protocols. Larry Malloy, his colleagues at NASA, and the contractors had no intentions of doing evil. Waivers were simply a way of saying that they believed that the risk is an *acceptable risk*.

The lifting of launch constraints and the issuance of waivers became the norm—standard operating procedure. Waivers became a way of life. If waivers were issued and the mission was completed successfully, then the same waivers would exist for the next flight and did not have to be brought up for discussion at the FRR meeting. The justification for the waivers seemed to be the similarity among flight launch conditions, temperature, and so on. Launching under similar conditions seemed to be important for the engineers at NASA and Thiokol because it meant that the forces acting on the O-rings were within their region of experience and could be correlated to existing data. The launch temperature effect on the O-rings was considered predictable and therefore constituted an acceptable risk to both NASA and Thiokol, thus perhaps eliminating costly program delays that would have resulted from having to redesign the O-rings. The completion of each shuttle mission added another data point to the region of experience, thus guaranteeing the same waivers on the next launch. Flying with acceptable risk became the norm in NASA's culture.

LAUNCH LIFTOFF SEQUENCE PROFILE: POSSIBLE ABORTS

During the countdown to liftoff, the launch team closely monitors weather conditions, not only at the launch site but also at touchdown sites, should the mission need to be prematurely aborted.

Dr. Feynman: "Would you explain why we are so sensitive to the weather?"

Mr. Moore (NASA's deputy administrator for space flight): "Yes, there are several reasons. I mentioned the return to the landing site. We need to have visibility if we get into a situation where we need to return to the landing site after launch, and the pilots and the commanders need to be able to see the runway and so forth. So, you need a ceiling limitation on it [i.e., weather].

"We also need to maintain specifications on wind velocity so we don't exceed crosswinds. Landing on a runway and getting too high of a crosswind may cause us to deviate off of the runway and so forth, so we have a crosswind limit. During ascent, assuming a normal flight, a chief concern is damage to tiles due to rain. We have had experiences in seeing what the effects of a brief shower can do in terms of the tiles. The tiles are thermal insulation blocks, very thick. A lot of them are very thick on the bottom of the orbiter. But if you have a raindrop and you are going at a very high velocity, it tends to erode the tiles, pock the tiles, and that causes us a grave concern regarding the thermal protection.

"In addition to that, you are worried about the turnaround time of the orbiters as well, because with the kind of tile damage that one could get in rain, you have an awful lot of work to do to go back and replace tiles back on the system. So, there are a number of concerns that weather enters into, and it is a major factor in our assessment of whether or not we are ready to launch."[10]

Approximately six to seven seconds prior to liftoff, the shuttle's main engines (liquid fuel) ignite. These engines consume one-half million gallons of liquid fuel. It takes nine hours prior to launch to fill the liquid-fuel tanks. At ignition, the engines are throttled up to 104 percent of rated power. Redundancy checks on the engines' systems are then made. The launch site ground complex and the orbiter's onboard computer complex check a large number of details and parameters about the main engines to make sure that everything is proper and that the main engines are performing as planned.

If a malfunction is detected, the system automatically goes into a shutdown sequence, and the mission is scrubbed. The primary concern at this point is to make the vehicle "safe." The crew remains on board and performs a number of functions to get the vehicle into a safe mode. These functions include making sure that all propellant and electrical systems are properly safed. Ground crews at the launch pad begin servicing the launch pad. Once the launch pad is in a safe condition, the hazard and safety teams begin draining the remaining liquid fuel out of the external tank.

If no malfunction is detected during this six-second period of liquid fuel burn, then a signal is sent to ignite the two SRBs, and liftoff occurs. For the next

---

[10] Ibid., p. 18.

two minutes, with all engines ignited, the shuttle goes through a Max Q, or high dynamic pressure phase, that exerts maximum pressure loads on the orbiter vehicle. Based on the launch profile, the main engines may be throttled down slightly during the Max Q phase to lower the loads.

After 128 seconds into the launch sequence, all of the solid fuel is expended and the SRB staging occurs. The SRB parachutes are deployed. The SRBs then fall back to Earth 162 miles from the launch site and are recovered for examination, cleaning, and reuse on future missions. The main liquid-fuel engines are then throttled up to maximum power. After 523 seconds into the liftoff, the external liquid-fuel tanks are essentially expended of fuel. The main engines are shut down. Ten to 18 seconds later, the external tank is separated from the orbiter and disintegrates on reentry into the atmosphere.

From a safety perspective, the most hazardous period is the first 128 seconds when the SRBs are ignited. Here's what Arnold Aldrich, manager of NASA's STS Program, Johnson Space Center, had to say:

Mr. Aldrich: "Once the shuttle system starts off the launch pad, there is no capability in the system to separate these [solid-propellant] rockets until they reach burnout. They will burn for two minutes and eight or nine seconds, and the system must stay together. There is not a capability built into the vehicle that would allow these to separate. There is a capability available to the flight crew to separate at this interface the orbiter from the tank, but that is thought to be unacceptable during the first stage when the booster rockets are on and thrusting. So, essentially the first two minutes and a little more of flight, the stack is intended and designed to stay together, and it must stay together to fly successfully."

Mr. Hotz: "Mr. Aldrich, why is it unacceptable to separate the orbiter at that stage?"

Mr. Aldrich: "It is unacceptable because of the separation dynamics and the rupture of the propellant lines. You cannot perform the kind of a clean separation required for safety in the proximity of these vehicles at the velocities and the thrust levels they are undergoing, [and] the atmosphere they are flying through. In that regime, it is the design characteristic of the total system."[11]

If an abort is deemed necessary during the first 128 seconds, the actual abort will not begin until *after* SRB staging has occurred, which is after 128 seconds into the launch sequence. Based on the reason and timing of an abort, options include those listed in Table II.

---

[11] Ibid., p. 51

TABLE II ABORT OPTIONS FOR SHUTTLE

Type of Abort	Landing Site
Once-around abort	Edwards Air Force Base
Transatlantic abort	DaKar
Transatlantic abort	Casablanca
Return to landing site	Kennedy Space Center

Arnold Aldrich was questioned on different abort profiles.

Chairman Rogers: "During the two-minute period, is it possible to abort through the orbiter?"

Mr. Aldrich: "You can abort for certain conditions. You can start an abort, but the vehicle won't do anything yet, and the intended aborts are built around failures in the main engine system, the liquid propellant systems and their controls. If you have a failure of a main engine, it is well detected by the crew and by the ground support, and you can call for a return-to-launch-site abort. That would be logged in the computer. The computer would be set up to execute it, but everything waits until the solids take you to altitude. At that time, the solids will separate in the sequence I described, and then the vehicle flies downrange some 400 miles, maybe 10 to 15 additional minutes, while all of the tank propellant is expelled through these engines.

"As a precursor to setting up the conditions for this return-to-launch-site abort to be successful towards the end of that burn downrange, using the propellants and the thrust of the main engines, the vehicle turns and actually points heads up back towards Florida. When the tank is essentially depleted, automatic signals are sent to close off the [liquid] propellant lines and to separate the orbiter, and the orbiter then does a similar approach to the one we are familiar with orbit back to the Kennedy Space Center for approach and landing."

Dr. Walker: "So, the propellant is expelled but not burned?"

Mr. Aldrich: "No, it is burned. You burn the system on two engines all the way down-range until it is gone, and then you turn around and come back because you don't have enough to burn to orbit. That is the return-to-launch-site abort, and it applies during the first 240 seconds of—no, 240 is not right. It is longer than that—the first four minutes, either before or after separation you can set that abort up, but it will occur after the solids separate, and if you have a main engine anomaly after the solids separate, at that time you can start the RTLS, and it will go through that same sequence and come back."

Dr. Ride: "And you can also only do an RTLS if you have lost just one main engine. So if you lose all three main engines, RTLS isn't a viable abort mode."

Mr. Aldrich: "Once you get through the four minutes, there's a period where you now don't have the energy conditions right to come back, and you have a forward abort, and Jesse mentioned the sites in Spain and on the coast of Africa. We have what is called a trans-Atlantic abort, and where you can use a very similar sequence to the one I just described. You still separate the solids, you still burn all the propellant out of the tanks, but you fly across and land across the ocean."

Mr. Hotz: "Mr. Aldrich, could you recapitulate just a bit here? Is what you are telling us that for two minutes of flight, until the solids separate, there is no practical abort mode?"

Mr. Aldrich: "Yes, sir."

Mr. Hotz: "Thank you."

Mr. Aldrich: "A trans-Atlantic abort can cover a range of just a few seconds up to about a minute in the middle where the across-the-ocean sites are effective, and then you reach this abort once-around capability where you go all the way around and land in California or back to Kennedy by going around the earth. And finally, you have abort-to-orbit where you have enough propulsion to make orbit but not enough to achieve the exact orbital parameters that you desire. That is the way that the abort profiles are executed.

"There are many, many nuances of crew procedure and different conditions and combinations of sequences of failures that make it much more complicated than I have described it."[12]

THE O-RING PROBLEM

There were two kinds of joints on the shuttle—field joints that were assembled at the launch site connecting together the SRB's cylindrical cases, and nozzle joints that connected the aft end of the case to the nozzle. During the pressure of ignition, the field joints could become bent such that the secondary O-ring could lose contact within an estimated 0.17 to 0.33 seconds after ignition. If the primary O-ring failed to seal properly before the gap within the joints opened up and the secondary seal failed, the results could be disastrous.

When the solid-propellant boosters are recovered after separation, they are disassembled and checked for damage. The O-rings could show evidence of coming into contact with heat. Hot gases from the ignition sequence could blow by the primary O-ring briefly before sealing. This blow-by phenomenon could last for only a few milliseconds before sealing and result in no heat damage to the O-ring. If the actual sealing process takes longer than expected, then charring and erosion of the O-rings can occur. This would be evidenced by gray or black soot and erosion to the O-rings. The terms used are "impingement erosion" and "bypass" erosion, with the latter identified also as "sooted blow-by."

---

[12] Ibid., pp. 51–52.

Roger Boisjoly of Thiokol describes blow-by erosion and joint rotation as follows:

> O-ring material gets removed from the cross section of the O-ring much, much faster than when you have bypass erosion or blow-by, as people have been terming it. We usually use the characteristic blow-by to define gas past it, and we use the other term [bypass erosion] to indicate that we are eroding at the same time. And so you can have blow-by without erosion, [and] you [can] have blow-by with erosion. . . .[13]
>
> At the beginning of the transient cycle [initial ignition rotation, up to 0.17 seconds] . . . [the primary O-ring] is still being attacked by hot gas, and it is eroding at the same time it is trying to seal, and it is a race between, will it erode more than the time allowed to have it seal.[14]

On January 24, 1985, STS 51-C [Flight No. 15] was launched at 51°F, which was the lowest temperature of any launch up to that time. Analyses of the joints showed evidence of damage. Black soot appeared between the primary and secondary O-rings. The engineers concluded that the cold weather had caused the O-rings to harden and move more slowly. This allowed the hot gases to blow by and erode the O-rings. This scorching effect indicated that low-temperature launches could be disastrous.

On July 31, 1985, Roger Boisjoly of Thiokol sent an interoffice memo to R. K. Lund, vice president for engineering at Thiokol:

> This letter is written to insure that management is fully aware of the seriousness of the current O-ring erosion problem in the SRM joints from an engineering standpoint.
>
> The mistakenly accepted position on the joint problem was to fly without fear of failure and to run a series of design evaluations which would ultimately lead to a solution or at least a significant reduction of the erosion problem. This position is now drastically changed as a result of the SRM 16A nozzle joint erosion which eroded a secondary O-ring with the primary O-ring never sealing.
>
> If the same scenario should occur in a field joint (and it could), then it is a jump ball as to the success or failure of the joint because the secondary O-ring cannot respond to the clevis opening rate and may not be capable of pressurization. The result would be a catastrophe of the highest order—loss of human life.
>
> An unofficial team [a memo defining the team and its purpose was never published] with [a] leader was formed on 19 July 1985 and was tasked with solving the problem for both the short and long term. This unofficial team is essentially nonexistent at this time. In my opinion, the team must be officially given the responsibility and the authority to execute the work that needs to be done on a non-interference basis (full time assignment until completed).

---

[13] Ibid., pp. 784–785.
[14] Ibid., p. 136.

It is my honest and very real fear that if we do not take immediate action to dedicate a team to solve the problem with the field joint having the number one priority, then we stand in jeopardy of losing a flight along with all the launch pad facilities.[15]

On August 9, 1985, a letter was sent from Brian Russell, manager of the SRM Ignition System, to James Thomas at the Marshall Space Flight Center. The memo addressed the following:

Per your request, this letter contains the answers to the two questions you asked at the July Problem Review Board telecon.

Question: If the field joint secondary seal lifts off the metal mating surfaces during motor pressurization, how soon will it return to a position where contact is re-established?

Answer: Bench test data indicate that the O-ring resiliency (its capability to follow the metal) is a function of temperature and rate of case expansion. MTI [Thiokol] measured the force of the O-ring against Instron plattens, which simulated the nominal squeeze on the O-ring and approximated the case expansion distance and rate.

At 100°F, the O-ring maintained contact. At 75°F, the O-ring lost contact for 2.4 seconds. At 50°F, the O-ring did not re-establish contact in 10 minutes at which time the test was terminated.

The conclusion is that secondary sealing capability in the SRM field joint cannot be guaranteed.

Question: If the primary O-ring does not seal, will the secondary seal seat in sufficient time to prevent joint leakage?

Answer: MTI has no reason to suspect that the primary seal would ever fail after pressure equilibrium is reached; i.e., after the ignition transient. If the primary O-ring were to fail from 0 to 170 milliseconds, there is a very high probability that the secondary O-ring would hold pressure since the case has not expanded appreciably at this point. If the primary seal were to fail from 170 to 330 milliseconds, the probability of the secondary seal holding is reduced. From 330 to 600 milliseconds the chance of the secondary seal holding is small. This is a direct result of the O-ring's slow response compared to the metal case segments as the joint rotates.[16]

At NASA, the concern for a solution to the O-ring problem became not only a technical crisis but also a budgetary crisis. In a July 23, 1985, memorandum from Richard Cook, program analyst, to Michael Mann, chief of the STS Resource Analysis Branch, the impact of the problem was noted:

Earlier this week you asked me to investigate reported problems with the charring of seals between SRB motor segments during flight operations. Discussions

---

[15] Ibid., pp. 691–692.
[16] Ibid., pp. 1568–1569.

with program engineers show this to be a potentially major problem affecting both flight safety and program costs.

Presently three seals between SRB segments use double O-rings sealed with putty. In recent Shuttle flights, charring of these rings has occurred. The O-rings are designed so that if one fails, the other will hold against the pressure of firing. However, at least in the joint between the nozzle and the aft segment, not only has the first O-ring been destroyed, but the second has been partially eaten away.

Engineers have not yet determined the cause of the problem. Candidates include the use of a new type of putty (the putty formerly in use was removed from the market by EPA [Environmental Protection Agency] because it contained asbestos), failure of the second ring to slip into the groove which must engage it for it to work properly, or new, and as yet unidentified, assembly procedures at Thiokol. MSC is trying to identify the cause of the problem, including on-site investigation at Thiokol, and OSF hopes to have some results from their analysis within thirty days. There is little question, however, that flight safety has been and is still being compromised by potential failure of the seals, and it is acknowledged that failure during launch would certainly be catastrophic. There is also indication that staff personnel knew of this problem sometime in advance of management's becoming apprised of what was going on.

The potential impact of the problem depends on the as yet undiscovered cause. If the cause is minor, there should be little or no impact on budget or flight rate. A worst case scenario, however, would lead to the suspension of shuttle flights, redesign of the SRB, and scrapping of existing stockpiled hardware. The impact on the FY 1987-8 budget could be immense.

It should be pointed out that Code M anagement [NASA's associate administrator for space flight] is viewing the situation with the utmost seriousness. From a budgetary standpoint, I would think that any NASA budget submitted this year for FY 1987 and beyond should certainly be based on a reliable judgment as to the cause of the SRB seal problem and a corresponding decision as to budgetary action needed to provide for its solution.[17]

On October 30, 1985, NASA launched Flight STS 61-A [Flight No. 22] at 75°F. This flight also showed signs of sooted blow-by, but the color was significantly blacker. Although there was some heat effect, there was no measurable erosion observed on the secondary O-ring. Since blow-by and erosion had now occurred at a higher launch temperature, the original premise that launches under cold temperatures were a problem was now being questioned. Table III shows the temperature at launch of all the shuttle flights up to this time and the O-ring damage, if any.

Management at both NASA and Thiokol wanted *concrete* evidence that launch temperature was directly correlated to blow-by and erosion. Other than

---

[17] Ibid., pp. 391–392.

TABLE III   EROSION AND BLOW-BY HISTORY (TEMPERATURE IN ASCENDING ORDER FROM COLDEST TO WARMEST)

Flight	Date	Temperature (°F)	Erosion Incidents	Blow-by Incidents	Comments
51-C	01/24/85	53	3	2	Most erosion any flight; blow-by secondary O-rings heated up
41-B	02/03/84	57	1		Deep, extensive erosion
61-C	01/12/86	58	1		O-rings erosion
41-C	04/06/84	63	1		O-rings heated but no damage
1	04/12/81	66			Coolest launch without problems
6	04/04/83	67			
51-A	11/08/84	67			
51-D	04/12/85	67			
5	11/11/82	68			
3	03/22/82	69			
2	11/12/81	70	1		Extent of erosion unknown
9	11/28/83	70			
41-D	08/30/84	70	1		
51-G	06/17/85	70			
7	06/18/83	72			
8	08/30/83	73			
51-B	04/29/85	75			
61-A	10/20/85	75		2	No erosion but soot between O-rings
51-1	08/27/85	76			
61	11/26/85	76			
41-G	10/05/84	78			
51-J	10/03/85	79			
4	06/27/82	80	No data; casing lost at sea		
51-F	07/29/85	81			

simply a gut feel, engineers were now stymied on how to show the direct correlation. NASA was not ready to cancel a launch simply due to an engineer's gut feel.

William Lucas, director of the Marshall Space Center, made it clear that NASA's manifest for launches would be adhered to. Managers at NASA were pressured to resolve problems internally rather than to escalate them up the chain of command. Managers became afraid to inform anyone higher up that they had a problem, even though they knew that one existed.

Richard Feynman, Nobel laureate and member of the Rogers Commission, concluded that a NASA official altered the safety criteria so that flights could be certified on time under pressure imposed by the leadership of William Lucas. Feynman commented:

> . . . They, therefore, fly in a relatively unsafe condition with a chance of failure of the order of one percent. Official management claims to believe that the probability of failure is a thousand times less.

Without concrete evidence of the temperature effect on the O-rings, the secondary O-ring was regarded as a redundant safety constraint, and the criticality factor was changed from C1 to C1R. Potentially serious problems were treated as anomalies peculiar to a given flight. Under the guise of anomalies, NASA began issuing waivers to maintain the flight schedules. Pressure was placed on contractors to issue closure reports. On December 24, 1985, L. O. Wear, NASA's SRM Program Office manager, sent a letter to Joe Kilminster, Thiokol's vice president for the Space Booster Program:

> During a recent review of the SRM Problem Review Board open problem list I found that we have 20 open problems, 11 opened during the past 6 months, 13 open over 6 months, 1 three years old, 2 two years old, and 1 closed during the past six months. As you can see our closure record is very poor. You are requested to initiate the required effort to assure more timely closures and the MTI personnel shall coordinate directly with the S&E personnel the contents of the closure reports.[18]

PRESSURE, PAPERWORK, AND WAIVERS

To maintain the flight schedule, critical issues such as launch constraints had to be resolved or waived. This would require extensive documentation. During the Rogers Commission investigation, it seemed that there had been a total lack of coordination between NASA's Marshall Space Center and Thiokol prior to the *Challenger* disaster. Joe Kilminster, Thiokol's vice president for the Space Booster Program, testified:

Mr. Kilminster: "Mr. Chairman, if I could, I would like to respond to that. In response to the concern that was expressed—and I had discussions with the team leader, the task force team leader, Mr. Don Kettner, and Mr. Russell and Mr. Ebeling. We held a meeting in my office and that was done in the October time period where we called the people who were in a support role to the task team, as well as the task force members themselves.

---

[18] Ibid., p. 1554.

"In that discussion, some of the task force members were looking to circumvent some of our established systems. In some cases, that was acceptable; in other cases, it was not. For example, some of the work that they had recommended to be done was involved with full-scale hardware, putting some of these joints together with various putty layup configurations; for instance, taking them apart and finding out what we could from that inspection process."

Dr. Sutter: "Was that one of these things that was outside of the normal work, or was that accepted as a good idea or a bad idea?"

Mr. Kilminster: "A good idea, but outside the normal work, if you will."

Dr. Sutter: "Why not do it?"

Mr. Kilminster: "Well, we were doing it. But the question was, can we circumvent the system, the paper system that requires, for instance, the handling constraints on those flight hardware items? And I said no, we can't do that. We have to maintain our handling system, for instance, so that we don't stand the possibility of injuring or damaging a piece of flight hardware.

"I asked at that time if adding some more people, for instance, a safety engineer—that was one of the things we discussed in there. The consensus was no, we really didn't need a safety engineer. We had the manufacturing engineer in attendance who was in support of that role, and I persuaded him that, typical of the way we normally worked, that he should be calling on the resources from his own organization, that is, in Manufacturing, in order to get this work done and get it done in a timely fashion.

"And I also suggested that if they ran across a problem in doing that, they should bubble that up in their management chain to get help in getting the resources to get that done. Now, after that session, it was my impression that there was improvement based on some of the concerns that had been expressed, and we did get quite a bit of work done. For your evaluation, I would like to talk a little bit about the sequence of events for this task force."

Chairman Rogers: "Can I interrupt? Did you know at that time it was a launch constraint, a formal launch constraint?"

Mr. Kilminster: "Not an overall launch constraint as such. Similar to the words that have been said before, each Flight Readiness Review had to address any anomalies or concerns that were identified at previous launches and in that sense, each of those anomalies or concerns were established in my mind as launch constraints unless they were properly reviewed and agreed upon by all parties."

Chairman Rogers: "You didn't know there was a difference between the launch constraint and just considering it an anomaly? You thought they were the same thing?"

Mr. Kilminster: "No, sir. I did not think they were the same thing."

Chairman Rogers: "My question is: Did you know that this launch constraint was placed on the flights in July 1985?"

Mr. Kilminster: "Until we resolved the O-ring problem on that nozzle joint, yes. We had to resolve that in a fashion for the subsequent flight before we would be okay to fly again."

Chairman Rogers: "So you did know there was a constraint on that?"

Mr. Kilminster: "On a one flight per one flight basis; yes, sir."

Chairman Rogers: "What else would a constraint mean?"

Mr. Kilminster: "Well, I get the feeling that there's a perception here that a launch constraint means all launches, whereas we were addressing each launch through the Flight Readiness Review process as we went."

Chairman Rogers: "No, I don't think—the testimony that we've had is that a launch constraint is put on because it is a very serious problem and the constraint means don't fly unless it's fixed or taken care of, but somebody has the authority to waive it for a particular flight. And in this case, Mr. Mulloy was authorized to waive it, which he did, for a number of flights before 51-L. Just prior to 51-L, the papers showed the launch constraint was closed out, which I guess means no longer existed. And that was done on January 23, 1986. Now, did you know that sequence of events?"

Mr. Kilminster: "Again, my understanding of *closing out,* as the term has been used here, was to close it out on the problem actions list, but not as an overall standard requirement. We had to address these at subsequent Flight Readiness Reviews to ensure that we were all satisfied with the proceeding to launch."

Chairman Rogers: "Did you understand the waiver process, that once a constraint was placed on this kind of a problem, that a flight could not occur unless there was a formal waiver?"

Mr. Kilminster: "Not in the sense of a formal waiver, no, sir."

Chairman Rogers: "Did any of you? Didn't you get the documents saying that?"

Mr. McDonald: "I don't recall seeing any documents for a formal waiver."[19]

MISSION 51-L

On January 25, 1986, questionable weather caused a delay of Mission 51-L to January 27. On January 26, the launch was reconfirmed for 9:37 a.m. on the 27th. However, on the morning of January 27, a malfunction with the hatch, combined with high crosswinds, caused another delay. All preliminary procedures had been completed and the crew had just boarded when the first problem appeared. A microsensor on the hatch indicated that the hatch was not shut securely. It turned out that the hatch was shut securely but the sensor had malfunctioned. Valuable time was lost in determining the problem.

After the hatch was finally closed, the external handle could not be removed. The threads on the connecting bolt were stripped and, instead of cleanly disengaging

---

[19] Ibid., pp. 1577–1578.

when turned, simply spun around. Attempts to use a portable drill to remove the handle failed. Technicians on the scene asked Mission Control for permission to saw off the bolt. Fearing some form of structural stress to the hatch, engineers made numerous time-consuming calculations before giving the go-ahead to cut off the bolt. The entire process consumed almost two hours before the countdown resumed.

However, the misfortunes continued. During the attempts to verify the integrity of the hatch and remove the handle, the wind had been steadily rising. Chief Astronaut John Young flew a series of approaches in the shuttle training aircraft and confirmed the worst fears of mission control. The crosswinds at the Cape were in excess of the level allowed for the abort contingency. The opportunity had been missed. The mission was then reset to launch the next day, January 28, at 9:38 a.m. Everyone was quite discouraged since extremely cold weather was forecast for Tuesday that could further postpone the launch.[20]

Weather conditions indicated that the temperature at launch could be as low as 26°F. This would be much colder and well below the temperature range that the O-rings were designed to operate in. The components of the solid rocket motors were qualified only to 40°F at the lower limit. Undoubtedly, when the sun came up and launch time approached, both the air temperature and vehicle would warm up, but there was still concern. Would the ambient temperature be high enough to meet the launch requirements? NASA's Launch Commit Criteria stated that no launch should occur at temperatures below 31°F. There were also worries over any permanent effects on the shuttle due to the cold overnight temperatures. NASA became concerned and asked Thiokol for their recommendation on whether or not to launch. NASA admitted under testimony that if Thiokol had recommended not launching, then the launch would not have taken place.

At 5:45 p.m. eastern standard time, a teleconference was held among the Kennedy Space Center, Marshall Space Flight Center, and Thiokol. Bob Lund, vice president for engineering, summarized the concerns of the Thiokol engineers that in Thiokol's opinion, the launch should be delayed until noontime or even later such that a launch temperature of at least 53°F could be achieved. Thiokol's engineers were concerned that no data were available for launches at this temperature of 26°F. This was the first time in 14 years that Thiokol had recommended not to launch.

The design validation tests originally done by Thiokol covered only a narrow temperature range. The temperature data did not include any temperatures below 53°F. The O-rings from Flight 51-C, which had been launched under cold conditions the previous year, showed very significant erosion. These were the only data available on the effects of cold, but all of the Thiokol engineers

---

[20] Hoover and Wallace, "Studies in Ethics, Safety and Liability for Engineers," pp. 3–4.

agreed that the cold weather would decrease the elasticity of the synthetic rubber O-rings, which in turn might cause them to seal slowly and allow hot gases to surge through the joint.[21]

Another teleconference was set up for 8:45 p.m. to invite more parties to be involved in the decision. Meanwhile, Thiokol was asked to fax all relevant and supporting charts to all parties involved in the 8:45 p.m. teleconference.

The following information was included in the pages that were faxed:

Blow-by History

SRM-15 Worst Blow-by
- Two case joints (80°), (110°) *Arc*
- Much worse visually than SRM-22

SRM-22 Blow-by
- Two case joints (30–40°)

SRM-13A, 15, 16A, 18, 23A, 24A
- Nozzle blow-by

Field Joint Primary Concerns—SRM-25
- A temperature lower than the current database results in changing primary O-ring sealing timing function
- SRM-15A—80° arc black grease between O-rings SRM-15B—110° arc black grease between O-rings
- Lower O-ring squeeze due to lower temp
- Higher O-ring shore hardness
- Thicker grease viscosity
- Higher O-ring pressure activation time
- If actuation time increases, threshold of secondary seal pressurization capability is approached.
- If threshold is reached then secondary seal may not be capable of being pressurized.

Conclusions

Temperature of O-ring is not only parameter controlling blow-by:
- SRM-15 with blow-by had an O-ring temp at 53°F.
- SRM-22 with blow-by had an O-ring temp at 75°F.
- Four development motors with no blow-by were tested at O-ring temp of 47° to 52°F.
- Development motors had putty packing which resulted in better performance.
- At about 50°F blow-by could be experienced in case joints.
- Temp for SRM-25 on 1-28-86 launch will be: 29°F 9 a.m.
  - 38°F 2 p.m.
- Have no data that would indicate SRM-25 is different than SRM-15 other than temp.

---

[21] Ibid., p. 4.

Recommendations
- O-ring temp must be ≥ 53°F at launch.
- Development motors at 47° to 52°F with putty packing had no blow-by.
- SRM-15 (the best simulation) worked at 53°F.
- Project ambient conditions (temp & wind) to determine launch time.

From NASA's perspective, the launch window was from 9:30 a.m. to 12:30 p.m. on January 28. This was based on weather conditions and visibility, not only at the launch site but also at the landing sites, should an abort be necessary. An additional consideration was the fact that the temperature might not reach 53°F prior to the launch window closing. Actually, the temperature at the Kennedy Space Center was not expected to reach 50°F until two days later. NASA was hoping that Thiokol would change its mind and recommend launch.

THE SECOND TELECONFERENCE

At the second teleconference, Bob Lund once again asserted Thiokol's recommendation not to launch below 53°F. NASA's Mulloy then burst out over the teleconference network: "My God, Morton Thiokol! When do you want me to launch—next April?"

NASA challenged Thiokol's interpretation of the data and argued that Thiokol was inappropriately attempting to establish a new Launch Commit Criterion just prior to launch. NASA asked Thiokol to reevaluate its conclusions. Crediting NASA's comments with some validity, Thiokol then requested a five- minute *off-line* caucus. In the room at Thiokol were 14 engineers, namely:

1. Jerald Mason, senior vice president, Wasatch Operations
2. Calvin Wiggins, vice president and general manager, Space Division
3. Joe C. Kilminster, vice president, Space Booster Programs
4. Robert K. Lund, vice president, Engineering
5. Larry H. Sayer, director, Engineering and Design
6. William Macbeth, manager, Case Projects, Space Booster Project
7. Donald M. Ketner, supervisor, Gas Dynamics Section and head Seal Task Force
8. Roger Boisjoly, member, Seal Task Force
9. Arnold R. Thompson, supervisor, Rocket Motor Cases
10. Jack R. Kapp, manager, Applied Mechanics Department
11. Jerry Burn, associate engineer, Applied Mechanics
12. Joel Maw, associate scientist, Heat Transfer Section
13. Brian Russell, manager, Special Projects, SRM Project
14. Robert Ebeling, manager, Ignition System and Final Assembly, SRB Project

There were no safety personnel in the room because nobody thought to invite them. The caucus lasted some 30 minutes. Thiokol (specifically Joe Kilminster) then returned to the teleconference stating that they were unable to sustain a valid argument that temperature affects O-ring blow-by and erosion. *Thiokol then reversed its position and was now recommending launch.*

NASA stated that the launch of the *Challenger* would not take place without Thiokol's approval. But when Thiokol reversed its position following the caucus and agreed to launch, NASA interpreted this as an acceptable risk. The launch would now take place.

Mr. McDonald (Thiokol): "The assessment of the data was that the data was not totally conclusive, that the temperature could affect everything relative to the seal. But there was data that indicated that there were things going in the wrong direction, and this was far from our experience base.

"The conclusion being that Thiokol was directed to reassess all the data because the recommendation was not considered acceptable at that time of [waiting for] the 53 degrees [to occur]. NASA asked us for a reassessment and some more data to show that the temperature in itself can cause this to be a more serious concern than we had said it would be. At that time Thiokol in Utah said that they would like to go off-line and caucus for about five minutes and reassess what data they had there or any other additional data.

"And that caucus lasted for, I think, a half hour before they were ready to go back on. When they came back on they said they had reassessed all the data and had come to the conclusions that the temperature influence, based on the data they had available to them, was inconclusive and therefore they recommended a launch."[22]

During the Rogers Commission testimony, NASA's Mulloy stated his thought process in requesting Thiokol to rethink their position:

General Kutyna: "You said the temperature had little effect?"

Mr. Mulloy: "I didn't say that. I said I can't get a correlation between O-ring erosion, blow-by and O-ring, and temperature."

General Kutyna: "51-C was a pretty cool launch. That was January of last year."

Mr. Mulloy: "It was cold before then but it was not that much colder than other launches."

General Kutyna: "So it didn't approximate this particular one?"

Mr. Mulloy: "Unfortunately, that is one you look at and say, aha, is it related to a temperature gradient and the cold. The temperature of the O-ring on 51-C, I believe, was 53 degrees. We have fired motors at 48 degrees."[23]

---

[22] *RPC*, p. 300.
[23] Ibid., p. 290

Mulloy asserted he had not pressured Thiokol into changing their position. Yet the testimony of Thiokol's engineers stated they believed they were being pressured.

Roger Boisjoly, one of Thiokol's experts on O-rings, was present during the caucus and vehemently opposed the launch. During testimony, Boisjoly described his impressions of what occurred during the caucus:

Mr. Boisjoly: "The caucus was started by Mr. Mason stating that a management decision was necessary. Those of us who were opposed to the launch continued to speak out, and I am specifically speaking of Mr. Thompson and myself because in my recollection, he and I were the only ones who vigorously continued to oppose the launch. And we were attempting to go back and rereview and try to make clear what we were trying to get across, and we couldn't understand why it was going to be reversed.

"So, we spoke out and tried to explain again the effects of low temperature. Arnie actually got up from his position which was down the table and walked up the table and put a quad pad down in front of the table, in front of the management folks, and tried to sketch out once again what his concern was with the joint, and when he realized he wasn't getting through, he just stopped.

"I tried one more time with the photos. I grabbed the photos and I went up and discussed the photos once again and tried to make the point that it was my opinion from actual observations that temperature was indeed a discriminator, and we should not ignore the physical evidence that we had observed.

"And again, I brought up the point that SRM-15 had a 110 degree arc of black grease, while SRM-22 had a relatively different amount, which was less and wasn't quite as black. I also stopped when it was apparent that I could not get anybody to listen."

Dr. Walker: "At this point did anyone else [i.e., engineers] speak up in favor of the launch?"

Mr. Boisjoly: "No, sir. No one said anything, in my recollection. Nobody said a word. It was then being discussed amongst the management folks. After Arnie and I had our last say, Mr. Mason said we have to make a management decision. He turned to Bob Lund and asked him to take off his engineering hat and put on his management hat. From this point on, management formulated the points to base their decision on. There was never one comment in favor, as I have said, of launching by any engineer or other nonmanagement person in the room before or after the caucus. I was not even asked to participate in giving any input to the final decision charts.

"I went back on the net with the final charts or final chart, which was the rationale for launching, and that was presented by Mr. Kilminster. It was handwritten on a notepad, and he read from that notepad. I did not agree with some of the statements that were being made to support the decision. I was never asked nor polled, and it was clearly a management decision from that point.

"I must emphasize, I had my say, and I never take any management right to take the input of an engineer and then make a decision based upon that input, and I truly believe that. I have worked at a lot of companies, and that has been done from time to time, and I truly believe that, and so there was no point in me doing anything any further [other] than [what] I had already attempted to do.

"I did not see the final version of the chart until the next day. I just heard it read. I left the room feeling badly defeated, but I felt I really did all I could to stop the launch. I felt personally that management was under a lot of pressure to launch, and they made a very tough decision, but I didn't agree with it.

"One of my colleagues who was in the meeting summed it up best. This was a meeting where the determination was to launch, and it was up to us to prove beyond a shadow of a doubt that it was not safe to do so. This is in total reverse to what the position usually is in a preflight conversation or a Flight Readiness Review. It is usually exactly opposite that."

Dr. Walker: "Do you know the source of the pressure on management that you alluded to?"

Mr. Boisjoly: "Well, the comments made over the net are what I felt. I can't speak for them, but I felt it. I felt the tone of the meeting exactly as I summed up, that we were being put in a position to prove that we should not launch rather than being put in the position and prove that we had enough data to launch."[24]

General Kutyna: "What was the motivation driving those who were trying to overturn your opposition?"

Mr. Boisjoly: "They felt that we had not demonstrated, or I had not demonstrated, because I was the prime mover in SRM-15. Because of my personal observations and involvement in the Flight Readiness Reviews, they felt that I had not conclusively demonstrated that there was a tie-in between temperature and blow-by.

"My main concern was if the timing function changed and that seal took longer to get there, then you might not have any seal left because it might be eroded before it seats. And then, if that timing function is such that it pushes you from the 170 millisecond region into the 330 second region, you might not have a secondary seal to pick up if the primary is gone. That was my major concern.

"I can't quantify it. I just don't know how to quantify that. But I felt that the observations made were telling us that there was a message there telling us that temperature was a discriminator, and I couldn't get that point across. I basically had no direct input into the final recommendation to launch, and I

---

[24] Ibid., pp. 793–794.

was not polled. "I think Astronaut Crippin hit the tone of the meeting exactly right on the head when he said that the opposite was true of the way the meetings were normally conducted. We normally have to absolutely prove beyond a shadow of a doubt that we have the ability to fly, and it seemed like we were trying to prove, have proved that we had data to prove that we couldn't fly at this time, instead of the reverse. That was the tone of the meeting, in my opinion."[25]

Jerald Mason, senior vice president at Thiokol's Wasatch Division, directed the caucus at Thiokol. Mason continuously asserted that a management decision was needed and instructed Bob Lund, vice president for engineering, to take off his engineering hat and put on his management hat. During testimony, Mason commented on his interpretation of the data:

Dr. Ride [a member of the Commission]: "You know, what we've seen in the charts so far is that the data was inconclusive and so you said go ahead."
Mr. Mason: ". . . I hope I didn't convey that. But the reason for the discussion was the fact that we didn't have enough data to quantify the effect of the cold, and that was the heart of our discussion. . . . We have had blow-by on earlier flights. We had not had any reason to believe that we couldn't experience it again at any temperature. . . ."[26]

At the end of the second teleconference, NASA's Hardy at Marshall Space Flight Center requested that Thiokol put their recommendation to launch in writing and fax it to both Marshall Space Flight Center and Kennedy Space Center. The list that follows is from a memo that was signed by Joe Kilminster, vice president for Thiokol's Space Booster Program, and faxed at 11:45 p.m. the night before the launch.

- Calculations show that SRM-25 O-rings will be 20° colder than SRM-15 O-rings.
- Temperature data not conclusive on predicting primary O-ring blow-by.

Engineering assessment is that:

- Colder O-rings will have increased effective durometer ("harder").
- "Harder" O-rings will take longer to "seat."
  - More gas may pass primary O-ring before the primary seal seats (relative to SRM-15).

---

[25] Ibid., p. 676.
[26] Ibid., p. 764.

- Demonstrated sealing threshold is three times greater than 0.038″ erosion experienced on SRM-15.
- If the primary seal does not seat, the secondary seal will seat.
  - Pressure will get to secondary seal before the metal parts rotate.
  - O-ring pressure leak check places secondary seal in outboard position, which minimizes sealing time.
- MTI recommends STS-51L launch proceed on 28 January 1986.
  - SRM-25 will not be significantly different from SRM-15.[27]

THE ICE PROBLEM

At 1:30 a.m. on the day of the launch, NASA's Gene Thomas, launch director, ordered a complete inspection of the launch site due to cold weather and severe ice conditions. The prelaunch inspection of the *Challenger* and the launch pad by the ice team was unusual, to say the least. The ice team's responsibility was to remove any frost or ice on the vehicle or launch structure. What they found during their inspection looked like something out of a science fiction movie. The freeze-protection plan implemented by Kennedy personnel had gone very wrong. Hundreds of icicles, some up to 16 inches long, clung to the launch structure. The handrails and walkways near the shuttle entrance were covered in ice, making them extremely dangerous if the crew had to make an emergency evacuation. One solid sheet of ice stretched from the 195-foot level to the 235-foot level on the gantry. However, NASA continued to cling to its calculations that there would be no damage due to flying ice shaken loose during the launch.[28] A decision was then made to delay the launch from 9:38 a.m. to 11:30 a.m. so that the ice on the launch pad could melt. The delay was still within the launch window of 9:30 a.m.–12:30 p.m.

At 8:30 a.m., a second ice inspection was made. Ice was still significantly present at the launch site. Robert Glaysher, vice president for orbital operations at Rockwell, stated that the launch was unsafe. Rockwell's concern was that falling ice could damage the heat tiles on the orbiter. This could have a serious impact during reentry.

At 10:30 a.m., a third ice inspection was made. Though some of the ice was beginning to melt, there was still significant ice on the launch pad. The temperature of the left SRB was measured at 33°F and the right booster was measured at 19°F. Even though the right booster was 34 degrees colder than Thiokol's original recommendation for a launch temperature (i.e., 53°F), no one seemed alarmed. Rockwell also agreed to launch, even though its earlier statement had been that the launch was unsafe.

---

[27] Ibid., p. 764.
[28] Hoover and Wallace, "Studies in Ethics, Safety and Liability for Engineers," p. 5.

Arnold Aldrich, manager of the STS Program at the Johnson Space Center, testified on the concern over the ice problem:

Mr. Aldrich: "Kennedy facility people at that meeting, everyone in that meeting, voted strongly to proceed and said they had no concern, except for Rockwell. The comment to me from Rockwell, which was not written specifically to the exact words, and either recorded or logged, was that they had some concern about the possibility of ice damage to the orbiter. Although it was a minor concern, they felt that we had no experience base launching in this exact configuration before, and therefore they thought we had some additional risk of orbiter damage from ice than we had on previous meetings, or from previous missions."

Chairman Rogers: "Did they sign off on it or not?"

Mr. Aldrich: "We don't *have* a sign-off at that point. It was not—it was not maybe 20 minutes, but it was close to that. It was within the last hour of launch."

Chairman Rogers: "But they still objected?"

Mr. Aldrich: "They issued what I would call a concern, a less than 100 percent concurrence in the launch. They did not say we do not want to launch, and the rest of the team overruled them. They issued a more conservative concern. They did not say don't launch."

General Kutyna: "I can't recall a launch that I have had where there was 100 percent certainty that everything was perfect, and everyone around the table would agree to that. It is the job of the launch director to listen to everyone, and it's our job around the table to listen and say there is this element of risk, and you characterize this as 90 percent, or 95, and then you get a consensus that that risk is an acceptable risk, and then you launch.

"So I think this gentleman is characterizing the degree of risk, and he's honest, and he had to say something."

Dr. Ride: "But one point is that their concern is a specific concern, and they weren't concerned about the overall temperature or damage to the solid rockets or damage to the external tank. They were worried about pieces of ice coming off and denting the tile."[29]

Following the accident, the Rogers Commission identified three major concerns about the ice-on-the-pad issue:

1. An analysis of all of the testimony and interviews established that Rockwell's recommendation on launch was ambiguous. The Commission found it difficult, as did Mr. Aldrich, to conclude that there was a

---

[29] Ibid., pp. 237–238.

no-launch recommendation. Moreover, all parties were asked specifically to contact Aldrich or Moore about launch objections due to weather. Rockwell made no phone calls or further objections to Aldrich or other NASA officials after the 9:00 a.m. Mission Management Team meeting and subsequent to the resumption of the countdown.

2. The Commission was also concerned about the NASA response to the Rockwell position at the 9:00 a.m. meeting. While it was understood that decisions have to be made in launching a Shuttle, the Commission was not convinced Levels I and II of NASA's management appropriately considered Rockwell's concern about the ice. However ambiguous Rockwell's position was, it was clear that they did tell NASA that the ice was an unknown condition. Given the extent of the ice on the pad, the admitted unknown effect of the Solid Rocket Motor and Space Shuttle Main Engines ignition on the ice, as well as the fact that debris striking the orbiter was a potential flight safety hazard, the Commission found the decision to launch questionable under those circumstances. In this situation, NASA appeared to be requiring a contractor to prove that it was not safe to launch, rather than proving it was safe. Nevertheless, the Commission had determined that the ice was not a cause of the 51-L accident and does not conclude that NASA's decision to launch specifically overrode a no-launch recommendation by an element contractor.

3. The Commission concluded that the freeze protection plan for launch pad 39B was inadequate. The Commission believed that the severe cold and presence of so much ice on the fixed service structure made it inadvisable to launch on the morning of January 28, and that margins of safety were whittled down too far.

It became obvious that NASA's management knew of the ice problem, but did they know of Thiokol's original recommendation not to launch and then its reversal? Larry Malloy, the SRB project manager for NASA, and Stanley Reinartz, NASA's manager of the Shuttle Office, both admitted that they told Arnold Aldrich, manager of the STS program, Johnson Space Center, about their concern for the ice problem, but there was no discussion about the teleconferences with Thiokol over the O-rings. It appeared that Malloy and Reinartz considered the ice as a potential problem whereas the O-rings constituted an acceptable risk. Therefore, only potential problems went up the chain of command, not the components of the "aggregate acceptable launch risk." It became common practice in FRR documentation to use the term "acceptable risk." This became the norm at NASA and resulted in insulating senior management from certain potential problems. The culture that had developed at NASA created the flawed decision-making process rather than an intent by individuals to withhold information and jeopardize safety.

THE ACCIDENT

Just after liftoff at 0.678 seconds into the flight, photographic data showed a strong puff of gray smoke spurting from the vicinity of the aft field joint on the right SRB. The two pad 39B cameras that would have recorded the precise location of the puff were inoperative. Computer graphic analysis of film from other cameras indicated the initial smoke came from the 270- to 310-degree sector of the circumference of the aft field joint of the right SRB. This area of the solid booster faced the external tank. The vaporized material streaming from the joint indicated there was incomplete sealing action within the joint.

Eight more distinctive puffs of increasingly blacker smoke were recorded between 0.836 and 2.500 seconds. The smoke appeared to puff upward from the joint. While each smoke puff was being left behind by the upward flight of the shuttle, the next fresh puff could be seen near the level of the joint. The multiple smoke puffs in this sequence occurred about four times per second, approximating the frequency of the structural load dynamics and resultant joint flexing. Computer graphics applied to NASA photos from a variety of cameras in this sequence again placed the smoke puffs' origin in the same 270- to 310-degree sector of the circumference as the original smoke spurt.

As the shuttle *Challenger* increased its upward velocity, it flew past the emerging and expanding smoke puffs. The last smoke was seen above the field joint at 2.733 seconds.

The black color and dense composition of the smoke puffs suggested that the grease, joint insulation, and rubber O-rings in the joint seal were being burned and eroded by the hot propellant gases.

At approximately 37 seconds, *Challenger* encountered the first of several high-altitude wind-shear conditions that lasted about 64 seconds. The wind shear created forces of relatively large fluctuations on the vehicle itself. These were immediately sensed and countered by the guidance, navigation, and control systems.

The steering system (thrust vector control) of the SRB responded to all commands and wind-shear effects. The wind shear caused the steering system to be more active than on any previous flight.

Both the *Challenger*'s main engines and the solid rockets operated at reduced thrust approaching and passing through the area of maximum dynamic pressure of 720 pounds per square foot. Main engines had been throttled up to 104 percent thrust, and the SRBs were increasing their thrust when the first flickering flame appeared on the right SRB in the area of the aft field joint. This first very small flame was detected on image-enhanced film at 58.788 seconds into the flight. It appeared to originate at about 305 degrees around the booster circumference at or near the aft field joint.

One film frame later from the same camera, the flame was visible without image enhancement. It grew into a continuous, well-defined plume at 59.262

seconds. At approximately the same time (60 seconds), telemetry showed a pressure differential between the chamber pressures in the right and left boosters. The right booster chamber pressure was lower, confirming the growing leak in the area of the field joint.

As the flame plume increased in size, it was deflected rearward by the aerodynamic slipstream and circumferentially by the protruding structure of the upper ring attaching the booster to the external tank. These deflections directed the flame plume onto the surface of the external tank. This sequence of flame spreading is confirmed by analysis of the recovered wreckage. The growing flame also impinged on the strut attaching the SRB to the external tank.

The first visual indication that swirling flame from the right SRB breached the external tank was at 64.660 seconds, when there was an abrupt change in the shape and color of the plume. This indicated that it was mixing with leaking hydrogen from the external tank. Telemetered changes in the hydrogen tank pressurization confirmed the leak. Within 45 milliseconds of the breach of the external tank, a bright, sustained glow developed on the black tiled underside of the *Challenger* between it and the external tank.

Beginning around 72 seconds, a series of events occurred extremely rapidly that terminated the flight. Telemetered data indicated a wide variety of flight system actions that supported the visual evidence of the photos as the shuttle struggled futilely against the forces that were destroying it.

At about 72.20 seconds, the lower strut linking the SRB and the external tank was severed or pulled away from the weakened hydrogen tank, permitting the right SRB to rotate around the upper attachment strut. This rotation was indicated by divergent yaw and pitch rates between the left and right SRBs.

At 73.124 seconds, a circumferential white vapor pattern was observed blooming from the side of the external tank bottom dome. This was the beginning of the structural failure of the hydrogen tank that culminated in the entire aft dome dropping away. This released massive amounts of liquid hydrogen from the tank and created a sudden forward thrust of about 2.8 million pounds, pushing the hydrogen tank upward into the intertank structure. About the same time, the rotating right SRB impacted the intertank structure and the lower part of the liquid oxygen tank. These structures failed at 73.137 seconds, as evidenced by the white vapors appearing in the intertank region.

Within milliseconds there was massive, almost explosive, burning of the hydrogen streaming from the failed tank bottom and the liquid oxygen breach in the area of the intertank.

At this point in its trajectory, while traveling at a Mach number of 1.92 at an altitude of 46,000 feet, the *Challenger* was totally enveloped in the explosive burn. The *Challenger*'s reaction control system ruptured, and a hypergolic burn of its propellants occurred, producing the oxygen-hydrogen flames. The reddish brown colors of the hypergolic fuel burn were visible on the edge of the main fireball. The orbiter, under severe aerodynamic loads, broke into several large sections, which emerged from the fireball. Separate sections that can be identified on film include the main engine/tail section with the engines still burning, one wing of the orbiter, and the forward fuselage trailing a mass of umbilical lines pulled loose from the payload bay.

The consensus of the Commission and participating investigative agencies was that the loss of the space shuttle *Challenger* was caused by a failure in the joint between the two lower segments of the right solid rocket motor. The specific failure was the destruction of the seals that were intended to prevent hot gases from leaking through the joint during the propellant burn of the rocket motor. The evidence assembled by the Commission indicates that no other element of the space shuttle system contributed to this failure.

In arriving at this conclusion, the Commission reviewed in detail all available data, reports, and records; directed and supervised numerous tests, analyses, and experiments by NASA, civilian contractors, and various government agencies; and then developed specific failure scenarios and the range of most probably causative factors.

The failure was due to a faulty design unacceptably sensitive to a number of factors. These factors were the effects of temperature, physical dimensions, the character of materials, the effects of reusability, processing, and the reaction of the joint to dynamic loading.

NASA AND THE MEDIA

Following the tragedy, many believed that NASA's decision to launch had been an attempt to minimize further ridicule by the media. Successful shuttle flights were no longer news because they were almost ordinary. However, launch aborts and delayed landings were more newsworthy because they were less common. The *Columbia* launch, which had immediately preceded the *Challenger* mission, had been delayed seven times. The *Challenger* launch had gone through four delays already. News anchor personnel were criticizing NASA. Some believed that NASA felt it had to do something quickly to dispel its poor public image.

The *Challenger* mission had had more media coverage and political ramifications than other recent missions. This would be the launch of the Teacher in Space Project. The original launch date of the *Challenger* had been scheduled just before President Reagan's State of the Union message, which was to be delivered the evening of January 28. Some believed that the president had intended to publicly praise NASA for the Teacher in Space Project and possibly even talk to Ms. McAuliffe live during his address. This would certainly have enhanced NASA's image. Following the tragedy, there were questions as to whether the White House had pressured NASA into launching the shuttle because of President Reagan's (and NASA's) love of favorable publicity. The Commission, however, found no evidence of White House intervention in the decision to launch.

FINDINGS OF THE COMMISSION

Determining the cause of an engineering disaster can take years of investigation. The *Challenger* disaster arose from many factors, including launch conditions, mechanical failure, faulty communication, and poor decision making. In the end, the last-minute decision to launch combined all possible factors into a lethal action.

The Commission concluded that the accident was rooted in history. The space shuttle's SRB problem began with the faulty design of its joint and increased as both NASA and contractor management first failed to recognize that they had a problem, then failed to fix it, and finally treated it as an acceptable flight risk.

Morton Thiokol, Inc., the contractor, did not accept the implication of tests early in the program that the design had a serious and unanticipated flaw. NASA did not accept the judgment of its engineers that the design was unacceptable, and as the joint problems grew in number and severity, NASA minimized them in management briefings and reports. Thiokol's stated position was that "the condition is not desirable but is acceptable."

Neither Thiokol nor NASA expected the rubber O-rings sealing the joints to be touched by hot gases of motor ignition, much less to be partially burned. However, as tests and then flights confirmed damage to the sealing rings, the reaction by both NASA and Thiokol was to increase the amount of damage considered "acceptable." At no time did management either recommend a redesign of the joint or call for the shuttle's grounding until the problem was solved.

The genesis of the *Challenger* accident—the failure of the joint of the right solid rocket motor—lay in decisions made in the design of the joint and in the failure by both Thiokol and NASA's Solid Rocket Booster project office to understand and respond to facts obtained during testing.

The Commission concluded that neither Thiokol nor NASA had responded adequately to internal warnings about the faulty seal design. Furthermore, Thiokol and NASA did not make a timely attempt to develop and verify a new seal after the initial design was shown to be deficient. Neither organization developed a solution to the unexpected occurrences of O-ring erosion and blow-by, even though this problem was experienced frequently during the shuttle flight history. Instead, Thiokol and NASA management came to accept erosion and blow-by as unavoidable and an acceptable flight risk. Specifically, the Commission found six things:

1. The joint test and certification program was inadequate. There was no requirement to configure the qualifications test motor as it would be in flight, and the motors were static tested in a horizontal position, not in the vertical flight position.
2. Prior to the accident, neither NASA nor Thiokol fully understood the mechanism by which the joint sealing action took place.

3. NASA and Thiokol accepted escalating risk apparently because they "got away with it last time." As Commissioner Feynman observed, the decision making was:

A kind of Russian roulette. . . . [The shuttle] flies [with O-ring erosion] and nothing happens. Then it is suggested, therefore, that the risk is no longer so high for the next flights. We can lower our standards a little bit because we got away with it last time. . . . You got away with it, but it shouldn't be done over and over again like that.

4. NASA's system for tracking anomalies for Flight Readiness Reviews failed in that, despite a history of persistent O-ring erosion and blow-by, flight was still permitted. It failed again in the strange sequence of six consecutive launch constraint waivers prior to 51-L, permitting it to fly without any record of a waiver, or even of an explicit constraint. Tracking and continuing only anomalies that are outside the database of prior flight allowed major problems to be removed from, and lost by, the reporting system.

5. The O-ring erosion history presented to Level I at NASA Headquarters in August 1985 was sufficiently detailed to require corrective action prior to the next flight.

6. A careful analysis of the flight history of O-ring performance would have revealed the correlation of O-ring damage and low temperature. Neither NASA nor Thiokol carried out such an analysis; consequently, they were unprepared to properly evaluate the risks of launching the 51-L mission in conditions more extreme than they had encountered before.

The Commission also identified a concern for the "silent" safety program. The Commission was surprised to realize after many hours of testimony that NASA's safety staff was never mentioned. No witness related the approval or disapproval of the reliability engineers, and none expressed the satisfaction or dissatisfaction of the quality assurance staff. No one thought to invite a safety representative or a reliability and quality assurance engineer to the January 27, 1986, teleconference between Marshall and Thiokol. Similarly, there was no safety representative on the Mission Management Team that made key decisions during the countdown on January 28, 1986.

The unrelenting pressure to meet the demands of an accelerating flight schedule might have been handled adequately by NASA if it had insisted on the exactingly thorough procedures that had been its hallmark during the *Apollo* program. An extensive and redundant safety program comprising interdependent safety, reliability, and quality assurance functions had existed during the lunar program to discover any potential safety problems. Between that period and 1986, however, the safety program had become ineffective. This loss of effectiveness seriously degraded the checks and balances essential for maintaining flight safety. On April 3, 1986, Arnold Aldrich, the Space Shuttle Program manager, appeared before the

Commission at a public hearing in Washington, D.C. He described five different communication or organization failures that affected the launch decision on January 28, 1986. Four of those failures related directly to faults within the safety program. These faults included a lack of problem reporting requirements, inadequate trend analysis, misrepresentation of criticality, and lack of involvement in critical discussions. A robust safety organization that was properly staffed and supported might well have avoided these faults, and thus eliminated the communication failures.

NASA had a safety program to ensure that the communication failures to which Mr. Aldrich referred did not occur. In the case of Mission 51-L, however, that program fell short.

The Commission concluded that there were severe pressures placed on the launch decision-making system to maintain a flight schedule. These pressures caused rational men to make irrational decisions.

With the 1982 completion of the orbital flight test series, NASA began a planned acceleration of the space shuttle launch schedule. One early plan contemplated an eventual rate of a mission a week, but realism forced several downward revisions. In 1985, NASA published a projection calling for an annual rate of 24 flights by 1990. Long before the *Challenger* accident, however, it was becoming obvious that even the modified goal of two flights a month was overambitious.

In establishing the schedule, NASA had not provided adequate resources. As a result, the capabilities of the launch decision-making system were strained by the modest nine-mission rate of 1985, and evidence suggested that NASA would not have been able to accomplish the 15 flights scheduled for 1986. These were the major conclusions of a Commission examination of the pressures and problems attendant upon the accelerated launch schedule:

1. The capabilities of the launch decision-making system were stretched to the limit to support the flight rate in winter 1985/1986. Projections into the spring and summer of 1986 showed a clear trend; the system, as it existed, would have been unable to deliver crew training software for scheduled flights by the designated dates. The result would have been an unacceptable compression of the time available for the crews to accomplish their required training.
2. Parts were in critically short supply. The shuttle program made a conscious decision to postpone spare parts procurements in favor of budget items of perceived higher priority. Lack of spare parts would likely have limited flight operations in 1986.
3. Stated manifesting policies were not enforced. Numerous late manifest changes (after the cargo integration review) had been made to both major payloads and minor payloads throughout the shuttle program:
   ● Late changes to major payloads or program requirements required extensive resources (money, manpower, facilities) to implement.

- If many late changes to "minor" payloads occurred, resources were quickly absorbed.
- Payload specialists frequently were added to a flight well after announced deadlines.
- Late changes to a mission adversely affected the training and development of procedures for subsequent missions.

4. The scheduled flight rate did not accurately reflect the capabilities and resources.
   - The flight rate was not reduced to accommodate periods of adjustment in the capacity of the workforce. There was no margin for error in the system to accommodate unforeseen hardware problems.
   - Resources were directed primarily toward supporting the flights; thus, not enough were available to improve and expand facilities needed to support a higher flight rate.

5. Training simulators may have been the limiting factor on the flight rate: The two simulators available at that time could not train crews for more than 12 to 15 flights per year.

6. When flights came in rapid succession, the requirements then current did not ensure that critical anomalies occurring during one flight would be identified and addressed appropriately before the next flight.

CHAIN-OF-COMMAND COMMUNICATION FAILURE

The Commission also identified a communication failure within the reporting structure at both NASA and Thiokol. Part of the problem with the chain-of-command structure was the idea of the proper reporting channel. Engineers report only to their immediate managers, while those managers report only to their direct supervisors. Engineers and managers believed in the chain-of-command structure; they felt reluctant to go above their superiors with their concerns. Boisjoly at Thiokol and Powers at Marshall felt that they had done all that they could as far as voicing their concerns. Anything more could have cost them their jobs. When questioned at the Rogers Commission hearing about why he did not voice his concerns to others, Powers replied, "That would not be my reporting channel." The chain-of-command structure dictated the only path that information could travel at both NASA and Thiokol. If information was modified or silenced at the bottom of the chain, there was not an alternate path for it to take to reach high-level officials at NASA. The Rogers Commission concluded that there was a breakdown in communication between Thiokol engineers and top NASA officials and faulted the management structure for not allowing important information about the SRBs to flow to the people who needed to know it. The Commission reported that the "fundamental problem was poor technical decision-making over a period of several years by top NASA and contractor personnel."

Bad news does not travel well in organizations like NASA and Thiokol. When the early signs of problems with the SRBs appeared, Thiokol managers did not believe that the problems were serious. Thiokol did not want to accept the fact that there could be a problem with its boosters. When Marshall received news of the problems, it considered it Thiokol's problem and did not pass the bad news upward to NASA headquarters. At Thiokol, Boisjoly described his managers as shutting out the bad news. He claims that he argued about the importance of the O-ring seal problems until he was convinced that "no one wanted to hear what he had to say." When Lund finally decided to recommend delay of the launch to Marshall, managers at Marshall rejected the bad news and refused to accept the recommendation not to launch. As with any information going up the chain of command at these two organizations, bad news was often modified so that it had less impact, perhaps skewing its importance.[30]

On January 31, 1986, President Ronald Reagan stated:

> The future is not free: the story of all human progress is one of a struggle against all odds. We learned again that this America, which Abraham Lincoln called the last, best hope of man on Earth, was built on heroism and noble sacrifice. It was built by men and women like our seven star voyagers, who answered a call beyond duty, who gave more than was expected or required and who gave it with little thought of worldly reward.[31]

EPILOGUE

Following the tragic accident, virtually every senior manager involved in the space shuttle *Challenger* decision-making processes, at both NASA and Thiokol, accepted early retirement. Whether this was the result of media pressure, peer pressure, fatigue, or stress. we can only postulate. The only true failures are the ones from which nothing is learned. Lessons on how to improve the risk management process were learned, unfortunately at the expense of human life.

On January 27, 1967, Astronauts Gus Grissom, Edward White, and Roger Chaffee were killed on board a test on *Apollo-Saturn 204*. James Webb, NASA's administrator at that time, was allowed by President Johnson to conduct an internal investigation of the cause. The investigation was primarily a technical investigation. NASA was fairly open with the media during the investigation. As a result of the openness, the credibility of the agency was maintained.

---

[30] "The *Challenger* Accident: Administrative Causes of the *Challenger* Accident," http://www. me.utexas.edu/~uer/challenger/chall3.html, pp. 8–9.

[31] "Transcript of the President's Eulogy for the Seven Challenger Astronauts." *New York Times*, February 1, 1986. Available at www.nytimes.com/1986/02/01/us/transcript-of-the-president-s-eulogy-for-the-seven-challenger-astronauts.html.

With the *Challenger* accident, confusion arose as to whether it had been a technical failure or a management failure. There was no question in anyone's mind that the decision-making process was flawed. NASA and Thiokol acted independently in their response to criticism. Critical information was withheld, at least temporarily, and this undermined people's confidence in NASA. The media, as might have been expected, began vengeful attacks on NASA and Thiokol.

Following the *Apollo-Saturn 204* fire, few changes were made in management positions at NASA. Those changes that did occur were the result of a necessity for improvement and where change was definitely warranted. Following the *Challenger* accident, almost every top management position at NASA underwent a change of personnel.

How an organization fares after an accident is often measured by how well it interfaces with the media. Situations such as the Tylenol tragedy (subject of another case study in this volume) and the *Apollo-Saturn 204* fire bore this out.

Following the accident and after critical data were released, papers were published showing that the O-ring data correlation was indeed possible. In one such paper, Frederick Lighthall showed that not only was a correlation possible, but the real problem may be a professional weakness shared by many people, but especially engineers, who are required to analyze technical data.[32] Lighthall's argument was that engineering curriculums might not provide engineers with strong enough statistical education, especially in covariance analysis. The Rogers Commission also identified this conclusion when they found that there were no engineers at NASA trained in statistical sciences.

Almost all scientific achievements require the taking of risks. The hard part is deciding which risk is worth taking and which is not. Every person who has ever flown in space, whether military or civilian, was a volunteer. They were all risk takers who understood that safety in space can never be guaranteed with 100 percent accuracy.

QUESTIONS

Following are a series of questions categorized according to the principles of risk management. There may not be any single right or wrong answer to these questions.

Risk Management Plan

1. Does it appear, from the data provided in the case, that a risk management plan was in existence?

---

[32] Frederick F. Lighthall, "Launching the Space Shuttle *Challenger*: Disciplinary Deficiencies in the Analysis of Engineering Data," *IEEE Transactions on Engineering Management* 38, no. 1 (February 1991): 63–74.

2. If such a plan did exist, then why wasn't it followed—or was it followed?
3. Is there a difference between a risk management plan, a quality assurance plan, and a safety plan, or are they the same?
4. Would there have been a better way to handle risk management planning at NASA assuming 16 flights per year, 25 flights per year, or as originally planned, 60 flights per year? Why is the number of flights per year critical in designing a formalized risk management plan?

Risk Identification

1. What is the difference between a risk and an anomaly? Who determines the difference?
2. Does there appear to have been a structured process in place for risk identification at either NASA or Thiokol?
3. How should problems with risk identification be resolved if there exist differences of opinion between the customer and the contractors?
4. Should senior management or sponsors be informed about all risks identified or just the overall "aggregate" risk?
5. How should one identify or classify the risks associated with using solid rocket boosters on manned spacecraft rather than the conventional liquid-fuel boosters?
6. How should one identify or classify trade-off risks, such as trading off safety for political acceptability?
7. How should one identify or classify the risks associated with pressure resulting from making promises that may be hard to keep?
8. Suppose that a risk identification plan had been established at the beginning of the space program when the shuttle was still considered an experimental design. If the shuttle is now considered an operational vehicle rather than an experimental design, could that affect the way that risks were identified to the point where the risk identification plan would need to be changed?

Risk Quantification

1. Given the complexity of the Space Shuttle Program, is it feasible and/or practical to develop a methodology for quantifying risks, or should each situation be addressed individually? Can we have both a quantitative and qualitative risk evaluation system in place at the same time?
2. How does one quantify the dangers associated with the ice problem?
3. How should risk quantification problems be resolved if there exist differences of opinion between the customer and the contractors?
4. If a critical risk is discovered, what is the proper way for the project manager to present to senior management the impact of the risk? How do you as a project manager make sure that senior management understand the ramifications?

5. How were the identified risks quantified at NASA? Is the quantification system truly quantitative or is it a qualitative system?
6. Were probabilities assigned to any of the risks? Why or why not?

Risk Response (Risk Handling)

1. How does an organization decide what is or is not an acceptable risk?
2. Who should have final say in deciding upon the appropriate response mechanism for a risk?
3. What methods of risk response were used at NASA?
4. Did it appear that the risk response method selected was dependent on the risk or on other factors?
5. How should an organization decide whether or not to accept a risk and launch if the risks cannot be quantified?
6. What should be the determining factors in deciding which risks are brought upstairs to the executive levels for review before selecting the appropriate risk response mechanism?
7. Why weren't the astronauts involved in the launch decision (i.e., the acceptance of the risk)? Should they have been involved?
8. What risk response mechanism did NASA administrators use when they issued waivers for the Launch Commit Criteria?
9. Are waivers a type of risk response mechanism?
10. Did the need to maintain a flight schedule compromise the risk response mechanism that would otherwise have been taken?
11. What risk response mechanism were managers at Thiokol and NASA using when they ignored the recommendations of their engineers?
12. Did the engineers at Thiokol and NASA do all they could to convince their own management that the wrong risk response mechanism was about to be taken?
13. When NASA pressed its contractors to recommend a launch, did NASA's risk response mechanism violate their responsibility to ensure crew safety?
14. When NASA discounted the effects of the weather, did NASA's risk response mechanism violate their responsibility to ensure crew safety?

Risk Control

1. How much documentation should be necessary for the tracking of a risk management plan? Can this documentation become excessive and create decision-making problems?
2. Risk management includes the documentation of lessons learned. In the case study, was there an audit trail of lessons learned, or was that audit trail simply protection memos?

3. How might Thiokol engineers have convinced both their own management and NASA to postpone the launch?

4. Should someone have stopped the *Challenger* launch, and, if so, how could this have been accomplished without risking one's job and career?

5. How might an engineer deal with pressure from above to follow a course of action that the engineer knows to be wrong?

6. How could the chains of communication and responsibility for the Space Shuttle Program have been made to function better?

7. Because of the ice problem, Rockwell could not guarantee the shuttle's safety but did nothing to veto the launch. Is there a better way for situations as this to be handled in the future?

8. What level of risk should have been acceptable for launch?

9. How should we handle situations where people in authority believe that the potential rewards justify what they believe to be relatively minor risks?

10. If you were on a jury attempting to place liability, whom would you say was responsible for the *Challenger* disaster?

Packer Telecom

BACKGROUND

The rapid growth of the telecom industry made it apparent to Packer's executives that risk management must be performed on all development projects. If Packer was late in the introduction of a new product, then market share would be lost. Furthermore, Packer could lose valuable opportunities to partner with other companies if Packer was regarded as being behind the learning curve with regard to new product development.

Another problem facing Packer was the amount of money being committed to R&D. Typical companies spend 8 to 10 percent of earnings on R&D, whereas in the telecom industry, the number may be as high as 15 to 18 percent. Packer was spending 20 percent on R&D, and only a small percentage of the projects that started out in the conceptual phase ever reached the commercialization phase, where Packer could expect to recover its R&D costs. Management attributed the problem to a lack of effective risk management.

THE MEETING

PM: "I have spent a great deal of time trying to benchmark best practices in risk management. I was amazed to find that most companies are in the same boat as us, with very little knowledge in risk management. From the limited results I have found from other companies, I have been able to develop a risk management template for us to use."

Sponsor: "I've read over your report and looked at your templates. You have words and expressions in the templates that we don't use here at Packer. This concerns me greatly. Do we have to change the way we manage projects to use these templates? Are we expected to make major changes to our existing project management methodology?"

PM: "I was hoping we could use these templates in their existing format. If the other companies are using these templates, then we should also. These templates also have the same probability distributions that other companies are using. I consider these facts equivalent to a validation of the templates."

Sponsor: "Shouldn't the templates be tailored to our methodology for managing projects and our life-cycle phases? These templates may have undergone validation, but not at Packer. The probability distributions are also based on someone else's history, not our history. I cannot see anything in your report that talks about the justification of the probabilities.

"The final problem I have is that the templates are based on history. It is my understanding that risk management should be forward looking, with an attempt at predicting the possible future outcomes. I cannot see any of this in your templates."

PM: "I understand your concerns, but I don't believe they are a problem. I would prefer to use the next project as a 'breakthrough project' using these templates. This will give us a good basis to validate the templates."

Sponsor: "I will need to think about your request. I am not sure that we can use these templates without some type of risk management training for our employees."

QUESTIONS

1. Can templates be transferred from one company to another, or should tailoring be mandatory?
2. Can probability distributions be transferred from one company to another? If not, then how do we develop a probability distribution?
3. How do you validate a risk management template?
4. Should a risk management template be forward looking?
5. Can employees begin using a risk management template without some form of specialized training?

 **Luxor Technologies**

Between 1992 and 1996, Luxor Technologies had seen their business almost quadruple in the wireless communications area. Luxor's success was attributed largely to the strength of its technical community, which was regarded as second to none. The technical community was paid very well and given the freedom to innovate. Even though Luxor's revenue came from manufacturing, Luxor was regarded by Wall Street as being a technology-driven company.

The majority of Luxor's products were based on low-cost, high-quality applications of the state-of-the-art technology rather than advanced state-of-the-art technological breakthroughs. Applications engineering and process improvement were major strengths at Luxor. Luxor possessed patents in technology breakthrough, applications engineering, and even process improvement. Luxor refused to license its technology to other firms, even if the applicant was not a major competitor.

Patent protection and design secrecy were of paramount importance to Luxor. In this regard, Luxor became vertically integrated, manufacturing and assembling all components of its products internally. Only off-the-shelf components were purchased. Luxor believed that if it were to use outside vendors for sensitive component procurement, they would have to release critical and proprietary data to the vendors. Since these vendors most likely also serviced Luxor's competitors, Luxor maintained the approach of vertical integration to maintain secrecy.

Being the market leader technically afforded Luxor certain luxuries. Luxor saw no need for expertise in technical risk management. In cases where the technical community was only able to achieve 75 to 80 percent of the desired specification limit, the product was released as it stood, accompanied by an announcement that there would be an upgrade the following year to achieve the remaining 20 to 25 percent of the specification limit, together with other features. Enhancements and upgrades were made on a yearly basis.

By the fall of 1996, however, Luxor's fortunes were diminishing. The competition was catching up quickly, thanks to major technological breakthroughs. Marketing estimated that, by 1998, Luxor would be a follower rather than a market leader. Luxor realized that something must be done, and quickly.

In January 1999, Luxor hired an expert in risk analysis and risk management to help it assess the potential damage to the firm and to assist in development of a mitigation plan. The consultant reviewed project histories and lessons learned on all projects undertaken from 1992 through 1998. The consultant concluded that the major risk to Luxor would be the technical risk and prepared Tables I and II. Table I shows the likelihood of a technical risk event occurring. The consultant identified the six most common technical risk events that could occur at Luxor over the next several years, based on the extrapolation of past and present data into the future. Table II shows the impact that a technical risk event could have on each project. Because of the high probability of state-of-the-art advancements needed in the future (i.e., 95 percent from Table I), the consultant identified the impact probabilities in Table II for both with and without state-of-the-art advancement needed.

Tables I and II confirmed management's fear that Luxor was in trouble. A strategic decision had to be made concerning the technical risks identified in Table I, specifically the first two risks. The competition had caught up to Luxor in applications engineering and was now surpassing Luxor in patents involving state-of-the-art advancements. From 1992 to 1998, time was considered a luxury for the technical community at Luxor. Now time was a serious constraint.

The strategic decision facing management was whether Luxor should struggle to remain a technical leader in wireless communications technology or simply

TABLE I   LIKELIHOOD OF A TECHNICAL RISK

Event	Likelihood Rating
State-of-the-art advance needed	0.95
Scientific research required(without advancements)	0.80
Concept formulation	0.40
Prototype development	0.20
Prototype testing	0.15
Critical performance demonstrated	0.10

TABLE II   IMPACT OF A TECHNICAL RISK EVENT

Event	Impact Rating	
	With State-of-the-Art Changes	**Without State-of-the-Art Changes**
Product performance not at 100% of specification	0.95	0.80
Product performance not at 75–80% of specification	0.75	0.30
Abandonment of project	0.70	0.10
Need for further enhancements	0.60	0.25
Reduced profit margins	0.45	0.10
Potential systems performance degradation	0.20	0.05

console itself with a future as a follower. Marketing was given the task of determining the potential impact of a change in strategy from a market leader to a market follower. The next list was prepared and presented to management by marketing:

1. The company's future growth rate will be limited.
2. Luxor will still remain strong in applications engineering but will need to outsource state-of-the-art development work.
3. Luxor will be required to provide outside vendors with proprietary information.
4. Luxor may no longer be vertically integrated (i.e., have backward integration).
5. Final product costs may be heavily influenced by the costs of subcontractors.
6. Luxor may not be able to remain a low-cost supplier.
7. Layoffs will be inevitable but perhaps not in the near term.
8. The marketing and selling of products may need to change. Can Luxor still market products as a low-cost, high-quality, state-of-the-art manufacturer?
9. Price-cutting by Luxor's competitors could have a serious impact on Luxor's future ability to survive.

The list presented by marketing demonstrated that there was a serious threat to Luxor's growth and even survival. Engineering then prepared a list of alternative courses of action that would enable Luxor to maintain its technical leadership position:

1. Luxor could hire away from the competition more staff personnel with pure and applied R&D skills. This would be a costly effort.

2. Luxor could slowly retrain part of its existing labor force using existing, experienced R&D personnel to conduct the training.
3. Luxor could fund seminars and university courses on general R&D methods as well as R&D methods for telecommunications projects. These programs were available locally.
4. Luxor could use tuition reimbursement funds to pay for distance learning courses (conducted over the Internet). These were full-semester programs.
5. Luxor could outsource technical development.
6. Luxor could purchase or license technology from other firms, including competitors. This assumed that competitors would agree to this at a reasonable price.
7. Luxor could develop joint ventures/mergers with other companies that, in turn, would probably require Luxor to disclose much of its proprietary knowledge.

With marketing's and engineering's lists before them, Luxor's management had to decide which path would be best for the long term.

QUESTIONS

1. Can the impact of one specific risk event, such as a technical risk event, create additional risks, which may or may not be technical risks? Can risk events be interrelated?
2. Does the list provided by marketing demonstrate the likelihood of a risk event or the impact of a risk event?
3. How does one assign probabilities to the marketing list?
4. The seven items in the list provided by engineering are all ways of mitigating certain risk events. If the company follows these suggestions, is it adopting a risk response mode of avoidance, assumption, reduction, or deflection?
5. Would you side with marketing or engineering? What should Luxor do at this point?

Altex Corporation

BACKGROUND

Following World War II, the United States entered into a Cold War with Russia. To win this Cold War, the United States had to develop sophisticated weapon systems with such destructive power that any aggressor knew that the retaliatory capability of the United States could and would inflict vast destruction.

Hundreds of millions of dollars were committed to ideas concerning technology that had not been developed as yet. Aerospace and defense contractors were growing without bounds, thanks to cost-plus-percentage-of-cost contract awards. Speed and technological capability were judged to be significantly more important than cost. To make matters worse, contracts were often awarded to the second or third most qualified bidder for the sole purpose of maintaining competition and maximizing the total number of defense contractors.

CONTRACT AWARD

During this period, Altex Corporation was elated when it learned that it had just been awarded the R&D phase of the Advanced Tactical Missile Program (ATMP). The terms of the contract specified that Altex had to submit to the Army, within 60 days after contract award, a formal project plan for the two-year ATMP effort. Contracts at that time did not require that a risk management plan be developed. A meeting was held with the project manager of R&D to assess the risks in the ATMP effort.

PM: "I'm in the process of developing the project plan. Should I also develop a risk management plan as part of the project plan?"

Sponsor: "Absolutely not! Most new weapon systems requirements are established by military personnel who have no sense of reality about what it takes to develop a weapon system based on technology that doesn't even exist yet. We'll be lucky if we can deliver 60–70 percent of the specification imposed on us."

PM: "But that's not what we stated in our proposal. I wasn't brought on board until after we won the award, so I wasn't privileged to know the thought process that went into the proposal. The proposal even went so far as to imply that we might be able to exceed the specification limits, and now you're saying that we should be happy with 60–70 percent."

Sponsor: "We say what we have to say to win the bid. Everyone does it. It is common practice. Whoever wins the R&D portion of the contract will also be first in line for the manufacturing effort and that's where the megabucks come from! If we can achieve 60–70 percent of specifications, it should placate the Army enough to give us a follow-on contract. If we told the Army the true cost of developing the technology to meet the specification limits, we would never get the contract. The program might even be canceled. The military people want this weapon system. They're not stupid! They know what is happening and they do not want to go to their superiors for more money until later on, downstream, after approval by Department of Defense and project kickoff. The government wants the lowest cost and we want long-term, follow-on production contracts, which can generate huge profits."

PM: "Aren't we simply telling lies in our proposal?"

Sponsor: "My engineers and scientists are highly optimistic and believe they can do the impossible. This is how technological breakthroughs are made. I prefer to call it 'overoptimism of technical capability' rather than 'telling lies.' If my engineers and scientists have to develop a risk management plan, they may become pessimistic, and that's not good for us!"

PM: "The problem with letting your engineers and scientists be optimistic is that they become reactive rather than proactive thinkers. Without proactive thinkers, we end up with virtually no risk management or contingency plans. When problems surface that require significantly more in the way of resources than we budgeted for, we will be forced to accept crisis management as a way of life. Our costs will increase and that's not going to make the Army happy."

Sponsor: "But the Army won't penalize us for failing to meet cost or for allowing the schedule to slip. If we fail to meet at least 60–70 percent of the specification limits, however, then we may well be in trouble. The Army knows there will be a follow-on contract request if we cannot meet specification limits. I consider 60–70 percent of the specifications to be the minimum acceptable limits for the Army. The Army wants the program kicked off right now.

"Another important point is that long-term contracts and follow-on production contracts allow us to build up a good working relationship with the Army. This is critical. Once we get the initial contract, as we did, the Army will always work with us for follow-on efforts. Whoever gets the R&D effort will almost always get the lucrative production contract. Military officers are under pressure to work with us because their careers may be in jeopardy if they have to tell their superiors that millions of dollars were awarded to the wrong defense contractor. From a career standpoint, the military officers are better off allowing us to downgrade the requirements than admitting that a mistake was made."

PM: "I'm just a little nervous managing a project that is so optimistic that major advances in the state of the art must occur to meet specifications. This is why I want to prepare a risk management plan."

Sponsor: "You don't need a risk management plan when you know you can spend as much as you want and also let the schedule slip. If you prepare a risk management plan, you will end up exposing a multitude of risks, especially technical risks. The Army might not know about many of these risks, so why expose them and open up Pandora's box? Personally, I believe that the Army does already know many of these risks, but does not want them publicized to their superiors.

"If you want to develop a risk management plan, then do it by yourself, and I really mean by yourself. Past experience has shown that our employees will be talking informally to Army personnel at least two to three times a week. I don't want anyone telling the customer that we have a risk management plan. The customer will obviously want to see it, and that's not good for us.

"If you are so incensed that you feel obligated to tell the customer what you're doing, then wait about a year and a half. By that time, the Army will have made a considerable investment in both us and the project, and they'll be locked into us for follow-on work. Because of the strategic timing and additional costs, they will never want to qualify a second supplier so late in the game. Just keep the risk management plan to yourself for now.

"If it looks like the Army might cancel the program, then we'll show them the risk management plan, and perhaps that will keep the program alive."

QUESTIONS

1. Why was a risk management plan considered unnecessary?
2. Should risk management planning be performed in the proposal stage or after contract award, assuming that it must be done?
3. Does the customer have the right to expect the contractor to perform risk analysis and develop a risk management plan if it is not called out as part of the contractual statement of work?

4. Would Altex have been more interested in developing a risk management plan if the project were funded entirely from within?

5. How effective will the risk management plan be if developed by the project manager without input from others?

6. Should the customer be allowed to participate in or assist the contractor in developing a risk management plan?

7. How might the Army have responded if it were presented with a risk management plan early during the R&D activities?

8. How effective is a risk management plan if cost overruns and schedule slippages are always allowed?

9. How can severe optimism or severe pessimism influence the development of a risk management plan?

10. How does one develop a risk management plan predicated upon needed advances in the state of the art?

11. Can the sudden disclosure of a risk management plan be used as a stopgap measure to prevent termination of a potentially failing project?

12. Can risk management planning be justified on almost all programs and projects?

Acme Corporation

BACKGROUND

Acme Corporation embarked on an optimistic project to develop a new product for the marketplace. Acme's scientific community made a technical breakthrough, and now the project appears to be in the development stage, more than being pure or applied research.

The product is considered to be high tech. If the product can be launched within the next four months, Acme expects to dominate the market for at least a year or so until the competition catches up. Marketing has stated that the product must sell for not more than $150 to $160 per unit to be the cost-focused market leader.

Acme uses a project management methodology for all multifunctional projects. The methodology has six life-cycle phases:

1. Preliminary planning
2. Detailed planning
3. Execution/design selection
4. Prototyping
5. Testing/buyoff
6. Production

At the end of each life-cycle phase, a gate/phase review meeting is held with the project sponsor and other appropriate stakeholders. Gate review meetings are formal meetings. The company has demonstrated success following this method-ology for managing projects.

At the end of the second life-cycle stage of this project, detailed planning, a meeting is held with just the project manager and the project sponsor. The purpose of the meeting is to review the detailed plan and identify any future problem areas that will require involvement by the project sponsor.

THE MEETING

Sponsor: "I simply do not understand this document you sent me titled 'Risk Management Plan.' All I see is a work breakdown structure with work pack-ages at level 5 of the WBS accompanied by almost 100 risk events. Why am I looking at more than 100 risk events? Furthermore, they're not categorized in any manner. Doesn't our project management methodology provide any guidance on how to do this?"

PM: "All of these risk events can and will impact the design of the final product. We must be sure we select the right design at the lowest risk. Unfortunately, our project management methodology does not include any provisions or guidance on how to develop a risk management plan. Perhaps it should."

Sponsor: "I see no reason for an in-depth analysis of 100 or so risk events. That's too many. Where are the probabilities and expected outcomes or damages?"

PM: "My team will not be assigning probabilities or damages until we get closer to prototype development. Some of these risk events may go away altogether."

Sponsor: "Why spend all of this time and money on risk identification if the risks can go away next month? You've spent too much money doing this. If you spend the same amount of money on all of the risk management steps, then we'll be way over budget."

PM: "We haven't looked at the other risk management steps yet, but I believe all of the remaining steps will require less than 10 percent of the budget we used for risk identification. We'll stay on budget."

QUESTIONS

1. Was the document given to the sponsor a risk management plan?
2. Did the project manager actually perform effective risk management?
3. Was the appropriate amount of time and money spent identifying the risk events?
4. Should one step be allowed to "dominate" the entire risk management process?
5. Are there any significant benefits to the amount of work already done for risk identification?
6. Should the 100 or so risk events identified have been categorized? If so, how?
7. Can probabilities of occurrence and expected outcomes (i.e., damage) be accu-rately assigned to 100 risk events?

8. Should a project management methodology provide guidance for the development of a risk management plan?
9. Given the life-cycle phases in the case study, in which phase would it be appropriate to identify the risk management plan?
10. What are your feelings on the project manager's comments that he must wait until the prototyping phase to assign probabilities and outcomes?

The Risk Management Department

BACKGROUND

In 1946, shortly after the end of World War II, Cooper Manufacturing Company was created. The company manufactured small appliances for the home. By 2010, Cooper Manufacturing had more than 30 manufacturing plants, all located in the United States. The business now included both small and large household appliances. Almost all of its growth came from acquisitions that were paid for out of cash flow and borrowing from the financial markets.

Cooper's strategic plan called for global expansion beginning in 2003. With this in mind and with large financial reserves, Cooper planned on acquiring five to six companies a year. This would be in addition to whatever domestic acquisitions were also available. Almost all of the acquisitions were manufacturing companies that produced products related to the household marketplace. However, some of the acquisitions included air conditioning and furnace companies as well as home security systems.

RISK MANAGEMENT DEPARTMENT

During the 1980s, when Cooper Manufacturing began its rapid acquisition approach, it established a Risk Management Department. The Risk Management Department reported to the chief financial officer (CFO) and was considered to be part of the financial discipline of the company. The overall objective of the

Risk Management Department was to coordinate the protection of the company's assets. The primary means by which this was done was through the implementation of loss prevention programs. The department worked very closely with other internal departments such as Environmental Health and Safety. Outside consultants were brought in as necessary to support these activities.

One method employed by the company to ensure the entire company's cooperation and involvement in the risk management process was to hold each manufacturing division responsible for any specific losses up to a designated self-insured retention level. If there was a significant loss, the division must absorb the loss and its impact on the division's bottom-line profit margin. This directly involved the division in both loss prevention and claims management. When a claim did occur, the Risk Management Department maintained regular contact with the division's personnel to establish protocol on the claim and cash reserves and ultimate disposition.

As part of risk management, the company purchased insurance above the designated retention levels. The insurance premiums were allocated to each division. The premiums were calculated based on sales volume and claims loss history, with the most significant percentage being allocated against claims loss history.

Risk management was considered an integral part of the due diligence process for acquisitions and divestitures. It began at the onset of the process rather than at the end and resulted in a written report and presentation to the senior levels of management.

A NEW RISK MATERIALIZES

The original intent of the Risk Management Department was to protect the company's assets, especially from claims and lawsuits. The department focused heavily on financial and business risks with often little regard for human assets. All of this was about to change.

The majority of Cooper's manufacturing processes were labor-intensive assembly line processes. Although Cooper modernized the plants with new equipment to support the assembly lines with hope of speeding up the work, the processes were still heavily labor intensive. The modernization of the plants did improve production. However, more people were getting injured and were out sick. Cooper's workers' compensation costs and health care premiums were sky-rocketing and taking an unexpected toll on the bottom line of the financial statements of many of the divisions.

Senior management recognized the gravity of the situation and asked the Risk Management Department to find ways to reduce injuries, lower the number of sick days that people were taking, and reduce workers' compensation costs. To do this, the Risk Management Department had to look at the way each worker performed his or her task and improve where possible the interaction between the workers and the equipment. The name of the department was then changed to Risk Management and Ergonomics.

ERGONOMICS

According to Wikipedia:

> Ergonomics is the science of designing the workplace environment to fit the user. Proper ergonomic design is necessary to prevent *repetitive strain injuries*, which can develop over time and can lead to long-term disability.
> The International Ergonomics Association defines ergonomics as follows:
>> Ergonomics (or human factors) is the scientific discipline concerned with the understanding of interactions among humans and other elements of a system, and the profession that applies theory, principles, data and methods to design in order to optimize human well-being and overall system performance.
>> Ergonomics is employed to fulfill the two goals of health and productivity. It is relevant in the design of such things as safe furniture and easy-to- use interfaces to machines.
>> Ergonomics is concerned with the "fit" between people and their technological tools and environments. It takes account of the user's capabilities and limitations in seeking to ensure that tasks, equipment, information and the environment suit each user.
>> To assess the fit between a person and the used technology, ergonomists consider the job (activity) being done and the demands on the user; the equipment used (its size, shape, and how appropriate it is for the task), and the information used (how it is presented, accessed, and changed). Ergonomics draws on many disciplines in its study of humans and their environments, including *anthropometry, biomechanics, mechanical engineering, industrial engineering, industrial design, kinesiology, physiology and psychology.*[1]

Ergonomics includes the fundamentals for the flexible workplace variability and compatibility with desk components that flex from individual work activities to team settings. Workstations provide supportive ergonomics for task-intensive environments.

Outside the discipline, the term "ergonomics" is generally used to refer to physical ergonomics as it relates to the workplace (as in, e.g., ergonomic chairs and *keyboards*). Ergonomics in the workplace has to do largely with the safety of employees, both long and short term. Ergonomics can help reduce costs by improving safety. This would decrease the money paid out in workers' compensation. For example, over 5 million workers sustain overextension injuries per year.

---

[1] Wikipedia contributors, "Human factors and ergonomics," Wikipedia, The Free Encyclopedia, https://en.wikipedia.org/w/index.php?title=Human_factors_and_ergonomics&oldid=754937541.

FIGURE V Ergonomics in the Workplace

Through ergonomics, workplaces can be designed so that workers do not have to overextend themselves and the manufacturing industry could save billions in workers' compensation (see Figure V).

Workplaces may either take the reactive or proactive approach when applying ergonomics practices. Reactive ergonomics is when something needs to be fixed and corrective action is taken. Proactive ergonomics is the process of seeking areas that could be improved and fixing the issues before they become a large problem. Problems may be fixed through equipment design, task design, or environmental design. Equipment design changes the actual, physical devices used by people. Task design changes what people do with the equipment. Environmental design changes the environment in which people work but not the physical equipment they use.

QUESTIONS

1. Was the original intent of creating the Risk Management Department correct in that it was designed to protect corporate assets? In other words, was this really risk management?
2. Are the new responsibilities of the department, specifically ergonomics, a valid interpretation of risk management?
3. Can the lowering of health care costs and workers' compensation costs be considered as a project?
4. How successful do you think Cooper was in lowering costs?

Part 12

CONFLICT MANAGEMENT

Conflicts can occur anywhere in the project and with anyone. Some conflicts are severe while others are easily solvable. In the past, project managers avoided conflicts when possible. Today, we believe that conflicts can produce beneficial results if the conflicts are managed correctly.

There are numerous methods available to project managers for the resolution of conflicts. The methods selected may vary depending on the severity of the conflict, the person with whom the conflict exists and his/her level of authority, the life-cycle phase of the project, the priority of the project, and the relative importance of the project as seen by senior management.

Facilities Scheduling at Mayer Manufacturing

Eddie Turner was elated with the good news that he was being promoted to section supervisor in charge of scheduling all activities in the new engineering research laboratory. The new laboratory was a necessity for Mayer Manufacturing. The engineering, manufacturing, and quality control directorates were all in desperate need of a new testing facility. Upper-level management felt that this new facility would alleviate many of the problems that previously existed.

The new organizational structure (as shown in Figure I) required a change in policy over use of the laboratory. The new section supervisor, on approval from his department manager, would have full authority for establishing priorities for the use of the new facility. The new policy change was a necessity because upper-level management felt that there would be inevitable conflict among manufacturing, engineering, and quality control.

After one month of operations, Eddie Turner was finding his job impossible, so he has a meeting with Gary Whitehead, his department manager.

Eddie: "I'm having a hell of a time trying to satisfy all of the department managers. If I give engineering prime-time use of the facility, then quality control and manufacturing say that I'm playing favorites. Imagine that! Even my own people say that I'm playing favorites with other directorates. I just can't satisfy everyone."

Gary: "Well, Eddie, you know that this problem comes with the job. You'll get the job done."

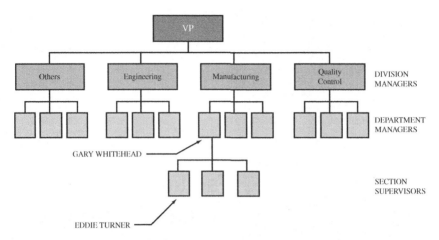

FIGURE I Mayer Manufacturing organizational structure

Eddie: "The problem is that I'm a section supervisor and have to work with department managers. These department managers look down on me like I'm their servant. If I were a department manager, then they'd show me some respect. What I'm really trying to say is that I would like you to send out the weekly memos to these department managers telling them of the new priorities. They wouldn't argue with you like they do with me. I can supply you with all the necessary information. All you'll have to do is to sign your name."

Gary: "Determining the priorities and scheduling the facilities is your job, not mine. This is a new position and I want you to handle it. I know you can because I selected you. I do not intend to interfere."

During the next two weeks, the conflicts got progressively worse. Eddie felt that he was unable to cope with the situation by himself. The department managers did not respect the authority delegated to him by his superiors. For the next two weeks, Eddie sent memos to Gary in the early part of the week asking whether Gary agreed with the priority list. There was no response to the two memos. Eddie then met with Gary to discuss the deteriorating situation.

Eddie: "Gary, I've sent you two memos to see if I'm doing anything wrong in establishing the weekly priorities and schedules. Did you get my memos?"

Gary: "Yes, I received your memos. But as I told you before, I have enough problems to worry about without doing your job for you. If you can't handle the work, let me know and I'll find someone who can."

Eddie returned to his desk and contemplated his situation. Finally, he made a decision. Next week he was going to put a signature block under his for Gary to sign, with carbon copies for all division managers. "Now let's see what happens," Eddie said.

QUESTIONS

1. What do you think most likely happened next?
2. What could Eddie do next if Gary refused to sign the documents?

Scheduling the Safety Lab

"Now see here, Tom, I understand your problem well," remarked Dr. Polly, director of the Research Laboratories. "I pay you a good salary to run the safety labs. That salary also includes doing the necessary scheduling to match our priorities. Now, if you can't handle the job, I'll get someone who can."

Tom: "Every Friday morning your secretary hands me a sheet with the listing of priorities for the following week. Once, just once, I'd like to sit in on the director's meeting and tell you people what you do to us in the safety lab when you continually shuffle around the priorities from week to week.

"On Friday afternoons, my people and I meet with representatives from each project to establish the following week's schedules."

Dr. Polly: "Can't you people come to an agreement?"

Tom: "I don't think you appreciate my problem. Two months ago, we all sat down to work out the lab schedule. Project X-13 had signed up to use the lab last week. Now, mind you, they had been scheduled for the past two months. But the Friday before they were to use it, your new priority list forced them to reschedule the lab at a later date, so that we could give the use of the lab to a higher-priority project. We're paying an awful lot of money for idle time and the redoing of network schedules. Only the project managers on the top-priority projects end up smiling after our Friday meetings."

Dr. Polly: "As I see your problem, you can't match long-range planning with the current priority list. I agree that it does create conflicts for you. But you have to remember that we, upstairs, have many other conflicts to resolve. I want that one solved at your level, not mine."

Tom: "Every project we have requires use of the safety lab. This is the basis for our problem. Would you consider letting us modify your priority list with regard to the safety lab?"

Dr. Polly: "Yes, but you had better have the agreement of all of the project managers. I don't want them coming to see me about your scheduling problems."

Tom: "How about if I let people do long-range scheduling for the lab, for three out of four weeks each month? The fourth week will be for the priority projects."

Dr. Polly: "That might work. You had better make sure that each project manager informs you immediately of any schedule slippages so that you can reschedule accordingly. From what I've heard, some of the project managers don't let you know until the last minute."

Tom: "That has been part of the problem. Just to give you an example, Project VX-161 was a top-priority effort and had the lab scheduled for the first week in March. I was never informed that they had accelerated their schedule by two weeks. They walked into my office and demanded use of the lab for the third week in February. Since they had the top priority, I had to grant them their request. However, Project BP-3 was planning on using the lab during that week and was bumped back three weeks. That cost them a pile of bucks in idle time pay and, of course, they're blaming me."

Dr. Polly: "Well, Tom, I'm sure you'll find a solution to your problem."

QUESTIONS

1. Should Dr. Polly have provided support?
2. What should Tom have done if he did not get the support he needed?

Telestar International

On November 15, 1998, the Department of Energy Resources awarded Telestar a $475,000 contract for the developing and testing of two waste treatment plants. Telestar had spent the better part of the last two years developing waste treatment technology under its own research and development (R&D) activities. This new contract would give Telestar the opportunity to break into a new field—that of waste treatment.

The contract was negotiated at a firm-fixed price. Any cost overruns would have to be incurred by Telestar. The original bid was priced out at $847,000. Telestar's management, however, wanted to win this one. The decision was made that Telestar would buy in at $475,000 so that it could at least get its foot into the new marketplace.

The original estimate of $847,000 was very rough because Telestar did not have any good man-hour standards, in the area of waste treatment, on which to base its man-hour projections. Corporate management was willing to spend up to $400,000 of its own funds in order to compensate the bid of $475,000.

By February 15, 1999, costs were increasing to such a point where overrun would be occurring well ahead of schedule. Anticipated costs to completion were now $943,000. The project manager decided to stop all activities in certain functional departments, one of which was structural analysis. The manager of the structural analysis department strongly opposed the closing out of the work order prior to the testing of the first plant's high-pressure pneumatic and electrical systems.

Structures manager: "You're running a risk if you close out this work order. How will you know if the hardware can withstand the stresses that will be imposed during the test? After all, the test is scheduled for next month and I can probably finish the analysis by then."

Project manager: "I understand your concern, but I cannot risk a cost overrun. My boss expects me to do the work within cost. The plant design is similar to one that we have tested before, without any structural problems being detected. On this basis I consider your analysis unnecessary."

Structures manager: "Just because two plants are similar does not mean that they will be identical in performance. There can be major structural deficiencies."

Project manager: "I guess the risk is mine."

Structures manager: "Yes, but I get concerned when a failure can reflect on the integrity of my department. You know, we're performing on schedule and within the time and money budgeted. You're setting a bad example by cutting off our budget without any real justification."

Project manager: "I understand your concern, but we must pull out all the stops when overrun costs are inevitable."

Structures manager: "There's no question in my mind that this analysis should be completed. However, I'm not going to complete it on my overhead budget. I'll reassign my people tomorrow. Incidentally, you had better be careful; my people are not very happy to work for a project that can be canceled immediately. I may have trouble getting volunteers next time."

Project manager: "Well, I'm sure you'll be able to adequately handle any future work. I'll report to my boss that I have issued a work stoppage order to your department."

During the next month's test, the plant exploded. Postanalysis indicated that the failure was due to a structural deficiency.

QUESTIONS

1. Who is at fault?
2. Should the structures manager have been dedicated enough to continue the work on his own?
3. Can a functional manager who considers his organization as strictly support still be dedicated to total project success?

The Problem with Priorities

For the past several years, Kent Corporation had achieved remarkable success in winning R&D contracts. The customers were pleased with the analytical capabilities of the R&D staff at Kent. Theoretical and experimental results were usually within 95 percent agreement. But many customers still felt that 95 percent was too low. They wanted 98 to 99 percent.

Kent updated its computer facility by purchasing a large computer. The increased performance with the new computer encouraged the R&D group to attempt to convert from two-dimensional to three-dimensional solutions to theoretical problems. Almost everyone except the director of R&D thought that this would give better comparison between experimental and theoretical data.

Kent Corporation had tried to develop the computer program for three-dimensional solutions with its own internal R&D programs, but the cost was too great. Finally, after a year of writing proposals, Kent Corporation convinced the federal government to sponsor the project. The project was estimated at $750,000, to begin January 2, and to be completed by December 20. Dan McCord was selected as project manager. Dan had worked with the electronic data processing department on other projects and knew the people and the man-hour standards.

Kent Corporation was big enough to support 100 simultaneous projects. With so many projects in existence at one time, continual reshuffling of resources was necessary. The corporation directors met every Monday morning to establish project priorities. Priorities were not enforced unless project and functional managers could not agree on the allocation and distribution of resources.

Because of the R&D director's persistence, the computer project was given a low priority. This posed a problem for Dan McCord. The computer department manager refused to staff the project with his best people. As a result, Dan was very skeptical about the success of the project.

In July, two other project managers held a meeting with Dan to discuss the availability of the new computer model.

"We have two proposals that we're favored to win, providing that we can state in our proposal that we have this new computer model available for use," remarked one of the project managers.

"We have a low priority and, even if we finish the job on time, I'm not sure of the quality of work because of the people we have assigned," said Dan.

"How do you propose we improve our position?" asked one project manager.

"Let's try to get in to see the director of R&D," asserted Dan.

"And what are we going to say in our defense?" asked the other project manager.

QUESTIONS

1. What can a project manager say in his/her defense?
2. How will you respond if nobody agrees with your defense?

Part 13

MORALITY AND ETHICS

When the survival of the firm is at stake, workers often make decisions that may violate moral and ethical principles. Some may view an action as a violation, whereas others may view it as an acceptable practice. Every day, people are placed in situations that may require a moral or ethical decision.

Some companies have found a solution to this problem by creating a standard practice manual or corporate credo that provides guidelines for how these decisions should be made. The guidelines identify the order in which certain stakeholders' interests should be satisfied.

The Project
Management
Lawsuit

BACKGROUND

A new president was hired and then restructured the company in a way that he thought would be better for project management. In doing this restructuring, the president violated a contractual agreement that had been made with the vice president of engineering hired three years earlier.

HIRING THE VICE PRESIDENT

In 2006, Phoenix Company hired Jim as vice president for engineering. As with all senior officers, the hiring process included a written contract that clearly stipulated the criteria for bonuses, stock options, severance packages, retirement packages, and golden parachutes.

Jim's bonus clause involved project management. All of the engineering projects would be headed up by project managers who would report directly to Jim. While part of Jim's bonus was based on overall corporate profitability, the selling price of the company' stock, and other factors, the major portion of the bonus was based on the profitability of the projects under Jim's direct control. Jim was experienced in project management and believed that this bonus plan would certainly be to his advantage.

From 2006 through 2008, Jim's bonuses were more than his actual salary, and the size of his bonus increased each year. The company was doing quite well, and Jim was pleased with his own performance and felt secure in his position with Phoenix Company.

HIRING A NEW PRESIDENT

In 2008, the president of Phoenix Company announced his retirement, to be effective the end of December 2008. The board of directors of Phoenix Company decided to look outside the company for a replacement and eventually hired a new president who had experience in project management. Initially, Jim viewed this as a positive factor, but this was about to change.

One of the first things that some executives do when taking over a company is to restructure the organization according to their desired span of control. This normally occurs within the first two months after a new president comes on board. Workers know that, if change will happen, it will be in the first two months.

The new president was knowledgeable in project management and had experience with the project management office (PMO). Phoenix Company had project management but did not have a PMO. The new president created a corporate PMO. Project managers in all divisions would no longer report to the division vice presidents and were assigned full time to the corporate PMO. The corporate PMO reported directly to the new president. With most PMOs, the project managers still report on a "solid line" to their respective division managers but report to the PMO on a "dotted line."

The president's decision to have the project managers permanently assigned to the PMO alienated the three divisions that had project managers, and the engineering division was particularly displeased. Previously, most of the project managers were in the engineering division. The engineering division no longer had control over the projects, even the projects that were mainly within engineering. The creation of the PMO had a serious impact on the bonuses of those divisions that lost their project managers. In effect, what the creation of the PMO did was to transfer profit-and-loss responsibility on the projects to the PMO. That meant that there would be no project profitability component as part of Jim's year-end bonus payout.

In 2009 and 2010, Jim's bonus payments were drastically reduced. In January of 2011, Jim resigned from the company and filed a lawsuit against Phoenix Company for the loss of part of his bonus payments over the previous two years. Jim and his attorney claimed that the creation of the corporate PMO and the transferral of profit-and-loss responsibility to the PMO in effect violated Jim's written agreement and affected his bonus.

QUESTIONS

1. Why do most executives hire into companies under written contracts rather than a one-page employment acceptance letter?
2. Does the president of a company have the right to restructure a company however he or she pleases?
3. Did the president have the right to transfer profit-and-loss responsibility from the functional divisions to the PMO?
4. Did Jim win or lose the lawsuit?

Managing Crisis Projects

BACKGROUND

Project managers have become accustomed to managing within a structure process such as an enterprise project management methodology. The statement of work has gone through several iterations and is clearly defined. A work breakdown structure exists, and everyone understands their roles and responsibilities as defined in the responsibility assignment matrix. All of this took time to do.

This is the environment we all take for granted. But now let's change the scenario a bit. The president of the company calls you into his office and informs you that several people have just died using one of your company's products. You are being placed in charge of this crisis project. The lobby of the building is swamped with the news media, all of whom want to talk to you to hear your plan for addressing the crisis. The president informs you that the media knows you have been assigned as the project manager, and a news conference has been set up for one hour from now. The president also says that he wants to see your plan for managing the crisis no later than 10:00 p.m. this evening. Where do you begin? What should you do first? Time is now an extremely inflexible constraint rather than merely a constraint that may be able to be changed. There is no time to perform all of the activities you are accustomed to doing. You may need to make hundreds if not thousands of decisions quickly, and many of these are decisions you never thought that you would have to make. This is crisis project management.

UNDERSTANDING CRISIS MANAGEMENT

The field of crisis management is generally acknowledged to have started in 1982 when seven people died after ingesting Extra-Strength Tylenol capsules that were laced with cyanide. Johnson & Johnson, the parent of Tylenol, handled the situation in a manner that became the standard for crisis management.

Today, crises are neither rare nor random. They are part of our everyday lives. Crises cannot always be foreseen or prevented, but when they occur, we must do everything possible to manage them effectively. We must also identify lessons learned and best practices so that mistakes are not repeated on future crises that will certainly occur.

Some crises are so well entrenched in our minds that they are continuously referenced in a variety of courses in business schools. Some crises that have become iconic in society are:

- Hurricane Katrina
- Mad cow disease
- The space shuttle *Challenger* explosion
- The space shuttle *Columbia* reentry disaster
- The Tylenol poisonings
- The Union Carbide chemical plant explosion in Bhopal, India
- The *Exxon Valdez* oil spill
- The Chernobyl nuclear disaster
- The Three Mile Island nuclear disaster
- The Russian submarine *Kursk* disaster
- The Enron and WorldCom bankruptcies

Some crises are the result of acts of God or natural disasters. The public is generally forgiving when these occur. Crisis management, however, deals primarily with man-made crises, such as product tampering, fraud, and environmental contamination. Unlike natural disasters, these man-made crises are not inevitable, and the general public knows this and is quite unforgiving. When the *Exxon Valdez* oil spill occurred, Exxon refused to face the media for five days. Eventually, Exxon blamed the ship's captain for the accident and also attacked the Alaska Department of the Environment for hampering its emergency efforts. Stonewalling the media and assuming a defensive posture created extensive negative publicity for Exxon.

Most companies neither have any processes in place to anticipate these crises, even though they perform risk management activities, nor know how to manage them effectively after they occur. When lives are lost because of man-made crises, the unforgiving public becomes extremely critical of the companies responsible for the crises. Corporate reputations are very fragile. Reputations that had taken years to develop can be destroyed in hours or days.

Some people contend that with effective risk management practices, these crises can be prevented. While it is true that looking at the risk triggers can prevent some crises, not all crises can be prevented. However, best practices in crisis management can be developed and implemented such that when a crisis occurs, we can prevent a bad situation from getting worse.

For some time, corporations in specific industries have found it necessary to simulate and analyze worst-case scenarios for their products and services. Product tampering would be an example. These worst-case scenarios have been referred to as contingency plans, emergency plans, or disaster plans. These scenarios are designed around "known unknowns" where at least partial information exists on what events could happen.

Crisis management requires a heads-up approach with a very quick reaction time combined with a concerted effort on the part of possibly all employees. In crisis management, decisions have to be made quickly, often without even partial information and perhaps before the full extent of the damages is known. Events happen so quickly and so unpredictably that it may be impossible to perform any kind of planning. Roles and responsibilities of key individuals may change on a daily basis. There may be very active involvement by a majority of the stakeholders, many of whom had previously been silent. Company survival could rest entirely on how well a company manages the crisis.

Crises can occur within any company, irrespective of its size. The larger the company involved in the crisis, the greater the media coverage. Also, crises can occur even when things are going extremely well. The management guru Peter Drucker noted that companies that have been overwhelmingly successful for a long time tend to become complacent, even though their initial assumptions and environmental conditions have changed. Under these conditions, crises are more likely to occur. Drucker calls this "the failure of success."

What is unfortunate is that most of the lessons learned will come from improper handling of the crisis. After reading each of the situations here, you will be asked a series of questions to determine if there were any common threads among the cases.

FORD VERSUS FIRESTONE

Product recalls are costly and embarrassing for the auto industry. Improper handling of a recall can have an adverse effect on consumer confidence and the selling price of the stock. Ford and tire manufacturer Firestone are still suffering from the repercussions of their handling of a product recall in 2000–2001.

In August of 2000, Firestone recalled 6.5 million tires in the United States primarily because of tread separation problems on Ford Explorers [sports utility vehicles (SUVs)]. The problems with the tires were known several years earlier. In 1997–1998, Saudi Arabia reported tread separation on the SUV Explorers. In August 1999, Firestone replaced the tires in Saudi Arabia. In February 2000,

Firestone replaced the tires in Malaysia and Thailand, and in May 2000, the tires were replaced in Venezuela.

Initially it was believed that the problem might be restricted to countries with hot climates and rough roads. However, by May 2000, the U.S. National Highway Traffic Safety Administration had received 90 complaints involving 27 injuries and four deaths. A U.S. recall of 6.5 million tires took place in August 2000.

Ford and Firestone adopted a unified response concerning the recall. Unfortunately, accidents continued after the recall. Ford then blamed Firestone for flaws in the tires, and Firestone blamed Ford for design flaws in the SUV Explorers. The Ford–Firestone relationship quickly deteriorated.

The finger-pointing between Ford and Firestone was juicy news for the media. Because neither company was willing to accept responsibility for its actions, probably because of pending lawsuits, consumer confidence in both companies diminished, as did their stock prices. Consumer sentiment was that financial factors were more important than consumer safety.

Ford's CEO, Jac Nasser, tried to allay consumer fears, but his actions did not support his words. In September of 2000, he refused to testify at the Senate and House Commerce Subcommittee on tire recall, stating that he was too busy. In October of 2000, Masatoshi Ono resigned as CEO of Bridgestone, Firestone's parent company. In October of 2001, Jac Nasser resigned. Both executives departed and left behind over 200 lawsuits filed against their companies.

LESSONS LEARNED

1. Early-warning signs appeared but were marginally addressed.
2. Each company blamed the other, leaving the public with the belief that neither company could be trusted with regard to public safety.
3. Actions must reinforce words; otherwise, the public will become nonbelievers.

AIR FRANCE CONCORDE CRASH

On July 25, 2000, an Air France Concorde flight crashed on takeoff, killing all 109 people on board and four people on the ground. Air France immediately grounded its entire Concorde fleet pending an accident investigation. In response to media pressure, Air France used its website for press releases, expressed sorrow and condolences from the company, and arranged for some financial consideration to be paid to the relatives of the victims prior to a full legal settlement. The chairman of Air France, Jean-Cyril Spinetta, visited the accident scene the day of the accident and later attended a memorial service for the victims.

Air France's handling of the crisis was characterized by fast and open communication with the media and sensitivity for the relatives of the victims. The selling price of the stock declined rapidly the day of the disaster but made a quick recovery.

British Airways (BA) also flew the Concorde but took a different approach immediately after the accident. BA waited a month before grounding all Concorde flights indefinitely, and only after the Civil Aviation Authority announced it would be withdrawing the Concorde's airworthiness certification. Eventually, the airworthiness certification was reinstated, but it took BA's stock significantly longer to recover its decline in price.

LESSONS LEARNED

1. Air France and British Airways took different approaches to the crisis.
2. The Air France chairman showed compassion by visiting the site of the disaster as quickly as possible and attending a memorial service for the victims. British Airways did neither, thus disregarding its social responsibility.

INTEL AND THE PENTIUM CHIP

Intel, the manufacturer of Pentium chips, suffered an embarrassing moment resulting in a product recall. While performing prime number calculations on 10-digit numbers, a mathematics professor discovered significant round-off errors using the Pentium chips. Intel believed that the errors were insignificant and would show up only in every few billion calculations. But the mathematician was performing billions of calculations, and the errors were now significant.

The professor informed Intel of the problem. Intel refused to take action on the problem, stating that these errors were extremely rare and would affect only a very small percentage of Pentium chip users. The professor went public with the disclosure of the error.

Suddenly, the percentage of people discovering the error was not as small as originally thought. Intel still persisted in its belief that the error affected only a small percentage of the population. Intel put the burden of responsibility on users to show that their applications necessitated a replacement chip. Protests from consumers grew stronger. Finally the company agreed to replace all chips, no questions asked, after IBM announced it would no longer use Pentium chips in its personal computers.

Intel created its own public relations nightmare. Its response was slow and insincere. Intel tried to solve the problem solely through technical channels and completely disregarded the human issue of the crisis. Telling people who work in hospitals or air traffic control that there is a flaw in their computer but it is insignificant is not an acceptable response. Intel spent more than $500 million on the recall, significantly more than the cost of an immediate replacement.

LESSONS LEARNED

1. Intel's inability to take immediate responsibility for the crisis and develop a crisis management plan made the situation worse.
2. Intel completely disregarded public opinion.
3. Intel failed to realize that a crisis existed.

RUSSIAN SUBMARINE *KURSK*

In August of 2000, the sinking of the nuclear-powered submarine *Kursk* resulted in the deaths of 118 crewmembers. Perhaps the crew could never have been saved, but the way the crisis was managed was a major debacle for both the Russian Navy and the Russian government.

Instead of providing honest and sincere statements to the media, the Russian Ministry of Defense tried to downplay the crisis by disclosing misleading information, telling the public that the submarine had run aground during a training exercise and that the crew was in no immediate danger. The ministry spread a rumor that there had been a collision with a NATO submarine. Finally the truth came out, and by the time the Russians sought assistance in mounting a rescue mission, it was too late.

Vladimir Putin, Russia's president, received enormous unfavorable publicity for his handling of the crisis. He was vacationing in southern Russia at the time and appeared on Russian television clad in casual clothes, asserting that the situation was under control. He then disappeared from sight for several days, which angered the public and the crew's family members, who believed it indicated his lack of desire to be personally involved in the crisis. When he finally visited the *Kursk*'s home base, he was greeted with anger and hostility.

LESSONS LEARNED

1. Lying to the public is unforgivable.
2. Russia failed to disclose the seriousness of the crisis.
3. Russia failed to ask other countries for assistance in a timely manner.
4. Russia demonstrated a lack of social responsibility by its president's refusal to appear at the site of the crisis and showed a lack of compassion for the victims and their families.

TYLENOL POISONINGS

In September 1982, seven people died after taking Extra-Strength Tylenol laced with cyanide.[1] All of the victims were relatively young. These deaths were the first ever to result from what came to be known as product tampering. All seven individuals died within a one-week time period. The symptoms of cyanide poisoning are rapid collapse and coma and are difficult to treat.

On the morning of September 30, 1982, reporters began calling the headquarters of Johnson & Johnson asking about information on Tylenol and Johnson & Johnson's reaction to the deaths. This was the first that Johnson & Johnson had heard about the deaths and the possible link to Tylenol.

With very little information available at that time and very little time to act, the crisis project was managed using three phases. The first phase was discovery, which included the gathering of all information from every possible source. The full complexity of the problem had to be known as well as the associated risks. Developing a recovery plan is almost impossible without having a complete understanding of the problem. The second phase was the assessment and quantification of the risks and the containment of potential damage. The third phase was the establishment of a recovery plan and risk mitigation. Unlike traditional "life-cycle" phases, which could be months or years in duration, these phases would be in hours or days.

From the start, the company found itself entering into a closer relationship with the press than it was accustomed to. James Burke, the CEO, quickly decided to elevate the management of the crisis to the corporate level, personally taking charge of the company's response and delegating responsibility for running the rest of the company to other members of the executive committee.

There were several reasons why Burke decided that it was necessary to take control of the situation himself. First, Burke believed that the crisis could become a national crisis with the future of self-medication at stake. Second, Burke recognized that the reputation of Johnson & Johnson was now at stake. Third, Burke knew that the McNeil Division, which manufactured Tylenol, might not be able to battle the crisis alone.

The fourth reason was the need for a Johnson & Johnson corporate spokesperson. James Burke was about to become the corporate spokesperson. One of Burke's first decisions was to completely cooperate with the news media. The general public, medical community, and Food and Drug Administration (FDA) were immediately notified.

Instead of providing incomplete information or only the most critical pieces and stonewalling the media, Burke provided all information available. He quickly

---

[1] For additional information on the Tylenol crises, see H. Kerzner, *Project Management Case Studies*, 2nd ed. (Hoboken, NJ: Wiley, 2006), pp. 487–514; also T. Moore, "The Fight to Save Tylenol," *Fortune*, November 29, 1982.

and honestly answered all questions from anyone. This was the first time that a corporate CEO had become so visible to the media and the public, and one of the few times that a CEO appeared on telelvision. James Burke spoke with an aura of trust.

Stakeholder Management

Burke had several options available to him in handling the crisis. Deciding which option to select would certainly be a difficult decision. Looking over Burke's shoulder were the stakeholders who would be affected by Johnson & Johnson's decision. Among the stakeholders were stockholders, lending institutions, employees, managers, suppliers, government agencies, and consumers.

Consumers: Consumers had the greatest stake in the crisis because their lives were on the line. Consumers must have confidence in the products they purchase and believe that they are safe to use as directed.

Stockholders: Stockholders had a financial interest in the selling price of the stock and the dividends. If the costs of removal and replacement or, in the worst-case scenario of product redesign, were substantial, it could lead to a financial hardship for some investors who were relying on the income.

Lending Institutions: Lending institutions provide loans and lines of credit. If the present and/or future revenue stream is impaired, then the funds available might be reduced and the interest rate charge could increase. The future revenue stream of the company's products could affect the quality rating of its debt.

Government: The primary concern of the government was in protecting public health. In this regard, government law enforcement agencies were committed to holding the responsible party accountable. Other government agencies would provide assistance in promoting and designing tamper-resistant packages in an effort to restore consumer confidence.

Management: Company management had the responsibility to protect the image of the company as well as its profitability. To do this, management must convince the public that it will take whatever steps are necessary to protect the consumer.

Employees: Employees have the same concerns as management but are also somewhat worried about possible loss of income or even employment.

Whatever decision Johnson & Johnson made was certain to displease at least some of the stakeholders. Therefore, how does a company decide which stakeholders' needs are more important? How does a company prioritize stakeholders?

For Jim Burke and the entire strategy committee, the decision was not very difficult—just follow the corporate credo. For more than 45 years, Johnson & Johnson had a corporate credo that clearly stated that the company's first priority is to the users of Johnson & Johnson's products and services. Everyone knew the credo, what it stood for, and the fact that it must be followed. The corporate credo guided the decision-making process, and everyone knew it without having to be told. The credo stated that the priorities, in order, were:

1. To the consumers
2. To the employees
3. To the communities being served
4. To the stockholder

When the crisis had ended, Burke recalled that no meeting had been convened for the first critical decision: to be open with the press and put the consumer's' interests first. "Every one of us knew what we had to do," he commented. "There was no need to meet. We had the credo philosophy to guide us."

Solution

The solution was new packaging. Tylenol capsules were reintroduced in November in triple-seal, tamper-resistant packaging, with the new packages beginning to appear on retail shelves in December. Despite the unsettled conditions at McNeil caused by the withdrawal of the Tylenol capsules in October, the company, with its new triple-sealed package, was the first in the industry to respond to the national mandate for tamper-resistant packaging and the new regulations from the FDA.

By Christmas week 1982, Tylenol had recovered 67 percent of its original market share. The product was coming back faster and stronger than the company had anticipated.

Tamper-Resistant Packaging

After the tampering in the Chicago area in the fall of 1982, the over-the-counter medications and the packaging industries examined a broad range of technologies designed to protect the consumer against product tamperings. Johnson & Johnson conducted an exhaustive review of virtually every viable technology. Ultimately, for Tylenol capsules, a triple-safety-sealed system was selected that consisted of glued flaps on the outer box, a tight printed plastic seal over the cap and neck of the bottle, and a strong foil seal over the mouth of the bottle.

The decision to reintroduce capsules was based on marketing research done during that time. This research indicated that the capsules remained the dosage form of choice for many consumers. Many felt that capsules were easier to swallow, and some felt that they provided more potent pain relief. While there was no basis in fact for the latter perception—tablets, caplets, and capsules are all equally effective—it was not an irrelevant consideration. "To the extent that some people think a pain reliever may be more powerful, a better result can often be achieved," said Burke. "This is due to a placebo effect but is nonetheless beneficial to the consumer."

Given these findings and Johnson & Johnson's confidence in the new triple-seal packaging system, the decision was made to reintroduce Extra-Strength Tylenol pain relievers in capsule form.

FOUR YEARS LATER: SECOND TYLENOL POISONINGS

On Monday afternoon, February 10, 1986, Johnson & Johnson was informed that Diane Elsroth, a young woman in Westchester County, New York, died of cyanide poisoning after ingesting Extra-Strength Tylenol capsules.[2] Johnson & Johnson immediately sent representatives to Yonkers to attempt to learn more and to assist in the investigation. Johnson & Johnson also began conferring by telephone with the FDA and the Federal Bureau of Investigation, both in Washington and at the respective field offices.

Johnson & Johnson endorsed the recommendation of the FDA and local authorities that people in the Bronxville/Yonkers, New York, area not take any of these capsules until the investigation was completed. Although from the outset Johnson & Johnson had no reason to believe that this was more than an isolated event, Johnson & Johnson concurred with the FDA recommendation that nationally no one take any capsules from the affected lot number ADF 916 until further notice. Consumers were asked to return products from this lot for credit or exchange. The tainted bottle was part of a batch of 200,000 packages shipped to retailers during the previous August, 95 percent of which had already been sold to consumers. Johnson & Johnson believed other people would have reported problems months before if the batch had been tainted either at the manufacturing plant or at distribution sites.

Once again Johnson & Johnson made its CEO, James Burke, its lead spokesperson. And once again the media treated Johnson & Johnson fairly because of the openness and availability of Burke and other Johnson & Johnson personnel. Johnson & Johnson responded rapidly and honestly to all information requests by the media, a lesson remembered from the 1982 Tylenol tragedy.

Withdrawal from the Direct-to-Consumer Capsule Business

Clearly, circumstances were different in 1986. In light of the Westchester events, Johnson & Johnson no longer believed that it could provide an adequate level of assurance of the safety of hollow-capsule products sold directly to consumers. For this reason, the decision was made to go out of that business. "We take this action with great reluctance and a heavy heart," Burke said in announcing the

---

[2] Robert D. McFadden, "Maker of Tylenol Discontinuing All Over-Counter Drug Capsules," *New York Times*, February 18, 1986. www.nytimes.com/1986/02/18/us/maker-of-tylenol-discontinuing-all-over-counter-drug-capsules.HTML?Pagewanted=All.

action. "But we cannot control random tampering with capsules after they leave our plant. Therefore, we feel our obligation to consumers is to remove capsules from the market to protect the public."[2] The company, Burke said, had "fought our way back" after the deaths of seven people who ingested cyanide-laced Tylenol capsules in 1982, and "we will do it again. We will encourage consumers to use either the solid tablets or caplets," he said. "The caplet is especially well-suited to serve the needs of capsule users. It is oval-shaped like a capsule, 35 percent smaller and coated to facilitate swallowing. It is also hard like a tablet and thus extremely difficult to violate without leaving clearly visible signs. We developed this dosage form as an alternative to the capsule. Since its introduction in 1984, it has become the analgesic dosage form of choice for many consumers."

"While this decision is a financial burden to us, it does not begin to compare to the loss suffered by the family and friends of Diane Elsroth," Burke said, his voice quavering as he referred to the woman who died. He expressed, on behalf of the company, "our heartfelt sympathy to Diane's family and loved ones."[3]

In abandoning the capsule business, Johnson & Johnson had taken the boldest option open to it in dealing with an attack on its prized Tylenol line. At the same time, the company used the publicity generated by the latest Tylenol scare as an opportunity to promote other forms of the drug, particularly its caplets. Johnson & Johnson initially balked at leaving the Tylenol capsule business, which represented about one-third of the $525 million in Tylenol sales in 1985. Burke said he was loath to take such a step, noting that if "we get out of the capsule business, others will get into it." Also, he said, pulling the capsules would be a "victory for terrorism."

People thought that the 1982 tampering incident meant an end to Tylenol. Now the same people believed that Tylenol would survive because it had demonstrated once that it could do so. If survival would be in caplets, then the industry would eventually follow. The FDA was planning to meet with industry officials to discuss what technological changes might be necessary to respond effectively to this problem. At stake was a reexamination of over-the-counter capsules that included dozens of products ranging from Contac decongestant to Dexatrim diet formula.

Conclusion

During congressional testimony, Burke said:

> I have been deeply impressed by the commitment and performance of government agencies, especially the FDA and the FBI. I cannot imagine how any organizations could have been more professional, more energetic or more rational in exercising their responsibilities to the American public.

---

[3] Ibid.

In addition, the media performed a critical role in telling the public what it needed to know in order to provide for its own protection. In the vast majority of instances, this was accomplished in a timely and accurate fashion. Within the first week following the Westchester incidents, polling revealed nearly 100 percent of consumers in the New York area were aware of the problem. I believe this is an example of how a responsible press can serve the public well-being.[4]

Accolades and Support

Once again, Johnson & Johnson received high marks, this time for the way it handled the second Tylenol tragedy. This was evident from the remarks of Robert Simonds, McNeil's vice president of marketing. "There's a real unity of purpose and positive attitude here at McNeil. Any reservations people had about the effect of removing capsules from the company's product line are long since resolved. We realized there was no good alternative" to the action announced by Johnson & Johnson chairman James E. Burke, he said, "as we simply could not guarantee the safety of the capsule form."

As in the aftermath of the 1982 tragedies, Mr. Simonds added, "everybody at McNeil pulled together." More than 300 McNeil employees manned the consumer phone lines, with four of them fielding calls in Spanish.

The employees were buoyed by unsolicited testimonials from consumers. One man in Savannah, Georgia, wrote Burke, "You and your people deserve the right to walk with pride!" Members of a fifth-grade class in Manchester, Missouri, wrote that they "will continue to support and buy your products."

Many of the correspondents did not know what they could do to help but clearly wanted to do something. A man in Yonkers, New York (near where the tampering occurred), for example, wrote, "In my small way I am donating a check in the amount of $10 to offset some of the cost of this kind of terrorism. I would like to donate it in the names of my two children, Candice and Jennifer. Now it [the capsule recall and discontinuation] will only cost Johnson & Johnson $149,999,990." (The check was returned with thanks.)

Nor were the testimonials and words of thanks just from consumers. The *New York Times* said that in dealing with a public crisis "in a forthright way and with his decision to stop selling Tylenol in capsule form, Mr. Burke is receiving praise from analysts, marketing experts and from consumers themselves."[5]

The *Cleveland Plain Dealer* said, "The decision to withdraw all over-the-counter capsule medications, in the face of growing public concern about the vulnerability of such products, was sadly but wisely arrived at."

---

[4] From J. E. Burke's testimony to the U.S. Senate Committee on Labor and Human Resources, February 28, 1986

[5] Steven Prokesch, "Man in the News; a Leader in a Crisis: James Edward Burke," *New York Times*, February 19, 1986. www.nytimes.com/1986/02/19/business/man-in-the-newws-a-leader-in-a-crisis-james-edward-burke.html.

The magazine *U.S. News & World Report* wrote: "No company likes bad news, and too few prepare for it. For dealing with the unexpected, they could take lessons from Johnson & Johnson."

And columnist Tom Blackburn in the *Miami News* put it this way: "Johnson & Johnson is in business to make money. It has done that very well. But when the going gets tough, the corporation gets human, and that makes it something special in the bloodless business world."

Perhaps most significantly, President Reagan, opening a meeting of the Business Council (comprised of corporate chief executives) in Washington, D.C., said, "Let me congratulate one of your members, someone who in recent days has lived up to the highest ideals of corporate responsibility and grace under pressure. Jim Burke of Johnson & Johnson, you have our deepest admiration."[6]

That kind of support from many sources has spurred McNeil, and the results are evident. "There is absolutely no doubt about it. We're coming back—again!"[7]

LESSONS LEARNED

1. On crisis projects, the (executive) project sponsor will be more actively involved and may end up performing as the project manager as well.
2. The project sponsor should function as the corporate spokesperson, responsible for all crisis communications. Strong communication skills are therefore mandatory.
3. Open and honest communications are essential.
4. The company must display a social consciousness as well as a sincere concern for people, especially victims and their families
5. Managing stakeholders with competing demands is essential.
6. The company, and especially the project sponsor, must maintain a close working relationship with the media.
7. A crisis committee should be formed and composed of the senior-most levels of management.
8. Corporate credos can shorten the response time during a crisis.
9. The company must be willing to seek help from all stakeholders and possibly also government agencies.
10. Corporate social responsibility must be a much higher priority than corporate profitability.

(continued)

---

[6] President Praises J&J Chairman in Tylenol Scare," Associated Press, February 19, 1986, www.apnewsarchive.com/1986/President-Praises-J-J-Chairman-in-Tylenol-Scare/id-a428ac5850e97f-6bada027adb37a2ca0.

[7] Johnson & Johnson Corporate Public Relations, "Tylenol Begins Making a Solid Recovery," *Worldwide*, 1986.

(continued)

11. The company, specifically the project sponsor, must appear at the scene of the crisis and demonstrate a sincere compassion for the families of those injured.
12. The company must try to prevent a bad situation from getting worse.
13. The company must manage the crisis as though all information is public knowledge.
14. The company must act quickly and with sincerity.
15. The company must assume responsibility for its products and services and its involvement in the crisis.

VICTIMS AND VILLAINS

The court of public opinion usually casts the deciding ballot as to whether the company involved in the crisis should be treated as a victim or a villain in the way it handled the crisis. The two determining factors are most often the company's demonstration of corporate social responsibility during the crisis and how well it dealt with the media.

During the Tylenol poisoning, Johnson & Johnson's openness with the media, willingness to accept full responsibility for its products, and rapid response to the crisis irrespective of the cost were certainly viewed favorably by the general public. Johnson & Johnson was viewed as a victim of the crisis.

Table I shows how the general public views company' performance during crises. The longer a crisis lasts, the greater the tendency that the company will be portrayed as a villain.

TABLE I   PUBLIC OPINION VIEW OF CRISIS MANAGEMENT

Crisis	Public Opinion View
Tylenol poisonings	Victim
Challenger explosion	Villain
Columbia reentry disaster	Villain
Exxon Valdez oil spill	Villain
Russian submarine *Kursk*	Villain
Ford and Firestone	Villains
Concorde: Air France	Victim
Concorde: British Airways	Villain
Intel and Pentium	Villain

LIFE-CYCLE PHASES

Crises can be shown to go through the life cycles illustrated in Figure I. Unlike traditional project management life-cycle phases, each of these phases can be measured in hours or days rather than in months. Unsuccessful management of any of these phases can lead to a corporate disaster.

Most crises are preceded by early-warning signs or risk triggers indicating that a crisis may occur. This is the early-warning phase. Typical warning signals might include violations of safety protocols during technology development, warnings from government agencies, public discontent, complaints from customers, and warnings/concerns from lower-level employees.

Most companies are poor at risk management, especially at evaluation of early-warning signs. Intel and the shuttle disasters were examples of this. Today, project managers are trained in the concepts of risk management, but specifically related to project management or product development. Once the product is commercialized, the most serious early-warning indicators can appear, and, by that time, the project manager may have been reassigned to another project. Someone else must then evaluate the early-warning signs.

Early-warning signs are indicators of potential risks. Because time and money are a necessity for evaluation of these indicators, the ability to evaluate all risks is precluded. Therefore, companies must be selective in the risks they consider.

The next life-cycle phase is understanding the problem causing the crisis. For example, during the Tylenol poisonings, once the deaths were related to the Tylenol capsules, the first concern was to discover whether the capsules were contaminated during the manufacturing process (i.e., an inside job) or during distribution and sales (i.e., an outside job). Without a fact-based understanding of the crisis, the media can formulate their own cause of the problem and pressure the company to follow the wrong path.

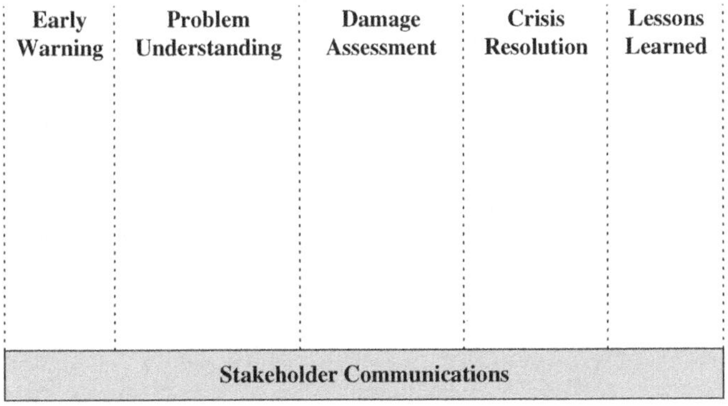

Early Warning	Problem Understanding	Damage Assessment	Crisis Resolution	Lessons Learned
	Stakeholder Communications			

FIGURE I Crisis management life-cycle phases

The third life-cycle phase is the damage assessment phase. The magnitude of the damage usually determines the method of resolution. Underestimating the magnitude of the damage and procrastination can cause the problem to escalate to a point where the cost of correcting the problem can grow by orders of magnitude. Intel found this out the hard way.

The crisis resolution stage is where the company announces its approach to resolve the crisis. The way the public views the company's handling of the crisis has the potential to make or break the company.

The final stage, lessons learned, mandates that companies learn not only from their own crises but also from how others handled crises. Learning from the mistakes of others is better than learning from one's own mistakes.

Perhaps the most critical component in Table I is stakeholder communications. When a crisis occurs, the assigned project manager may need to communicate with stakeholders who previously were of minor importance, such as the media and government agencies, all of whom may have competing interests. These competing interests mandate that project managers understand stakeholder needs and objectives and also possess strong communication skills, conflict resolution skills, and negotiation skills.

QUESTIONS

1. When a crisis project occurs, who should be the leader of the crisis team?
2. Will there be a crisis committee or a crisis project sponsor?
3. How important is effective communication during a crisis?
4. How important is stakeholder relations management during a crisis?
5. Should a company immediately assume responsibility for a crisis?
6. How important is response time when a crisis occurs?
7. How important is it to show compassion for people who may have been injured?
8. How important is it to maintain a paperwork trail?
9. How important is it to capture lessons learned?

Is It Fraud?

BACKGROUND

Paul was a project management consultant and often helped the Judge Advocate General's office (JAG) by acting as an expert witness in lawsuits filed by the U.S. government against defense contractors. While most lawsuits were based upon unacceptable performance by the contractors, this lawsuit was different; it was based upon supposedly superior performance.

MEETING WITH COLONEL JENSEN

Paul sat in the office of Army colonel Jensen listening to the colonel's description of the history behind this contract. Colonel Jensen stated:

> We have been working with the Welton Company for almost ten years. This contract was one of several contracts we have had with them over the years. It was a one year contract to produce 1500 units for the Department of the Navy. Welton told us during contract negotiations that they needed two quarters to develop their manufacturing plans and conduct procurement. They would then ship the Navy 750 units at the end of the third quarter and the remaining 750 units at the end of the fourth quarter. On some other contracts, manufacturing planning and procurement was done in less than one quarter. On other contracts similar to this one, the Navy would negotiate a firm-fixed-price contract because the risk to both the buyer and seller was quite low. The government's proposal statement of

work also stated that this would be a firm-fixed-price contract. But during final contract negotiations, Welton became adamant in wanting this contract to be an incentive-type contract with a bonus for coming in under budget and/or ahead of schedule. We were somewhat perplexed about why they wanted an incentive contract. Current economic conditions in the United States were poor during the time we did the bidding and companies like Welton were struggling to get government contracts and keep their people employed. Under these conditions, we believed that they would want to take as long as possible to finish the contract just to keep their people working.

Their request for an incentive contract made no sense to us, but we reluctantly agreed to it. We often change the type of contract based on special circumstances. We issued a fixed-price-incentive-fee contract with a special incentive clause for a large bonus should they finish the work early and ship all 1,500 units to the Navy. The target cost for the contract, including $10 million in procurement, was $35 million with a sharing ratio of 90%–10% and a profit target of $4 million. The point of total assumption was at a contract price of $43.5 million.

Welton claimed that they finished their procurement and manufacturing plans in the first quarter of the year. They shipped the Navy 750 units at the end of the second quarter and the remaining 750 units at the end of the third quarter. According to their invoices, which we audited, they spent $30 million in labor in the first nine months of the contract and $10 million in procurement. The government issued them checks totaling $49.5 million. That included $43.5 million plus the incentive bonus of $6 million for early delivery of the units.

The JAG office believes that Welton took advantage of the Department of the Navy when [they] demanded and received a fixed-price-incentive-fee contract. We want you to look over their proposal and what they did on the contract and see if anything looks suspicious.

CONSULTANT'S AUDIT

The first thing that Paul did was to review the final costs on the contract.

Labor:	$30,000.000
Material:	$10,000,000
	$40,000,000
Cost overrun:	$ 5,000,000
Welton's cost:	$ 500,000
Final profit:	$ 3,500,000

Welton completed the contract exactly at the contract price ceiling, also the point of total assumption, of $43.5 million.

The cost overrun of $5 million was entirely in labor. Welton originally expected to do the job in 12 months for $25 million in labor. That amounted to an average monthly labor expenditure of $2,083,333. But Welton actually spent

$30 million in labor over nine months, which amounted to an average monthly labor cost of $3,333,333. Welton was spending about $1.25 million more per month than planned for during the first nine months. Welton explained that part of the labor overrun was due to overtime and using more people than anticipated. It was pretty clear in Paul's mind what Welton had done. Welton overspent the labor by $5 million, and only $500,000 of the overrun was paid by Welton because of the sharing ratio. In addition, Welton received a $6 million bonus for early delivery. Simply stated, Welton received $6 million for a $500,000 investment.

Paul knew that believing this to be true was one thing, but being able to prove this in court would require more supporting information. His next step was to read the proposal that Welton submitted. On the bottom of the first page of the proposal was a paragraph entitled "Truth of Negotiations," which stated that everything in the proposal was the truth. The letter was signed by a senior officer at Welton.

Paul then began reading the management section of the proposal. In the management section, Welton bragged about previous contracts almost identical to this one with the Department of the Navy and other government organizations. Welton also stated that most of the people used on this contract had worked on the previous contracts. Paul found other statements in the proposal that implied that the manufacturing plans for this contract were similar to those of other contracts, and Paul now wondered why two quarters were needed to develop the manufacturing plans for this project. Paul was now convinced that something was wrong.

QUESTIONS

1. What information does Paul have to support his belief that something is wrong?
2. Knowing that you are not an attorney, does it appear from a project management perspective that sufficient information exists for a possible lawsuit to recover all or part of the incentive bonus for early delivery?
3. How do you think this case study ended? (It is a factual case, and the author was the consultant.)

The Management Reserve

BACKGROUND

A project sponsor forces the project management to include a management reserve in the cost of a project. However, the project sponsor intends to use the management reserve for his own "pet" project, and this creates problems for the project manager.

SOLE-SOURCE CONTRACT

The Structural Engineering Department at Avcon, Inc., made a breakthrough in the development of a high-quality, low-weight composite material. Avcon believed that the new material could be manufactured inexpensively and its clients would benefit by lowering their manufacturing and shipping costs.

News of the breakthrough spread through the industry. Avcon was asked by one of its most important clients to submit an unsolicited proposal for design, development, and testing of products for the client using the new material. Jane would be the project manager. She had worked with the client previously as the project manager on several other projects that were considered successes.

MEETING WITH TIM

Because of the relative newness of the technology, both Avcon and the client understood that this could not be a firm-fixed-price contract. They ultimately agreed to a cost-plus-incentive-fee contract type. However, the target costs still had to be determined.

Jane worked with all of the functional managers to determine what their efforts would be on this contract. The only unknown was the time and cost needed for structural testing. Structural testing would be done by the Structural Engineering Department, which was responsible for making the technical breakthrough.

Tim was head of the Structural Engineering Department. Jane set up a meeting to discuss the cost of testing on this project. During the meeting, Tim replied: "A full test matrix will cost about $100,000. I believe that we should price out the full test matrix and also include a management reserve of at least $100,000 should anything go wrong."

Jane was a little perplexed about adding in a management reserve. Tim was usually right on the money on his estimates, and Jane knew from previous experience that a full test matrix might not be needed. But Tim was the subject matter expert, and Jane reluctantly agreed to include in the contract a management reserve of $100,000. As Jane was about to exit Tim's office, Tim said: "Jane, I had requested to be your project sponsor on this effort, and management has given me the okay. You and I will be working together on this effort. As such, I would like to see all of the cost figures before submitting the final bid to the client."

REVIEWING COST FIGURES

Jane had worked with Tim before, but not in a situation where Tim would be the project sponsor. However, it was common on some contracts that lower and middle levels of management would assume the sponsorship role rather than having all sponsorship at the top of the organization. Jane met with Tim and showed him the following information, which would appear in the proposal:

- Sharing ratio: 90–10%
- Contract cost target: $800,000
- Contract profit target: $50,000
- Management reserve: $100,000
- Profit ceiling: $70,000
- Profit floor: $35,000

Tim looked at the numbers and Jane could see that he was somewhat unhappy.

Tim then said, "Jane, I do not want to identify to the client that we have a management reserve. Let's place the management reserve in with the $800,000 and change the target cost to $900,000. I know that the cost baseline should not include the management reserve, but in this case I believe it is necessary to do so."

Jane knew that the cost baseline of a project does not include the management reserve, but there was nothing she could do; Tim was the sponsor and had the final say. Jane simply could not understand why Tim was trying to hide the management reserve.

EXECUTION BEGINS

Tim instructed Jane to include in the structural test matrix work package the entire management reserve of $100,000. Jane knew from previous experience that a full test matrix was not required and that the typical cost of this work package should be between $75,000 and $90,000. Establishing a work package of $200,000 meant that Tim had complete control over the management reserve and how it would be used.

Jane was now convinced that Tim had a hidden agenda. Unsure what to do next, Jane contacted a colleague in the Project Management Office. The colleague informed Jane that Tim had tried unsuccessfully to get some of his pet projects included in the portfolio of projects, but management refused to include any of those projects in the budget for the portfolio.

It was now clear what Tim was asking Jane to be part of and why Tim had requested to be the project sponsor. Tim was forcing Jane to violate PMI's Code of Ethics and Professional Conduct.

QUESTIONS

1. Why did Tim want to add in a management reserve?
2. Why did Tim want to become the project sponsor?
3. Are Tim's actions a violation of the Code of Ethics and Professional Conduct?
4. If Jane follows Tim instructions, is Jane also in violation of the Code of Ethics and Professional Responsibility?
5. What are Jane's options if she decides not to follow Tim's instructions?

Part 14

MANAGING SCOPE CHANGES

Scope changes on a project can occur regardless of how well the project is planned or executed. Scope changes can be the result of something that was omitted during the planning stage, due to changing customer requirements, or due to changes in technology that have taken place.

The two most commonly used methods for scope change control are (1) allowing continuous scope changes to occur but under the guidance of the configuration management process and (2) clustering all scope changes together to be accomplished later as an enhancement project. There are risks and rewards in each of these methods. The decision of when to select one over the other is not always black or white, but more so a gray area.

Denver International Airport

BACKGROUND

How does one convert a $1.2 billion project into a $5.0 billion project? It's easy. Just build a new airport in Denver. The decision to replace Denver's Stapleton Airport with Denver International Airport (DIA) was made by well-intentioned city officials. The city of Denver would need a new airport eventually, and it seemed like the right time to build an airport that would satisfy Denver's needs for at least 50 to 60 years. DIA could become the benchmark for other airports to follow.

A summary of the critical events is listed next:

1985: Denver Mayor Federico Pena and Adams County officials agree to build a replacement for Stapleton International Airport.

Project estimate: $1.2 billion

1986: Peat Marwick, a consulting firm, is hired to perform a feasibility study including projected traffic. Results indicate that, depending on the season, as many as 50 percent of the passengers would change planes. The new airport would have to handle this smoothly. United and Continental object to the idea of building a new airport, fearing the added cost burden.

May 1989: Denver voters pass an airport referendum.

Project estimate: $1.7 billion

March 1993: Denver Mayor Wellington Webb announces the first delay. Opening day would be postponed from October 1993 to December 1993. (Federico Pena becomes Secretary of Transportation under Clinton.)

Project estimate: $2.7 billion

October 1993: Opening day is to be delayed to March 1994. There are problems with the fire and security systems in addition to the inoperable baggage handling system.

Project estimate: $3.1 billion

December 1993: The airport is ready to open, but without an operational baggage handling system. Another delay is announced.

February 1994: Opening day is to be delayed to May 15, 1994, because of baggage handling system.

May 1994: Airport misses the fourth deadline.

August 1994: DIA finances a backup baggage handling system. Opening day is delayed indefinitely.

Project estimate: $4 billion plus

December 1994: Denver announces that DIA may have been built over an old Native American burial ground. An agreement is reached to bless the site.

AIRPORTS AND AIRLINE DEREGULATION

Prior to the Airline Deregulation Act of 1978, airline routes and airfare were established by the Civil Aeronautics Board (CAB). Airlines were allowed to charge whatever they wanted for airfare, if approved by CAB. The cost of additional aircraft was eventually passed on to the consumer. Prior to the Act, the high cost for airfare restricted travel to businesspeople and wealthy individuals who could afford it.

Before deregulation, increases in passenger travel were moderate. Most airports were already underutilized, and growth was achieved by adding terminals or runways on existing airport sites. The need for new airports was not deemed critical for the near term.

Following deregulation, the airline industry had to prepare for open market competition. This meant that airfares were expected to decrease dramatically. Airlines began purchasing large numbers of planes, and most routes were fair game. Airlines had to purchase more planes and fly more routes in order to remain profitable. The increase in passenger traffic was expected to come from the average person who could finally afford air travel.

Deregulation made it clear that airport expansion would be necessary. While airport management conducted feasibility studies, the recession of 1979 to 1983 occurred. Several airlines, such as Braniff, filed for bankruptcy protection under Chapter 11, and the airline industry headed for consolidation through mergers and leveraged buyouts.

Cities took a wait-and-see attitude rather than risk billions in new airport development. Noise abatement policies, environmental protection acts, and land acquisition were viewed as headaches. The only major airport that had been built in the previous twenty years was Dallas–Ft. Worth, which was completed in 1974.

DOES DENVER NEED A NEW AIRPORT?

In 1974, even prior to deregulation, Denver's Stapleton Airport was experiencing such rapid growth that Denver's Regional Council of Governments concluded that Stapleton would not be able to handle the traffic expected by the year 2000. Modernization of Stapleton could have extended the inevitable problem to 2005. But were the headaches with Stapleton better cured through modernization or by building a new airport? There was no question that insufficient airport capacity would cause Denver to lose valuable business. Because Denver is 500 miles from any other major city, air travel was very important to it.

In 1988, Denver's Stapleton International Airport ranked as the fifth busiest in the country, with 30 million passengers. The busiest airports were Chicago, Atlanta, Los Angeles, and Dallas–Ft. Worth. By the year 2000, Denver anticipated 66 million passengers, just below Dallas–Ft. Worth's 70 million and Chicago's 83 million estimates.

Delays at Denver's Stapleton Airport caused major delays at all other airports. By one estimate, bad weather in Denver caused up to $100 million in lost income to the airlines each year because of delays, rerouting, canceled flights, putting travelers into hotels overnight, employee overtime pay, and passengers switching to other airlines.

United Airlines and Continental comprised 80 percent of all flights in and out of Denver. Table I shows the service characteristics of United and Continental between December 1993 and April 1994. Table II shows all of the airlines serving Denver as of June 1994. Exhibit III shows the cities serviced from Denver. It should be obvious that delays in Denver could cause delays in each of these cities. Figure I shows the top 10 domestic passenger origin-destination markets from Denver Stapleton.

Stapleton was ranked as one of the 10 worst air traffic bottlenecks in the United States. Even low clouds at Denver Stapleton could bring delays of 30 to 60 minutes.

Stapleton had two parallel north–south runways that were close together. During bad weather where instrument landing conditions exist, the two runways were considered as only one. This drastically reduced takeoffs and landings each hour. The new airport would have three north–south runways initially with a master plan calling for eight eventually. This would triple or quadruple instrument flights occurring at the same time to 104 aircraft per hour. Stapleton could handle only 30 landings per hour under instrument conditions with a *maximum* of 80 aircraft per hour during clear weather.

TABLE I   SERVICE CHARACTERISTICS: UNITED AIRLINES AND CONTINEN-TAL AIRLINES, DECEMBER 1993 AND APRIL 1994

	Enplaned Passengers[a]	Scheduled Seats[b]	Boarding Load Factor	Scheduled Departures[b]	Average Seats per Departure
December 1993					
United Airlines	641,209	1,080,210	59%	7,734	140
United Express	57,867	108,554	53%	3,582	30
Continental Airlines	355,667	624,325	57%	4,376	143
Continental Express	52,680	105,800	50%	3,190	33
Other	2_3_6_,_7_5_1	3_5_7_,_2_1_4	66%	2_,_8_5_1	125
Total	1,344,174	2,276,103	59%	21,733	105
April 1994					
United Airlines	717,093	1,049,613	68%	7,743	136
United Express	44,451	92,880	48%	3,395	27
Continental Airlines	275,948	461,168	60%	3,127	147
Continental Express	24,809	92,733	27%	2,838	33
Other	2_3_4_,_0_9_1	3_5_4_,_9_5_0	66%	2_,_8_3_3	125
Total	1,296,392	2,051,344	63%	19,936	103

[a] Airport management records.

[b] Official Airline Guides, Inc. (online database), for periods noted.

TABLE II   AIRLINES SERVING DENVER, JUNE 1994

Major/National Airlines	Regional/Commuter Airlines
America West Airlines	Air Wisconsin (United Express)[b]
American Airlines	Continental Express
Continental Airlines	GP Express Airlines
Delta Air Lines	Great Lakes Aviation (United Express)
Markair	Mesa Airlines (United Express)
Midway Airlines	Midwest Express[b]
Morris Air[a]	
Northwest Airlines	**Cargo Airlines**
TransWorld Airlines	Airborne Express
United Airlines	Air Vantage
USAir	Alpine Air
	American International Airways Ameriflight
Charter Airlines	Bighorn Airways
Aero Mexico	Burlington Air Express
American Trans Air	Casper Air
Casino Express	Corporate Air
Express One	DHL Worldwide Express

Great American	Emery Worldwide
Private Jet	Evergreen International Airlines EWW Airline/Air Train
Sun Country Airlines	Federal Express Kitty Hawk
Foreign Flag Airlines (scheduled)	Majestic Airlines
Martinair Holland	Reliant Airlines
Mexicana de Aviacion	United Parcel Service
	Western Aviators

[a] Morris Air was purchased by Southwest Airlines in December 1993. The airline announced that it would no longer serve Denver as of October 3, 1994.

[b] Air Wisconsin and Midwest Express have both achieved the level of operating revenues needed to qualify as a national airline as defined by the Federal Aviation Administration. However, for purposes of this report, these airlines are referred to as regional airlines.

Source: Airport management, June 1994.

TABLE III   TOP 10 DOMESTIC PASSENGER ORIGIN-DESTINATION MARKETS AND AIRLINE SERVICE, STAPLETON INTERNATIONAL AIRPORT (FOR THE 12 MONTHS ENDED SEPTEMBER 30, 1993)

City of Origin or Destination[a]	Air Miles from Denver	% Certificated Airline Passengers	Average Daily Nonstop Departures[b]
1. Los Angeles[c]	849	6.8	34
2. New York[d]	1,630	6.2	19
3. Chicago[e]	908	5.6	26
4. San Francisco[f]	957	5.6	29
5. Washington, D.C.[g]	1,476	4.9	12
6. Dallas–Fort Worth	644	3.5	26
7. Houston[h]	864	3.2	15
8. Phoenix	589	3.1	19
9. Seattle	1,019	2.6	14
10. Minneapolis	693	2.3	16
Cities listed		43.8	210
All others		56.2	241
Total		100.0	451

[a] Top 10 cities based on total inbound and outbound passengers (on large certificated airlines) at Stapleton International Airport in 10 percent sample for the 12 months ended September 30, 1993.

[b] Official Airline Guides, Inc. (on-line database), April 1994. Includes domestic flights operated at least four days per week by major/national airlines and excludes the activity of foreign-flag and commuter/regional airlines.

[c] Los Angeles International, Burbank–Glendale–Pasadena, John Wayne (Orange County), Ontario International, and Long Beach Municipal Airports.

[d] John F. Kennedy International, LaGuardia, and Newark International Airports.

[e] Chicago-O'Hare International and Midway Airports.

[f] San Francisco, Metropolitan Oakland, and San Jose International Airports.

[g] Washington Dulles International, Washington National, and Baltimore/Washington International Airports.

[h] Houston Intercontinental and William P. Hobby Airports.

Sources: U.S. Department of Transportation/Air Transport Association of America, "Origin-Destination Survey of Airline Passenger Traffic, Domestic," third quarter 1993, except as noted.

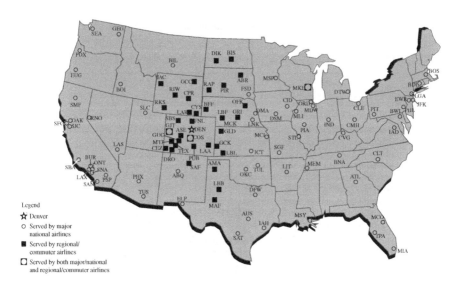

FIGURE I   U.S. airports served nonstop from Denver
Source: Official Airline Guides, Inc. (online database), June 1994.

The runway master plan called for ten 12,000-foot and two 16,000-foot runways. By opening day, three north–south and one east–west 12,000-foot runways would be in operation, and one of the 16,000-foot north–south runways would be operational shortly thereafter.

The airfield facilities also included a 327-foot Federal Aviation Administration (FAA) air traffic control tower (the nation's tallest) and base building structures. The tower's height allowed controllers to visually monitor runway thresholds as much as three miles away. The runway/taxiway lighting system, with lights embedded in the concrete pavement to form centerlines and stopbars at intersections, would allow air traffic controllers to signal pilots to wait on taxiways and cross active runways and to lead them through the airfield in poor visibility.

Due to shifting winds, runway operations were shifted from one direction to another. At the new airport, the changeover would require four minutes as opposed to the 45 minutes at Stapleton.

Sufficient spacing was provided for in the concourse design such that two FAA Class 6 aircraft (i.e., 747-XXX) could operate back to back without impeding each other.

Even when two aircraft (one from each concourse) had pushed back at the same time, there could still exist room for a third FAA Class 6 aircraft to pass between them.

City officials believed that Denver's location, being equidistant from Japan and Germany, would allow twin-engine, extended-range transports to reach both countries nonstop. The international opportunities were there. Between late 1990 and early 1991, Denver was entertaining four groups of leaders per month from Pacific Rim countries to look at DIA's planned capabilities.

In the long term, Denver saw the new airport as a potential hub for Northwest or USAir. This would certainly bring more business to Denver. Very few airports in the world can boast multiple hubs.

THE ENPLANED PASSENGER MARKET

Perhaps the most critical parameter that illustrates the necessity for a new airport is the enplaned passenger market. (An enplaned passenger is one who gets on a flight, either an origination flight or connecting flight.)

Table IV identifies the enplaned passengers for individual airlines servicing Denver Stapleton for 1992 and 1993.

Connecting passengers were forecast to decrease about 1 million between 1993 and 1995 before returning to a steady growth of 3.0 percent per year, totaling 8,285,500 in 2000. As a result, the number of connecting passengers was forecast

TABLE IV ENPLANED PASSENGERS BY AIRLINE, 1992–1993, STAPLETON INTERNATIONAL AIRPORT

Enplaned Passengers	1992	1993
United	6,887,936	7,793,246
United Express[a]	470,841	578,619
	7,358,777	8,371,865
Continental	5,162,812	4,870,861
Continental Express	514,293	532,046
	5,677,105	5,402,907
American Airlines	599,705	563,119
America West Airlines	176,963	156,032
Delta Air Lines	643,644	634,341
MarkAir	2,739	93,648
Northwest Airlines	317,507	320,527
TransWorld Airlines	203,096	182,502
USAir	201,949	197,095
Other	256,226	398,436
	2,401,829	2,545,700
Total	15,437,711	16,320,472

[a] Includes Mesa Airlines, Air Wisconsin, Great Lakes Aviation, and Westair Airlines.

Source: Department of Aviation management records.

to represent a smaller share (46 percent) of total enplaned passengers at the airport in 2000 than in 1993 (50 percent). Total enplaned passengers at Denver were forecast to increase from 16,320,472 in 1993 to 18,161,000 in 2000—an average increase of 1.5 percent per year (decreasing slightly from 1993 through 1995, then increase 2.7 percent per year after 1995).

The increase in enplaned passengers would necessitate an increase in the number of aircraft departures. Since landing fees are based on aircraft landed weight, more arrivals and departures will generate more landing fee revenue. Since airport revenue is derived from cargo operations as well as passenger activities, it was important to recognize that enplaned cargo would also be expected to increase.

LAND SELECTION

The site selected was a 53-square-mile area 18 miles northeast of Denver's business district. The site would be larger than the Chicago O'Hare and Dallas–Ft. Worth airports combined. State law prohibited political entities from annexing land without the consent of its residents (the land was in Adams County). Before the vote was taken, Adams County and Denver negotiated an agreement limiting noise and requiring the creation of a buffer zone to protect surrounding residents. The agreement also included continuous noise monitoring as well as limits on such businesses as airport hotels that could be in direct competition with existing services provided in Adams County. The final part of the agreement limited DIA to such businesses as airline maintenance, cargo, small-package delivery, and other such airport-related activities.

With those agreements in place, Denver annexed 45 square miles and purchased an additional 8 square miles for noise buffer zones. Denver rezoned the buffer area to prohibit residential development within a 65 LDN (Level Day/Night) noise level. (LDN is a weighted noise measurement intended to determine perceived noise in both day and night conditions.) Adams County enacted even stiffer zoning regulations, calling for no residential development with an LDN noise level of 60.

Most of the airport land embodied two ranches. About 550 people were relocated. The site had overhead power lines and gas wells, which were relocated or abandoned. The site lacked infrastructure development, and there were no facilities for providing water, power, sewage disposal, or other such services.

FRONT RANGE AIRPORT

Located 2.5 miles southeast of DIA is Front Range Airport, which had been developed to relieve Denver's Stapleton Airport of most nonairline traffic operations.

Information in the section "Land Selection" is from David A. Brown, "Denver Aims for Global Hub Status with New Airport Under Construction," *Aviation Week and Space Technology,* March 11, 1991, pp. 42–44.

As a satellite airport to DIA, Front Range Airport was offering six aviation business services by 1991:

1. Air cargo and air freight, including small package services (This is direct competition for DIA.)
2. Aircraft manufacturing
3. Aircraft repair (This is direct competition for DIA.)
4. Fixed base operators to service general (and corporate) aviation
5. Flight training
6. Military maintenance and training

The airport was located on a 4,800-acre site and was surrounded by a 12,000-acre industrial park. The airport was owned and operated by Adams County, which had completely different ownership from DIA. By 1991, Front Range Airport had two east–west runways: a 700-foot runway for general aviation use and an 8,000-foot runway to be extended to 10,000 feet. By 1992, the general plans called for two more runways to be built, both north–south. The first runway would be 10,000 feet initially with expansion capability to 16,000 feet to support wide-body aircraft. The second runway would be 7,000 feet to service general aviation.

Opponents of DIA contended that Front Range Airport could be enlarged significantly, thus reducing pressure on Denver's Stapleton Airport, and that DIA would not be necessary at that time. Proponents of DIA argued that Front Range should be used to relieve pressure on DIA if and when DIA became a major international airport as all expected. Both sides were in agreement that, initially, Front Range Airport would be a competitor to DIA.

AIRPORT DESIGN

The Denver International Airport was based on a "home-on-the-range" design. The city wanted a wide-open entry point for visitors. In spring of 1991, the city began soliciting bids.

To maintain a distinctive look that would be easily identified by travelers, a translucent tentlike roof was selected. The roof was made of two thicknesses of translucent, Teflon-coated glass fiber material suspended from steel cables hanging from the structural supports. The original plans for the roof called for a conventional design using 800,000 tons of structural steel. The glass fiber roof would require only 30,000 tons of structural steel, thus providing substantial savings on construction costs. The entire roof would permit about 10 percent of the sunlight to shine through, thus providing an open, outdoor-like atmosphere.

The master plan for the airport called for four concourses, each with a maximum of 60 gates. However, only three concourses would be built initially, and none would be full size. The first, Concourse A, would have 32 airline gates and six commuter gates. This concourse would be shared by Continental and any future international carriers. Continental had agreed to give up certain gate positions if requested to do so in order to accommodate future international operations. Continental was the only long-haul international carrier, with one daily flight to London. Shorter international flights were to Canada and Mexico. Concourses B and C would each have 20 gates initially for airline use plus six commuter gates. Concourse B would be the United Concourse. Concourse C would be for all carriers other than Continental or United.

All three concourses would provide a total of 72 airline gates and 18 commuter gates. This would be substantially less than what the original master plan called for.

Although the master plan identified 60 departure gates for each concourse, cost became an issue. The first set of plans identified 106 departure gates (not counting commuter gates) and was then scaled down to 72 gates. United Airlines originally wanted 45 departure gates but settled for 20. The recession was having its effect.

The original plans called for a train running through a tunnel beneath the terminal building and the concourses. The train would carry 6,000 passengers per hour. Road construction on and adjacent to the airport was planned to take one year. Runway construction was planned to take one year but was deliberately scheduled for two years in order to save on construction costs.

The principal benefits of the new airport compared to Stapleton were:

- A significantly *improved airfield configuration* that allowed for triple simultaneous instrument landings in all weather conditions, improved efficiency and safety of airfield operations, and reduced taxiway congestion.
- *Improved efficiency in the operation of the regional airspace,* which, coupled with the increased capacity of the airfield, was supposed to significantly reduce aircraft delays and airline operating costs both at Denver and system-wide.
- *Reduced noise impacts* resulting from a large site that was situated in a relatively unpopulated area.
- A *more efficient terminal/concourse/apron layout* that minimized passenger walking distance, maximized the exposure of concessions to passenger flows, provided significantly greater curbside capacity, and allowed for the efficient maneuvering of aircraft in and out of gates.
- *Improved international facilities* including longer runway lengths for improved stage length capability for international flights and larger Federal Inspection Services facilities for greater passenger processing capability.
- *Significant expansion capability* of each major functional element of the airport.
- *Enhanced efficiency of airline operations* as a result of new baggage handling, communications, deicing, fueling, mail sorting, and other specialty systems.

One of the problems with the airport design related to the high wind shears that would exist where the runways were placed. This could eventually become a serious issue.

PROJECT MANAGEMENT

The city of Denver selected two companies to assist in the project management process. The first was Greiner Engineering, an engineering, architecture, and airport planning firm. The second company was Morrison-Knudsen Engineering, which is a design-construct firm. The city of Denver and Greiner/Morrison-Knudsen Engineering would function as the project management team (PMT) responsible for schedule coordination, cost control, information management, and administration of approximately 100 design contracts, 160 general contractors, and more than 2,000 subcontractors.

In the selection of architects, it became obvious that there would be a split between those who would operate the airport and the city's aspirations. Airport personnel were more interested in an easy-to-clean airport and convinced the city to hire a New Orleans–based architectural firm with which Stapleton personnel had worked previously. The city wanted a "thing of beauty" rather than an easy-to-clean venue.

In an unusual division of responsibilities, the New Orleans firm was contracted to create standards that would unify the entire airport and to take the design of the main terminal only through schematics and design development, at which point it would be handed off to another firm. This sharing of the wealth with several firms would later prove more detrimental than beneficial.

The New Orleans architectural firm complained that the direction given by airport personnel focused on operational issues rather than aesthetic values. Furthermore, almost all decisions seemed to be made in reaction to maintenance or technical issues. This created a problem for the design team because the project's requirements specified that the design reflect a signature image for the airport, one that would capture the uniqueness of Denver and Colorado.

The New Orleans team designed a stepped-roof profile supported by an exposed truss system over a large central atrium, resembling the structure of train sheds. The intent was to bring the image of railroading, which was responsible for Denver's early growth, into the jet age.

The mayor, city council, and others were concerned that the design did not reflect a $2 billion project. A blue-ribbon commission was formed to study the matter. The city council eventually approved the design.

Financial analysis of the terminal indicated that the roof design would increase the cost of the project by $48 million and would push the project off schedule. A second architectural firm was hired. The final design was a peaked roof with Teflon-coated fabric designed to bring out the image of the Rocky Mountains. The second architectural firm had the additional responsibility to take the project from design development through to construction. The cost savings from the new design was so substantial that the city upgraded the floor finish in the terminal and doubled the size of the parking structure to 12,000 spaces.

The effectiveness of the PMT was being questioned. The PMT failed to sort out the differences between the city's aspirations and the operators' maintenance orientation. It failed to detect the cost and constructability issues with the first design even though both PMT partners had vast in-house expertise. The burden of responsibility was falling on the shoulders of the architects. The PMT also did not appear to be aware that the first design might not have met the project's standards.

Throughout the design battle, no one heard from the airlines. Continental and United controlled 80 percent of the flights at Stapleton. Yet the airlines refused to participate in the design effort, hoping the project would be canceled. The city ordered the design teams to proceed for bids without any formal input from the users.

With a recession looming and Continental fighting for survival, the city needed the airlines to sign on. To entice the airlines to participate, the city agreed to a stunning range of design changes while assuring the bond rating agencies that the 1993 opening date would be kept. Continental convinced Denver to move the international gates away from the north side of the main terminal to terminal A and to build a bridge from the main terminal to terminal A. This duplicated the function of a below-ground people-mover system. A basement was added the full length of the concourses. A second level was added to the service cores located between gates.

United's changes were more significant. It widened concourse B by eight feet to accommodate two moving walkways in each direction. It added a second level of service cores and had the roof redesigned to provide a clerestory of natural light.

Most important, United wanted a destination-coded vehicle (DCV) baggage handling system where bags could be transferred between gates in less than 10 minutes, thus supporting short turnaround times. The DCV was to be on Concourse B (United) only. Within a few weeks, DIA proposed that the baggage handling system be extended to the entire airport. Yet even with these changes in place, United and Continental *still* did not sign a firm agreement with DIA, thus keeping bond interest expense at a higher-than-anticipated level. Some people contended that United and Continental were holding DIA hostage.

From a project management perspective, there was no question that disaster was on the horizon. Nobody knew what to do about the DCV system. The risks were unknown. Nobody realized the complexity of the system, especially the software requirements. By one account, the launch date should have been delayed by at least two years. The contract for DCV hadn't been awarded yet, and terminal construction was already under way. Everyone wanted to know why the design (and construction) was not delayed until after the airlines had signed on.

How could DIA install and maintain the terminal's baseline design without having a design for the baggage handling system? Everyone felt that what they were now building would have to be ripped apart.

There were going to be massive scope changes. DIA management persisted in its belief that the airport would open on time. Work in process was now $130 million per month. Acceleration costs, because of the scope changes, would be $30 to $40 million. Three shifts were running at DIA with massive overtime. People were getting burned out to the point where they couldn't continue.

To reduce paperwork and maintain the schedule, architects became heavily involved during the construction phase, which was highly unusual. The PMT seemed to be abdicating control to the architects who would be responsible for coordination. The trust that had developed during the early phases began evaporating.

Even the car rental companies got into the act. They balked at the fees for their in-terminal location and said that servicing within the parking structures was inconvenient. They demanded and finally received a separate campus. Passengers would now be forced to take shuttle buses out of the terminal complex to rent or return vehicles.

THE BAGGAGE HANDLING SYSTEM

DIA's $200 million baggage handling system was designed to be state of the art. Conventional baggage handling systems are manual. Each airline operates its own system. DIA opted to buy a single system and lease it back to the airlines. In effect, it would be a one-baggage-system-fits-all configuration.

The system would contain 100 computers, 56 laser scanners, conveyor belts, and thousands of motors. As designed, the system would contain 400 fiberglass carts, each carrying a single suitcase through 22 miles of steel tracks. Operating at 20 miles per hour, the system could deliver 60,000 bags per hour from dozens of gates. United was worried that passengers would have to wait for luggage since several of its gates were more than a mile from the main terminal. The system design was for the luggage to go from the plane to the carousel in eight to 10 minutes. The luggage would reach the carousel before the passengers.

The baggage handling system would be centered on track-mounted cars propelled by linear induction motors. The cars slow down, but don't stop, as a conveyor ejects bags onto their platform. During the induction process, a scanner reads the bar-coded label and transmits the data through a programmable logic controller to a radio frequency identification tag on a passing car. At this point, the car knows the destination of the bag it is carrying, as does the computer software that routes the car to its destination. To illustrate the complexity of the situation, consider 4,000 taxicabs in a major city, all without drivers, being controlled by a computer through the city streets.

EARLY RISK ANALYSIS

Construction began in 1989 without a signed agreement from Continental and United. By March 1991, the bidding process was in full swing for the main terminal, concourses, and tunnel. Preliminary risk analysis involved three areas: cost, human resources, and weather.

1. *Cost.* The grading of the terminal area was completed at about $5 million under budget, and the grading of the first runway was completed at about $1.8 million under budget. This led management to believe that the original construction cost estimates were accurate. Also, many of the construction bids being received were below the city's own estimates.

2. *Human resources.* The economic recession hit Denver a lot harder than the rest of the nation. DIA was at that time employing about 500 construction workers. By late 1992, it was anticipated that 6,000 construction workers would be needed. Although more than 3,000 applications were on file, there remained the question of available, qualified labor. If the recession were to be prolonged, the lack of qualified suppliers could be an issue as well.

3. *Bad weather.* Bad weather, particularly in the winter, was considered to be the greatest risk to the schedule. Fortunately, the winters of 1989–1990 and 1990–1991 were relatively mild, which gave promise to future mild winters. Actually, more time was lost due to bad weather in the summer of 1990 than in either of the two previous winters.

MARCH 1991

By early March 1991, Denver had already issued more than $900 million in bonds to begin construction of the new airport. Denver planned to issue another $500 million in bonds the following month. Standard & Poor's (S&P) lowered the rating on the DIA bonds from BBB to BBB–, just a notch above the junk grade rating. This could prove to be extremely costly to DIA because any downgrading in bond quality ratings would force DIA to offer higher yields on its new bond offerings, thus increasing the yearly interest expense.

Denver was in the midst of a mayoral race. Candidates were calling for the postponement of the construction, not only because of the lower ratings, but also because Denver *still* did not have a firm agreement with either Continental or United Airlines that they would use the new airport. The situation became more intense because three months earlier, in December of 1990, Continental had filed for bankruptcy protection under Chapter 11. Fears existed that Continental might drastically reduce the size of its hub at DIA or even pull out altogether.

Denver estimated that cancellation or postponement of the new airport would be costly. The city had $521 million in contracts that could not be canceled.

Approximately $22 million had been spent in debt service for the land, and $38 million in interest on the $470 million in bond money was already spent. The city would have to default on more than $900 million in bonds if it could not collect landing fees from the new airport. The study also showed that a two-year delay would increase the total cost by $2 billion to $3 billion and increase debt service to $340 million per year. It now appeared that the point of no return was at hand.

Fortunately for DIA, Moody's Investors Service, Inc. did *not* lower its rating on the $1 billion outstanding of airport bonds. Moody's confirmed its conditional Baa1 rating, which was slightly higher than the S&P rating of BBB–. Moody's believed that the DIA effort was a strong one and that even at depressed airline traffic levels, DIA would be able to service its debt for the scaled-back airport. Had both Moody's and S&P lowered their ratings together, DIA's future might have been in jeopardy.

APRIL 1991

Denver issued $500 million in serial revenue bonds with a maximum yield of 9.185 percent for bonds maturing in 2023. A report by Fitch Investors Service estimated that the airport was ahead of schedule and 7 percent below budget. The concerns of the investor community seemed to have been tempered despite the bankruptcy filing of Continental Airlines. However, there was still concern that no formal agreement existed between DIA and either United Airlines or Continental Airlines.

MAY 1991

The city of Denver and United Airlines finally reached a tentative agreement. United would use 45 of the potential 90 to 100 gates at Concourse B. This would be a substantial increase from the 26 gates DIA had originally thought that United would require. The 50 percent increase in gates would also add 2,000 reservations jobs. United also expressed an interest in building a $1 billion maintenance facility at DIA employing 6,000 people.

United stated later that the agreement did not constitute a firm commitment but was contingent on legislative approval of a tax incentive package of $360 million over 30 years plus $185 million in financing and $23 million in tax exemptions. United would decide by the summer in which city the maintenance facility would be located. United reserved the right to renegotiate the hub agreement if DIA was not chosen as the site for the maintenance facility.

Some people believed that United had delayed signing a formal agreement until it was in a strong bargaining position. With Continental in bankruptcy and DIA beyond the point of no return, United was in a favorable position to demand tax incentives of $200 million in order to keep its hub in Denver and build a maintenance facility. The state legislature would have to be involved in approving the incentives. United Airlines ultimately located the $1 billion maintenance facility at the Indianapolis Airport.

AUGUST 1991

Hotel developers expressed concern about building at DIA, which is 26 miles from downtown; in contrast Stapleton is eight miles from downtown Denver. DIA officials initially planned for a 1,000-room hotel attached to the airport terminal, with another 300 to 500 rooms adjacent to the terminal. The 1,000-room hotel had been scaled back to 500 to 700 rooms and was not likely to be ready when the airport was scheduled to open in October 1993. Developers had expressed resistance to building close to DIA unless industrial and office parks were also built near the airport. Even though ample land existed, developers were putting hotel development on the back burner until after 1993.

NOVEMBER 1991

Federal Express and United Parcel Service (UPS) planned to move cargo operations to the smaller Front Range Airport rather than to DIA. The master plan for DIA called for cargo operations to be at the northern edge of DIA, thus increasing the time and cost for deliveries to Denver. Shifting operations to Front Range Airport would certainly have been closer to Denver but would have alienated northern Adams County cities that counted on an economic boost in their areas. Moving cargo operations would have been in violation of the original agreement between Adams County and Denver for the annexation of the land for DIA.

The cost of renting at DIA was estimated at $0.75 per square foot, compared to $0.25 per square foot at Front Range. DIA would have higher landing fees of $2.68 per 1,000 pounds compared to $2.15 for Front Range. UPS demanded a cap on landing fees at DIA if another carrier were to go out of business. Under the UPS proposal, area landholders and businesses would set up a fund to compensate DIA if landing fees were to exceed the cap. Cargo carriers at Stapleton were currently paying $2 million in landing fees and rental of facilities per year.

As the dogfight over cargo operations continued, the FAA issued a report calling for cargo operations to be colocated with passenger operations at the busier metropolitan airports. This included both full cargo carriers as well as passenger cargo (i.e., "belly cargo") carriers. Proponents of Front Range argued that the report didn't preclude the use of Front Range because of its proximity to DIA.

DECEMBER 1991

United Airlines formally agreed to a 30-year lease for 45 gates at Concourse B. With the firm agreement in place, the DIA revenue bonds shot up in price almost

TABLE V   AIRLINE AGREEMENTS

Airline	Term (Years)	Number of Gates
American Airlines	5	3
Delta Air Lines[a]	5	4
Frontier Airlines	10	2
MarkAir	10	5
Northwest Airlines	10	2
TransWorld Airlines	10	2
USAir[a]	5	2
Total		20

[a] The city has entered into use and lease agreements with these airlines. The USAir lease is for one gate on Concourse C, and USAir has indicated its interest in leasing a second gate on Concourse C.

$30 per $1,000 bond. Earlier in the year, Continental signed a five-year lease agreement.

Other airlines also agreed to service DIA. Table V sets forth the airlines that either executed use and lease agreements for, or indicated an interest in leasing, the 20 gates on Concourse C on a first-preferential-use basis.

JANUARY 1992

BAE Automated Systems was selected to design and build the baggage handling system. The airport had been under construction for three years before BAE was brought on board. BAE agreed to do eight years of work in two years to meet the October 1993 opening date.

JUNE 1992

DIA officials awarded a $24.4 million contract for the new airport's telephone services to U.S. West Communication Services. Officials of DIA had considered controlling its own operations through shared tenant service, which would allow the airport to act as its own telephone company. All calls would be routed through an airport-owned computer switch. By grouping tenants together into a single shared entity, the airport would be in a position to negotiate discounts with long-distance providers, thus enabling cost savings to be passed on to the tenants.

By one estimate, the city would generate $3 million to $8 million annually in new, nontax net revenue by owning and operating its own telecommunication network. Unfortunately, DIA officials did not feel that sufficient time existed for them to operate their own system. The city of Denver was unhappy over this lost income.

SEPTEMBER 1992

By September 1992, the city had received $501 million in FAA grants and $2.3 billion in bonds with interest rates of 9.0 to 9.5 percent in the first issue to 6 percent in the latest issue. The decrease in interest rates due to the recession was helpful to DIA. The rating agencies also increased the city's bond rating one notch.

The FAA permitted Denver to charge a $3 departure tax at Stapleton with the income earmarked for construction of DIA. Denver officials estimated that over 34 years, the tax would generate $2.3 billion.

The cities bordering the northern edge of DIA (where the cargo operations were to be located) teamed up with Adams County to file lawsuits against DIA in its attempt to relocate cargo operations to the southern perimeter of DIA. This relocation would appease the cargo carriers and hopefully end the yearlong battle with Front Range Airport. The Adams County commissioner contended that relocation would violate the Clean Air Act and the National Environmental Policy Act and would be a major deviation from the original airport plan approved by the FAA.

OCTOBER 1992

The city issued $261 million of Airport Revenue Bonds for the construction of facilities for United Airlines. (See Appendix A at the end of this case.)

MARCH 1993

The city of Denver announced that the launch date for DIA would be pushed back to December 18 rather than the original October 30 date in order to install and test all of the new equipment. The city wanted to delay the opening until late in the first quarter of 1994 but deemed it too costly because the airport's debt would have to be paid without an adequate stream of revenue. The interest on the bond debt was now at $500,000 per day.

The delay to December 18 angered the cargo carriers. This would be their busiest time of the year, usually twice their normal cargo levels, and a complete revamping of their delivery service would be needed to compensate for the delay. The Washington-based Air Freight Association urged the city to allow the cargo carriers to fly out of Stapleton through the holiday period.

By March 1993, Federal Express, Airborne Express, and UPS (reluctantly) had agreed to house operations at DIA after the city pledged to build facilities for them at the south end of the airport. Negotiations were also under way with Emery Worldwide and Burlington Air Express. The "belly" carriers, Continental and United, had already signed on.

UPS had wanted to create a hub at Front Range Airport. If Front Range were a cargo-only facility, it would free up UPS from competing with passenger traffic for runway access even though both Front Range and DIA were in the same air traffic control pattern. UPS stated that it would not locate a regional hub at DIA. This would mean the loss of a major development project that would have attracted other businesses that relied on UPS delivery.

For UPS to build a regional hub at Front Range would have required the construction of a control tower and enlargement of the runways, both requiring federal funds. The FAA refused to free up the funds needed for this construction, largely due to a lawsuit by United Airlines and environmental groups.

United had an ulterior motive in filing their lawsuit. Adams County officials repeatedly stated that they had no intention of building passenger terminals at Front Range. However, if federal funds were given to Front Range, a commercial airline could not be prevented from setting up shop there. The threat to United was that low-cost carriers such as Southwest Airlines might elect to operate out of Front Range rather than DIA. Because costs were fixed, fewer passengers traveling through DIA meant less profit for the airlines. United simply did not want any airline activities removed from DIA.

AUGUST 1993

Plans for a train to connect downtown Denver to DIA were under way. A $450,000 feasibility study and federal environmental assessment were being conducted, with the results due November 30, 1993. Union Pacific had spent $350,000 preparing a design for the new track, which could be constructed in 13 to 16 months. The major hurdle would be the financing, which was estimated between $70 million and $120 million, based on hourly trips or 20-minute trips. The more frequent the trips, the higher the cost.

The feasibility study also considered the possibility of baggage check-in at each of the stops. This would require financial support and management assistance from the airlines.

SEPTEMBER 1993

Denver officials disclosed plans for transferring airport facilities and personnel from Stapleton to DIA. The move would be stage-managed by Larry Sweat, a retired military officer who had coordinated troop movements for Operation Desert Shield. Bechtel Corporation would be responsible for directing the transport and setup of machinery, computer systems, furniture, and service equipment, all of which had to be accomplished overnight since the airport had to be operational again in the morning.

OCTOBER 1993

DIA, which was already $1.1 billion over budget, was to be delayed again. The new opening date would be March 1994. The city blamed the airlines for the delays, citing the numerous scope changes required. Even the fire safety system hadn't been completed.

Financial estimates became troublesome. Airlines would have to charge a $15 per person tax, the highest in the nation. Fees and rent charged the airlines would triple from $74 million at Stapleton to $247 million at DIA.

JANUARY 1994

Front Range Airport and DIA were considering the idea of being designated as one system by the FAA. Front Range could legally be limited to cargo only. This would also prevent low-cost carriers from paying lower landing fees and rental space fees at Front Range.

FEBRUARY 1994

Southwest Airlines said that it would not service DIA. As a low-cost, no-frills carrier, Southwest wanted to keep its airport fees below $3 a passenger. Current projections indicated that DIA would have to charge between $15 and $20 per passenger in order to service its debt. This was based on a March 9 opening day.

Continental announced that it would provide a limited number of low-frills service flights in and out of Denver. Furthermore, Continental said that because of the high landing fees, it would cancel 23 percent of its flights through Denver and relocate some of its maintenance facilities.

United Airlines expected its operating cost to be $100 million more per year at DIA than at Stapleton. With the low-cost carriers either pulling out or reducing service to Denver, United was under less pressure to lower airfares.

MARCH 1994

The city of Denver announced the fourth delay in opening DIA, from March 9 to May 15. The cost of the delay, $100 million, would be paid mostly by United and Continental. As of March, only Concourse C, which housed the carriers other than United and Continental, was granted a temporary certificate of occupancy by the city.

As the finger-pointing began, blame for this delay was given to the baggage handling system, which was experiencing late changes, restricted access flow, and a slowdown in installation and testing. A test by Continental Airlines indicated that only 39 percent of baggage was delivered to the correct location. Other problems also existed. As of December 31, 1993, there were 2,100 design changes. The city of Denver had taken out insurance for construction errors and omissions. The city's insurance claims cited failure to coordinate design of the ductwork with ceiling and structure, failure to properly design the storm draining systems for the terminal to prevent freezing, failure to coordinate mechanical and structural designs of the terminal, and failure to design an adequate subfloor support system.

Consultants began identifying potential estimating errors in DIA's operations. The runways at DIA were six times longer than the runways at Stapleton, but DIA had purchased only 25 percent more equipment. DIA's cost projections would be $280 million for debt service and $130 million for operating costs, for a total of $410 million per year. The total cost at Stapleton was $120 million per year.

APRIL 1994

DIA began having personnel problems. According to DIA's personnel officer, Linda Rubin Royer, relocating the city's airport 17 miles away from the site of the old Stapleton Airport was creating serious problems. One of the biggest issues was the additional 20-minute drive that employees had to bear. To resolve this problem, she proposed a car/van pooling scheme and tried to get the city bus company to transport people to and from the new airport. There was also the problem of transferring employees to similar jobs elsewhere if they truly disliked working at DIA. The scarcity of applicants was creating a problem as well.

MAY 1994

S&P lowered the rating on DIA's outstanding debt to the noninvestment grade of BB, citing the problems with the baggage handling system with no immediate cure in sight. Denver was currently paying $33.3 million per month to service debt. Stapleton was generating $17 million per month, and United Airlines had agreed to pay $8.8 million in cash for the next three months only. That left a current shortfall of $7.5 million each month that the city would have to fund. Beginning in August 1994, the city would be burdened with $16.3 million in debt financing each month.

BAE personnel began to complain that they were pressured into doing the impossible. The only other system of this type in the world was in Frankfurt, Germany. That system required six years to install and two years to debug. BAE was asked to do it all in two years.

BAE underestimated the complexity of the routing problems. During trials, cars crashed into one another, luggage was dropped at the wrong location, cars that were needed to carry luggage were routed to empty waiting pens, and some cars traveled in the wrong direction. Sensors became coated with dirt, throwing the system out of alignment, and luggage was dumped prematurely because of faulty latches that jammed cars against the side of a tunnel. By the end of May, BAE was conducting a worldwide search for consultants who could determine what was going wrong and how long it would take to repair the system.

BAE conducted an end-of-month test with 600 bags. Outbound (terminal to plane), the sort accuracy was 94 percent; inbound, the accuracy was 98 percent.

The system had a zero downtime for both inbound and outbound testing. The specification requirements called for 99.5 percent accuracy.

BAE hired three technicians from Germany's Logplan, which helped solve similar problems with the automated system at Frankfurt. With no opening date set, DIA contemplated opening the east side of the airport for general aviation and air cargo flights. That would begin generating at least some revenue.

JUNE 1994

The cost for DIA was now approaching $3.7 billion, and jokes about DIA appeared everywhere. One common joke was that when you fly to Denver, you will have to stop in Chicago to pick up your luggage. Others included the abbreviation DIA as part of the joke.

The people who were not laughing at these jokes were the concessionaires, including about 50 food service operators, who had been forced to rehire, retrain, and reequip, at considerable expense. Several small businesses were forced to call it quits because of the eight-month delay. Red ink was flowing despite the fact that the $45-a-square-foot rent would not have to be paid until DIA officially opened. Several of the concessionaires had requested that the rent be cut by $10 a square foot for the first six months or so after the airport opened. A merchant's association was formed at DIA to fight for financial compensation.

THE PROJECT'S WORK BREAKDOWN STRUCTURE

The city had managed the design and construction of the project by grouping design and construction activities into seven categories, or *areas:*

Area #0 Program management/preliminary design

Area #1 Site development

Area #2 Roadways and on-grade parking

Area #3 Airfield

Area #4 Terminal complex

Area #5 Utilities and specialty systems

Area #6 Other

Since the fall of 1992, the project budget had increased by $224 million (from $2.7 billion to over 2.9 billion), principally as a result of scope changes.

- Structural modifications to the terminal buildings (primarily in the Landside Terminal and Concourse B) to accommodate the automated baggage system

- Changes in the interior configuration of Concourse B
- Increases in the scope of various airline tenant finished, equipment, and systems, particularly in Concourse B
- Grading, drainage, utilities, and access costs associated with the relocation of air cargo facilities to the south side of the airport
- Increases in the scope and costs of communication and control systems, particularly premises wiring
- Increases in the costs of runway, taxiway, and apron paving and change orders as a result of changing specifications for the runway lighting system
- Increased program management costs because of schedule delays

Yet even with all of these design changes, the airport was ready to open except for the baggage handling system.

JULY 1994

The Securities and Exchange Commission (SEC) disclosed that DIA was one of 30 municipal bond issuers under investigation for improper contributions to the political campaigns of Pena and his successor, Mayor Wellington Webb. Citing public records, Pena was said to have received $13,900, and Webb's campaign fund increased by $96,000. The SEC said that the contributions might have been in exchange for the right to underwrite DIA's municipal bond offerings. Those under investigation included Merrill Lynch, Goldman Sachs & Co., and Lehman Brothers, Inc.

AUGUST 1994

Continental confirmed that as of November 1, 1994, it would reduce its flights out of Denver from 80 to 23. At one time, Continental had 200 flights out of Denver.

Denver announced that it expected to sell $200 million in new bonds. Approximately $150 million would be used to cover future interest payments on existing DIA debt and to replenish interest and other money paid due to the delayed opening.

Approximately $50 million would be used to fund the construction of an interim baggage handling system of the more conventional tug-and-conveyor type. The interim system would require 500 to 600 people rather than the 150 to 160 people needed for the computerized system. Early estimates said that the conveyor belt/tug-and-cart system would be at least as fast as the system at Stapleton and would be built using proven technology and off-the-shelf parts. However, modifications would have to be made to both the terminal and the concourses.

United Airlines asked for a 30-day delay in approving the interim system for fear that it would not be able to satisfy the airline's requirements. The original lease agreement with DIA and United stipulated that, on opening day, there would be a fully operational automated baggage handling system in place. United had 284 flights a day out of Denver and had to be certain that the interim system would support a 25-minute turnaround time for passenger aircraft.

The city's District Attorney's Office said it was investigating accusations of falsified test data and shoddy workmanship at DIA. Reports had come in regarding fraudulent construction and contracting practices. No charges were filed at that time.

DIA began repairing cracks, holes, and fissures that had emerged in the runways, ramps, and taxiways. Officials said that the cracks were part of the normal settling problems and might require maintenance for years to come.

United Airlines agreed to invest $20 million and act as the project manager to the baggage handling system at Concourse B. DIA picked February 28, 1995, as the new opening date as long as either the primary or secondary baggage handling system was operational.

UNITED BENEFITS FROM CONTINENTAL'S DOWNSIZING

United had been building up its Denver hub since 1991, increasing its total departures 9 percent in 1992, 22 percent in 1993, and 9 percent in the first six months of 1994. Stapleton is United's second largest connecting hub after Chicago O'Hare (ORD), ahead of San Francisco (SFO), Los Angeles (LAX), and Washington Dulles (IAD) International Airports, as shown in Figure II.

In response to the downsizing by Continental, United is expected to absorb a significant portion of Continental's Denver traffic by means of increased load factors and increased service (i.e., capacity), particularly in larger markets where significant voids in service might be left by Continental. United served 24 of the 28 cities served by Continental from Stapleton in June, 1994, with about 79 percent more total available seats to those cities—23,937 seats provided by United compared with 13,400 seats provided by Continental. During 1993, United's average load factor from Denver was 63 percent, indicating that, with its existing service and available capacity, United had the ability to absorb many of the passengers abandoned by Continental. In addition, United had announced plans to increase service at Denver to 300 daily flights by the end of the calendar year.

As a result of its downsizing in Denver, Continental was forecasted to lose more than 3.9 million enplaned passengers from 1993 to 1995—a total decrease of 80 percent. However, this decrease was expected to be largely offset by the forecasted 2.2 million increase in enplaned passengers by United and 1.0 million by the other airlines, resulting in a total of 15,877,000 enplaned passengers at Denver

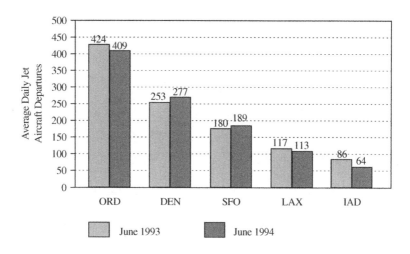

Note: Does not include activity by United Express.

FIGURE II Comparative United Airlines service at hub airports, June 1983 and June 1994 (activity by United Express not included)

Source: Official Airline Guides, Inc. (www.oag.com), for periods shown.

in 1995. As discussed earlier, it was assumed that, in addition to a continuation of historical growth, United and the other airlines would pick up much of the traffic abandoned by Continental through a combination of added service, larger average aircraft size, and increased load factors.

From 1995 to 2000, the increase in total enplaned passengers was based on growth rates of 2.5 percent per year in originating passengers and 3.0 percent per year in connecting passengers. Between 1995 and 2000, United's emerging dominance at the airport (with almost twice the number of passengers of all other airlines combined) should have resulted in somewhat higher fare levels in the Denver markets, and therefore dampened traffic growth. As shown in Figure III, of the 18.2 million forecasted enplaned passengers in 2000, United and United Express together was forecasted to account for 70 percent of total passengers at the airport—up from about 51 percent in 1993—while Continental's share, including GP Express, was forecasted to be less than 8 percent—down from about 33 percent in 1993.

Total connecting passengers at Stapleton increased from about 6.1 million in 1990 to about 8.2 million in 1993—an average increase of about 10 percent per year. The number of connecting passengers was forecast to decrease in 1994 and 1995, as a result of the downsizing by Continental, and then return to

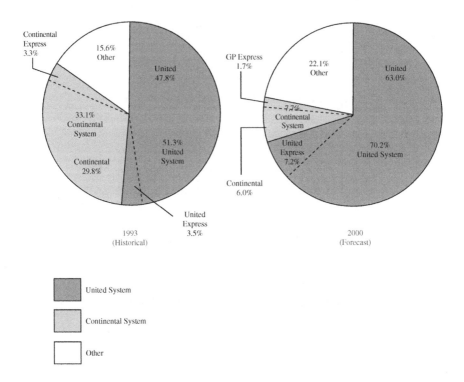

FIGURE III Enplaned passenger market shares at Denver airports
Source: 1993: Airport Management Records

steady growth of 3.0 percent per year through 2000, reflecting expected growth in passenger traffic nationally and a stable market share by United in Denver. Airline market share of connecting passengers in 1993 and 1995 are shown in Figure IV.

SEPTEMBER 1994

Denver began discussions with cash-strapped MarkAir of Alaska to begin service at DIA. For an undercapitalized carrier, the prospects of tax breaks, favorable rents, and a $30 million guaranteed city loan were enticing.

DIA officials estimated an $18 per person charge on opening day. Plans to allow only cargo carriers and general aviation to begin operations at DIA were canceled.

Total construction cost for the main terminal exceeded $455 million (including the parking structure and the airport office building). (See Table VI.)

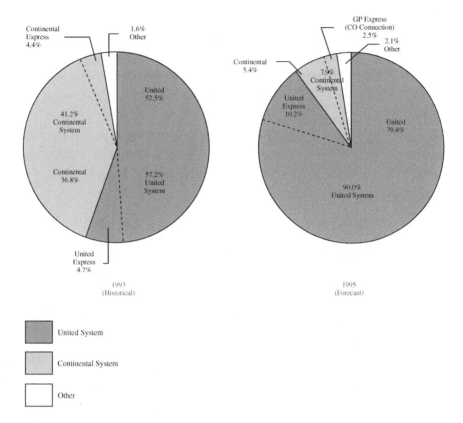

FIGURE IV Connecting passenger market shares at Denver Airports

Source: 1993: Airport Management Records and U.S. Department of Transportation

TABLE VI **TOTAL CONSTRUCTION COSTS FOR DENVER AIRPORT**

General site expenses, commission	$ 38,667,967
Sitework, building excavations	15,064,817
Concrete	89,238,296
Masonry	5,501,608
Metals	40,889,411
Carpentry	3,727,408
Thermal, moisture protection	8,120,907
Doors and windows	13,829,336
Finishes	37,025,019

(continued)

EXHIBIT VI   *(Continued)*

Specialties	2,312,691
Building equipment	227,720
Furnishings	3,283,852
Special construction	39,370,072
Conveying systems	23,741,336
Mechanical	60,836,566
Electrical	73,436,575
Total	$455,273,581

OCTOBER 1994

A federal grand jury convened to investigate faulty workmanship and falsified records at DIA. The faulty workmanship had resulted in falling ceilings, buckling walls, and collapsing floors.

NOVEMBER 1994

The baggage handling system was working, but only in segments. Frustration still existed in not being able to get the whole system to work at the same time. The problem appeared to be with the software required to get computers to talk to computers. The fact that a mere software failure could hold up Denver's new airport for more than a year put into question the project's risk management program. Jerry Waddles was the risk manager for Denver. He left that post to become risk manager for the State of Colorado. Eventually the city found an acting risk manager, Molly Austin Flaherty, to replace Waddles, but for the most part, DIA construction over the past several months had continued without a full-time risk manager. The failure of the baggage handling system had propelled DIA into newspaper headlines around the country. The SEC had launched a probe into whether Denver officials had deliberately deceived bondholders about how equipment malfunctions would affect the December 19, 1993, opening. The allegations were made by Denver's KCNC-TV. Internal memos indicated that by the summer of 1993 city engineers believed it would take at least until March 1994 to get the system working. However, Mayor Webb did not announce the delayed opening until October 1993. The SEC was investigating whether the last postponement misled investors holding $3 billion in airport bonds.

Under a new agreement, the city agreed to pay BAE an additional $35 million for modifications *if* the system was working for United Airlines by February 28, 1995. BAE would then have until August 1995 to complete the rest of the system for the other tenants. If the system was not operational by February 28, the city could withhold payment of the $35 million.

BAE lodged a $40 million claim against the city, alleging that the city caused the delay by changing the system's baseline configuration after the April 1, 1992, deadline. The city filed a $90 million counterclaim, blaming BAE for the delays.

The lawsuits were settled out of court when BAE agreed to pay $12,000 a day in liquidated damages dating from December 19, 1993, to February 28, 1995, or approximately $5 million. The city agreed to pay BAE $6.5 million to cover some invoices submitted by BAE for work already done to repair the system.

Under its DIA construction contract, BAE's risks were limited. BAE's liability for consequential damages resulting from its failure to complete the baggage handling system on time was capped at $5 million. BAE had no intention of being held liable for changes to the system. The system as it was at the time was not the system that BAE had been hired to install.

Additional insurance policies also existed. Builder's risk policies generally pay damages caused by defective parts or materials, but so far none of the parts used to construct the system had been defective. BAE was also covered for design errors or omissions. The unknown risk at that point was who would be responsible if the system worked for Concourse B (i.e., United) but then failed when it was expanded to cover all concourses.

A study was under way to determine the source of respiratory problems suffered by workers at the construction site. The biggest culprit appeared to be the use of concrete in a confined space.

The city and DIA were also protected from claims filed by vendors whose businesses were put on hold because of the delays under a hold-harmless agreement in the contracts. However, the city had offered to permit the concessionaires to charge higher fees and also to extend their leases for no charge to make up for lost income due to the delays.

DECEMBER 1994

The designer of the baggage handling system was asked to reexamine the number of bags per minute that the BAE system was required to accommodate as per the specifications. The contract called for departing luggage to Concourse A to be delivered at a peak rate of 90 bags per minute. The designer estimated peak demand at 25 bags per minute. Luggage from Concourse A was contracted for at 223 bags per minute, but, again, the designer calculated peak demand at a lower rate of 44 bags per minute.

AIRPORT DEBT

By December 1994, DIA was more than $3.4 billion in debt, as shown in Table VII.

TABLE VII   OUTSTANDING DEBT AT DENVER AIRPORT

Series 1984 bonds	$ 103,875,000
Series 1985 bonds	175,930,000
Series 1990A bonds	700,003,843
Series 1991A bonds	500,003,523
Series 1991D bonds	600,001,391
Series 1992A bonds	253,180,000
Series 1992B bonds	315,000,000
Series 1992C bonds	392,160,000
Series 1992D–G bonds	135,000,000
Series 1994A bonds	257,000,000
	$3,432,153,757

AIRPORT REVENUE

Airports generally have two types of contracts with their tenants. The first type is the residual contract where the carriers guarantee that the airport will remain solvent. Under this contract, the carriers absorb the majority of the risk. The airport maintains the right to increase rents and landing fees to cover operating expenses and debt coverage. The second type of contract is the compensatory contract where the airport is at risk. DIA has a residual contract with its carriers.

Airports generate revenue from several sources. The most common breakdown includes landing fees and rent from the following entities: airline carriers, passenger facilities, rental car agencies, concessionary stores, food and beverage services, retail shops, and parking garages. Retail shops and other concessionary stores also pay a percentage of sales.

AIRLINE COSTS PER ENPLANED PASSENGER

Revenues derived from the airlines are often expressed on a per enplaned passenger basis. The average airline cost per enplaned passenger at Stapleton in 1993 was $5.02. However, this amount excludes costs related to major investments in terminal facilities made by United Airlines in the mid-1980s and, therefore, understates the true historical airline cost per passenger.

Average airline costs per enplaned passenger at the airport in 1995 and 2000 are forecast to be as shown in Table VIII.

TABLE VIII   TOTAL AVERAGE AIRLINE COSTS PER ENPLANED PASSENGER

Year	Current Dollars	1990 Dollars
1995	$18.15	$14.92
2000	17.20	11.62

The forecasted airline costs per enplaned passenger at the airport are considerably higher than costs at Stapleton today and are the highest of any major airport in the United States. (The cost per enplaned passenger at Cleveland Hopkins is $7.50.) The relatively high airline cost per passenger is attributable, in part, to (1) the unusually large number of tenant finishes, equipment, and systems costs being financed as part of the project relative to other airport projects and (2) delayed costs incurred since the original opening date for purposes of the Plan of Financing (January 1, 1994).

The city estimates that, as a result of the increased capacity and efficiency of the airfield, operation of the airport will result in annual delay savings to the airlines of $50 million to $100 million per year (equivalent to about $3 to $6 per enplaned passenger), and that other advanced technology and systems incorporated into the design of the airport will result in further operational savings. In the final analysis, the cost effectiveness of operating at the airport is a judgment that must be made by the individual airlines in deciding to serve the Denver market.

It is assumed for the purposes of this analysis that the city and the airlines will resolve the current disputes regarding cost allocation procedures and responsibility for delay costs and that the airlines will pay rates generally in accordance with the procedures of the use and lease agreements as followed by the city and as summarized in the accompanying exhibits.

FEBRUARY 28, 1995

The airport opened as planned on February 28, 1995. However, several problems became apparent. First, the baggage handling system did have "bad days." Passengers traveling to and from Denver felt more comfortable carrying bags than having them transferred by the computerized baggage handling system. Large queues began to form at the end of the escalators in the main terminal going down to the concourse trains. The trains were not running frequently enough, and the number of cars in each train did not appear to be sufficient to handle the necessary passenger traffic.

The author flew from Dallas–Ft. Worth to Denver in one hour and 45 minutes. It then took one hour and 40 minutes to catch the airport shuttles (which stop at all the hotels) and arrive at the appropriate hotel in downtown Denver. Passengers began to balk at the discomfort of the remote rental car facilities, the additional $3 tax per day for each rental car, and the fact that the nearest gas station was 15 miles away. How does one return a rental car with a full tank of gas?

Departing passengers estimated it would take two hours to drive to the airport from downtown Denver, unload luggage, park their automobile, check in, and take the train to the concourse.

Faults in the concourse construction were becoming apparent. Tiles that were supposed to be ⅝ inch thick were found to be ½ inch thick. Tiles began to crack. During rainy weather, rain began seeping in through the ceiling.

QUESTIONS

1. Is the decision to build a new airport at Denver strategically a sound decision?
2. Perform an analysis for strengths, weaknesses, opportunities, and threats (SWOT) on the decision to build DIA.
3. Who are the stakeholders and what are their interests or objectives?
4. Did the airlines support the decision to build DIA?
5. Why was United opposed to expansion at Front Range Airport?
6. Why was the new baggage handling system so important to United?
7. Is DIA a good strategic fit for Continental?
8. What appears to be the single greatest risk in the decision to build DIA?
9. United is a corporation in business to make money. How can United issue tax-free municipal bonds?
10. What impact do the rating agencies (i.e., Moody's and Standard & Poor's) have in the financing of the airport?
11. According to the prospectus, the DIA bonds were rated as BBB by Standard & Poor's Corporation. Yet, at the same time, the city of Denver was given a rating of AA. How can this be?
12. What is the function of the project management team (PMT) and why were two companies involved?
13. When did the effectiveness of the project management team begin to be questioned?
14. Did it sound as though the statement of work/specifications provided by the city to the PMT was "vague" for the design phase?
15. During the design phase, contractors were submitting reestimates for work, 30 days after their original estimates, and the new estimates were up to $50 million larger than the prior estimate. Does this reflect on the capabilities of the PMT?
16. Should the PMT be qualified to perform risk analyses?
17. Why were the architects coordinating the changes at the construction site?
18. Should the PMT have been replaced?
19. Do scope changes reflect on the ineffectiveness of a PMT?
20. Why did United Airlines decide to act as the project manager for the baggage handling system on Concourse B?

REFERENCES

David A. Brown, "Denver Aims for Global Hub Status with New Airport Under Construction," *Aviation Week & Space Technology*, March 11, 1991, pp. 42–45.

"Satellite Airport to Handle Corporate, General Aviation for Denver Area," *Aviation Week & Space Technology*, March 11, 1991, pp. 44–45.

"Denver to Seek Bids This Spring for Wide-Open Terminal Building," *Aviation Week & Space Technology*, March 11, 1991, p. 50.

"Denver City Council Supports Airport Despite Downgrade," *The Wall Street Journal*, March 20, 1991, p. A1D.

"Denver Airport Bonds' Rating Is Confirmed by Moody's Investors," *The Wall Street Journal*, March 22, 1991, p. C14.

"Bonds for Denver Airport Priced to Yield up to 9.185%," *New York Times*, April 10, 1991, p. D16.

Marj Charlier, "Denver Reports a Tentative Agreement with United over Hub at New Airport," *The Wall Street Journal*, May 3, 1991, p. B2.

Brad Smith, "New Airport Has Its Ups and Downs," *Los Angeles Times*, July 9, 1991, p. A5.

Christopher Wood, "Hotel Development at New Airport Not Likely Until After '93," *Denver Business Journal*, August 2, 1991, p. 8S.

Christopher Wood, "FAA: Link Air Cargo, Passengers," *Denver Business Journal*, November 1–7, 1991, p. 3.

Christopher Wood, "Airport May Move Cargo Operations, Offer Reserve Funds," *Denver Business Journal*, December 6–12, 1991, pp. 1, 34.

"UAL in Accord on Denver," *The New York Times*, December 7, 1991, p. 39L.

Thomas Fisher, "Projects Flights of Fantasy," *Progressive Architecture* (March 1992), p. 103.

Tom Locke, "Disconnected," *Denver Business Journal*, June 12–18, 1992, p. 19.

"Big Ain't Hardly the Word for It," *ENR*, September 7, 1992, pp. 28–29.

Christopher Wood, "Adams Seeks Action," *Denver Business Journal*, September 4–10, 1992, pp. 1, 13.

"Denver Airport Rises under Gossamer Roof," *The Wall Street Journal*, November 17, 1992, p. B1.

Mark B. Solomon, "Denver Airport Delay Angers Cargo Carriers," *Journal of Commerce*, March 17, 1993, p. 3B.

"Denver Airport Opening Delayed Until December," *Aviation Week & Space Technology*, May 10, 1993, p. 39.

Aldo Svaldi, "DIA Air Train Gathering Steam as Planners Shift Possible Route," *Denver Business Journal*, August 27–September 2, 1993, p. 74.

Dirk Johnson, "Opening of New Denver Airport Is Delayed Again," *The New York Times*, October 26, 1993, p. A19.

"Denver's Mayor Webb Postpones Opening International Airport," *The Wall Street Journal*, October 26, 1993, p. A9.

"An Airport Comes to Denver," *Skiing* (December 1993), p. 66.

Ellis Booker, "Airport Prepares for Takeoff," *Computerworld*, January 10, 1994.

Aldo Svaldi, "Front Range, DIA Weigh Merging Airport Systems," *Denver Business Journal*, January 21–27, 1994, p. 3.

Don Phillips, "$3.1 Billion Airport at Denver Preparing for a Rough Takeoff," *The Washington Post*, February 13, 1994, p. A10.

"New Denver Airport Combines Several State-of-the-Art Systems," *Travel Weekly*, February 21, 1994, p. 20.

Steve Munford, "Options in Hard Surface Flooring," *Buildings* (March 1994), p. 58.

Mars Charles, "Denver's New Airport, Already Mixed in Controversy, Won't Open Next Week," *The Wall Street Journal*, March 2, 1994, pp. B1, B7.

"Denver Grounded for Third Time," *ENR*, March 7, 1994, p. 6.

Shannon Peters, "Denver's New Airport Creates HR Challenges," *Personnel Journal* (April 1994), p. 21.

Laura Del Rosso, "Denver Airport Delayed Indefinitely," *Travel Weekly*, May 5, 1994, p. 37.

"DIA Bond Rating Cut," *Aviation Week & Space Technology*, May 16, 1994, p. 33.

Robert Scheler, "Software Snafu Grounds Denver's High-Tech Airport," *PC Week*, May 16, 1994, p. 1.

John Dodge, "Architects Take a Page from Book on Denver Airport-Bag System," *PC Week*, May 16, 1994, p. 3.

Jean S. Bozman, "Denver Airport Hits Systems Layover," *Computerworld*, May 16, 1994, p. 30.

Richard Woodbury, "The Bag Stops Here," *Time*, May 16, 1994, p. 52.

"Consultants Review Denver Baggage Problems," *Aviation Week & Space Technology*, June 6, 1994, p. 38.

"Doesn't It Amaze? The Delay that Launched a Thousand Gags," *Travel Weekly*, June 6, 1994, p. 16.

Michael Romano, "This Delay Is Costing Business a Lot of Money," *Restaurant Business*, June 10, 1994, p. 26.

Scott Armstrong, "Denver Builds New Airport, Asks 'Will Planes Come?'" *The Christian Science Monitor*, June 21, 1994, p. 1.

Benjamin Weiser, "SEC Turns Investigation to Denver Airport Financing," *The Washington Post*, July 13, 1994, p. D1.

Bernie Knill, "Flying Blind at Denver International Airport," *Material Handling Engineering* (July 1994), p. 47.

Keith Dubay, "Denver Airport Seeks Compromise on Baggage Handling," *American Banker Washington Watch*, July 25, 1994, p. 10.

Dirk Johnson, "Denver May Open Airport in Spite of Glitches," *The New York Times*, July 27, 1994, p. A14.

Jeffrey Leib, "Investors Want a Plan," *The Denver Post*, August 2, 1994, p. A1.

Marj Charlier, "Denver Plans Backup Baggage System for Airport's Troubled Automated One," *The Wall Street Journal*, August 5, 1994, p. B2.

Louis Sahagun, "Denver Airport to Bypass Balky Baggage Mover," *Los Angeles Times*, August 5, 1994, p. A1.

Len Morgan, "Airports Have Growing Pains," *Flying* (August 1994), p. 104.

Adam Bryant, "Denver Goes Back to Basics for Baggage," *The New York Times*, August 6, 1994, pp. 5N, 6L.

"Prosecutors Scrutinize New Denver Airport," *The New York Times*, August 21, 1994, p. 36L.

Kevin Flynn, "Panic Drove New DIA Plan," *Rocky Mountain News*, August 7, 1994, p. 5A.

David Hughes, "Denver Airport Still Months from Opening," *Aviation Week & Space Technology*, August 8, 1994, p. 30.

"Airport May Open in Early '95," *Travel Weekly*, August 8, 1994, p. 57.

Michael Meyer and Daniel Glick, "Still Late for Arrival," *Newsweek*, August 22, 1994, p. 38.

Andrew Bary, "A $3 Billion Joke," *Barron's*, August 22, 1994, p. MW10.

Jean Bozman, "Baggage System Woes Costing Denver Airport Millions," *Computerworld* (August 22, 1994), p. 28.

Edward Phillips, "Denver, United Agree on Baggage System Fixes," *Aviation Week & Space Technology*, August 29, 1994.

Glenn Rifkin, "What Really Happened at Denver's Airport," *Forbes*, August 29, 1994, p. 110.

Andrew Bary, "New Denver Airport Bond Issue Could Face Turbulence from Investors," *Barron's*, August 29, 1994, p. MW9.

Andrew Bary, "Denver Airport Bonds Take Off as Investors Line Up for Higher Yields," *Barron's*, August 29, 1994, p. MW9.

Susan Carey, "Alaska's Cash-Strapped MarkAir Is Wooed by Denver," *The Wall Street Journal*, September 1, 1994, p. B6.

Dana K. Henderson, "It's in the Bag(s)," *Air Transport World* (September 1994), p. 54.

Dirk Johnson, "Late Already, Denver Airport Faces More Delays," *The New York Times*, September 25, 1994, p. 26L.

Gordon Wright, "Denver Builds a Field of Dreams," *Building Design and Construction* (September 1994), p. 52.

Alan Jabez, "Airport of the Future Stays Grounded," *Sunday Times*, October 9, 1994, Features Section.

Jean Bozman, "United to Simplify Denver's Troubled Baggage Project," *Computerworld*, October 10, 1994, p. 76.

"Denver Aide Tells of Laxity in Airport Job," *The New York Times*, October 17, 1994, p. A12.

Brendan Murray, "In the Bags: Local Company to Rescue Befuddled Denver Airport," *Marietta Daily Journal*, October 21, 1994, p. C1.

Joanne Wojcik, "Airport in Holding Pattern, Project Is Insured, but Denver to Retain Brunt of Delay Costs," *Business Insurance*, November 7, 1994, p. 1.

James S. Russell, "Is This Any Way to Build an Airport?" *Architectural Record* (November 1994), p. 30.

Appendix A
Municipal Bond Prospectus _____

$261,415,000

City and County of Denver, Colorado 6.875% Special Facilities Airport Revenue
   Bonds

(United Airlines Project) Series 1992A

Date: October 1, 1992

Due: October 1, 2032 Rating: Standard & Poor's BBB–

Moody's Baa2

INTRODUCTION

This official statement* is provided to furnish information in connection with the
sale by the City and County of Denver, Colorado (the "City") of 6.875% Special
Facilities Airport Revenue Bonds (United Airlines Project) series 1992A in the
aggregate principal amount of $261,415,000 (the "Bonds"). The bonds will be
dated, mature, bear interest, and be subject to redemption prior to maturity as
described herein.

   The Bonds will be issued pursuant to an Ordinance of the City and County of
Denver, Colorado (the "Ordinance").

   The proceeds received by the City from the sale of the Bonds will be used
to acquire, construct, equip, or improve (or a reimbursement of payments for the
acquisition, construction, equipping, or improvement of) certain terminals, Concourse B, aircraft maintenance, ground equipment maintenance, flight kitchen,
and air freight facilities (the "Facilities") at the new Denver International Airport
(the "New Airport").

---

*Only excerpts from the prospectus are included here.

The City will cause such proceeds to be deposited, distributed, and applied in accordance with the terms of a Special Facilities and Ground Lease, dated as of October 1, 1992 (the "Lease") between United Airlines and the City. Under the Lease, United has agreed to make payments sufficient to pay the principal, premium, if any, and interest on the Bonds. Neither the Facilities nor the ground rental payments under the Lease are pledged as security for the payment of principal, premium, if any, and interest on the bonds.

AGREEMENT BETWEEN UNITED AND THE CITY

On June 26, 1991, United and the City entered into an agreement followed by a second agreement on December 12, 1991, which, among other things, collectively provide for the use and lease by United of certain premises and facilities at the New Airport. In the United Agreement, United agrees among other things, to (1) support the construction of the New Airport, (2) relocate its present air carrier operations from Stapleton to the New Airport, (3) occupy and lease certain facilities at the New Airport, including no less than 45 gates on Concourse B within two years of the date of beneficial occupancy as described in the United Agreement, and (4) construct prior to the date of beneficial occupancy, a regional reservation center at a site at Stapleton. In conjunction with the execution of the United Agreement, United also executes a 30-year use and lease agreement. United has agreed to lease, on a preferential use basis, Concourse B, which is expected to support 42 jet aircraft with up to 24 commuter aircraft parking positions at the date of beneficial occupancy, and, on an exclusive use basis, certain ticket counters and other areas in the terminal complex of the New Airport.

THE FACILITIES

The proceeds of the bonds will be used to finance the acquisition, construction, and equipping of the Facilities, as provided under the Lease. The Facilities will be located on approximately 100 acres of improved land located within the New Airport, which United will lease from the City. The Facilities will include an aircraft maintenance facility capable of housing ten jet aircraft, a ground equipment support facility with 26 maintenance bays, an approximately 55,500-square-foot air freight facility, and an approximately 155,000-square-foot flight kitchen. Additionally, the proceeds of the Bonds will be used to furnish, equip, and install certain facilities to be used by United in Concourse B and in the terminal of the New Airport.

REDEMPTION OF BONDS

The Bonds will be subject to optional and mandatory redemption prior to maturity in the amounts, at the times, at the prices, and in the manner as provided in the

Ordinance. If less than all of the Bonds are to be redeemed, the particular Bonds to be called for redemption will be selected by lot by the Paying Agent in any manner deemed fair and reasonable by the Paying Agent.

The bonds are subject to redemption prior to maturity by the City at the request of United, in whole or in part, by lot, on any date on or after October 1, 2002, from an account created pursuant to the Ordinance used to pay the principal, premium, if any, and interest on the Bonds (the "Bond Fund") and from monies otherwise available for such purpose. Such redemptions are to be made at the applicable redemption price shown below as a percentage of the principal amount thereof, plus interest accrued to the redemption date:

Redemption Period	Optional Redemption Price
October 1, 2002 through September 30, 2003	102%
October 1, 2003 through September 30, 2004	101%
October 1, 2004 and thereafter	100%

The Bonds are subject to optional redemption prior to maturity, in whole or in part by lot, on any date, upon the exercise by United of its option to prepay Facilities Rentals under the Lease at a redemption price equal to 100% of the principal amount thereof plus interest accrued to the redemption date, if one or more of the following events occurs with respect to one or more of the units of the Leased Property:

(a) the damage or destruction of all or substantially all of such unit or units of the Leased Property to such extent that, in the reasonable opinion of United, repair and restoration would not be economical and United elects not to restore or replace such unit or units of the Leased Property; or,

(b) the condemnation of any part, use, or control of so much of such unit or units of the Leased Property that such unit or units cannot be reasonably used by United for carrying on, at substantially the same level or scope, the business theretofore conducted by United on such unit or units.

In the event of a partial extraordinary redemption, the amount of the Bonds to be redeemed for any unit of the Leased Property with respect to which such prepayment is made shall be determined as set forth below (expressed as a percentage of the original principal amount of the Bonds) plus accrued interest on the Bonds to be redeemed to the redemption date of such Bonds provided that the amount of Bonds to be redeemed may be reduced by the aggregate principal amount (valued at par) of any Bonds purchased by or on behalf of United and delivered to the Paying Agent for cancellation:

Terminal Concourse B Facility	Aircraft Maintenance Facility	Ground Equipment Maintenance Facility	Flight Kitchen	Air Freight Facility
20%	50%	10%	15%	5%

The Bonds shall be subject to mandatory redemption in whole prior to maturity, on October 1, 2023, at a redemption price equal to 100% of the principal amount thereof, plus accrued interest to the redemption date if the term of the Lease is not extended to October 1, 2032, in accordance with the provisions of the Lease and subject to the conditions in the Ordinance.

LIMITATIONS

Pursuant to the United Use and Lease Agreement, if costs at the New Airport exceed $20 per revenue enplaned passenger, in 1990 dollars, for the preceding calendar year, calculated in accordance with such agreement, United can elect to terminate its Use and Lease Agreement. Such termination by United would not, however, be an event of default under the Lease.

If United causes an event of default under the Lease and the City exercises its remedies thereunder and accelerates Facilities Rentals, the City is not obligated to relet the Facilities. If the City relets the Facilities, it is not obligated to use any of the payments received to pay principal, premium, if any, or interest on the Bonds.

APPLICATION OF THE BOND PROCEEDS

It is estimated that the proceeds of the sale of the Bonds will be applied as follows:

Cost of Construction...	$226,002,433
Interest on Bonds During Construction	22,319,740
Cost of Issuance Including Underwriters' Discount	1,980,075
Original Issue Discount	11,112,742
Principal Amount of the Bonds	$261,415,000

TAX COVENANT

Under the terms of the lease, United has agreed that it will not take or omit to take any action with respect to the Facilities or the proceeds of the bonds (including any investment earnings thereon), insurance, condemnation, or any other proceeds derived in connection with the Facilities, which would cause the interest on the Bonds to become included in the gross income of the Bondholder for federal income tax purposes.

OTHER MATERIAL COVENANTS

United has agreed to acquire, construct, and install the Facilities to completion pursuant to the terms of the Lease. If monies in the Construction Fund are insufficient to pay the cost of such acquisition, construction, and installation in full, then United shall pay the excess cost without reimbursement from the City, the Paying Agent, or any Bondholder.

United has agreed to indemnify the City and the Paying Agent for damages incurred in connection with the occurrence of certain events, including without limitation, the construction of the Facilities, occupancy by United of the land on which the Facilities are located, and violation by United of any of the terms of the Lease or other agreements related to the Leased Property.

During the Lease Term, United has agreed to maintain its corporate existence and its qualifications to do business in the state. United will not dissolve or otherwise dispose of its assets and will not consolidate with or merge into another corporation provided, however, that United may, without violating the Lease, consolidate or merge into another corporation.

ADDITIONAL BONDS

At the request of United, the City may, at its option, issue additional bonds to finance the cost of special Facilities for United upon the terms and conditions in the Lease and the Ordinance.

THE GUARANTY

Under the Guaranty, United will unconditionally guarantee to the Paying Agent, for the benefit of the Bondholders, the full and prompt payment of the principal, premium, if any, and interest on the Bonds, when and as the same shall become due whether at the stated maturity, by redemption, acceleration, or otherwise. The obligations of United under the Guaranty are unsecured, but are stated to be absolute and unconditional, and the Guaranty will remain in effect until the entire principal, premium, if any, and interest on the Bonds has been paid in full or provision for the payment thereof has been made in accordance with the Ordinance.

Part 15

WAGE AND SALARY ADMINISTRATION

It is very difficult for the true benefits of project management to be realized unless project management is integrated into the wage and salary administration program. Some companies view project management as a career path position while others view it simply as a part-time profession.

The situation becomes even more complex when dealing with functional employees who report to multiple bosses. When employees are notified that they are being assigned to a new project, their first concern is: What is in it for them? How will they be evaluated? How will their boss know whether or not they did a good job? Project managers must have either a formal or informal input into the employee's performance review.

Photolite Corporation (A)

Photolite Corporation is engaged in the sale and manufacture of cameras and photographic accessories. The company was founded in Baltimore in 1980 by John Benet. After a few rough years, the company began to flourish, with the majority of its sales coming from the military. By 1985, sales had risen to $5 million.

By 2015, sales had increased to almost $160 million. However, in 1996, competition from larger manufacturers and from some Japanese and German imports made itself felt on Photolite's sales. The company did what it could to improve its product line, but due to lack of funds, it could not meet the competition head on. The company was slowly losing its market share and was approached by several larger manufacturers as to the possibility of a merger or acquisition. Each offer was turned down.

During this time period, several meetings took place with department heads and product managers regarding the financial health of Photolite. At one of the more recent meetings, John Benet expressed his feelings in this manner:

> I have been offered some very attractive buyouts, but frankly the companies that want to acquire us are just after our patents and processes. We have a good business, even though we are experiencing some tough times. I want our new camera lens project intensified. The new lens is just about complete, and I want it in full-scale production as soon as possible! Harry Munson will be in charge of this project as of today, and I expect everyone's full cooperation. This may be our last chance for survival.

With that, the meeting was adjourned.

PROJECT INFORMATION

The new lens project was an innovation that was sure to succeed if followed through properly. The innovation was a lens that could be used in connection with sophisticated camera equipment. It was more intense than the wide-angle lens and had no distortion. The lens was to be manufactured in three different sizes, enabling the lens to be used with the top-selling cameras already on the market. The lens would be operable not only with the camera equipment manufactured by Photolite but also with that of their competitors.

Management was certain that if the manufactured lens proved to be as precise as the prototypes, the Central Intelligence Agency and possibly government satellite manufacturers would be their largest potential customers.

THE PROJECT OFFICE

Harry Munson was a young project manager, 29 years of age, who had both sales and engineering experience, in addition to an MBA degree. He had handled relatively small projects in the past and realized that this was the most critical project, not only to his career but also for the company's future.

Project management was still relatively new at Photolite, having been initiated only 15 months earlier. Some of the older department heads were very much against letting go of their subordinates for any length of time, even though it was only a sharing arrangement. This was especially true of Herb Wallace, head of the manufacturing division. He felt his division would suffer in the long run if any of his people were to spend much time on projects and reporting to another manager or project leader.

Harry Munson went directly to the personnel office to review the personnel files of available people from the manufacturing division. There were nine folders available for review. Harry had expected to see at least 20 folders but decided to make the best of the situation. Harry was afraid that it was Herb Wallace's influence that had reduced the number of files down to nine.

Harry had several decisions to make before looking at the folders. He felt that it was important to have a manufacturing project engineer assigned full time to the project rather than having to negotiate for part-time specialists who would have to be shared with other projects. The ideal manufacturing project engineer would have to coordinate activity in production scheduling, quality control, manufacturing engineering, procurement, and inventory control. Because project management had only recently been adopted, there were no individuals qualified for this position. This project would have to become the training ground for development of a manufacturing project engineer.

Due to the critical nature of the project, Harry realized that he must have the most competent people on his team. He could always obtain specialists on a part-time basis, but his choice for the project engineering slot would have to be not only the best person available but someone who would be willing to give as much extra time as the project demanded for at least the next 18 months. After all, the project engineer would also be the assistant project manager since only the project manager and project engineer would be working full time on the project. Now Harry was faced with the problem of trying to select the individual who would be best qualified for this slot. He decided to interview each of the potential candidates, in addition to analyzing their personnel files.

QUESTIONS

1. What would be the ideal qualifications for the project engineering slot?
2. What information should Harry look for in the personnel files?
3. Harry decided to interview potential candidates after reviewing the files. This is usually a good idea, because the files may not address all of his concerns. What questions should Harry ask during the interviews? Why is he interviewing candidates? What critical information may not appear in the personnel files?

Photolite
Corporation (B)

A meeting was held between Jesse Jaimeson, the director of personnel, and Ronald Ward, the wage and salary administrator. The purpose of the meeting was to discuss the grievances by the functional employees that Photolite's current employee evaluation procedures are inadequate for an organization that supports a project management structure.

Jesse Jaimeson: "Ron, we're having a lot of trouble with our functional employees over their evaluation procedures. The majority of the complaints stem from situations where the functional employee works closely with the project manager. If the functional manager does not track the work of this employee closely, then the functional manager must rely heavily upon the project manager for information during employee evaluation."

Ron Ward: "There aren't enough hours in a day for a functional manager to keep close tabs on all of his or her people, especially if those people are working in a project environment. Therefore, the functional manager will ask the project manager for evaluation information. This poses several problems. First, there are always situations where functional and project management disagree as to either direction or quality of work. The functional employee has a tendency of bending toward the individual who signs his or her promotion and evaluation form. This can alienate the project manager into recommending a poor evaluation regardless of how well the functional employee performs.

"In the second situation, the functional employee will spend most of this time working by her or himself, with very little contact with the project manager. In this case, the project manager tends to give an average evaluation, even if the employee's performance is superb. This could result from a situation where the employee has perhaps only a one to two week effort on a given project. This doesn't give that employee enough time to get to know anybody.

"In the third situation, the project manager allows personal feelings to influence his or her decision. A project manager who knows an employee personally might be tempted to give a strong or weak recommendation, regardless of the performance. When personalities influence the evaluation procedure, chaos usually results."

Jaimeson: "There's also a problem if the project manager makes an overly good recommendation to a functional manager. If the employee knows that he or she has received a good appraisal for work done on a given project, that employee feels that he or she should be given an above-average pay increase or possibly a promotion. Many times this puts severe pressure upon the functional manager. We have one functional manager here at Photolite who gives only average salary increases to employees who work a great deal of time on one project, perhaps away from view of the functional manager. In this case, the functional manager claims that he cannot give the individual an above-average evaluation because he hasn't seen him enough. Of course, this is the responsibility of the functional manager.

"We have another manager who refuses to give employees adequate compensation if they are attached to a project that could eventually grow into a product line. His rationale is that if the project grows big enough to become a product line, then the project will have its own cost center account and the employee will then be transferred to the new cost center. The functional manager thus reserves the best salary increases for those employees who he feels will stay in his department and make him look good."

Ward: "Last year we had a major confrontation on the Coral Project. The Coral Project manager took a grade 5 employee and gave him the responsibilities of a grade 7 employee. The grade 5 employee did an outstanding job and naturally expected a large salary increase or even a promotion. Unfortunately, the functional manager gave the employee an average evaluation and argued that the project manager had no right to give the employee this added responsibility without first checking with the functional manager. We're still trying to work this problem out. It could very easily happen again."

Jaimeson: "Ron, we have to develop a good procedure for evaluating our employees. I'm not sure if our current evaluation form is sufficient. Can we develop multiple evaluation forms, one for project personnel and another one for nonproject personnel?"

Ward: "That might really get us in trouble. Suppose we let each project manager fill out a project evaluation form for each functional employee who works more than, say, 60 hours on a given project. The forms are then given to the functional manager. Should the project manager fill out these forms at project termination or when the employee is up for evaluation?"

Jaimeson: "It would have to be at project termination. If the evaluation were made when the employee is up for promotion and the employee is not promoted, then that employee might slack off on the job if he or she felt that the project manager rated him or her down. Of course, we could always show the employee the project evaluation sheets, but I'm not sure that this would be the wise thing to do. This could easily lead into a situation where every project manager would want to see these forms before staffing a project. Perhaps these forms should be solely for the functional manager's use."

Ward: "There are several problems with this form of evaluation. First, some of our functional employees work on three or four projects at the same time. This could be a problem if some of the evaluations are good while others are not. Some functional people are working on departmental projects and, therefore, would receive only one type of evaluation. And, of course, we have the people who charge to our overhead structure. They also would have one evaluation form."

Jaimeson: "You know, Ron, we have both exempt and nonexempt people charging to our projects. Should we have different evaluation forms for these people?"

Ward: "Probably so. Unfortunately, we're now using just one form for our exempt, nonexempt, technical, and managerial personnel. We're definitely going to have to change. The question is how to do it without disrupting the organization."

Jaimeson: "I'm dumping this problem into your lap, Ron. I want you to develop an equitable way of evaluating our people here at Photolite Corporation, and I want you to develop the appropriate evaluation forms. Just remember one thing—I do not want to open Pandora's box. We're having enough personnel problems as it is."

QUESTIONS

1. Can a company effectively utilize multiple performance evaluation forms within an organization? What are the advantages and disadvantages?
2. If we use only one form, what information should be evaluated so as to be equitable to everyone?
3. If multiple evaluation forms are used, what information should go into the form filled out by the project manager?
4. What information can and cannot a project manager evaluate effectively? Could it depend on the project manager's educational background and experience?

Photolite Corporation (C)

After more than two months of effort, Ron Ward, the wage and salary administrator for Photolite Corporation, was ready to present his findings on the most equitable means of evaluating personnel who are required to perform in a project management organizational structure. Jesse Jaimeson, the director of personnel, was eagerly awaiting the results.

Ron Ward: "Well, Jesse, after two months of research and analysis, we've come to some reasonable possibilities. My staff looked at the nine basic performance appraisal techniques. They are (1) essay appraisal, (2) graphic rating scale, (3) field review, (4) forced choice rating, (5) critical incident appraisal, (6) management by objectives, (7) the work-standards approach, (8) ranking methods, and (9) assessment centers.

(Exhibit I contains a brief description of each technique.)

EXHIBIT I. BASIC APPRAISAL TECHNIQUES

Essay Appraisal

This technique asks raters to write a short statement covering a particular employee's strengths, weaknesses, areas for improvement, potential, and so on.

(continued)

(continued)

This method is often used in the selection of employees when written recommendations are solicited from former employers, teachers, or supervisors. The major problem with this type of appraisal is the extreme variability in length and content, which makes comparisons difficult.

Graphic Rating Scale

A typical graphic rating scale assesses a person on the quality and quantity of his or her work and on a variety of other factors that vary with the specific job. Usually included are personal traits such as flexibility, cooperation, level of self-motivation, and organizational ability. The graphic rating scale results in more consistent and quantifiable data, though it does not provide the depth of the essay appraisal.

Field Review

As a check on reliability of the standards used among raters, a systematic review process may be utilized. A member of the personnel or central administrative staff meets with small groups of raters from each supervisory unit to go over ratings for each employee to identify areas of dispute and to arrive at an agreement on the standards to be utilized. This group judgment technique tends to be more fair and valid than individual ratings, but is considerably more time-consuming.

Forced-Choice Rating

There are many variations of this method, but the most common version asks raters to choose from among groups of statements those that best fit the person being evaluated and those that least fit. The statements are then weighted and scored in much the same way psychological tests are scored. The theory behind this type of appraisal is that since the rater does not know what the scoring weight of each statement is, he or she cannot play favorites.

Critical Incident Appraisal

Supervisors are asked to keep a record on each employee and to record actual incidents of positive and negative behavior. While this method is beneficial in that it deals with actual behavior rather than abstractions, it is time-consuming for the supervisor, and the standards of recording are set by the supervisor.

Management by Objectives

In this approach, employees are asked to set, or help set, their own performance goals. This approach has considerable merit in its involvement of the individual in setting the standards by which he or she will be judged and the emphasis on results rather than on abstract personality characteristics.

Work-Standards Approach

Instead of asking each employee to set his or her own performance standards, many organizations set measured daily work standards. The work-standards technique establishes work and staffing targets aimed at increasing productivity. When realistically used and when standards are fair and visible, it can be an effective type of performance appraisal. The most serious problem is that of comparability. With different standards for different people, it is difficult to make comparisons for the purposes of promotion.

Ranking Methods

For purposes of comparing people in different units, the best approach appears to be a ranking technique involving pooled judgment. The two most effective ranking methods include alternation-ranking and paired-comparison ranking. Essentially, supervisors are asked to rank who is "most valuable."

Assessment Centers

Assessment centers are coming into use more for the prediction and assessment of future potential. Typically, individuals from different areas are brought together to spend two or three days working on individual and group assignments. The pooled judgment of observers leads to an order-of-merit ranking of participants. The greatest drawback to this system is that it is very time-consuming and costly.

"We tried to look at each technique objectively. Unfortunately, many of my people are not familiar with project management and, therefore, had some difficulties. We had no so-called standards of performance against which we could evaluate each technique. We therefore listed the advantages and disadvantages that each technique would have if utilized in a project management structure.

Jesse Jaimeson: "I'm not sure of what value your results are in this case because they might not directly apply to our project management organization."

Ward: "In order to select the technique most applicable to a project management structure, I met with several functional and project managers as to the establishment of a selection criteria. The functional managers felt that conflicts were predominant in a project organization, and that these conflicts could be used as a comparison. I therefore decided to compare each of the appraisal techniques to

TABLE I   RATING EVALUATION TECHNIQUES AGAINST TYPES OF CONFLICT

Type of Conflict	Essay Appraisal	Graphic Rating Scale	Field Review	Forced-Choice Review	Critical Incident Appraisal	Management by Objectives	Work Standards Approach	Ranking Medthods	Assessment Center
Conflict over schedules	●	●		●	●		●	●	
Conflict over priorities	●	●		●	●		●	●	
Conflict over technical issues	●			●			●		
Conflict over administration	●	●	●	●			●	●	●
Personality conflict	●	●		●			●		
Conflict over cost	●		●	●	●		●	●	●

Note: Circles indicate areas of difficulty.

the most commonly mentioned conflicts that exist in project management organizational forms. The comparison is shown in Table I.

"Analysis of Table I shows the management by objectives (MBO) technique to be the most applicable system. There are several factors supporting this conclusion.

"The *essay appraisal* technique appears in most performance appraisals and is characterized by a lack of standards. As a result, it tends to be subjective and inconsistent.

"The *graphic rating scale* technique is marked by checking boxes and does not have the flexibility required by the constantly changing dynamic structure required in project management.

"The *field review* system probably would account for the majority of performance appraisal problems. However, it is costly and provides for another management overlay as well as an additional cost and time factors.

"The *forced-choice rating* technique has the same problems as the essay technique with the added problem of being inflexible.

"*Critical incident appraisal* centers on the individual's performance and does not take into account decisions made by one's superiors or the problems beyond the individual's control. Again, it is time-consuming.

"*Management by objectives.* allows all parties—the project manager, the functional manager, and the employee—to share and to participate in the appraisal. It epitomizes the systems approach since it allows for objectives modification without undue or undeserved penalty to the employee. Finally, it uses objective data and downplays subjective data.

"The *work-standards approach* lends itself easily to technical projects. Though not usually recognized formally, it is probably the most common project management performance appraisal technique. However, it is not flexible and downplays the effect of personality conflicts with little employee input.

"The *ranking method* allows for little individual input. Most conflict possibilities are maximized with this technique.

"The *assessment centers* method cannot be used on site and is very costly. It is probably most applicable (if not the best technique) for selecting project management human resources.

"In summary, MBO appears to be the best technique for performance appraisal in a project management organization."

Jaimeson: "Your conclusions lead me to believe that the MBO appraisal technique is applicable to all project management appraisal situations and should be recommended. However, I do have a few reservations. A key point is that the MBO approach does not eliminate, or even minimize, the problems inherent in project and matrix management organizations. MBO provides the technique through which human resources can be fairly appraised (and, of course, rewarded and punished). MBO has the weakness that it prohibits individual input and systems that employ poorly trained appraisers and faulty follow-up techniques. Of course, such weaknesses would kill any performance appraisal system. The MBO technique most exemplifies the systems approach and, even with its inherent weaknesses, should be considered when the systems approach to management is being employed."

Ward: "There is another major weakness that you have omitted. What about those situations where the employee has no say in setting the objectives? I'm sure we have project managers, as well as functional managers, who will do all of the objective-setting themselves."

Jaimeson: "I'm sure this situation either exists now or will eventually exist. But that's not what worries me. If we go to an MBO approach, how will it affect our current evaluation forms? We began this study to determine the best appraisal method for our organization. I've yet to see any kind of MBO evaluation form that can be used in a project management environment. This should be our next milestone."

QUESTIONS

1. Do you agree with the results in Exhibit II? Why or why not? Defend your answers.
2. Are there any other techniques that may be better?

Photolite Corporation (D)

Ron Ward, the wage and salary administrator for Photolite Corporation, met with Jesse Jaimeson, the director of personnel, to discuss their presentation to senior management for new evaluation techniques in the recently established matrix organization.

Jesse Jaimeson: "I've read your handout on what you're planning to present to senior management, and I feel a brief introduction should also be included. [See Exhibit I.] Some of these guys have been divorced from lower-level appraisals for over 20 years. How do you propose to convince these guys?"

EXHIBIT I. RECOMMENDED APPROACH

I. Prework

Employee and manager record work to be done using goals, work plans, position guide.

Employee and manager record measurements to be used.

Note: This may not be possible at this time since we are in the middle of a cycle. For 1999 only, the process will start with the employees submitting a list of their key tasks (i.e., job description) as they see it. Manager will review that list with the employee.

(continued)

(continued)

II. Self-Appraisal
- Employee submits self-appraisal for key tasks.
- It becomes part of the record.

III. Managerial Appraisal
- Manager evaluates each task.
- Manager evaluates total effort.
- Skills displayed are recorded.
- Development effort required is identified.

Note: Appraisals should describe what happened, both good and bad.

IV. Objective Review
- Employee relations reviews the appraisal.
  - Assure consistent application of ratings.
  - Assist in preparation, if needed.
  - Be a sounding board.

V. One-over-One Review
- Managerial perspective is obtained.
- A consistent point of view should be presented.

VI. Appraisal Discussion
- Discussion should be participative.
- Differences should be reconciled. If this is not possible, participants must agree to disagree.
- Work plans are recycled.
- Career discussion is teed up.
- Employee and manager commit to development actions.

VII. Follow-up
- Checkpoints on development plan allow for this follow-up.

Ron Ward: "We do have guidelines for employee evaluation and appraisal. These include:

   A. To record an individual's specific accomplishments for a given period of time.
   B. To formally communicate to the individual on four basic issues:
     1. What is expected of him/her (in specifics).
     2. How he/she is performing (in specifics).

    3. What his/her manager thinks of his/her performance (in specifics).

    4. Where he/she could progress within the present framework.

C. To improve performance.

D. To serve as a basis for salary determination.

E. To provide a constructive channel for upward communication.

"Linked to the objectives of the performance appraisal, we must also consider some of the possible negative influences impacting on a manager involved in this process. Some of these factors could be:

- A manager's inability to control the work climate.
- A normal dislike to criticize a subordinate.
- A lack of communication skills needed to handle the employee interview.
- A dislike for the general mode in the operation of the business.
- A mistrust of the validity of the appraisal instrument.

"To determine the magnitude of management problems inherent in the appraisal of employees working under the matrix concept, the above-mentioned factors could be increased four or five times, the multiplier effect being caused by the fact that an employee working under the project/matrix concept could be working on as many as four or five projects during the appraisal period, thereby requiring all the project managers and the functional manager to input their evaluations regarding a subordinate's performance and the appraisal system itself."

Jaimeson: "Of course, managers cannot escape making judgments about subordinates. Without these evaluations, Photolite would be unable to adequately administer its promotion and salary policies. But in no instance can a performance appraisal be a simple accept or reject concept involving individuals. Unlike the quality appraisal systems used in accepting or rejecting manufactured units, our personnel appraisal systems must include a human factor. This human factor must take us beyond the scope of job objectives into the values of an individual's worth, human personality, and dignity. It is in this vein that any effective personnel appraisal system must allow the subordinate to participate fully in the appraisal activities."

Ward: "Prior to a couple of years ago, this was a major problem within Photolite. Up to that time, all appraisals were based on the manager or managers assessing an individual's progress toward goals that had been established and passed on to subordinates. Although an employee meeting was held to discuss the outcome of an employee's appraisal, in many instances it was one-sided, without meaningful participation by the person being reviewed. Because of such a system, many employees began to view the appraisal concept as inconsistent and without true concern for the development of the individual. This also led many to believe that promotions and salary increases were based on favoritism rather than merit.

"Problems inherent in these situations are compounded in the matrix organization when an individual is assigned to several projects with varying degrees of importance placed on each project, but knowing that each project manager will contribute to the performance appraisal based on the success of their individual projects. Such dilemmas can only be overcome when the individual is considered as the primary participating party in the appraisal process and the functional manager coordinates and places prime responsibility on the subordinate contributor in the project for which prime interest has been focused by the company. Other project contributions are then considered, but on a secondary basis."

Jaimeson: "Although we have discussed problems that are inherent in a matrix organization and can be compounded by the multiple performance determination, a number of positives can also be drawn from such a work environment. It is obvious, based on its design, that a project/matrix organization demands new attitudes, behavior, knowledge, and skills. This in turn has substantial implications for employee selection, development, and career progression. The ultimate success of the individual and the project depends largely on the ability of the organization to help people learn how to function in new ways.

"The matrix organization provides an opportunity for people to develop and grow in ways and rates not normally possible in the more traditional functional organizational setting. Although the project/matrix organization is considered to be high tension in nature, it places greater demands on people but offers greater development and career opportunities than does the functional organization.

"Because of the interdependencies of projects in a matrix, increased communications and contact between people is necessary. This does not mean that in a functional organization interdependency and communication are not necessary. What it does say, however, is that in a functional setting, roles are structured so that individuals can usually resolve conflicting demands by talking to their functional manager. In a matrix, such differences would be resolved by people from different functions who have different attitudes and orientations."

Ward: "From the very outset, organizations such as Photolite ran into conflict between projects involving such items as:

- Assignment of personnel to projects
- Manpower costs
- Project priority
- Project management status (as related to functional managers)
- Overlap of authority and power in the matrix

"If not adequately planned for in advance, these factors could be significant factors in the performance appraisal of matrix/project members. However, where procedures exist to resolve authority and evaluation conflicts, a more equitable performance appraisal climate exists. Unfortunately, such a climate rarely exists in any functioning organization.

"With the hope of alleviating such problems, my group has redefined its approach to exempt performance appraisals. [See Exhibits I and II.] This approach is based on the management by objectives technique. This approach allows both management and employees to work together in establishing performance goals.

"Beyond this point of involvement, employees also perform a self-evaluation of their performance, which is considered a vital portion of the performance appraisal. Utilization of this system also opens up communication between management and the employee, thereby allowing two-way communication to become a natural item. Although it is hoped that differences can be reconciled, if this cannot occur, the parties involved have at least established firm grounds on which to disagree. These grounds are not hidden to either and the employee knows exactly how his/her performance appraisal was determined."

Jaimeson: "Okay, I'm convinced we're talking the same language. We won't have any problem convincing these people of what we're trying to do."

EXHIBIT II. PERFORMANCE SUMMARY

When writing the overall statement of performance:

- Consider the degree of difficulty of the work package undertaken in addition to the actual results.
- Reinforce performance outcomes that you would like to see in the future by highlighting them here.
- Communicate importance of missed targets by listing them here.
- Let employees know the direction that performance is taking so that they can make decisions about effort levels, skill training emphasis, future placement possibilities, and so on.

When determining the overall rating number:

- Choose the paragraph that best describes performance in total, then choose the number that shades the direction it leans.
- Use the individual task measurements plus some weighting factor—realistically, some projects are worth more than others and should carry more weight.
- Again, consider the degree of difficulty of the work package undertaken.

(continued)

(continued)

Strong points are:

- Demonstrated in the accomplishment of the work.
- Found in the completion of more than one project.
- Relevant—avoid trivia.
- Usually not heard well by employees.
- Good subjects for sharpening and growing.

Areas requiring improvement usually:

- Show up in more than one project.
- Are known by subordinate.
- Limit employee effectiveness.
- Can be improved to some degree.

Areas of disagreement:

- Can be manager or subordinate initiated.
- Need not be prepared in advance.
- Require some effort on both parts before recording.
- Are designed to keep problems from hiding beneath the surface.

Your review of the self-appraisal may reveal some disagreement. Discuss this with the employee before formally committing it to writing.

QUESTIONS

1. If you were an executive attending this briefing, how would you react?
2. Are there any additional questions that need to be addressed?

First Security Bank of Cleveland

The growth rate of First Security of Cleveland had caused several executives to do some serious thinking about whether the current organizational structure was adequate for future operations. The big question was whether the banking community could adapt to a project management structure.

Tom Hood had been the president of First Security for the past 10 years. He had been a pioneer in bringing computer technology into the banking industry. Unfortunately, the size and complexity of the new computer project created severe integration problems, problems with which the current traditional organization was unable to cope. What was needed was a project manager who could drive the project to success and handle the integration of work across functional lines.

Tom Hood met with Ray Dallas, one of the bank's vice presidents, to discuss possible organizational restructuring.

Tom Hood: "I've looked at the size and complexity of some twenty projects that First Security did last year. Over 50 percent of these projects required interaction between four or more departments."

Ray Dallas: "What's wrong with that? We're growing and our problems are likewise becoming more complex."

Hood: "It's the other 50 percent that worry me. We can change our organizational structure to adapt to complex problem solving and integration. But what happens when we have a project that stays in one functional department? Who's going to drive it home? I don't see how we can tell a functional manager that he or she is a support group in one organizational form and a project manager in the other and have both organizational forms going on at the same time.

"We can have either large, complex projects or small ones. The small ones will be the problem. They can exist in one department or be special projects assigned to one person or a task force team. This means that if we incorporate project management, we'll have to live with a variety of structures. This can become a bad situation. I'm not sure that our people will be able to adapt to this changing environment."

Dallas: "I don't think it will be as bad as you make it. As long as we clearly define each person's authority and responsibility, we'll be all right. Other industries have done this successfully. Why can't we?"

Hood: "There are several questions that need answering. Should each project head be called a project manager, even if the project requires only one person? I can see our people suddenly becoming title-oriented. Should all project managers report to the same boss, even if one manager has 30 people working on the project and the other manager has none? This could lead to power struggles. I want to avoid that because it can easily disrupt our organization."

Dallas: "The problem you mentioned earlier concerns me. If we have a project that belongs in one functional department, the ideal solution is to let the department manager wear two hats, the second one being project manager. Disregarding for the moment the problem that this manager will have in determining priorities, to whom should he or she report to as to the status of the work? Obviously, not to the director of project management."

Hood: "I think the solution must be that all project managers report to one person. Therefore, even if the project stays in one functional department, we'll still have to assign a project manager Under project management organizational forms, functional managers become synonymous with resource managers. It is very dangerous to permit a resource manager to act also as a project manager. The resource manager might consider the project as being so important that he or she will commit all the department's best people to it and make it into a success at the expense of all the department's other work. That would be like winning a battle but losing the war."

Dallas: "You realize that we'll need to revamp our wage and salary administration program if we go to project management. Evaluating project managers might prove difficult. Regardless of what policies we establish, there are still going to be project managers who try to build empires, thinking that their progress is dependent upon the number of people they control. Project management will definitely give some people the opportunity to build a empire. We'll have to watch that closely."

Hood: "Ray, I'm a little worried that we might not be able to get good project managers. We can't compete with the salaries the project managers get in other industries such as engineering, construction, or computers. Project management cannot be successful unless we have good managers at the controls. What's your feeling on this?"

Dallas: "We'll have to promote from within. That's the only viable solution. If we try to make project management salaries overly attractive, we'll end up throwing the organization into chaos. We must maintain an adequate salary structure so that people feel that they have the same opportunities in both project management and the functional organization. Of course, we'll still have some people who will be more title-oriented than money-oriented, but at least each person will have the same opportunity for salary advancement."

Hood: "See if you can get some information from our personnel people on how we could modify our salary structure and what salary levels we can pay our project managers. Also, check with other banks and see what they're paying their project managers. I don't want to go into this blind and then find out that we're setting the trend for project management salaries. Everyone would hate us. I'd rather be a follower than a leader in this regard."

QUESTIONS

1. What are the major problems identified in the case?
2. What are your solutions to the above question and problems?

Jackson Industries

"I wish the hell that they had never invented computers," remarked Tom Ford, president of Jackson Industries. "This damn computer has been nothing but a thorn in our side for the past 10 years. We're gonna resolve this problem now. I'm through watching our people fight with one another. We must find a solution to this problem."

In 2002, Jackson Industries decided to purchase a mainframe computer, primarily to handle the large, repetitive tasks found in the accounting and finance functions of the organization. It was only fitting, therefore, that control of the computer came under the director of finance, Al Moody. For two years, operations went smoothly. In 2004, the computer department was reorganized in three sections: scientific computer programming, business computer programming, and systems programming. The reorganization was necessary because the computer department had grown into the fifth largest department, employing some 30 people, and was experiencing some severe problems working with other departments.

After the reorganization, Ralph Gregg, the computer department manager, made the following remarks in a memo distributed to all personnel:

> The computer department has found it increasingly difficult to work with engineering and operations functional departments, which continue to permit their personnel to write and document their own computer programs. In order to maintain some degree of consistency, the computer department will now assume the responsibility for writing all computer programs. All requests should be directed

533

to the department manager. My people are under explicit instructions that they are to provide absolutely no assistance to any functional personnel attempting to write their own programs without authorization from me. Company directives in this regard will be forthcoming.

The memo caused concern among the functional departments. If engineering wanted a computer program written, it would now have to submit a formal request and then have the person requesting the program spend a great deal of time explaining the problem to the scientific programmer assigned to this effort. The department managers were reluctant to have their people "waste time" in training scientific programmers to be engineers. The computer department manager countered this argument by stating that once the programmer was fully familiar with the engineering problem, then the engineer's time could be spent more fruitfully on other activities until the computer program was ready for implementation. This same problem generated more concern by department managers when they were involved in computer projects that required integration among several departments. Although Jackson Industries operated on a traditional structure, the new directive implied that the computer department would be responsible for managing all projects involving computer programming even if they crossed into other departments. Many people looked on this as a "baby" project management structure within the traditional organization.

In June 2006, Al Moody and Ralph Gregg met to discuss the deterioration of working relationships between the computer department and other organizations.

Al Moody: "I'm getting complaints from the engineering and operations departments that they can't get any priorities established on the work to be done in your group. What can we do about it?"

Ralph Gregg: "I set the priorities as I see fit, for what's best for the company. Those guys in the engineering and operations have absolutely no idea how long it takes to write, debug, and document a computer program. Then they keep feeding me this crap about how their projects will slip if this computer program isn't ready on time. I've told them what problems I have, and yet they still refuse to let me participate in the planning phase of their activities."

Al Moody: "Well, you may have a valid gripe there. I'm more concerned about this closed shop you've developed for your department. You've built a little empire down there, and it looks like your people are unionized where the rest of us are not. Furthermore, I've noticed that your people have their own informal organization and tend to avoid socializing with the other employees. We're supposed to be one big, happy family, you know. Can't you do something about that?"

Ralph Gregg: "The problem belongs to you and Tom Ford. For the last three years, the average salary increase for the entire company has been 7.5 percent, and our department has averaged a mere 5 percent because you people upstairs do not feel as though we contribute anything to company profits. My scientific programmers feel that they're doing engineering work and that they're making the same contribution to profits as is the engineers. Therefore, they should be on the engineering pay structure and receive an 8 percent salary increase."

Al Moody: "You could have given your scientific programmers more money. You had a budget for salary increases, the same as everyone else."

Ralph Gregg: "Sure I did. But my budget was less than everyone else's. I could have given the scientific people 7 percent and everyone else 3 percent. That would be an easy way to tell people that we think they should look for another job. My people do good work and do, in fact, contribute to profits. If Tom Ford doesn't change his impression of us, then I expect to lose some of my key people. Maybe you should tell him that."

Al Moody: "Between you and me, all of your comments are correct. I agree with your concerns. But my hands are tied, as you know.

"We are contemplating the installation of a management information system for all departments and, especially, for executive decision making. Tom is contemplating creating a new position, Director of Information Services. This would move the computer out of a department under finance and up to the directorate level. I'm sure this would have an impact on yearly salary increases for your people.

"The problem that we're facing involves the managing of projects under the new directorate. It looks like we'll have to create a project management organization just for this new directorate. Tom likes the traditional structure and wants to leave all other directorates intact. We know that this new directorate will have to integrate the new computer projects across multiple departments and divisions. Once we solve the organizational structure problem, we'll begin looking at implementation. Got any good ideas about the organizational structure?"

Ralph Gregg: "You bet I do. Make me director and I'll see that the work gets done."

QUESTIONS

1. What are the basic problems in the case?
2. What is the rationale for the problem?
3. How does Gregg know what is best for the company?
4. Are the computer groups taking on too much power?
5. Does the president understand computers?
6. Will the creation of a management information system group solve the salary disparity problem?
7. Was there good corporate direction?
8. What are your recommendations?

Part 16

TIME MANAGEMENT

Managing projects within time, cost, and performance is easier said than done. The project management environment is extremely turbulent and is composed of numerous meetings, report writing, conflict resolution, continuous planning and replanning, customer communications, and crisis management.

To manage all of these activities requires that the project manager and team members effectively manage their time each day. Some people are morning people and soon learn they are more productive in the morning than afternoon. Others are afternoon people. Knowing your own energy cycle is important. Also, good project managers realize that not all of the activities that they are asked to do are their responsibility.

Time Management Exercise

Effective time management is one of the most difficult chores facing even the most experienced managers. For a manager who manages well-planned repetitive tasks, effective time management can be accomplished without very much pain. But for a project manager who must plan, schedule, and control resources and activities on unique, one-of-a-kind projects or tasks, effective time management may not be possible because of the continuous stream of unexpected problems that develop.

This exercise is designed to make you aware of the difficulties of time management both in a traditional organization and in a project environment. Before beginning the exercise, you must make the following assumptions concerning the nature of the project:

- You are the project manager on a project for an outside customer.
- The project is estimated at $3.5 million with a time span of two years.
- The two-year time span is broken down into three phases: Phase I—one year, beginning February 1; Phase II—six months; Phase III—six months. You are now at the end of Phase I. (Phases I and II overlap by approximately two weeks. You are now in the Monday of the next to the last week of Phase I.) Almost all of the work has been completed.
- Your project employs 35 to 60 people, depending on the phase that you are in.

539

- You, as the project manager, have three full-time assistant project managers who report directly to you in the project office; an assistant project manager each for engineering, cost control, and manufacturing. (Material procurement is included as part of the responsibilities of the manufacturing assistant project manager.)
- Phase I appears to be proceeding within the time, cost, and performance constraints.
- You have a scheduled team meeting for each Wednesday from 10 A.M. to 12 noon. The meeting will be attended by all project office team members and the functional team members from all participating line organizations. Line managers are not team members and therefore do not show up at team meetings. It would be impossible for them to show up at the team meetings for all projects and still be able to function as line managers. Even when requested, they may not show up at team meetings because it is not effective time management for them to show up for a two-hour meeting simply to discuss 10 minutes of business. (Disregard the possibility that a team meeting agenda could resolve this problem.)

It is now Monday morning and you are home eating breakfast, waiting for your car pool to pick you up. As soon as you enter your office, you will be informed about problems, situations, tasks, and activities that have to be investigated. Your problem will be to accomplish effective time management for this entire week based on the problems and situations that occur.

You will take each day one at a time. You will be given ten problems and/or situations that will occur for each day and the time necessary for resolution. You must try to optimize your time for each of the next five days and get the maximum amount of productive work accomplished. Obviously, the word "productive" can take on several meanings. You must determine what is meant by "productive work." For the sake of simplicity, let us assume that your energy cycle is such that you can do eight hours of productive work in an eight-hour day. You do not have to schedule idle time, except for lunch. However, you must be aware that in a project environment, the project manager occasionally must take care of all work that line managers, line personnel, and even executives do not feel like accomplishing. Following the 10 tasks for each day, you will find a worksheet that breaks down each day into half-hour blocks between 9:00 A.M. and 5:00 P.M. Your job will be to determine which of the tasks you wish to accomplish during each half-hour block. The following assumptions are made in scheduling work:

- Because of car pool requirements, overtime is not permitted.
- Family commitments for the next week prevent work at home. Therefore, you will not schedule any work after 5:00 P.M.
- You as project manager are advised of the 10 tasks as soon as you arrive at work.

The first step in the solution to the exercise is to establish the priorities for each activity based on:

- *Priority A:* This activity is urgent and must be completed today. (However, some A priorities can be postponed until the team meeting.)
- *Priority B:* This activity is important but not necessarily urgent.
- *Priority C:* This activity can be delayed, perhaps indefinitely.

Fill in the space after each activity as to the appropriate priority. Next, determine which of the activities you have time to accomplish for this day. You have either seven or seven and one-half hours to use for effective time management, depending on whether you want a half-hour or a full hour for lunch.

You have choices as to how to accomplish each of the activities. These choices are:

- You can do the activity *yourself* (symbol = Y).
- You can *delegate* the responsibility to one of your assistant project managers (symbol = D). If you use this technique, you can delegate only one hour's worth of *your* work to each of your assistants without incurring a penalty. The key word here is that you are delegating *your* work. If the task that you wish to delegate is one that the assistant project manager would normally perform, then it does *not* count toward the one hour's worth of your work. This type of work is transmittal work and will be discussed later. For example, if you wish to delegate five hours of work to one of your assistant project managers and four of those hours are activities that would normally be his responsibility, then no penalty will be assessed. You are actually transmitting four hours and delegating one. You may assume that whatever work you assign to an assistant project manager will be completed on the day it is assigned, regardless of the priority.
- Many times, project managers and their team are asked to perform work that is normally the responsibility of someone else, say, an executive or a line manager. As an example, a line employee states that he doesn't have sufficient time to write a report and he wants you to do it, since you are the project manager. These types of requests can be returned to the requestor since they normally do not fall within the project manager's responsibilities. You may, therefore, select one of these four choices:
  - You can *return* the activity request back to the originator, whether line manager, executive, or subordinate, since it is not your responsibility (symbol = R). Of course, you might want to do this activity, if you have time, in order to build up goodwill with the requestor.
  - Many times, work that should be requested of an assistant project manager is automatically sent to the project manager. In this case, the project manager will automatically *transmit* this work to the appropriate

assistant project manager (symbol = T). As before, if you feel that you have sufficient time available or if your assistants are burdened, you may wish to do the work yourself. Work that is normally the responsibility of an assistant project manager is transmitted, not delegated. Thus you can transmit four hours of work (T) and still delegate one hour of work (D) to the same assistant project manager without incurring any penalty.

- You can *postpone* work from one day to the next (symbol = P). As an example, you decide that you want to accomplish a given Monday activity but do not have sufficient time. You can postpone the activity until Tuesday. If you do not have sufficient time on Tuesday, you may then decide to transmit (T) the activity to one of your assistants, delegate (D) the activity to one of your assistants, return (R) the activity to the requestor, or postpone (P) the activity another day. Postponing activities can be a trap. On Monday, you decide to postpone a category B priority. On Tuesday, the activity may become a category A priority, and you have no time to accomplish it. If you make a decision to postpone an activity from Monday to Tuesday and find that you have made a mistake by not performing this activity on Monday, you *cannot* go back in time and correct the situation.

- You can simply consider the activity as unnecessary and *avoid* doing it (symbol = A).

After you have decided which activities you will perform each day, place them in the appropriate time slot based on your own energy cycle. Later we discuss energy cycles and the order of the activities accomplished each day. You will find one worksheet for each day. The worksheets follow the 10 daily situations and/or problems.

Repeat the procedure for each of the five days. Remember to keep track of the activities that are carried over from the previous days. Several of the problems can be resolved by more than one method. If you are thoroughly trapped between two or more choices on setting priorities or modes of resolution, then write a note or two to justify your answer in space beneath each activity.

SCORING SYSTEM

Briefly look at the work plan for one of the days. Under the column labeled "Priority," the 10 activities for each day are listed. You must first identify the priorities (A, B, or C as described on page 541) for each activity. Next, under the column labeled "Method," you must select the method of accomplishment according to the list of options below the work plan. At the same time, you must fill in the activities you wish to perform yourself under the "Accomplishment" column in the appropriate time slot because your method for accomplishment may be dependent on whether you have sufficient time to accomplish the activity.

Notice that there is a space provided for you to keep track of activities that have been carried over. This means that if you have three activities on Monday's list that you wish to carry over until Tuesday, then you must turn to Tuesday's work plan and record these activities so that you will not forget.

You will not score any points until you complete Friday's work plan. Using the scoring sheets that follow Friday's work plan, you can return to the daily work plans and fill in the appropriate points. You will receive either positive points or negative points for each decision that you make. Negative points should be subtracted when calculating totals.

After completing the work plans for all five days, fill in the summary work plan on page 556 and be prepared to answer the summary questions.

You will not be told at this time how the scoring points will be awarded because it may affect your answers.

Monday's Activities

Activity	Description	Priority
1.	The detailed schedules for Phase II must be updated prior to Thursday's meeting with the customer. (Time = 1 hr)	_____
2.	The manufacturing manager calls you and states that he cannot find a certain piece of equipment for tomorrow's production run test. (Time = ½ hr)	_____
3.	The local university has a monthly distinguished lecturer series scheduled for 3–5 P.M. today. You have been directed by the vice president to attend and hear the lecture. The company will give you a car. Driving time to the university is one hour. (Time = 3 hrs)	_____
4.	A manufacturer's representative wants to call on you today to show you why his product is superior to the one that you are now using. (Time = ½ hr)	_____
5.	You must write a two-page weekly status report for the vice president. Report is due on his desk by 1:00 P.M. Wednesday. (Time = 1 hr)	_____
6.	A vice president calls you and suggests that you contact one of the other project managers about obtaining a uniform structure for the weekly progress reports. (Time = ½ hr)	_____
7.	A functional manager calls to inform you that, due to a schedule slippage on another project, your beginning milestones on Phase II may slip to the right because his people will not be available. He wants to know if you can look at the detailed schedules and modify them. (Time = 2 hrs)	_____
8.	The director of personnel wants to know if you have reviewed the three resumes that he sent you last week. He would like your written comments by quitting time today. (Time = 1 hr)	_____
9.	One of your assistant project managers asks you to review a detailed Phase III schedule that appears to have errors. (Time = 1 hr)	_____
10.	The procurement department calls with a request that you tell them approximately how much money you plan to spend on raw materials for Phase III. (Time = ½ hr)	_____

Day: _____Monday_____

WORK PLAN

Priority			Method		Accomplishment		
Activity	Priority	Points	Method of Accomplishment	Points	Time	Activity	Points
1					9:00–9:30		
2					9:30–10:00		
3					10:00–10:30		
4					10:30–11:00		
5					11:00–11:30		
6					11:30–12:00		
7					12:00–12:30		
8					12:30–1:00		
9					1:00–1:30		
10					1:30–2:00		
	Total		Total		2:00–2:30		
					2:30–3:00		
					3:00–3:30		
					3:30–4:00		
					4:00–4:30		
					4:30–5:00		
						Total	

Methods of accomplishment:
Y = you
D = delegate
T = transmit
R = return
A = avoid
P = postpone

Activities Postponed Until Today	Today's Priority

Points	
Priority points	
Methods points	
Accomplishment points	
Today's points total	

Tuesday's Activities

Activity	Description	Priority
11.	A functional manager calls you wanting to know if his people should be scheduled for overtime next week. (Time = ½ hr)	_____
12.	You have a safety board meeting today from 1–3 P.M. and must review the agenda. (Time = 2½ hrs)	_____
13.	Because of an impending company cash flow problem, your boss has asked you for the detailed monthly labor expenses for the next three months. (Time = 2 hrs)	_____
14.	The vice president has just called to inform you that two congressmen will be visiting the plant today and you are requested to conduct the tour of the facility from 3–5 P.M. (Time = 2 hrs)	_____
15.	You have developed a new policy for controlling overtime costs on Phase II. You must inform your people by memo, phone, or team meeting. (Time = ½ hr)	_____
16.	You must sign and review 25 purchase order requisitions for Phase III raw materials. It is company policy that the project manager sign all forms. Almost all of the items require a three-month lead time. (Time = 1 hr)	_____
17.	The engineering division manager has asked you to assist one of his people this afternoon in the solution of a technical problem. You are not required to do this. It would be a personal favor for the engineering manager, a man to whom you reported for the six years that you were an engineering functional manager. (Time = 2 hrs)	_____
18.	The data processing department manager informs you that the company is trying to eliminate unnecessary reports. He would like you to tell him which reports you can do without. (Time = ½ hr)	_____
19.	The assistant project manager for cost informs you that he does not know how to fill out the revised corporate project review form. (Time = ½ hr)	_____
20.	One of the functional managers wants an immediate explanation of why the scope of effort for Phase II was changed this late into the project and why he wasn't informed. (Time = 1 hr)	_____

WORK PLAN

Day: Tuesday

Priority			Method		Accomplishment		
Activity	Priority	Points	Method of Accomplishment	Points	Time	Activity	Points
11					9:00–9:30		
12					9:30–10:00		
13					10:00–10:30		
14					10:30–11:00		
15					11:00–11:30		
16					11:30–12:00		
17					12:00–12:30		
18					12:30–1:00		
19					1:00–1:30		
20					1:30–2:00		
					2:00–2:30		
Total			Total		2:30–3:00		
					3:00–3:30		
					3:30–4:00		
					4:00–4:30		
					4:30–5:00		
					Total		

Methods of accomplishment:
Y = you
D = delegate
T = transmit
R = return
A = avoid
P = postpone

Activities Postponed Until Today	Today's Priority

Points	
Priority points	
Methods points	
Accomplishment points	
Today's points total	

Wednesday's Activities

Activity	Description	Priority
21.	A vice president calls you stating that he has just read the rough draft of your Phase I report and wants to discuss some of the conclusions with you before the report is submitted to the customer on Thursday. (Time = 2 hrs)	_____
22.	The reproduction department informs you that it is expecting the final version of the in-house quarterly report for your project by noon today. The report is on your desk waiting for final review. (Time = 1 hr)	_____
23.	The manufacturing department manager calls to say that the department may have to do more work than initially defined in Phase II. A meeting is requested. (Time = 1 hr)	_____
24.	Quality control sends you a memo stating that, unless changes are made, it will not be able to work with the engineering specifications developed for Phase III. A meeting will be required with all assistant project managers in attendance. (Time = 1 hr)	_____
25.	A functional manager calls to tell you that the raw data from yesterday's tests are terrific and invites you to come up to the laboratory and see the results yourself. (Time = 1 hr)	_____
26.	Your assistant project manager is having trouble resolving a technical problem. The functional manager wants to deal with you directly. This problem must be resolved by Friday or else a major Phase II milestone might slip. (Time = 1 hr)	_____
27.	You have a technical interchange meeting with the customer scheduled for 1–3 P.M. on Thursday and must review the handout before it goes to publication. The reproduction department has requested at least 12 hours' notice. (Time = 1 hr)	_____
28.	You have a weekly team meeting from 10 A.M. to 12 noon (Time = 2 hrs)	_____
29.	You must dictate minutes to your secretary concerning your weekly team meeting, which is held on Wednesday 10 A.M. to 12 noon (Time = ½ hr)	_____
30.	A new project problem has occurred in the manufacturing area, and your manufacturing functional team members are reluctant to make a decision. (Time = 1 hr)	_____

WORK PLAN

Day: _____Wednesday_____

Priority			Method		Accomplishment		
Activity	Priority	Points	Method of Accomplishment	Points	Time	Activity	Points
21					9:00–9:30		
22					9:30–10:00		
23					10:00–10:30		
24					10:30–11:00		
25					11:00–11:30		
26					11:30–12:00		
27					12:00–12:30		
28					12:30–1:00		
29					1:00–1:30		
30					1:30–2:00		
	Total		Total		2:00–2:30		
					2:30–3:00		
					3:00–3:30		
					3:30–4:00		
					4:00–4:30		
					4:30–5:00		
					Total		

Methods of accomplishment:
Y = you
D = delegate
T = transmit
R = return
A = avoid
P = postpone

Activities Postponed Until Today	Today's Priority

Points	
Priority points	
Methods points	
Accomplishment points	
Today's points total	

Thursday's Activities

Activity	Description	Priority
31.	The electrical engineering department informs you that it has completed some Phase II activities ahead of schedule and wants to know if you wish to push any other activities to the left. (Time = 1 hr)	_____
32.	The assistant project manager for cost informs you that the corporate overhead rate is increasing faster than anticipated. If this continues, severe cost overruns will occur in Phases II and III. A schedule and cost review is necessary. (Time = 2 hrs)	_____
33.	Your insurance agent is calling to see if you wish to increase your life insurance. (Time = ½ hr)	_____
34.	You cannot find one of last week's manufacturing line manager's technical reports regarding departmental project status. You'll need it for the customer technical interchange meeting. (Time = ½ hr)	_____
35.	One of your car pool members wants to talk to you concerning next Saturday's golf tournament. (Time = ½ hr)	_____
36.	A functional manager calls to inform you that, due to a change in his division's workload priorities, people with the necessary technical expertise may not be available for next week's Phase II tasks. (Time = 2 hrs)	_____
37.	An employee calls you stating that he is receiving conflicting instructions from one of your assistant project managers and his line manager. (Time = 1 hr)	_____
38. .	The customer has requested bimonthly instead of monthly team meetings for Phase II. You must decide whether to add an additional project office team member to support the added workload. (Time = ½ hr)	_____
39.	Your secretary reminds you that you must make a presentation to the Rotary Club tonight on how your project will affect the local economy. You must prepare your speech. (Time = 2 hrs)	_____
40.	The bank has just called you concerning your personal loan. The information is urgent to get loan approval in time. (Time = ½ hr)	_____

Day: _____Thursday_____

WORK PLAN

Priority			Method		Accomplishment		
Activity	Priority	Points	Method of Accomplishment	Points	Time	Activity	Points
31					9:00–9:30		
32					9:30–10:00		
33					10:00–10:30		
34					10:30–11:00		
35					11:00–11:30		
36					11:30–12:00		
37					12:00–12:30		
38					12:30–1:00		
39					1:00–1:30		
40					1:30–2:00		
	Total		Total		2:00–2:30		
					2:30–3:00		
					3:00–3:30		
					3:30–4:00		
					4:00–4:30		
					4:30–5:00		
						Total	

Methods of accomplishment:
Y = you
D = delegate
T = transmit
R = return
A = avoid
P = postpone

Activities Postponed Until Today	Today's Priority

Points	
Priority points	
Methods points	
Accomplishment points	
Today's points total	

Friday's Activities

Activity	Description	Priority
41.	An assistant project manager has asked for your solution to a recurring problem. (Time = ½ hr)	_____
42.	A functional employee is up for a merit review. You must fill out a brief checklist form and discuss it with the employee. The form must be on the functional manager's desk by next Tuesday. (Time = ½ hr)	_____
43.	The personnel department wants you to review the summer vacation schedule for your project office personnel. (Time = ½ hr)	_____
44.	The vice president calls you into his office stating that he has seen the excellent test results from this week's work and feels that a follow-on contract should be considered. He wants to know if you can develop reasonable justification for requesting a follow-on contract at this early date. (Time = 1 hr)	_____
45.	The travel department says that you'll have to make your own travel arrangements for next month's trip to one of the customers, since you are taking a planned vacation trip in conjunction with the customer visit. (Time = ½ hr)	_____
46.	The personnel manager has asked if you would be willing to conduct a screening interview for an applicant who wants to be an assistant project manager. The applicant will be available this afternoon 1–2 p.m. (Time = 1 hr)	_____
47.	Your assistant project manager wants to know why you haven't approved his request to take MBA courses this quarter. (Time = ½ hr)	_____
48.	Your assistant project manager wants to know if he has the authority to visit vendors without informing procurement. (Time = ½ hr)	_____
49.	You have just received your copy of *Engineering Review Quarterly* and would like to look it over. (Time = ½ hr)	_____
50.	You have been asked to make a statement before the grievance committee (this Friday, 10 a.m.–noon) because a functional employee has complained about working overtime on Sunday mornings. You'll have to be in attendance for the entire meeting. (Time = 2 hrs)	_____

WORK PLAN

Day: _____Friday_____

Priority			Method		Accomplishment		
Activity	Priority	Points	Method of Accomplishment	Points	Time	Activity	Points
41					9:00–9:30		
42					9:30–10:00		
43					10:00–10:30		
44					10:30–11:00		
45					11:00–11:30		
46					11:30–12:00		
47					12:00–12:30		
48					12:30–1:00		
49					1:00–1:30		
50					1:30–2:00		
					2:00–2:30		
Total			Total		2:30–3:00		
					3:00–3:30		
					3:30–4:00		
					4:00–4:30		
					4:30–5:00		
					Total		

Methods of accomplishment:
Y = you
D = delegate
T = transmit
R = return
A = avoid
P = postpone

Activities Postponed Until Today	Today's Priority

Points	
Priority points	
Methods points	
Accomplishment points	
Today's points total	

RATIONALE AND POINT AWARDS

In the answers that follow, your recommendations may differ from those of the author because of the type of industry or the nature of the project. You will be given the opportunity to defend your answers at a later time.

a. If you selected the correct priority according to the table on pages 543–552, employ the following system for awarding points:

Priority	Points
A	10
B	5
C	3

b. If you selected the correct accomplishment mode according to the table on pages 543–552, employ the following system for assigning points:

Method of Accomplishment	Points
Y	10
T	10
P	8
D	8
A	6

c. You will receive 10 bonus points for each correctly postponed or delayed activity accomplished during the team meeting.

d. You will receive 5 points for each half-hour time slot in which you perform a priority A activity (one that is correctly identified as priority A).

e. You will receive a 10-point penalty for any activity that is split.

f. You will receive a 20-point penalty for each priority A or B activity not accomplished by you or your team by Friday at 5:00 p.m.

Activity	Rationale
1.	The updating of schedules, especially for Phase II, should be of prime importance because of the impact on functional resources. These schedules can be delegated to assistant project managers. However, with a team meeting scheduled for Wednesday, it should be an easy task to update the schedules when all of the players are present. The updating of the schedules should not be delayed until Thursday. Sufficient time must be allocated for close analysis and reproduction services.
2.	This must be done immediately. Your assistant project manager for manufacturing should be able to handle this activity.
3.	You must handle this yourself.
4.	Here, we assume that the representative is available only today. The assistant project managers can handle this activity. This activity may be important if you were unaware of this vendor's product.
5.	This could be delegated to your assistants provided that you allow sufficient time for personal review on Wednesday.
6.	Delaying this activity for one more week should not cause any problems. This activity can be delegated.
7.	You must take charge at once.
8.	Even though your main concern is the project, you still must fulfill your company's administrative requirements.
9.	This can be delayed until Wednesday's team meeting, especially since these are Phase III schedules. However, there is no guarantee that line people will be ready or knowledgeable to discuss Phase III this early. You will probably have to do this yourself.
10.	The procurement request must be answered. Your assistant project manager for manufacturing should have this information available.
11.	This is urgent and should not be postponed until the team meeting. Good project managers will give functional managers as much information as possible as early as possible for resource control. This task can be delegated to the assistant project managers, but it is not recommended.
12.	This belongs to the project manager. The agenda review and the meeting can be split, but it is not recommended.
13.	This must be done immediately. The results could severely limit your resources (especially if overtime would normally be required). Although your assistant project managers will probably be involved, the majority of the work is yours.
14.	Most project managers hate a request like this but know that situations such as these are inevitable.
15.	Project policies should be specified by project managers themselves. Policy changes should be announced as early as possible. Team meetings are appropriate for such actions.
16.	Obviously, you must do this task yourself. Fortunately, there is sufficient time if the lead times are accurate.
17.	The priority of this activity is actually your choice, but an A priority is preferred if you have time. This activity cannot be delegated.
18.	This activity must be done, but the question is when. Parts of this task can be delegated, but the final decision must be made by you, the project manager.

19. Obviously you must do this yourself. Your priority, of course, depends on the deadline on the corporate project review form.

20. You must perform this activity immediately.

21. Top-level executives from both the customer and contractor often communicate project status among themselves. Therefore, since the conclusions in the report reflect corporate policy, this activity should be accomplished immediately.

22. The reproduction department considers each job as a project. Therefore you should try not to violate their milestones. This activity can be delegated, depending on the nature of the report.

23. This could have a severe impact on your program. Although you could delegate this to one of your assistants, you should do it yourself because of the ramifications.

24. This must be done, and the team meeting is the ideal place.

25. You personally should give the functional manager the courtesy of showing you his outstanding results. However, it is not a high priority and could even be delegated or postponed since you'll see the data eventually.

26. The question here is the importance of the problem. The problem must be resolved by Thursday in case an executive meeting needs to be scheduled to establish company direction. Waiting until the last minute can be catastrophic here.

27. You as project manager should personally review all data presented to the customer. Check Thursday's schedule. Did you forget the interchange meeting?

28. This is your show.

29. This should be done immediately. Nonparticipants need to know the project status. The longer you wait, the greater the risk that you will neglect something important. This activity can be delegated, but it is not recommended.

30. You may have to solve this yourself even though you have an assistant project manager for manufacturing. The decision may affect the schedule and milestones.

31. Activities such as this do not happen very often. But when they do, you should make the most of them, as fast as you can. These are gold mine activities. They can be delegated but not postponed.

32. If this activity is not accomplished immediately, the results can be catastrophic. Regardless of your first inclination to delegate, this activity should be done by you, project manager, yourself.

33. This activity can be postponed or even avoided, if necessary.

34. Obviously, if the report is that important, then your assistant project managers should have copies of the report, and the activity can be delegated.

35. This activity should be discussed in the car pool, not on company time.

36. This is extremely serious. The line manager would probably prefer to work directly with you on this problem.

37. This is an activity that you should handle. Transmitting this to one of your assistants may aggravate the situation further. Although it is possible that this activity could be postponed, it is highly unlikely that time would smooth out the conflict.

38. This is a decision for you as project manager. Extreme urgency may not be necessary.

39. Project managers also have a social responsibility.

40. The solution to this activity is up for grabs. Most companies realize that employees occasionally need company time to complete personal business.

(continued)

41. Why is he asking you about a recurring problem? How did he solve it last time? Let him do it again.

42. You must do this personally, but it can wait until Monday.

43. This activity is not urgent and can be accomplished by your assistant project managers.

44. This could be your lucky day.

45. Although most managers would prefer to delegate this activity to their secretaries, it is really the responsibility of the project manager since it involves personal business.

46. This is an example of an administrative responsibility that is required of all personnel regardless of the job title or management level. This activity must be accomplished today, if time permits.

47. Although you might consider this as a B priority or one that can be postponed, you must remember that your assistant project manager considers this as an A priority and would like an answer today. You are morally obligated to give him the answer today.

48. Why can't he get the answer himself? Whether you handle this activity or not might depend on the priority and how much time you have available.

49. How important is it for you to review the publication?

50. This is mandatory attendance on your behalf. You have total responsibility for all overtime scheduled on your project. You may wish to bring one of your assistant project managers with you for moral support.

Now take the total points for each day and complete the following table:

Summary Work Plan	
Day	**Points**
Monday	
Tuesday	
Wednesday	
Thursday	
Friday	
Total	

CONCLUSIONS AND SUMMARY QUESTIONS

1. Project managers have a tendency to want to carry the load themselves, even if it means working 60 hours a week. You were told to do everything within your normal working day. But, as a project manager, you probably have a natural tendency to want to postpone some work until a later date so that you can do it yourself. Doing the activities when they occur, even through transmittal or delegation, is probably the best policy. You might wish to do the same again at a later time and see if you can beat your present score. Only this time, try to do as many tasks as possible on each day, even if it means delegation.

2. Several of the activities were company, not project, requests. Project managers have a tendency to avoid administrative responsibilities unless they deal directly with their project. This process of project management "tunnel vision" can lead to antagonism and conflicts if the project manager does not develop the proper attitude, which can easily trickle down to his assistants as well.

3. Several of the activities could have been returned to the requestor. However, in a project environment where the project manager cannot be successful without the functional manager's support, most project managers would never turn away a line employee's request for assistance.

4. Make a list of the activities where your answers differ from those of the answer key and where you feel that there exists sufficient justification for your interpretation.

5. Quite often self-productivity can be increased by knowing one's own energy cycle. Are your more important meetings in the mornings or afternoons? What time of day do you perform your most productive work? When do you do your best writing? Does your energy cycle vary according to the day of the week?

Part 17

INDUSTRY SPECIFIC: CONSTRUCTION

Many project management situations or problems are somewhat complex and involve many interacting factors, all contributing to a common situation. For example, poor planning on a project may appear on the surface to be a planning issue, whereas the real problem may be the corporate culture, lack of line management support, or poor employee morale. The case studies in this chapter involve interacting factors.

Robert L. Frank Construction Company

It was Friday afternoon, a late November day in 2013, and Ron Katz, a purchasing agent for Robert L. Frank Construction, pored over the latest earned value measurement reports. The results kept pointing out the same fact: The Lewis project was seriously over budget. Man-hours expended to date were running 30 percent over the projection, and, despite this fact, the project was not progressing sufficiently to satisfy the customer. Material deliveries had experienced several slippages, and the unofficial indication from the project scheduler was that, due to delivery delays on several of the project's key items, it was no longer possible to meet the scheduled completion date of the coal liquefaction pilot plant

Katz was completely baffled. Each day for the past few months as he reviewed the daily printout of project time charges, he would note that the purchasing and expediting departments were working on the Lewis project, even though it was not an unusually large project, dollarwise, for Frank. Two years earlier, Frank was working on a $300 million contract, a $100 million contract, and a $50 million contract concurrently with the Frank Chicago purchasing department responsible for all the purchasing, inspection, and expediting on all three contracts. The Lewis project was the largest project in house and was valued at only $90 million. What made this project so different from previous contracts and caused such problems? There was little Katz felt that he could do to correct the situation. All that he could do was to understand what had occurred in an effort to prevent a recurrence. He began to write his man-hour report as requested by the project manager for the next day.

COMPANY BACKGROUND

Robert L. Frank Construction Company was an engineering and construction firm serving the petroleum, petrochemical, chemical, iron and steel, mining, pharmaceutical, and food-processing industries from its corporate headquarters in Chicago, Illinois, and its worldwide offices. Its services include engineering, purchasing, inspection, expediting, construction, and consultation.

Frank's history began in 1947 when Robert L. Frank opened his office. In 1955, a corporation was formed, and by 1960, the company had completed contracts for the majority of the American producers of iron and steel. In 1962, an event occurred that was to have a large impact on Frank's future. This was the merger of Wilson Engineering Company, a successful refinery concern, with Robert L. Frank, now a highly successful iron and steel concern. This merger greatly expanded Frank's scope of operations and brought with it a strong period of growth. Several offices were opened in the United States in an effort to better handle the increase in business. Future expansions and mergers enlarged the Frank organization to the point where it had 15 offices or subsidiaries located throughout the United States and 20 offices worldwide. Through its first 20 years of operations, Frank had more than 2,500 contracts for projects having an erected value of over $1 billion.

Frank's organizational structure has been well suited to the type of work undertaken. The projects Frank contracted for typically had a time constraint, a budget constraint, and a performance constraint. They all involved an outside customer, such as a major petroleum company or a steel manufacturer. Upon acceptance of a project, a project manager was chosen (and usually identified in the proposal). The project manager would head up the project office, typically consisting of the project manager, one to three project engineers, a project control manager, and project secretaries. The project team also included the necessary functional personnel from the engineering, purchasing, estimating, cost control, and scheduling areas. Figure I is a simplified depiction. Of the functional areas, the purchasing department is somewhat unique in its organizational structure. The purchasing department is organized on a project management basis much as the project as a whole would be organized. Within the purchasing department, each project had a project office that included a project purchasing agent, one or more project expeditors, and a project purchasing secretary. Within the purchasing department, the PPA had line authority over only the project expeditor(s) and project secretary. However, for the PPA to accomplish his goals, the various functions within the purchasing department had to commit sufficient resources. Figure II illustrates the organization within the purchasing department.

HISTORY OF THE LEWIS PROJECT

Since 2008, the work backlog at Frank has been steadily declining. The Rovery Project, valued at $600 million, had increased company employment sharply

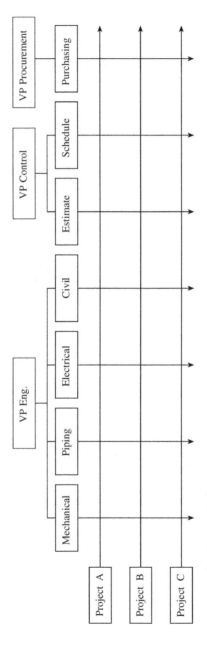

FIGURE I Frank organization

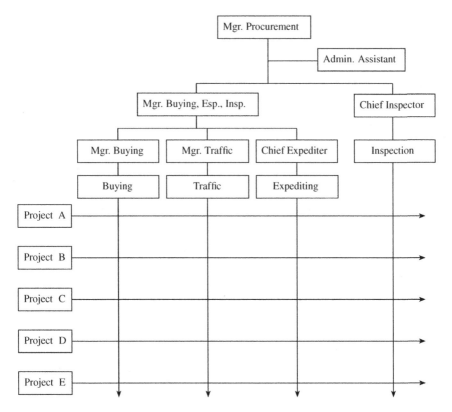

FIGURE II   Frank purchasing organization

since its inception in 2007. In fact, the engineering on the Rovery project was such a large undertaking that in addition to the Chicago office's participation, two other U.S. offices, the Canadian office, and the Italian subsidiary were heavily involved. However, since the Rovery project completion in 2011, Frank did not receive enough new work to support its workforce, thus necessitating recent layoffs of engineers, including a few project engineers.

Company officials were very disturbed with the situation. Frank's company policy was to "maintain an efficient organization of sufficient size and resources, and staffed by people with the necessary qualifications, to execute projects in any location for the industries served by Frank." However, the recent downturn in business meant that there was not enough work even with the reduction in employees. Further cutbacks would jeopardize Frank's prospects of obtaining future large projects as prospective clients look to contractors with a sufficient staff of qualified people to accomplish their work. By contrast, supporting employees out of overhead was not the way to do business either. It became increasingly important to cut the fat out of the proposals being submitted for possible projects. Despite this, new projects were few and far between, and the projects that were received were small in scope and dollar value and therefore did not provide work for very many employees.

When rumors of a possible construction project for a new coal liquefaction pilot plant started circulating, Frank officials were extremely interested in bidding for the work. It was an excellent prospect for two reasons. Besides Frank's desperate need for work, the Lewis chemical process used in the pilot plant would benefit Frank in the long run by the development of state-of-the-art technology. If the pilot plant project could be executed successfully, when it came time to construct the full-scale facility, Frank would have the inside track as it had already worked with the technology. The full-scale facility offered prospects exceeding the Rovery project, Frank's largest project to date. Top priority was therefore put on obtaining the Lewis project. It was felt that Frank had a slight edge due to successful completion of a Lewis project six years earlier. The proposal submitted to Lewis contained estimates for material costs, man-hours, and the fee. Any changes in scope after contract award would be handled by change order to the contract. Both Lewis and Frank had excellent scope change control processes as part of their configuration management plans. The functional department affected would submit an estimate of extra man-hours involved to the project manager, who would review the request and submit it to the client for approval. Frank's preference was for cost-plus-fixed-fee contracts.

One of the unique aspects stated in the Lewis proposal was the requirement for participation by both of Frank Chicago's operating divisions. Previous Frank contracts were well suited to either Frank's Petroleum and Chemical Division or the Iron and Steel Division. However, due to the unusual chemical process, one that starts with coal and ends up with a liquid energy form, one of the plant's three units was well suited to the Petroleum and Chemical Division and one was well suited to the Iron and Steel Division. The third unit was an off-site unit and was not of particular engineering significance.

The award of the contract six weeks later led to expectations by most Frank personnel that the company's future was back on track. The project began inauspiciously. The project manager was a well-liked, easygoing sort who had been manager of several Frank projects. The project office included three of Frank's most qualified project engineers.

In the purchasing department, the project purchasing agent (PPA) assigned to the project was Frank's most experienced PPA. Bill Hall had just completed his assignment on the Rovery Project and had done well, considering the magnitude of the job. The project had its problems, but they were small in comparison to the achievements. Bill had alienated some of the departments slightly, but that was to be expected. Purchasing upper management was somewhat dissatisfied with him in that, due to the size of the project, he didn't always use the normal Frank purchasing methods; rather, he used whatever method he felt was in the best interest of the project. Also, after the Rovery project, a purchasing upper management reshuffling left him in the same position but with less power and authority rather than receiving a promotion he had felt he had earned. As a result, he began to subtly criticize the purchasing management. This action caused upper management to hold him in less than high regard, but, at the time of the Lewis project, Hall was the best person available.

Due to the lack of float in the schedule and the early field start date, it was necessary to *fast start* the Lewis project. All major equipment was to be purchased within the first three months. This, with few exceptions, was accomplished. The usual problems occurred, such as late receipt of requisition from engineering and late receipt of bids from suppliers.

One of the unique aspects of the Lewis project was the requirement for purchase order award meetings with vendors. Typically, Frank would hold award meetings with vendors of major equipment, such as reactors, compressors, large process towers, or large pumps. However, almost each time Lewis approved purchase of a mechanical item or vessel, it requested that the vendor come in for a meeting. Even if the order was for an on-the-shelf stock pump or small drum or tank, a meeting was held. Initially, the purchasing department meeting attendees included the project purchasing agent, the buyer, the manager of the traffic department, the chief expeditor, and the chief inspector. Engineering representatives included the responsible engineer and one or two of the project engineers. Other Frank attendees were the project control manager and the scheduler. Quite often, these meetings would accomplish nothing except the reiteration of what had been included in the proposal or what could have been resolved with a phone call or even e-mail. The PPA was responsible for issuing meeting notes after each meeting.

One day at the end of the first three-month period, the top-ranking Lewis representative met with Larry Broyles, the Frank project manager.

Lewis rep: "Larry, the project is progressing, but I'm a little concerned. We don't feel like we have our finger on the pulse of the project. The information we are getting is sketchy and untimely. What we would like to do is meet with Frank every Wednesday to review progress and resolve problems."

Larry: "I'd be more than happy to meet with any of the Lewis people because I think your request has a lot of merit."

Lewis rep: "Well, Larry, what I had in mind was a meeting between all the Lewis people, yourself, your project office, the project purchasing agent, his assistant, and your scheduling and cost control people."

Larry: "This sounds like a pretty involved meeting. We're going to tie up a lot of our people for one full day a week. I'd like to scale this thing down. Our proposal took into consideration meetings, but not to the magnitude we're talking about."

Lewis rep: "Larry, I'm sorry, but we're footing the bill on this project and we've got to know what's going on."

Larry: "I'll set it up for this coming Wednesday."

Lewis rep: "Good."

The required personnel were informed by the project manager that effective immediately, meetings with the client would be held weekly. However, Lewis was dissatisfied with the results of the meetings, so the Frank project manager informed his people that a premeeting would be held each Tuesday to prepare the Frank portion of the Wednesday meeting. All of the Wednesday participants attended the Tuesday premeetings.

Lewis requests for additional special reports from the purchasing department were given in to without comment. The PPA and his assistants (project started with one and expanded to four) were devoting a great majority of their time to special reports and putting out fires instead of being able to track progress and prevent problems. For example, recommended spare parts lists were normally required from vendors on all Frank projects. Lewis was no exception. However, after the project began, Lewis decided it wanted the spare parts recommendations early into the job. Usually spare parts lists are left for the end of an order. For example, on a pump with 15-week delivery, normally Frank would pursue the recommended spare parts list three to four weeks prior to shipment, as it would tend to be more accurate. This improved accuracy was due to the fact that at this point in the order, all changes probably had been made. In the case of the Lewis project, spare parts recommendations had to be expedited from the day the material was released for fabrication. Changes could still be made that could dramatically affect the design of the pump. Thus, a change in the pump after receipt of the spare parts list would necessitate a new spare parts list. The time involved in this method of expediting the spare parts list was much greater than the time involved in the normal Frank method. Added to this situation was Lewis's request for a fairly involved biweekly report on the status of spare parts lists on all the orders. In addition, a full-time spare parts coordinator was assigned to the project.

The initial lines of communication between Frank and Lewis were well defined. The seven in-house Lewis representatives occupied the area adjacent to the Frank project office. (See Figure III.) At first, all communications from Lewis were channeled through the Frank project office to the applicable functional employee. In the case of the purchasing department, the Frank project office would channel Lewis requests through the purchasing project office. Responses or return communications followed the reverse route. Soon the volume of communications increased to the point where response time was becoming unacceptable. In several special cases, an effort was made to cut this response time. Larry

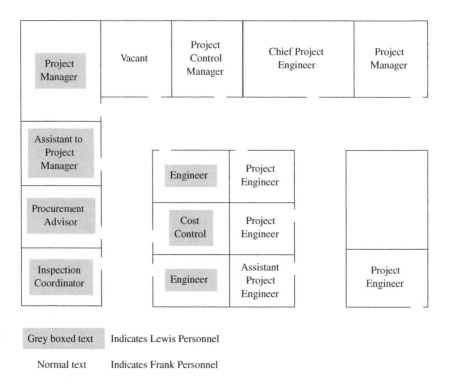

Project Manager	Vacant	Project Control Manager	Chief Project Engineer	Project Manager

Assistant to Project Manager				
Procurement Advisor	Engineer	Project Engineer		
	Cost Control	Project Engineer		
Inspection Coordinator	Engineer	Assistant Project Engineer	Project Engineer	

Grey boxed text	Indicates Lewis Personnel
Normal text	Indicates Frank Personnel

FIGURE III   Floor plan—Lewis project teams

Broyles told the Lewis team members to call or go see the functional person (i.e., buyer or engineer) for the answer. However, this practice soon became the rule rather than the exception. Initially, the project office was kept informed of these conversations, but this soon stopped. The Lewis personnel had integrated themselves into the Frank organization to the point where they had become part of the organization.

The project continued, and numerous problems cropped up. Vendors' material delays occurred, companies with Frank purchase orders went bankrupt, and progress was not to Lewis's satisfaction. Upper management soon became aware of the problems on this project due to its sensitive nature, and the Lewis project was now receiving much more intense involvement by senior management than it had previously. Upper management sat in on the weekly meetings in an attempt to pacify Lewis. Further problems plagued the project. Purchasing management, in an attempt to placate Lewis, replaced the project purchasing agent. Ron Katz, a promising young MBA graduate, had five years of experience as an assistant to several of the project purchasing agents. He was most recently a PPA on a fairly small project that had been very successful. Purchasing upper management thought that this move was a good one, for two reasons. First, it would remove Bill Hall from the project as PPA. Second, by appointing Ron Katz, Lewis would be pacified, as Katz was a promising talent with a successful project under his belt.

However, under direction of Katz, the project still experienced problems in the purchasing area. Revisions by engineering to material already on order caused serious delivery delays. Recently requisitioned material could not be located with an acceptable delivery promise. Katz and purchasing upper management, in an attempt to improve the situation, assigned more personnel to the project, personnel who were more qualified than the positions dictated. Buyers and upper-level purchasing officials were sent on trips to vendors' facilities, trips that were normally handled by traveling expediters. In the last week the Lewis representative met with the project manager, Broyles:

Lewis rep: "Larry, I've been reviewing these man-hour expenditures, and I'm disturbed by them."

Larry: "Why's that?"

Lewis rep: "The man-hour expenditures are far outrunning project progress. Three months ago, you reported that the project completion percentage was 30 percent, but according to my calculations, we've used 47 percent of the man-hours. Last month, you reported 40 percent project completion, and I show a 60 percent expenditure of man-hours."

Larry: "Well, as you know, due to problems with vendors' deliveries, we've really had to expedite intensively to try to bring them back in line."

Lewis rep: "Larry, I'm being closely watched by my people on this project, and a cost or schedule overrun not only makes Frank look bad, it makes *me* look bad."

Larry: "Where do we go from here?"

Lewis rep: "What I want is an estimate from your people on what is left, man-hour wise. Then I can sit down with my people and see where we are."

Larry: "I'll have something for you the day after tomorrow."

Lewis rep: "Good."

The functional areas were requested to provide this information, which was reviewed and combined by the project manager and submitted to Lewis for approval. Lewis's reaction was unpleasant, to say the least. The estimated man-hours in the proposal were now insufficient. The revised estimate was for almost 40 percent over the proposal. The Lewis representative immediately demanded an extensive report on the requested increase. In response, the project manager requested man-hour breakdowns from the functional areas. Purchasing was told to do a purchase order by purchase order breakdown of expediting and inspection man-hours. The buying section had to break down the estimate of the man-hours needed to purchase each requisition, many of which were not even issued. Things appeared to have gone from bad to worse.

The Lyle
Construction Project

At 6:00 P.M. on Thursday in late October 2008, Don Jung, an Atlay Company project manager assigned to the Lyle contract, sat in his office thinking about the comments brought up during a meeting with his immediate superior earlier that afternoon. During that meeting, Fred managers, criticized Don for not promoting a cooperative attitude Franks, the supervisor of project between him and the functional managers. Fred had had a high-level meeting with the vice presidents in charge of the various functional departments (i.e., engineering, construction, cost control, scheduling, and purchasing) earlier that day. One of these vice presidents, John Mabby (head of the purchasing department), had indicated that his department, according to his latest projections, would overrun its man-hour allocation by 6,000 hours. Bob Stewart, the PPA assigned to the Lyle Project, had relayed this to Don twice in the past, but Don had not seriously considered the problem with the overrun possibility because some of the purchasing was now going to be done by the subcontractor at the job site (who had enough available man-hours to cover this additional work). John complained that, even though the subcontractor was doing some of the purchasing in the field, his department still would overrun its man-hour allocation. He also indicated to Fred that Don had better do something about this man-hour problem now. At this point in the meeting, the vice president of engineering, Harold Mont, stated that he had experienced the same problem in that Don seemed to ignore their requests for additional man-hours. Also at this meeting, the various vice presidents indicated that Don had not been operating within the established standard company procedures. In an effort to make up for time lost due to initial delays that occurred in the process development stage of this project, Don and his project team had been getting various functional people working on the contract to cut corners and in many cases to buck the standard operating procedures of their respective functional departments to save time. His actions and the actions of his project team were alienating the vice presidents in charge of the functional departments. During this meeting, Fred received a good deal of criticism due to this fact. He was also told that Don had better shape up, because it was the consensus opinion of these vice presidents that his method of operating might seriously hamper the project's ability to finish on time and within budget. It was very important that this job be completed in accordance with the Lyle requirements since Frank would be building two more similar plants within the next 10 years. A good effort on this job could further enhance Atlay's chances for being awarded the next two jobs.

Fred Franks related these comments and a few of his own to Don Jung. Fred seriously questioned Don's ability to manage the project effectively and told him so. However, Fred was willing to allow Don to remain on the job if he would begin to operate in accordance with the various functional departments' standard operating procedures and if he would listen and be more attentive to the comments from the various functional departments and do his best to cooperate with them in the best interests of the company and the project itself.

INCEPTION OF THE LYLE PROJECT

In April of 2009, Bob Briggs, Atlay's vice president of sales, was notified by Lyle's vice president of operations, Fred Wilson, that Atlay had been awarded the $600 million contract to design, engineer, and construct a polypropylene plant in Louisiana. Bob immediately notified Atlay's president and other high- level officials in the organization. (See Figure I.) He then contacted Fred Franks in order to finalize the members of the project team. Briggs wanted George Fitz, who was involved in developing the initial proposal, to be the project manager. However, George was in the hospital and would be essentially out of action for another three months. Atlay then had to scramble to appoint a project manager, since Lyle wanted to conduct a kickoff meeting in a week with all the principals present.

One of the persons most available for the position of project manager was Don Jung. Don had been with the company for about 15 years. He had started with the company as a project engineer and then was promoted to the position of manager of computer services. He was in charge of computer services for six months until he had a confrontation with Atlay's upper management regarding the policies under which the computer department was operating. He had served the company in two other functions since that time—his most recent position was as a senior project engineer on a small project that was handled out of the Houston

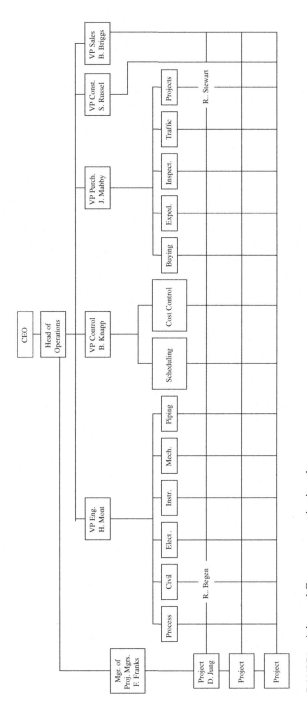

FIGURE I Atlay and Company organization chart

office. One big plus was the fact that Don knew Lyle's Fred Wilson personally since they belonged to many of the same community organizations. It was decided that Don would be the project manager and John Neber, an experienced project engineer, would be assigned as the senior project engineer. The next week was spent advising Don regarding the contents of the proposal and determining the rest of the members to be assigned to the project team.

A week later, Lyle's contingent arrived at Atlay's headquarters. (See Figure II.) Atlay was informed that Steve Zorn would be the assistant project manager on this job for Lyle. The position of project manager would be left vacant for the time being. The rest of Lyle's project team was then introduced. Lyle's project team consisted of individuals from various Lyle divisions around the country, including Texas, West Virginia, and Philadelphia. Many of the Lyle project team members had met each other for the first time only two weeks ago.

During this initial meeting, Fred Wilson emphasized that it was essential that this plant be completed on time since their competitor was in the process of preparing to build a similar facility in the same general location. The first plant finished would most likely be the one that would establish control for polypropylene material over the southwestern U.S. market. Fred Wilson felt that Lyle had a six-week head start over its competitor at the moment and would like to increase that difference, if at all possible. He then introduced Lyle's assistant project manager who completed the rest of the presentation.

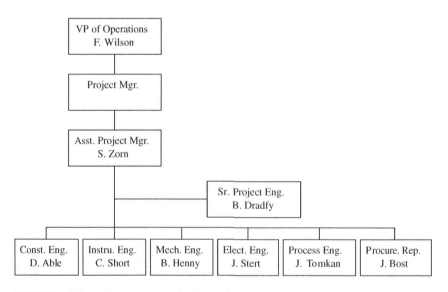

FIGURE II Lyle project team organizational chart

At this initial meeting, the design package was handed over to Atlay's Don Jung so that the process engineering stage of the project could begin. This package was, according to the inquiry letter, so complete that all material requirements for this job could be placed within three months after project award (since very little additional design work was required by Atlay on this project). Two weeks later, Don contacted the lead process engineer on the project, Raphael Begen. He wanted to get Raphael's opinion regarding the condition of the design package.

Begen: "Don, I think you have been sold a bill of goods. This package is in bad shape."

Jung: "What do you mean, this package is in bad shape? Lyle told us that we would be able to have all the material on order within three months since this package was in such good shape."

Begen: "Well in my opinion, it will take at least six weeks to straighten out the design package. Within three months from that point you will be able to have all the material on order."

Jung: "What you are telling me then is that I am faced with a six-week schedule delay right off the bat due to the condition of the package."

Begen: "Exactly."

Don Jung went back to his office after his conversation with the lead process engineer. He thought about the status of his project. He felt that Raphael Begen was being overly pessimistic and that the package wasn't really all that bad. Besides, a month shouldn't be too hard to make up if the engineering section would do its work quicker than normal and if purchasing would cut down on the amount of time it takes to purchase materials and equipment needed for this plant.

CONDUCT OF THE PROJECT

The project began on a high note. Two months after contract award, Lyle sent in a contingent of representatives. These representatives would be located at Atlay's headquarters for the next eight to 10 months. Don Jung had arranged to have the Lyle offices set up on the other side of the building away from his project team. At first there were complaints from Lyle's assistant project manager regarding the physical distance that separated Lyle's project team and Atlay's project team. However, Don assured him that there just wasn't any available space that was closer to the Atlay project team other than the one they were now occupying.

The Atlay project team operating within a matrix organizational structure plunged right into the project. (See Figure III.) They were made aware of the delay that was incurred at the onset of the job (due to the poor design package) by Don Jung. His instructions to them were to cut corners whenever doing so might result in time savings. They were also to suggest to members of the functional

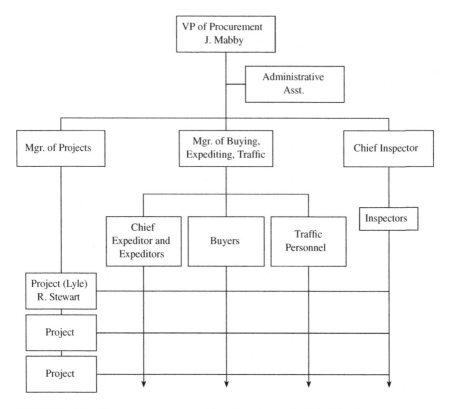

FIGURE III   Atlay Company procurement department organizational chart

departments that were working on this project methods that could possibly result in quicker turnaround of the work required of them. The project team coerced the various engineering departments into operating outside of their normal procedures due to the special circumstances surrounding this job. For example, the civil engineering section prepared a special preliminary structural steel package, and the piping engineering section prepared preliminary piping packages so that the purchasing department could go out on inquiry immediately. Normally the purchasing department would have to wait for formal take-offs from both of these departments before it could send out inquiries to potential vendors. Operating in this manner could result in some problems, however. For example, the purchasing department might arrange for discounts from vendors based on the quantity of structural steel estimated during the preliminary take-off. After the formal take-off was been done by the civil engineering section (which would take about a month), they might find out that they underestimated the quantity of structural steel required on the project by 50 tons. This was damaging, because knowing that an additional 50 tons of structural steel was required might have aided the purchasing department in securing an additional discount of $.20 per pound (or $160,000 discount for 400 tons of steel).

In an effort to make up for lost time, the project team convinced the functional engineering departments to use catalog drawings or quotation information whenever they lacked engineering data on a particular piece of equipment. The engineering section leaders pointed out that this procedure could be very dangerous and could result in additional work and further delays to the project. If, for example, the dimensions for the scale model being built were based on preliminary information without the benefit of having certified vendor drawings in house, then the scale for that section of the model might be off. When the certified data prints were received later and it was apparent that the dimensions were incorrect, that portion of the model might have to be rebuilt entirely. This would further delay the project. However, if the information did not change substantially, the company could save approximately a month in engineering time. Lyle was advised in regard to the risks and potential benefits involved when Atlay operated outside of its normal operating procedure. Steve Zorn informed Don Jung that Lyle was willing to take these risks in an effort to make up for lost time. The Atlay project team then proceeded accordingly.

The method that the project team was utilizing appeared to be working. It seemed as if the work was being accomplished at a much quicker pace than what was initially anticipated. The only snag in this operation occurred when Lyle had to review/approve something. Drawings, engineering requisitions, and purchase orders would sit in the Lyle area for about two weeks before Lyle personnel would review them. Half of the time these documents were returned two weeks later with a request for additional information or with changes noted by some of Lyle's engineers. Then the Atlay project team would have to review the comments/changes, incorporate them into the documents, and resubmit them to Lyle for review/approval. They would then sit for another week in that area before finally being reviewed and eventually returned to Atlay with final approval. It should be pointed out that the contract procedures stated that Lyle would have only five days to review/approve the various documents being submitted to it. Don Jung felt that part of the reason for this delay had to do with the fact that all the Lyle team members went back to their homes for the weekends. Their routine was to leave around 10:00 A.M. on Friday and return around 3:00 P.M. on the following Monday. Therefore, essentially two days of work by the Lyle project team out of the week were lost. Don reminded Steve Zorn that, according to the contract, Lyle was to return documents that needed approval within five days after receiving them. He also suggested that if the Lyle project team would work a full day on Monday and Friday, it would probably increase the speed at which documents were being returned. However, neither corrective action was undertaken by Lyle's assistant project manager, and the situation failed to improve. All the time the project team had saved by cutting corners was now being wasted, and further project delays seemed inevitable. In addition, other problems were being encountered during the interface process between the Lyle and Atlay project team members. It seems that the Lyle project team members (who were on temporary loan to Steve from various functional departments within the Lyle organization) were more concerned with producing a perfect end product. They did not seem to realize that their actions, as well as the actions of the Atlay project team, had a significant impact on this particular project. They did not seem aware that they were also constrained by time and cost as well as performance. Instead, they had a very relaxed and informal operating procedure. Many of the changes made by Lyle were given to Atlay verbally. They explained to the Atlay project team members that written confirmation of the changes was unnecessary because "we are all working on the same team." Many significant changes in the project were made when a Lyle engineer was talking directly to an Atlay engineer. The Atlay engineer would then incorporate the changes into the drawings he was working on, and sometimes he failed to inform his project engineer about the changes. Because of this informal way of operating, there were instances in which Lyle was dissatisfied with Atlay because changes were not being incorporated or were not made in strict accordance with Lyle's requests. Steve called Don into his office to discuss this problem:

Steve: "Don, I've received complaints from my personnel regarding your teams inability to follow through and incorporate Lyle's comments/changes accurately into the drawings."

Don: "Steve, I think my staff has been doing a fairly good job of incorporating your team's comments/changes. You know the whole process would work a lot better, however, if you would send us a letter detailing each change. Sometimes my engineers are given two different instructions regarding the scope of the change recommended by your people. For example, one of your people will tell our process engineer to add a check valve to a specific process line and another would tell him that check valves are not required in that service."

Steve: "Don, you know that if we documented everything that was discussed between our two project teams, we would be buried in paperwork. Nothing would ever get accomplished. Now, if you get two different instructions from my project team, you should advise me accordingly so that I can resolve the discrepancy. I've decided that since we seem to have a communication problem regarding engineering changes, I want to set up a weekly engineering meeting for every Thursday. These meetings should help to cut down on the misunderstandings as well as keep us advised of your progress in the engineering area of this contract without the need of a formal status report. I would like all members of your project staff present at these meetings."

Don: "Will this meeting be in addition to our overall progress meetings that are held on Wednesdays?"

Steve: "Yes. We will now have two joint Atlay/Lyle meetings a week—one discussing overall progress on the job and one specifically aimed at engineering."

On the way back to his office, Don Jung thought about the request for an additional meeting. That meeting will be a waste of time, he thought, just as the Wednesday meeting currently is. It will just take away another day from the Lyle project team's available time for approving drawings, engineering, requisitions, and purchase orders. Now there are three days during the week where at least a good part of the day is taken up by meetings, in addition to a meeting with his project team on Mondays in order to freely discuss the progress and problems of the job without intervention by Lyle personnel. A good part of his project team's time, therefore, was now being spent preparing for and attending meetings during the course of the week. "Well," Don rationalized, "they are the client, and if they desire a meeting, then I have no alternative but to accommodate them."

JUNG'S CONFRONTATION

When Don Jung returned to his desk he saw a message stating that John Mabby, vice president of procurement, had called. Don returned his call and found out that John requested a meeting. A meeting was set up for the following day. At 9:00 A.M. the next day, Don was in John's office. John was concerned about the unusual procedures that were being utilized on this project. It seems as if he had had a rather lengthy discussion with Bob Stewart, the PPA assigned to the Lyle project. During the course of that conversation, it became apparent that this particular project was not operating within the normal procedures established for the purchasing department. This deviation from normal procedures was the result of instructions given by Don Jung to Bob Stewart. This upset John, since he felt that Don Jung have discussed these deviations with him prior to his instructing Bob to proceed in this manner:

Mabby: "Don, I understand that you advised my PPA to work around the procedures that I established for this department so that you could possibly save time on your project."
Jung: "That's right, John. We ran into a little trouble early in the project and started running behind schedule, but by cutting corners here and there, we've been able to make up some of the time."
Mabby: "Well, I wish you had contacted me first regarding this situation. I have to tell you, however, that if I had known about some of these actions, I would never have allowed Bob Stewart to proceed. I've instructed Stewart that from now on, he is to check with me prior to going against our standard operating procedure."

Jung: "But John Stewart has been assigned to me for this project? Therefore, I feel that he should operate in accordance with my requests, whether they are within your procedures or not."

Mabby: "That's not true. Stewart is in my department and works for me. I am the one who reviews him, approves the size of his raise, and decides if and when he gets a promotion. I have made that fact very clear to Stewart, and I hope I've made it very clear to you, also. In addition, I hear that Stewart has been predicting a 6,000 man-hour overrun for the purchasing department on your project. Why haven't you submitted an additional change request to the client?"

Jung: "Well, if what Stewart tells me is true, the main reason that your department is short man-hours is because the project manager who was handling the initial proposal, George Fitz, underestimated your requirements by 7,000 man-hours. Therefore, from the very beginning you were short man-hours. Why should I be the one that goes to the client and tells him that we blew our estimate when I wasn't even involved in the proposal stage of this contract? Besides, we are taking away some of your duties on this job, and I personally feel that you won't even need those additional 6,000 man-hours."

Mabby: "Well, I have to attend a meeting with your boss, Fred Franks, tomorrow, and I think I'll talk to him about these matters."

Jung: "Go right ahead. I'm sure you'll find out that Fred stands behind me 100 percent."

Part 18

INDUSTRY SPECIFIC: DISNEY THEME PARKS

Every company has some unique characteristics that it wishes its employees to possess to be successful. For Disney theme parks, the skill is in Imagineering project management. But even some of the best-managed companies in the world, such as Disney, still succumb to risks, mainly from enterprise environmental factors that can influence decisions made on projects. Although we all want to believe that we have some understanding of these enterprise environmental factors, the impacts can be severe when a variety of international cultures are involved.

Disney (A): Imagineering Project Management

INTRODUCTION

Not all project managers are happy with their jobs, and they often believe that changing industries might help. Some want to manage "the world's greatest construction projects" while others want to design the next-generation cell phone or mobile device. However, the project managers who probably are the happiest are the Imagineering project managers who work for the Walt Disney Company, even though they probably could earn higher salaries elsewhere on projects that have profit and loss statements. Three Imagineering project managers—John Hench, Claude Coats, and Martin Sklar—retired with a combined 172 years of Imagineering project management work experience with the Walt Disney Company. But how many project managers in other industries truly understand what skills are needed to be successful as an Imagineering project manager? Is it possible that many Imagineering project management skills are applicable to other industries and we do not recognize it?

The *PMBOK® Guide* is, as the name implies, just a guide. Each company may have unique or specialized skills needed for the projects it undertakes above and beyond what is included in the *PMBOK® Guide.* Even though the principles of the *Guide* apply to Disney's theme park projects, other skills are needed that are significantly different from much of the material taught in traditional project management courses. Perhaps the most common skills among all Imagineering project managers are brainstorming, problem solving, decision making, and thinking in three rather than two dimensions. While many of these skills are not taught in depth in traditional project management programs, they may very well be necessities for *all* project managers. Yet most of us may not recognize this fact.

WALT DISNEY IMAGINEERING

Walt Disney Imagineering (also known as WDI or simply Imagineering) is the design and development arm of the Walt Disney Company, responsible for the creation and construction of Disney theme parks worldwide. Founded by Walt Disney to oversee the production of Disneyland Park, the company was originally known as WED Enterprises, from the initials meaning "Walter Elias Disney," the company founder's full name.[1]

The term "Imagineering" was introduced in the 1940s by Alcoa to describe its blending of imagination and engineering, and used by Union Carbide in an in-house magazine in 1957, with an article by Richard F. Sailer called "BRAIN-STORMING IS IMAGINation enginEERING." Disney filed for a copyright for the term in 1967, claiming first use of the term in 1962. Imagineering is responsible for designing and building Disney theme parks, resorts, cruise ships, and other entertainment venues at all levels of project development. Imagineers possess a broad range of skills and talents, and thus over 140 different job titles fall under the banner of Imagineering, including illustrators, architects, engineers, lighting designers, show writers, graphic designers, and many more.[2] It could be argued that all Imagineers are project managers and all project managers at WDI are Imagineers. Most Imagineers work from the company's headquarters in Glendale, California, but are often deployed to satellite branches within the theme parks for long periods of time.

PROJECT DELIVERABLES

All I want you to think about is when people walk through or have access to anything you design, I want them, when they leave, to have smiles on their faces. Just remember that. It's all I ask of you as a designer.

—Walt Disney

Parts of this case study have been adapted from Wikipedia contributors, "Walt Disney Imagineering," *Wikipedia, The Free Encyclopedia,* https://en.wikipedia.org/w/index.php?title=Walt_Disney_Imagineering&oldid=758012775

[1] Alex Wright, *Imagineers: The Imagineering Field Guide to Magic Kingdom at Walt Disney World* (New York: Disney Editions, 2005).

[2] Ibid.

Unlike traditional projects where the outcome of a project is a hardware or software deliverable, Imagineering project outcomes for theme park attractions are visual stories. The entire deliverable is designed to operate in a controlled environment where every component has a specific meaning and contributes to part of telling a story. It is visual storytelling. Unlike traditional movies or books that are two dimensional, theme parks and the accompanying characters come to life in three dimensions. Most project managers do not see themselves as storytellers.

The intent of a theme park attraction is to remove people from reality once they enter the attraction and make them believe that they are living out a story and possibly interacting with their favorite characters. Theme park visitors of all ages are made to feel that they are participants in the story rather than just observers.

Some theme parks are composed of rides that appeal to just one of your senses; Disney's attractions, in contrast, appeal to several senses, thus leaving a greater impact when people exit the attraction. "People must learn how to see, hear, smell, touch and taste in new ways."[3] Everything is designed to give people an experience. In the ideal situation, people are made to believe that they are part of the story. When new attractions are launched, Imagineers pay attention to guests' faces as they come off of a ride. This is important for continuous improvement efforts.

THE IMPORTANCE OF CONSTRAINTS

Most project management courses emphasize that there are three constraints on projects, namely time, cost, and scope. Although these constraints exist for Imagineering projects as well, there are three other theme park constraints that are often considered more important than time, cost, and scope. The additional constraints are safety, quality, and aesthetic value.

Safety, quality, and aesthetic value are all interrelated constraints. Disney will never sacrifice safety. It is first and foremost the primary constraint. All attractions operate every few minutes 365 days each year and must satisfy the strictest of building codes. Some rides require special effects, such as fire, smoke, steam, and water. All of this is accomplished with safety in mind. Special effects include fire that actually does not burn, simulated fog that people can breathe safely, and explosions that do not destroy anything. Another special effect is the appearance of bubbling molten lava that is actually cool to the touch.

Reliability and maintainability are important quality attributes for all project managers but are of critical importance for the Imagineers. In addition to fire, smoke, stream, and water, there are a significant number of moving parts in each

---

[3] John Hench with Peggy Van Pelt, *Designing Disney: Imagineering and the Art of the Show* (New York: Disney Editions, 2008), p. 2

attraction. Reliability considers how long something will perform without requiring maintenance. Maintainability concerns how quickly repairs can be made. Attractions are designed with consideration given to component malfunctions and ways to minimize the down time. Some people may have planned their entire vacation around the desire to see specific attractions, and if these attractions are down for repairs for a lengthy time, park guests will be unhappy.

BRAINSTORMING

With traditional projects, brainstorming may be measured in hours or days. Members of the brainstorming group are few in number and may include marketing for the purpose of identifying the need for a new product or enhancement to an existing product and technical personnel to state how long it takes and the approximate cost. Quite often, traditional project managers may not be assigned and brought on board until after the project has been approved, added into the queue, and after the statement of work (SOW) is well defined. At Disney's Imagineering organization, brainstorming may be measured in years and a multitude of Imagineering personnel will participate, including the project managers.

Attractions at most traditional amusement parks are designed by engineers and architects. Imagineering brainstorming at Disney is done by storytellers who must visualize their ideas in both two and three dimensions. Brainstorming could very well be the most critical skill for an Imagineer. It requires that Imagineers put themselves in the guests' shoes and think like children and as well as adults in order to see what the visitors will see. Those who design an attraction must know the primary audience.

Brainstorming can be structured or unstructured. Structured brainstorming could entail thinking up an attraction based on a newly released animated or nonanimated Disney movie. Unstructured brainstorming is usually referred to as "blue sky" brainstorming. Several sessions may be required to come up with the best idea because people need time to brainstorm. Effective brainstorming mandates that people be open-minded to all ideas. And even if everyone agrees on the idea, Imagineers always ask, "Can we make it even better?" Unlike traditional brainstorming, it may take years before an idea comes to fruition at the Imagineering Division.

Imagineering brainstorming must focus on a controlled themed environment where every component is part of telling the story. Critical questions must be addressed and answered as part of Imagineering brainstorming:

- How much space will I have for the attraction?
- How much time will the guests need to feel the experience?
- Will the attraction be seen on foot or using people movers?
- What colors should we use?
- What music should we use?

- What special effects and/or illusions must be in place?
- Does technology exist for the attraction, or must new technology be created?
- What landscaping and architecture will be required?
- What other attractions precede this attraction or follow it?

Before brainstorming is completed, the team must consider the cost. Regardless of the technology, can we afford to build it? This question must be addressed during structured and blue-sky brainstorming sessions.

GUIDING PRINCIPLES

If I could pick any job here, I'd move my office to the Imagineering building and immerse myself in all that lunacy and free-thinking.
—Michael D. Eisner, former CEO, Walt Disney

When developing new concepts and improving existing attractions, Imagineers are governed by a few key principles. Often new concepts and improvements are created to fulfill specific needs and to make the impossible appear possible. Many ingenious solutions to problems are Imagineered in this way, such as the ride vehicle of the attraction Soarin' Over California. The Imagineers knew they wanted guests to experience the sensation of flight but weren't sure how to accomplish the task of loading the people onto a ride vehicle in an efficient manner where everyone had an optimal viewing position. One day an Imagineer found an Erector set in his attic and was able to envision and design a ride vehicle that would effectively simulate hang gliding.[4]

Imagineers are also known for returning to ideas for attractions and shows that, for whatever reason, never came to fruition. It could be years later when they revisit the ideas. These ideas are often reworked and appear in a different form—like the Museum of the Weird, a proposed walk-through wax museum that eventually became the Haunted Mansion.[5]

Finally, there is the principle of "blue-sky speculation," a process where Imagineers generate ideas with no limitations. The custom at Imagineering has been to start the creative process with what is referred to as "eyewash"—the boldest, wildest, best idea a person can come up with, presented in absolutely convincing detail. Many Imagineers consider this to be the true beginning of the design process and operate under the notion that if it can be dreamed of, it can be built.[6] Disney believes that everyone can brainstorm and that everyone wants to

---

[4] George Scribner and Jerry Rees (directors), *Disneyland: Secrets, Stories, and Magic* (DVD). Walt Disney Video, 2007.

[5] Ibid.

[6] Karal Ann Marling, *Designing Disney's Theme Parks: The Architecture of Reassurance* (New York: Flammarion, 1997).

contribute to the brainstorming process. No ideas are bad ideas. Effective brainstorming sessions neither evaluate nor criticize ideas. They are recorded and may be revisited years later.

Imagineers are always seeking to improve on their work—what Walt Disney called "plussing." He firmly believed that "Disneyland will never be completed as long as there's imagination left in the world," meaning there is always room for innovation and improvement.[7] Ideas and eventually future attractions can also come from the animated films produced by the Walt Disney Company or other film studios.

> The brainstorming subsides when the basic idea is defined, understood, and agreed upon by all group members. It belongs to all of us, keeping strong a rich heritage left to us by Walt Disney. Teamwork is truly the heart of Imagineering . . .
>
> In that spirit, though Imagineering is a diverse collection of architects, engineers, artists, support staff members, writers, researchers, custodians, schedulers, estimators, machinists, financiers, model-makers, landscape designers, special effects and lighting designers, sound technicians, producers, carpenters, accountants, and filmmakers—we all have the honor of sharing the same unique title. Here, you will find only Imagineers.[8]

IMAGINEERING INNOVATIONS

Over the years, WDI has been granted over 115 patents in areas such as ride systems, special effects, interactive technology, live entertainment, fiber optics, and advanced audio systems.[9] WDI is responsible for technological advances such, as the Circle-Vision 360° film technique and the FastPass virtual queuing system.

Imagineering must find a way to blend technology with the story. Imagineering is perhaps best known for its development of Audio-Animatronics, a form of robotics created for use in shows and attractions in the theme parks that allowed Disney to animate things in three dimensions instead of just two dimensions. The idea sprang from Disney's fascination with a mechanical bird he purchased in New Orleans, which eventually led to the development of the attraction the Enchanted Tiki Room. The Tiki Room, which debuted in 1963 and featured singing audio-animatronic birds, was the first to use such technology. The 1964 World's Fair featured an audio-animatronic figure of Abraham Lincoln that actually stood up and delivered part of the Gettysburg Address (which incidentally had just passed its centennial at the time) for the "Great Moments With Mr. Lincoln" figure exhibit, the first human Audio-Animatronic.[10]

---

[7] Scribner and Rees, *Disneyland.*

[8] Disney Book Group, *Walt Disney Imagineering* (New York: Disney Editions, 1996), p. 21.

[9] Walt Disney Imagineering website, www.imagineeringdisney.com.

[10] Ibid.

Today, audio-animatronics are featured prominently in many popular Disney attractions, including Pirates of the Caribbean, the Haunted Mansion, the Hall of Presidents, Country Bear Jamboree, Star Tours: The Adventures Continue, and Muppet*Vision 3D. Guests also have the opportunity to interact with some Audio-Animatronic characters, such as Lucky the Dinosaur, WALL-E, and Remy from *Ratatouille*. The next wave of audio-animatronic development focuses on completely independent figures, or "autonomatronics." Otto, the first autonomatronic figure, is capable of seeing, hearing, sensing a person's presence, having a conversation, and even sensing and reacting to guests' emotions.

STORYBOARDING

Most traditional project managers may be unfamiliar with the use of storyboarding as applied to projects. At Disney Imagineering, it is an essential part of the project. Ideas at Imagineering begin as a two-dimensional vision drafted on a piece of white paper. Storyboards, which are graphic organizers in the form of illustrations or images displayed in sequence for the purpose of pre-visualizing the relationship between time and space in the attraction, assist the Imagineers in seeing the entire attraction. Storyboards also are used in motion pictures, animation, motion graphics, and interactive media. They provide a visual layout of events as they are to be seen by the guests. The storyboarding process, in the form it is known today, was developed at Walt Disney Productions during the early 1930s, after several years of similar processes being in use at Walt Disney and other animation studios.

A storyboard is essentially a large comic of the attraction produced beforehand to help the Imagineers visualize the scenes and find potential problems before they occur. Storyboards also help estimate the cost of the overall attraction and save development time. Storyboards can be used to identify where changes to the music are needed to fit the mood of the scene. Often storyboards include arrows or instructions that indicate movement. When animation and special effects are part of the attraction, the storyboarding stage may be followed by simplified mock-ups called "animatics" to give a better idea of how the scene will look and feel with motion and timing. At its simplest, an animatic is a series of still images edited together and displayed in sequence with a rough dialogue and/or rough sound track added to the sequence of still images (usually taken from a storyboard) to test whether the sound and images are working together effectively.

The storyboarding process can be very time-consuming and intricate. Today, storyboarding software is available to speed up the process.

MOCK-UPS

Once brainstorming has been completed, mock-ups of the idea are created. Mock-ups are common to some other industries, such as construction. Simple mock-ups can be made from paper, cardboard, Styrofoam, plywood, or metal.

The modelmaker is the first Imagineer to make a concept real. The art of bringing a two-dimensional design into three dimensions is one of the most important and valued steps in the Imagineering process. Models enable the Imagineer to visualize, in miniature, the physical layout and dimensions of a concept, and the relationships of show sets or buildings as they will appear.

As the project evolves, so too do the models that represent it. Once the project team is satisfied with the arrangements portrayed on massing models, small-scale detailed-oriented study models are begun. This reflects the architectural styles and colors for the project.

Creating a larger overall model, based upon detailed architectural and engineering drawings, is the last step in the model-building process. This show model is the exact replica of the project as it will be built, featuring the tiniest of details, including building exteriors, landscape, color schemes, the complete ride layout, vehicles, show sets, props, figures and suggested lighting and graphics.[11]

Computer models of the complete attraction, including the actual ride, are next. They are computer generated so that the Imagineers can see what the final product looks like from various positions without actually having to build a full-scale model. Computer models, similar to CAD/CAM modeling, can show in three dimensions the layout of all of the necessary electrical, plumbing, HVAC, special effects, and other equipment.

AESTHETICS

Imagineers view the aesthetic value of an attraction in a controlled theme environment as a constraint. This aesthetic constraint is more of a passion for perfection than the normal constraints that most project managers are familiar with.[12]

Aesthetics are the design elements that identify the character and the overall theme and control the environment and atmosphere of each setting. This includes color, landscaping, trees, colorful flowers, architecture, music, and special effects. Music must support the mood of the ride. The shape of the rocks used in the landscape is also important. Pointed or sharp rocks may indicate danger whereas rounded or smooth rocks may represent safety. Everything in the attraction is there for the purpose of reinforcing a story. Imagineers go to minute levels of detail for everything needed to support the story without overwhelming the viewers with too many details. Details that are contradictory can leave the visitors confused about the meaning of the story.

A major contributor to the aesthetics of the attraction are the special effects. Special effects are created by "Illusioneering," which is a subset of Imagineering.

---

[11] Disney Book Group, *Walt Disney Imagineering: A Behind the Scenes Look at Making the Magic Real*, p. 72.

[12] Some people argue that the aesthetics focus more on creating a controlled environment than on reality, thus controlling your imagination.

Special effects can come in many different forms. Typical projected special effects can include:

- Steam, smoke clouds, drifting fog, swirling effects
- Erupting volcano, flowing lava
- Lightning flashes and strikes, sparks
- Water ripple, reflection, waterfall, flows
- Rotating and tumbling images
- Flying, falling, rising, moving images
- Moving images with animated sections
- Kaleidoscopic projections
- Liquid projections, bubbles, waves
- Aurora borealis, lumia, abstract light effects
- Twinkling stars (when fiber optics cannot be used, such as on rear-projection screen)
- Spinning galaxies in perspective, comets, rotating space stations, pulsars, meteor showers, shooting stars, and any astronomical phenomena
- Fire, torches, forest fire
- Expanding rings
- Ghosts, distorted images
- Explosions, flashes[13]

Perhaps the most important contributor to the aesthetic value of an attraction is color. Traditional project managers rely on sales or marketing personnel to select the colors for a deliverable. At Imagineering, it is done by the Imagineers. Color is a form of communication. Even the colors of the flowers and the landscaping are critical. People feel emotions from certain colors, either consciously or subconsciously. Imagineers treat color as a language. Some colors catch the eye quickly, and we focus our attention on it. "We must ask not only how colors work together, but how they make the viewer feel in a given situation. . . . It is the Imagineer's job to understand how colors work together visually and why they can make guests feel better."[14]

"White represents cleanliness and purity, and in many European and North American cultures . . . is the color most associated with weddings, and with religious ceremonies such as christenings. Silver-white suggest joy, pleasure and

---

[13] See "Bill Novey and the Business of Theme Park Special Effects," http://blooloop.com/feature/disney-imagineering-bill-novey-and-the-business-of-theme-park-special-effects-2/. The paper provides an excellent summary of various special effects used by Illusioneers. In addition to the projected special effects, the paper also describes laser effects, holographic images, floating images, mirror gags, gas discharge effects, and fiber optics.

[14] Hench with Van Pelt, *Designing Disney*, p. 104.

delight. In architecture and interior design, white can be monotonous if used over large areas."[15] "We have created an entire color vocabulary at Imagineering, which includes colors and patterns we have found that stir basic human instincts – including that of survival."[16]

Aesthetics also impacts the outfits and full-body costumes of the cast members who are part of the attraction. The outfits that the cast members wear must support the attraction. Unlike animation, where there are no physical limitations to a character's identity or mobility, people may have restricted motion once in the costume. Care must be taken that the colors used in the full-body costumes maintain the character's identity without conflicting with the background colors used in the attraction. Even the colors in the rest rooms must fit the themed environment.

Imagineers also try to address queue design by trying to make it a pleasant experience. As people wait in line to see an attraction, aesthetics can introduce them to the theme of the attraction. The aesthetics must also consider the time it takes people to go from attraction to attraction as well as what precedes this attraction and what follows it. "For transition to be smooth, there must be a blending of themed foliage, color, sound, music, and architecture. Even the soles of your feet feel a change in the paving explicitly and tell you something new is on the horizon."[17]

THE ART OF THE SHOW

Over the years, Imagineering has conceived a whole range of retail stores, galleries, and hotels that are designed to be experienced and to create and sustain a very specific mood. For example, the mood of Disney's Contemporary Resort could be called "the hello futuristic optimism," and it is readily apparent, given the resort's A-frame structure, futuristic building techniques, modern décor, and the monorail gliding quietly through the lobby every few minutes. Together, these details combine to tell the story of the hotel.[18]

Imagineering is, first and foremost, a form of storytelling, and visiting a Disney theme park should feel like entering a show. Extensive theming, atmosphere, and attention to detail are the hallmarks of the Disney experience. The mood is distinct and identifiable, the story made clear by details and props. Pirates of the Caribbean evokes a "rollicking buccaneer adventure," according to Imagineering Legend John Hench,[19] whereas the Disney Cruise Line's ships create an elegant seafaring atmosphere. Even the shops and restaurants within the theme parks tell

---

[15] Ibid., p.135.

[16] Disney Book Group, *Walt Disney Imagineering*, p. 94.

[17] Ibid., p. 90.

[18] Marling, *Designing Disney's Theme Parks*.

[19] Hench and Van Pelt, *Designing Disney*, p. 56.

stories. Every detail is carefully considered, from the menus to the names of the dishes to the cast members' costumes.[20] Disney parks are meant to be experienced through all senses—for example, as guests walk down Main Street, U.S.A., they are likely to smell freshly baked cookies, a small detail that enhances the story of small town America at the turn of the nineteenth century.

The story of Disney theme parks is often told visually, and the Imagineers design the guest experience in what they call "The Art of the Show." John Hench was fond of comparing theme park design to moviemaking and often used film-making techniques, such as forced perspective, in the Disney parks.[21] Forced perspective is a design technique in which the designer plays with the scale of an object in order to affect the viewer's perception of the object's size. One of the most dramatic examples of forced perspective in the Disney parks is Cinderella's Castle. The scale of architectural elements is much smaller in the upper reaches of the castle compared to the foundation, making it seem significantly taller than its actual height of 189 feet.[22]

THE POWER OF ACKNOWLEDGMENT

Project managers like to be told that they have done a good job. It is a motivational force encouraging them to continue performing well. However, acknowledgment does not have to come with words; it can come from results. At Disney's Imagineering Division, the fact that more than 132,500,000 visitors passed through the gates of the 11 Disney theme parks in 2013 is probably the greatest form of acknowledgment. The Walt Disney Company does acknowledge some Imagineers in other ways. Disney established a society entitled "Imagineering Legends." Three of their most prominent Imagineering Legends are John Hench (65 years with Disney), Claude Coats (54 years with Disney), and Martin Sklar (53 years with Disney). The contributions of these three Imagineers appear throughout the Disney theme park attractions worldwide. The goal of all Imagineers at Disney may very well be the acknowledgment of becoming an Imagineering Legend.

THE NEED FOR ADDITIONAL SKILLS

All projects have special characteristics that may mandate a unique set of project management skills above and beyond what we teach using the *PMBOK® Guide*. Some of the additional skills that Imagineers may need are summarized next.

- The ability to envision a story
- The ability to brainstorm

---

[20] Hench and Van Pelt, *Designing Disney.*

[21] Ibid., p. 74.

[22] Wright, *Imagineers.*

- The ability to create a storyboard and build mock-ups in various stages of detail
- A willingness to work with a multitude of disciplines in a team environment
- An understanding of theme park design requirements
- Recognizing that the customers and stakeholders range from toddlers to senior citizens
- An ability to envision the attraction through the eyes and shoes of the guests
- An understanding of the importance of safety, quality, and aesthetic value as additional competing constraints
- A passion for aesthetic details
- An understanding of the importance of colors and the relationship between colors and emotions
- An understanding of how music, animatronics, architecture, and landscaping must support the story

Obviously, this list is not all-inclusive, but it does show that not everyone can be an Imagineer for Disney. These skills also apply to many of the projects that most project managers are struggling with. Learning and applying these skills could very well make all of us better project managers.

QUESTIONS

1. Why do most project managers not recognize that they either need or can use the skills required to perform as an Imagineering project manager?
2. What is the fundamental difference between a ride and an attraction?
3. What are some of the differences between traditional brainstorming and Imagineering brainstorming?
4. How many project constraints are there on a traditional theme park attraction?
5. How would you prioritize the constraints?
6. Why is it necessary to consider cost before the Imagineering brainstorming sessions are completed?
7. What is Audio-Animatronics?
8. What is storyboarding, and how is it used on Disney projects?
9. What is meant by "project aesthetics," and how might it apply to projects other than at Disney?

Disney (B): Imagineering Project Management in Action—The Haunted Mansion

INTRODUCTION

The Haunted Mansion attraction opened to the public August 9, 1969. One week after opening, more than 82,000 guests had seen the attraction. During the first busy season, the time to stand in the queue to see the attraction was three to four hours. Eventually, an army of diehard fans claimed that the Haunted Mansion was their favorite attraction. Today, stores and websites are dedicated to the sale of souvenirs of the Haunted Mansion and its inhabitants.

WHY STUDY THE HAUNTED MANSION?

Some projects have unique characteristics that can make them more difficult to manage than other projects. Projects that involve imagination and creativity fall into this category. Years ago, project managers believed that, if you understood the concepts of project management, you could work in just about any industry. But today, we recognize the importance of these unique characteristics that may make changing industries more complex.

Disney's Haunted Mansion opened to guests in 1969, the same year that the Project Management Institute (PMI) was formed. The Haunted Mansion attraction was completed without the use of the *PMBOK® Guide* or Project Management Professionals, since these did not appear until the mid-1980s. Many of the individuals assigned to the Haunted Mansion Project were the most creative and inventive people in the world. So, how did project management take place on such an endeavor as the Haunted Mansion? What were some of the unique characteristics needed for the project?

In the Disney (A) Case Study, we identified some of the characteristics that differentiated Imagineering project managers from traditional project managers. We will now look more closely at Imagineering project management in action using Disney's Haunted Mansion Project.

The literature abounds with both authorized and unauthorized stories of Walt Disney's Haunted Mansion. Unfortunately, all of the versions do not directly discuss project management, thus mandating some assumptions and interpretation. The comparison of Imagineering project management with traditional project management and the accompanying conclusions are solely the author's interpretation and may not necessarily represent Disney's conclusions. The material in this case study was extracted from numerous sources that are referenced throughout this case study.

CONSTRAINTS

All project have constraints. For almost 50 years, project managers were taught to focus on the triple constraints of time, cost, and scope. But for the projects at the Disney theme parks, the constraints of safety, quality, and aesthetic value also must be included.

The *PMBOK® Guide* did not begin discussing the importance of competing constraints until the fourth edition was released in 2008. Prior to that time, only the importance of the traditional triple constraints were discussed. Yet even as early as the 1950s with the design of the Disneyland theme park, Disney understood the importance of competing constraints and the fact that they must be prioritized.

The most important constraint at Disney was, and still is, the safety of the guests. This constraint is never sacrificed at Disney. In the author's opinion, quality and aesthetic value were probably tied for second and third behind safety. If trade-offs had to be made on certain attractions, it appears that the trade-offs took place on time, cost, and scope but not on safety, aesthetic value, or quality. Today safety, aesthetic value, and quality are attributes of the Disney image. The importance of these constraints is discussed further in this case study.

LIFE-CYCLE PHASES

When companies strive for some degree of project management maturity, they usually begin with the creation of an enterprise project management methodology

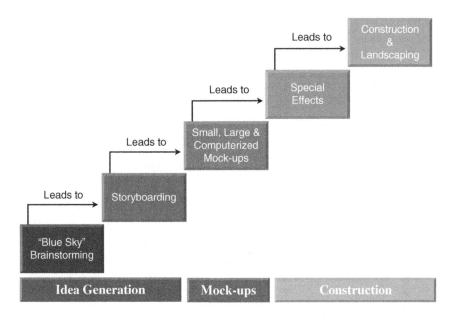

FIGURE I Typical life-cycle phases

developed around required life-cycle phases. The literature does not identify any project management methodology or identify any life-cycle phases for Disney's theme park attractions. However, the literature does identify many of the various steps in creating an attraction. From these steps, we can assume that typical life-cycle phases might appear such as shown in Figure I. Some of the detailed steps that are performed in each life-cycle phase are described in the Disney (A) Case Study.

The life-cycle phases shown in Figure I appear as sequential phases. However, in reality, many of the phases can overlap. As an example, special effects activities can take place in any or all of the life-cycle phases, including construction.

THE SCOPE CONSTRAINT

Most project managers are accustomed to having a well-defined SOW at the onset of a project. The SOW serves as the scope constraint. Even though the SOW may be highly narrative, the accompanying work breakdown structure and specifications can provide significant detail to support the narrative SOW.

Well-defined SOWs are based on a well-defined business case where the concept for the project is understood. If the concept is not well understood, then the

SOW may not appear until the end of the concept development or idea generation life-cycle phase in Figure I.

On a project such as the Haunted Mansion, we must remember that, first of all, it is an Imagineering project, and Imagineering efforts will continue throughout the life of the project and beyond due to continuous improvement efforts. Expecting a well-defined SOW at the beginning of a project like the Haunted Mansion, and having it remain unchanged throughout the project, is highly unlikely. The SOW is most likely a constantly evolving document, possibly finalized as the opening day of the attraction approaches.

To understand the complexity of creating a formal SOW, we must first look at the questions that had to be addressed during the concept development phase of the Haunted Mansion Project. Typical questions include:

- Should the attraction be based on a single ghost story concept or several stories?
- Should it be a scary or humorous attraction?
- What should the Haunted Mansion look like?
- What colors and type of landscaping should be used?
- Should it be a walk-through attraction or a ride using a people-mover?
- If it is a ride, how many people can we have in the people-mover at one time?
- How long should it take to go through the attraction?
- How many ghostly special effects will be needed?
- Will a script be needed to accompany some of the special effects?
- Will we need a host to guide people through the ghostly attraction?
- If a host is needed, will a live person or a ghost do the hosting?
- Will there be eerie music to accompany the special effects?
- Does technology exist for the ghostly images, or must new technology be created?
- How much should be budgeted for the attraction?

These questions were not easy to answer at the onset of the project and were also influenced by who was working on the project at that time. Walt Disney assigned some of his seasoned veterans to the project. Many team members had worked with him for decades and had been highly creative on other projects. They brought with them their own unique ideas that often created ego problems. If people were reassigned during the Haunted Mansion Project, which they were, their replacements came with their own ideas, and then the answers for many of the listed questions could change.

To understand the complexities of creating a SOW for the Haunted Mansion, many of these questions could not be answered until the project was well under way. The questions were, as expected, interrelated and not always easy to answer even in later life-cycle stages. The answer to one question could cause the answers to several other questions to change. If the answers to some questions could not be made until well into the project, then there could be a significant amount of scope changes.

SCOPE CHANGES

To understand the interrelatedness of the questions and how scope changes can occur even near the completion of the project, consider the Hatbox Ghost special effect. The Hatbox Ghost was a character that was planned for the Haunted Mansion at Disneyland but was removed shortly after the attraction's debut. Located in the ride's attic scene, the figure was described as "an elderly ghost in a cloak and top hat, leaning on a cane with a wavering hand and clutching a hatbox in the other."[1]

The idea behind the Hatbox Ghost was for his head to vanish from atop his shoulders and reappear inside his hatbox,[2] in time with an adjacent bride figure's beating heart. According to Imagineer Chris Merritt in an interview with Doom-Buggies.com, the effect was never completely successful due to the illusion's close proximity to the ride vehicles:

> The gag was based purely on lighting. The ghost's head was illuminated by black lighting. A light inside the hatbox he held would rhythmically illuminate and hide the head in the hatbox, while, in tandem, the actual head on the ghost's shoulders would be hidden by extinguishing the black lighting.

The Hatbox Ghost illusion was installed inside the Haunted Mansion and in place for cast member (park employee) previews on the nights of August 7 and 8, 1969.[3] Almost immediately, it became apparent that the effect had failed, as ambient light in the attraction's attic scene prevented the specter's face from disappearing fully, even when its designated spotlight was turned off. Attempts were made to remedy technical problems, but the effect was not convincing enough, and the ghost was decommissioned after a few months. This was just one example of how things can change well into the project.

---

[1] DoomBuggies.com, "Hats Off: The Secret of the Attic." www.doombuggies.com/myths2.php 1.

[27] Davelandweb.com, "Daveland Hatbox Ghost Movie."

[3] Jason Surrell, *The Haunted Mansion: From the Magic Kingdom to the Movies* (2nd edition; New York: Disney Editions, 2009), p. 86; Doombuggies.com, "History of the Haunted Mansion." www.doombuggies.com/history6.php.

THE TIME CONSTRAINT

Walt Disney thought up the idea for the Haunted Mansion in the early 1950s. It took almost 18 years for the idea to become reality. To understand the time constraints and complexity of the project, including the interrelatedness of the questions asked previously, we should look at a brief history of the attraction.

The Haunted Mansion is a haunted house dark ride located at Disneyland, Magic Kingdom (Walt Disney World), and Tokyo Disneyland. Phantom Manor, a significantly reimagined version of the Haunted Mansion, is located exclusively in Disneyland Paris. Another Disney attraction involving the supernatural and set in a mansion, Mystic Manor, has opened at Hong Kong Disneyland. The Haunted Mansion features a ride-through tour in Omnimover (or people-mover) vehicles called "Doom Buggies," preceded by a walk-through show in the queue. The attraction utilizes a range of technology, from centuries-old theatrical effects to modern special effects and spectral Audio-Animatronics.

The attraction predates Disneyland, to when Walt Disney hired the first of his Imagineers. The first known illustration of the park showed a Main Street setting, green fields, a western village, and a carnival. Disney Imagineering Legend Harper Goff developed a black-and-white sketch of a crooked street leading away from Main Street by a peaceful church and graveyard, with a run-down manor perched high on a hill that towered over Main Street.

Walt Disney assigned Imagineer Ken Anderson to create a story around Goff's idea. Plans were made to build a New Orleans–themed land in the small transition area between Frontierland and Adventureland. Weeks later, New Orleans Square appeared on the souvenir map and promised a thieves' market, a pirate wax museum, and a haunted house walk-through. Anderson studied New Orleans and old plantations and came up with a drawing of an antebellum manor overgrown with weeds, dead trees, swarms of bats, and boarded doors and windows topped by a screeching cat as a weather vane.

Walt Disney, however, did not like the idea of a run-down building in his pristine park. He visited the Winchester Mystery House in San Jose, California, and was captivated by the massive mansion with its stairs to nowhere, doors that opened to walls and holes, and elevators. When the decision was made to begin full-scale development of the Haunted Mansion, Imagineer Marc Davis asked Disney if he wanted the house to look scary. Disney replied:

> No, I want the lawn beautifully manicured. I want beautiful flowers. I want the house well-painted and well cared for so that people would know that we took care of thing in the park, and it's a clean, good park for families to come and have a good time. You can put all the spider webs inside that you want, I don't care about that . . . but the outside has to be pristine and clean at all times.[4]

---

[4] Alice Davis, panel presentation, "Spirited Seance and Haunting Seminar panel" at "The Haunted Mansion 40th Anniversary Merchandise Event." August 9, 2009, Disneyland.

Anderson came up with several possible stories for the mansion. Some of the stories included:

- A wedding gone awry when a ghost suddenly appears and kills the groom. The man that eventually appears hanging from the ceiling in the attic could be the bride's husband.
- Similar to the above story, a ghost appears and kills the groom. The bride then commits suicide and appears hanging in the attic.
- A newly married bride discovers that her husband is really a blood-thirsty pirate. The pirate kills his bride in a jealous rage, but her ghost returns to haunt him. He could not live with himself for what he did to his true love, so he hangs himself in the attic rafters.
- Another story focused on calling the Haunted Mansion "Bloodmere Manor," which may have involved more bloody scenes and body parts. People would look as if they had been violently murdered. The story would end with the Headless Horseman in the graveyard.

The number one rule in Imagineering is that the attraction must tell a story. Unfortunately, no one could agree on what the story should be or whether the attraction could be described by just one story. In the meantime, other Imagineers were developing illusions for the Haunted House without having any story to go by. It appeared that no firm SOW existed other than the fact that the Haunted House attraction would eventually be built. There were still too many questions that were unanswered.

In 1961, handbills announcing a 1963 opening of the Haunted Mansion were given out at Disneyland's main entrance. Construction began a year later, and the exterior was completed in 1963. The Haunted Mansion was actually a replica of a preexisting building. Even though the façade of the Haunted Mansion was completed, the project was put on hold because of Disney's involvement in the 1964–1965 New York World's Fair. Similar to what happens in most companies when priorities change, all of the resources that were assigned to the Haunted Mansion project were reassigned to efforts to support the World's Fair. Changes in priorities caused efforts on the Haunted Mansion to wax and wane over the years.

In 1963, inspired by Disney, Marty Sklar, former vice chairman and principal creative executive at Walt Disney Imagineering, created a sign inviting ghosts to continue practicing their trade in active retirement in the Haunted Mansion. The sign is shown in Figure II.[5] The intent of the sign was to keep people focused on the fact that the Haunted Mansion would eventually be built even though the sign hung for many years in front of an empty building. The public's perception of the

---

[5] The actual sign was not on a tombstone as depicted here. For a picture of the actual sign, see Jeff Baham, *Walt Disney's Haunted Mansion* (n.p.: Theme Park Press, 2014), p.44.

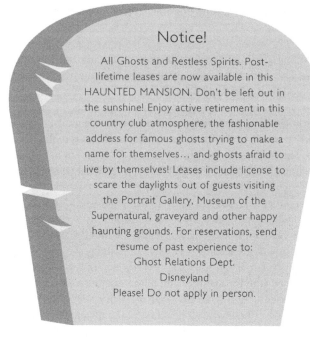

FIGURE II The Invitation

abandoned construction project took on a life of its own even though there was still no story line to accompany the abandoned building.

The sign became quite popular. Some Disney literature stated:

> The world's greatest collection of "actively retired" ghosts will soon call this Haunted Mansion "home." Walt Disney and his "Imagineers" are now creating 1001 eerie illusions. Marble busts will talk. Portraits that appear "normal" one minute will change before your eyes. And, of course, ordinary ghost tricks (walking through solid walls, disappearing at the drop of a sheet) will also be seen . . . and felt. Here will live famous and infamous ghosts, ghosts trying to make a name for themselves . . . and ghosts afraid to live by themselves![6]

When the project started up again in 1966, a new team of Imagineers was assigned—the fourth Imagineering team to work on the Haunted Mansion project. Marc Davis and Claude Coats were responsible for the continuity of the ride and the backgrounds. The responsibility for the special effects was placed

---

[6] 1966 Guide to Disneyland souvenir book.

in the hands of two Imagineers who were also referred to as Illusioneers, Rolly Crump and Yale Gracey. Crump was an artist who loved stage magic and illusions, and Gracey was an animator, mechanical genius, and considered the father of Illusioneering. Disney could select from an army of talented people in various organizations. He had a knack of putting people "in conflict" together and telling them to work as a team, knowing full well that the results would be exceptional, despite the ego problems that typically exist with highly talented teams.

Following Walt Disney's death in 1966, many of the Imagineers clashed over the direction of the project. Imagineer Xavier Atencio was brought on board to put together a coherent story. Without such a story for direction, there was a fear that the Haunted Mansion would simply be a multitude of special effects and illusions. Even with Atencio's focus, there was still the question of whether the Haunted Mansion should be a scary attraction or not.

Disney's original dream was to scare people, but in a pleasant sort of way. That meant that there would be no oozing of blood, missing eye sockets, gory body parts, or horrifying decaying bodies that some guests might see as offensive. The decision was made that the animation should focus on the lighter or cartoon-like tone of a Haunted Mansion rather than the scarier tone. The Imagineers also decided that, instead of looking like an "old spook house" that may have been Ken Anderson's original thoughts, the Haunted Mansion would be full of illusions.

The Haunted Mansion's long development was rife with discarded story concepts, disagreement on the type of scenes and effects to be used, conflicts over how many viewers should be carted through the attraction per hour—even as basic an idea as whether the attraction should be scary or not. Egos were bruised, tempers flared, and at the end of the day, it just seemed that there were "too many cooks in the kitchen," as Imagineer Marc Davis often recalled. [7]

Making the attraction cartoonlike rather than scary and having the outside of the Haunted Mansion pristine was certainly in line with Disney's original idea for the attraction. But how do you then make the Haunted Mansion structure look somewhat scary? John Hench was regarded as the color expert at Disney's Imagineering Division. According to Hench:

> We wanted to create an imposing southern-style house that would look old, but not in ruins. So we painted it a cool off-white with dark, cold blue-grey accents in shadowed areas such as the porch ceilings and wrought iron details. To accentuate the eerie, deserted feelings, I had the underside of exterior details painted the same dark color, creating exaggerated, unnaturally deep cast shadows, since we associate dark shadows with things hidden, or half-hidden. The shadow treatment enhanced the structure's other worldliness.[8]

---

[7] Baham, *Walt Disney's Haunted Mansion*, p. xiv.

[8] Hench, *Designing Disney*, p. 116.

There was still another critical decision that had to be made. Should the attraction be a walk-through or a ride? There were pros and cons to each approach. With a walk-through, it would be easier to create a single story line for the entire attraction. Believing that Ken Anderson's approach in the 1950s of a single story and a walk-through attraction would be selected, the Imagineers created some illusions where the guests could be more actively involved with the ghosts. However, walk-throughs required a live host as a tour guide, the speed of the tour might be difficult to control, and there would always be the risk of vandalism or damage to props and equipment in the attraction.

The decision was made to go with a ride. This meant that, instead of a single story line that would work with a walk-through, the ride would have several stories. Story lines would be needed for each of the ghosts. The tour guide could now be one of the dastardly ghosts. The total attraction, and each individual story, had to be unique and with some degree of weirdness.

ADDITIONAL TIME CONSTRAINTS

For most project managers, "time management" refers to the duration of the project, which for the Haunted Mansion would be 18 years from Disney's original concept to the date when the attraction was opened to the public. But for the Imagineers, there were two other time management issues once the decision was made that this would be a ride rather than a walk-through:

1. How much time will people need to view each of the scenes?
2. How many people can we service each hour?

Writer Bob Thomas interviewed Carl Walker from Disneyland and Dick Irvine, representing WED, the early name of the Imagineering Division:

> "Then there was the matter of how to conduct people through the ride," said Walker. "At first, it might be a walk-through, with 30 on a conducted tour. But that was difficult to manage, and besides, people don't scare as easily in crowds. So we made it a ride-through, with three people in a car—their crypt so to speak," said Irvine. "The cars could be programmed to face the right direction, tilt back and keep moving. They provided the capacity we need for rides at Disneyland— 2,300 per hour." [9]

The people-mover system was called the "Doom Buggies." It was a modification of the Omnimover system that Disney used in the 1964 New York World's

---

[9] Quoted in Bob Thomas, "Dave McIntyre's Front Row," *San Diego Evening Tribune*, August 19, 1969.

Fair. Since the guests were seated, the Imagineers could force them to watch precisely what was intended. The Doom Buggies were programmed to control the angle (i.e., line of sight) at which the guests would see the set without being able to see the supporting animatronics. It also kept the guests at a distance where they could not touch any of the props used in the scenes. Like other attractions at Disney's theme parks, the Haunted Mansion had now become a controlled environment for the guests.

This concept also allowed the designers to place infrastructure elements of the attraction, such as lighting and projectors, behind, above, or below the vehicles without concern for having the attraction's illusions revealed to the guests. The system consists of a chain of vehicles operating on a track, usually hidden beneath the floor. The chain of vehicles maintains constant motion at a specific speed, thus controlling the viewing time. The duration of the rides varies from 5:50–8:20 minutes, at a maximum speed of 3 miles per hour.[10]

One of the features that differentiates this system from other people-mover systems is the ability of the vehicle to be rotated to a predetermined orientation. In addition to the main ride rails, each vehicle also has two control rails attached to a wheel. One controls swiveling, allowing the vehicle to face in any direction at any point on the track. The other allows the vehicle to tilt in relation to the inclining and declining portions of the track.

Because the entire attraction is in a controlled environment, the Imagineers can control what the guests see. The Imagineers can make it appear that a ghost is in the Doom Buggy with guests.

THE COST CONSTRAINT

The Haunted Mansion was completed at a cost of $7 million. In today's dollars, that would be equivalent to approximately $50 million. When the Haunted Mansion was built, mainframe computers were just entering the marketplace. Cost control software did not exist, and all cost control was done manually.

There is a misconception that, when imagination and creativity are allowed to run wild on projects, budgets must be unlimited. That certainly is not the case. The Walt Disney Company monitors all costs. Budgets for each attraction are established during the idea/conceptual phase.

The larger the project, the greater the chance for scope changes and increases in the budget. Unfortunately, the literature does not provide any information related to the original budget or the number of scope changes. The cost of each attraction at Disney theme parks is generally regarded as proprietary knowledge.

---

[10] The Doom Buggies at Walt Disney World can handle 3,200 guests per hour traveling at about 1.4 miles per hour. At Disneyland, because of the shorter track, only 2,618 guests can be handled.

THE SAFETY CONSTRAINT

Safety is the main concern at Disney parks. As mentioned, Disney himself wanted the Haunted Mansion to be scary but in a pleasant sort of way. The term "scared to death" could lead to wrongful-death lawsuits.

Since the Haunted Mansion is a controlled environment, repeat visitors recognized the predictability of the attraction. An attempt was made to have some cast members dress up in a knight's armor suit and wield an axe (actually made of rubber). Some people were very frightened, and there were complaints. Disney parks discontinued this practice.

THE AESTHETICS CONSTRAINT

Aesthetics and quality go hand in hand. All Disney theme park attractions must be aesthetically appealing to the guests. The Haunted Mansion was no exception. The Imagineers were be able to convert ideas into reality. Imagineers, and especially Illusioneers, are often considered dreamers, inventors, and even mad scientists. They must have an obsessive commitment to detail and quality.

Nightly maintenance takes place in the Haunted Mansion to make sure that every prop is in place. All of the props are real. Some of the props were used elsewhere; for example, the pipe organ was used in the film *20,000 Leagues under the Sea.*

Even the cobwebs and dust must be in place. A liquid cobweb spinner makes the cobwebs. There must also be a proper amount of dust (which is actually a rubber cement that cannot induce allergies in guests).

A tour of the Haunted Mansion includes:

- The Grounds
- The Foyer
- The Stretching Room
- The Portrait Corridor
- The Library and Music Room
- The Endless Staircase
- The Endless Hallway
- The Conservatory
- The Corridor of Doors
- The Séance Circle
- The Grand Hall
- The Attic
- The Graveyard
- The Crypt

When the decision was made to have a not-so-scary Haunted Mansion, Imagineers/Illusioneers Gracey and Crump read ghost stories and watched ghost movies to decide what type of ghosts they could create. The men created numerous effects and often left the special effects running all night long. The night cleaning crew were often spooked and complained to management, which asked the Imagineers not to scare off the cleaning crew.

But instead of leaving the lights on and the special effects off, the two Imagineers decided to connect their special effects to a motion-detector switch. When the duo came to work in the morning, they found a broom left in the middle of their studios. The Imagineers had to clean their studios by themselves from that point on, as management told them that the night cleaning crew were never coming back.

Special effects in each location support the aesthetic constraints. Some of the special effects include:

- Digital projections
- Computer-controlled effects
- Audio-Animatronics
- Holograms (although they were not used)
- Special lighting
- Real props

Many of the special effects and illusions are based on Pepper's Ghost, an illusion that dates back to the 1800s. Pepper's Ghost is an illusion technique that has been used in theatres, haunted houses, dark rides, and magic tricks. It uses plate glass, Plexiglas, or plastic film and special lighting techniques to make objects seem to appear or disappear, become transparent, or morph into something else. It is named after John Henry Pepper, who popularized the effect.

For the illusion to work, two rooms are required. The viewer must be able to see into the main room but not into the hidden room. The edge of the glass separating the rooms is sometimes hidden by a cleverly designed pattern in the floor.

The hidden room must be an identical mirror image of the main room, so that its reflected image matches the main rooms; this approach is useful in making objects seem to appear or disappear. This illusion can also be used to make one object or person reflected in the mirror seem to morph into another behind the glass (or vice versa). The hidden room may instead be painted black, with only light-colored objects in it. In this case, when light is cast on the room, only the light objects reflect the light and look like ghostly translucent images superimposed in the visible room.

In the Haunted Mansion at Disneyland, Walt Disney World, and Tokyo Disneyland, the glass is vertical to guests rather than in the normal angled position, to reflect animated props below and above guests that create the appearance of three-dimensional, translucent "ghosts" that appear to dance through the ballroom and interact with props in the physical ballroom. The apparitions appear and disappear when the lights on the animations turn on and off.

Some of the special effects created for the Haunted Mansion include:

- A ghost host
- Exploding ghosts
- Talking and singing statues
- Furniture that comes to life
- A man made of dripping wax
- A grandfather clock that looks like a coffin
- A graveyard band of ghosts playing music
- A pet cemetery
- An invisible ghost horse with only a saddle and reins
- A ghost poetess creating a poem
- Dancing ghosts
- Ghosts that fade in and out
- Ghosts that suddenly become headless
- A ghost playing a piano
- Portraits that change from reality to the supernatural
- Wallpaper patterned with monster faces
- Hanging ghosts
- A crypt that plays music if you touch the instruments

Some of these special effects are discussed below in more detail.[11]

GHOST HOST

The Ghost Host is one of the first characters guests meet at the Haunted Mansion, so to speak. He remains invisible throughout the tour, guiding "foolish mortals" with an ominous voice. The voice is that of Paul Frees, a popular Disneyland announcer and vocal talent (well-known as the voice of the Pillsbury Doughboy, Ludwig von Drake, and Boris Badenov in the popular cartoon series *The Adventures of Rocky and Bullwinkle*). Frees's gleefully sardonic narration often features death-related puns and maniacal laughter. In the Stretching Room scene near the beginning of the tour, it is revealed that he committed suicide by hanging himself from the rafters in the cupola.

AGING MAN

Above the fireplace in the foyer of the Walt Disney World and Tokyo Haunted Mansions is a portrait of a former owner of the house. The painting gradually changes from a handsome blue-eyed, black-haired young man to a withered, balding old man and finally, to a decaying skeleton. This portrait can also be found in

---

[11] For more information on the special effects, see Wikipedia contributors, "Haunted Mansion," Wikipedia, The Free Encyclopedia, https://en.wikipedia.org/w/index.php?title=Haunted_Mansion&oldid=756609637.

the changing portrait hallway of the Disneyland Haunted Mansion, but it morphs from the young man to the skeleton with flashes of lightning.

CHANGING PORTRAIT CHARACTERS

Lightning flashes transform the paintings at the Disneyland and Walt Disney World Haunted Mansions from benign to frightening. The portraits consist of:

- A beautiful young princess reclining on a couch who changes into a werecat.
- A gallant knight (identified as "The Black Prince" in concept art) atop a rearing horse, who both become skeletal.
- A handsome young man who decays into a ghastly corpse.
- The beautiful, red-haired Medusa, who becomes a hideous Gorgon.

STRETCHING PORTRAIT CHARACTERS

These characters are depicted in the portraits of the Stretching Room:

- A balding man with a brown mustache and beard, dressed in a black tailcoat, a white shirt, a red sash, and a black bowtie. When the portrait stretches, it is revealed that he is not wearing pants (only red and white–striped boxer shorts), and he is standing atop a lit keg of dynamite. In an early attraction script, which named the characters in the stretching portraits, he was an ambassador named Alexander Nitrokoff, who came to the Mansion one night "with a bang."
- Constance Hatchaway, an old woman holding a rose and smiling. When the portrait stretches, it is revealed that she is seated on top of the tombstone of her late husband, George Hightower, who is depicted as a marble bust with his head split by an ax. The ghost of Constance as a young woman is later seen in the attic.
- A brown-haired man with his arms crossed, dressed in a brown suit and wearing a brown derby hat. When the portrait stretches, it is revealed that he is sitting on the shoulders of another man, who is sitting on the shoulders of another man who is waist deep in quicksand.
- A pretty young brunette woman holding a pink parasol. When the portrait stretches, it is revealed that she is balancing on a fraying tightrope above the gaping jaws of an alligator.

COFFIN OCCUPANT

In the center of the Conservatory is a large coffin occupied by a possessed corpse attempting to break out. He calls for help in the voice of a feeble old man, and his skeletal hands can be seen attempting to pry open the nailed-down coffin lid. He is voiced by Xavier Atencio, who wrote the attraction's script.

MADAME LEOTA

Madame Leota is one of the iconic characters of the ride. She is the spirit of a psychic medium, conducting an otherworldly séance in an attempt to summon spirits and assist them in materializing. Her ghostly head appears within a crystal ball on a table in the middle of her dark chamber, from which she speaks her incantations. Musical instruments and furniture levitate and make noises in response. Imagineer Leota Toombs was chosen for the face of the medium in the crystal ball. Toombs also plays the Ghost Hostess who appears at the end of the attraction, though it is unknown whether she and Madame Leota are meant to be the same character.

In 2002, a tombstone for Madame Leota debuted at Walt Disney World's Mansion. The epitaph reads: "Dear sweet Leota, beloved by all. In regions beyond now, but having a ball."

Madame Leota summons the Mansion's restless spirits and encourages them to appear by reciting suitable surreal incantations.

DUELISTS

The ghosts of two top hat—wearing gentlemen emerge from paintings of themselves and shoot each other.

HITCHHIKING GHOSTS

The Hitchhiking Ghosts—"The Prisoner," "The Skeleton," and "The Traveler"—are often considered to be the mascots of The Haunted Mansion. They alone have the most merchandise, including pins, stuffed toys, action figures, and bobble heads. The Hitchhiking Ghosts are a tongue-in-cheek send-up of urban legends involving phantom hitchhikers. They are seen standing together inside a crypt, thumbs extended. They hitch a ride with guests traveling in Doom Buggies and appear alongside them in mirrors. "They have selected you to fill our quota, and they'll haunt you until you return," says the Ghost Host. In 2011 at Walt Disney World's Haunted Mansion, the mirror scene was updated with digital effects that enable the ghosts to interact with the guests.

Fans often refer to the Hitchhiking Ghosts as "Gus" (Prisoner), "Ezra" (Skeleton), and "Phineas" (Traveler). These names first appeared in fan fiction created by cast members who worked at the Walt Disney World Haunted Mansion. Since then, the names have appeared on merchandise for the characters and in various media licensed by Disney.

REFERENCED CHARACTERS

On numerous tombstones and crypts at the Disneyland, Walt Disney World, and Tokyo Haunted Mansions (and in the Servants Quarters of Walt Disney World's

Haunted Mansion) are the names of characters who may or may not appear in the attraction. Most of the names are actually tributes to Imagineers who were involved in the creation of the attraction.

Outside each Mansion are crypts labeled with pun-based names. At Tokyo, they are identified as "Restless Spirits."

- Asher T. Ashes (Ashes to ashes)
- Bea Witch (Bewitch)
- Clare Voince (Clairvoyance)
- C. U. Later (See you later)
- Dustin T. Dust (Dust to dust)
- G. I. Missyou (Gee, I miss you)
- Hail N. Hardy (Hale and hearty)
- Hal Lusinashun (Hallucination)
- Hap A. Rition (Apparition)
- Harry After (Hereafter)
- Hobb Gobblin (Hobgoblin)
- L. Beback (I'll be back)
- Emma Spook (I am a spook)
- M. Mortal (I am mortal) or (immortal)
- M. Ready (I am ready)
- Trudy Departed (I truly departed)
- Trudy Dew (I truly do)
- Levi Tation/Lev Itation (Levitation)
- Love U. Trudy (Love you truly)
- Manny Festation (Manifestation)
- Metta Fisiks (Metaphysics)
- M. T. Tomb (Empty tomb)
- Paul Tergyst (Poltergeist)
- Pearl E. Gates (Pearly gates)
- Ray. N. Carnation (Reincarnation)
- Rustin Peece (Rest in peace)
- Rusty Gates (Rusty gates)
- Theo Later (See you later)
- U. R. Gone (You are gone)
- Wee G. Bord (Ouija board)

SPECIAL EFFECTS AND MUSIC

The special effects were groundbreaking for the time. They included an attic with the ghost of a spurned bride, a crypt and a cemetery, halls that appear endless, and the mystical fortune teller Madame Leota, who appears as a disembodied head inside a crystal ball with musical instruments floating in the air around her. Finally, the guests are shown that a "hitchhiking ghost" has hopped into the Doom Buggy with them.

Although the setting is spooky, the mood is kept light by the upbeat "Grim Grinning Ghosts" music that plays throughout the ride. The music was composed by Buddy Baker and the lyrics were written by Xavier Atencio. The deep voice of Thurl Ravenscroft sings as part of a quartet of singing busts in the graveyard scene. Ravenscroft's face is used as well as it is projected onto the bust with a detached head.

CONTINUOUS IMPROVEMENTS

All theme park attractions undergo continuous improvement efforts. According to Bob Zalk, Walt Disney Imagineer and show producer:

> The idea of going back into an iconic attraction and adding, changing, adjusting, removing elements—the standards are extremely high when you reach the finish line. We have to deliver. Unlike new attractions, re-imagining an established attraction carries with it its own sense of history and tradition that the entire team has to take into account. It's a big challenge, but an exciting one.[12]

QUESTIONS

1. What are the primary differences between traditional projects and the Haunted Mansion Project?
2. Why aren't all of the project's constraints of equal importance?
3. Why was it impossible to prepare a clearly defined statement of work at the beginning of the Haunted Mansion Project?
4. In the list of questions that had to be addressed at project initiation, which three questions were probably most critical for SOW preparation? (Note: There can be several answers to this question. What is important is the justification behind the three answers selected.)
5. Why did the Haunted Mansion Project take 18 years from concept to completion?
6. Why did Walt Disney not want the exterior of the Haunted Mansion to look like a traditional haunted house?
7. Most Disney attractions tell a story. Why was it so difficult to create a single story for the Haunted Mansion?
8. Why do some people, such as Imagineers, often have ego issues?
9. Why was the Haunted Mansion attraction created as a "controlled" ride?

---

[12] Quoted in Susan Veness, *The Hidden Magic of Walt Disney World* (Avon, Mass.: Adams Media, 2009), p. 24.

Disney (C): Disney Theme Parks and Enterprise Environmental Factors

PMI first began promoting the term "enterprise environmental factors" (EEF) in 2004 with the third edition of the *PMBOK® Guide*. This was 12 years after Euro Disney had opened. It is interesting to look at the launch of Euro Disney, now called Disneyland Paris, by analyzing the EEFs that existed at that time, even though they were not referred to by that phrase.

UNDERSTANDING ENTERPRISE ENVIRONMENTAL FACTORS

EEFs are conditions that exist now or in the future and may or may not have an impact on the project. If there is an impact, it can occur at any time over the life of the project. EEFs can influence how the project will be managed, whether changes in scope or quality are required, and whether the project is viewed as a success. These factors can include the state of the economy, current and future legislation, politics, influence of labor unions, competitive forces in play, and cultural issues. The factors can also change after the project is completed and turn an initially successful project into a failure.

Project planning is generally based on history, especially past successes. EEFs are assumptions and predictions about both the present and the future and therefore are directly related to risk management activities.

It is generally the responsibility of senior management, the project sponsor, or the governance committee to identify the EEFs. The factors may be listed as EEFs in the project's business case, or they may appear as assumptions concerning the business environment. EEFs are generally interpretations by one person or a consensus of several people including experts. One person may see a factor as having a favorable impact on the project whereas another person may see it as an unfavorable condition. Simply stated, EEFs are subject to interpretation as well as misinterpretation by the people requesting and funding the project. The impact can be devastating unless corrections and changes can be made quickly. Even some of the best-managed companies, such as the Walt Disney Company, can be impacted by unanticipated changes in EEFs.

ENTERPRISE ENVIRONMENTAL FACTORS AND CULTURE

Perhaps the most important EEF to be considered in the Walt Disney Company's decision to expand globally was the impact of multinational cultures. Expanding onto foreign soil (i.e., outside of the United States) would be a challenge. Disney theme parks would have to be integrated culturally and socially with host countries and their neighbors.

The company understood the American culture, and foreign visitors who came to Disneyland and Walt Disney World understood that they were visiting an American theme park. But how would people react to an American theme park on foreign soil? What might happen if the park did not adhere to the cultural and social norms of the host country? How much of a change would be necessary from the way that theme parks are managed in the United States?

ENTERPRISE ENVIRONMENTAL FACTORS AND COMPETING CONSTRAINTS

Before discussing Euro Disney, it is important to understand competing constraints. Project managers and executives can take some actions to alleviate the impact of unfavorable EEFs. Although project managers and sponsors cannot eliminate all EEFs, options may be available to lessen the impact. The actions we take almost always mandate trade-offs on competing constraints, thus perhaps making it impossible to meet all of the constraints. In such cases, the constraints may have to be prioritized to guide in the order of the trade-offs. A trade-off may result in a schedule elongation or significant cost overrun.

Although most companies focus on the triple constraints of time, cost, and scope, Disney's theme parks in the United States have six constraints; time, cost, scope, safety, aesthetic value, and quality. Although not discussed in the literature, it appears that safety is, understandably, the most important constraint at Disney parks, followed by aesthetic value and quality. These three important constraints very rarely undergo trade-offs because they have a direct bearing on Disney's image and reputation. All trade-offs appear to be on time, cost, and scope.

When expanding onto foreign soil, EEFs may also impose constraints on culture and even social behavior. Cultural and social constraints can involve designing the settings to be more aligned to local architecture, serving foods preferred by the population of the host country, and specifying dress codes that are acceptable to the general population. Host countries may not want on their soil a theme park that tries to "Americanize" the guests.

All of the constraints may be interrelated when trade-offs are necessary. For example, the Walt Disney Company maintains blueprints for attractions erected at Walt Disney World and Disneyland. Having to modify the attraction to be more aligned with the local architecture of the host country can require the creation of new blueprints, thus possibly increasing the project's cost and lengthening the schedule.

THE DECISION TO BUILD EURO DISNEY

In 1984, the Walt Disney Company made the decision to build a theme park in Europe by 1992. It wanted to build a large, state-of-the-art theme park, a decision that eventually led to "budget breaker" scope changes. Many of the changes were last-minute changes made by Michael D. Eisner, Disney's chief executive officer (CEO).

History has shown that large projects, especially those designed around state-of-the-art technology, are prone to large cost overruns. The baggage handling system at Denver International Airport (Chapter 14) and the Iridium Project (Chapter 9), which was designed to create a worldwide wireless handheld mobile phone system with the ability to communicate anywhere in the world at any time, are two examples where the cost overruns were in the billions of dollars.

Highly optimistic financial projections were established for Euro Disney based on the expectation of 11 million visitors the first year and 16 million visitors yearly after the turn of the twenty-first century. Nine years earlier, when Tokyo Disneyland opened on April 15, 1983, more than 13,000 visitors entered the park. Within the same year, Tokyo Disneyland broke the attendance record for theme parks with a one-day attendance of 93,000 visitors. Within four years, they again broke the one-day record with 111,500 visitors.

The Walt Disney Company viewed the Euro Disney theme park as a potentially profitable revenue generator for decades since the company would have a leisure and entertainment monopoly in Europe. The definition of a monopoly is when there are no rivals and high barriers, especially financial, to prevent others from entering the same market.

SITE SELECTION

The Walt Disney Company considered approximately 1,200 locations in Europe, and everyone wanted to host Euro Disney. The locations included Portugal, Spain, France, Italy, and Greece. Part of Disney's selection criteria included a warm climate, good weather, a centralized location, and available land for further growth. The list was narrowed down to four locations: two in Spain and two in France. Spain had better weather, but France had a denser population. The final decision was to build Euro Disney east of Paris in a newly built suburb, Marne-la-Vallée, 20 miles from the heart of Paris. This meant that 17 million people were less than a two-hour drive from Euro Disney, 68 million people within a four-hour drive, 110 million people within a six-hour drive, and another 310 million people less than two hours by air. In addition, tourists from around the world frequently visited Paris.

Because it would have a monopoly on theme parks and its success of its smaller $1.4 billion Tokyo Disneyland, the Walt Disney Company decided to build a much larger state-of-the-art theme park in Paris. It needed approximately $5 billion to build Euro Disney. More than $1 billion was provided by the French government in the belief that Euro Disney would create 30,000 jobs. At the time (1992), Europe was in a recession. France had an unemployment rate near 14 percent. Also, France expected a large percentage of the projected 11 million visitors to the theme park each year to be foreign, thus bringing revenue into the country.

The French were willing to make concessions to acquire the theme park. The land was provided at a low price of $7,500 per acre. Euro Disney would be built in a plot of 1,945 acres in the center of a 4,400-acre site. The French would pay for new road construction and provide water, sewage, gas, electricity, and other necessary services, such as a subway and train system.

PROJECT FINANCING

To satisfy the French government's legal requirements and to limit the financial exposure to the Walt Disney Company, a new company was formed: Euro Disney S.C.A. Unlike Tokyo Disneyland, where Japan's Oriental Land Company owned and operated the park and paid Disney royalties, Euro Disney S.C.A. would be a publicly held company. Disney would own a maximum of 49 percent of the new company, and Europeans would own at least 51 percent.

Euro Disney S.C.A. was set up using a project financing model. Project financing involves the establishment of a legally independent project company, Euro Disney S.C.A., where the providers of funds are repaid out of cash flow and earnings and where the assets of the new company, and only the new company, are used as collateral for the loans. Debt repayment would come from Euro Disney S.C.A. rather than from any other entity. In case of a default on the bank loans, lenders could take legal action against Euro Disney S.C.A. but not the Walt Disney Company.

A risk with project financing is that the capital assets may have a limited life. This constraint often makes it difficult to get lenders to agree to long-term financial arrangements. With Euro Disney, the attractions in the park would have to undergo continuous improvement and new attractions added. If there is insufficient cash flow to fund growth, the company would have to incur additional debt.

Another critical issue with project financing, especially for high-technology projects, is that the projects are generally long term. It may be several years before service will begin, and in terms of technology, this can be an eternity. Project financing is often considered a bet on the future. And if the project were to fail, the company may be worth nothing after liquidation.

European banks, looking at the financial success of Disneyland, Walt Disney World, and Tokyo Disneyland, rushed in to provide the Walt Disney Company with whatever funding was necessary in the way of construction loans. More than 60 banks entered into loan agreements. The company negotiated a deal whereby it paid only $160 million to help fund a $5 billion theme park. The Walt Disney Company would collect hundreds of millions of dollars each year in royalty payments, even if the theme park lost money. Following a royalty agreement similar to Tokyo Disneyland, the Walt Disney Company would receive 10 percent royalties on admissions; 5 percent royalties on food, beverage, and merchandise sales; a management fee equivalent to 3 percent of revenue; a licensing fee for using the Disney name and characters; 5 percent of gross revenues of themed hotels; and 49 percent of the profits.[1] The company also received royalties from companies that sponsored specific rides. For its $160 million investment, the Walt Disney Company estimated that its profits would be $230 to $600 million the first year and $300 million to $1 billion in the second year. In exchange for royalty payments, Disney provided expertise in theme park management and allowed Euro Disney to use the trademarked Disney characters and Disney Imagineering's intellectual property.

The total price of $5 billion for a theme park would certainly serve as an impediment to prevent rivals from entering the market. The quality and aesthetic value of the Walt Disney Company's products and services, its reputation as one of the world's leaders in leisure and entertainment, and its uniqueness characterized by its brand name made it appear on the surface to be a monopoly. The Walt Disney Company believed that the attendance projections were correct: 11 million people would visit the first year, with 16 million per year by the turn of the twenty-first century.

---

[1] Tokyo Disneyland was a licensing agreement between the Walt Disney Company and the Oriental Land Company. Disney received royalty payments but did not share in the profits. Euro Disney was more of a joint venture where Disney would receive royalty payments and a percentage of the profits. These agreements are discussed in the Disney (D) Case Study.

Much of the Walt Disney Company's thinking was predicated on its phenomenal past success with Disneyland (1955), Walt Disney World (1970), and Tokyo Disneyland (1983). With all three theme parks, the company adapted correctly to most of the EEFs it considered and the impact the EEFs might have on project success. Construction cost of Tokyo Disneyland was $1.4 billion and the debt, which was 80 percent of the construction cost, was paid off in three years. The question, of course, was whether these same EEFs and assumptions considered in Tokyo Disneyland were transferable to and appropriate for the European marketplace.

Unlike Disneyland and Walt Disney World locations, weather in Tokyo was an issue for Disney. When people in Japan showed that they were willing to brave the cold and snow to enjoy the theme park, Disney was convinced that the Europeans would follow suit.

Disneyland and Walt Disney World were based on an American Disney philosophy. The Japanese wanted an American-style theme park, not one customized to the Japanese culture. The younger Japanese wanted American-style food. However, some Japanese-style restaurants were built for patrons who preferred traditional Japanese meals.

Blinded by the success of Tokyo Disneyland, the Walt Disney Company believed that it could introduce the same American Disney philosophy into Europe without significant changes. However, would the EEFs related to culture be the same for the European marketplace? Once inside the park, would Europeans accept being Americanized?

UNDERSTANDING CULTURAL DIFFERENCES

Perhaps the Walt Disney Company's greatest mistake was not fully understanding the cultural differences between the Japanese and the Europeans, primarily the French. This mistake would have a significant impact on Euro Disney's revenue stream when the park opened. Some of the critical differences are shown in Table I. These are the differences based on the year when the parks were opened.

In defense of Disney's actions, which some argue was not enough to manage the cultural differences effectively, it is important to recognize that Euro Disney is an American theme park and Disney's actions were to protect its image, brand names, and reputation. Expecting Disney to make major cultural changes and alter the image of the park would be a mistake.

LAND DEVELOPMENT

Euro Disney eventually opened in April of 1992. The CEO of Walt Disney Company Michael Eisner commented:

> To all who come to this happy place, welcome. Once upon a time. . . . A master storyteller, Walt Disney, inspired by Europe's best loved tales, used his own

TABLE I   CULTURAL DIFFERENCES BETWEEN JAPAN AND FRANCE

Factor	Japan	France
Economy	Booming	In a recession
Per-capita income	Increasing	Decreasing
Time spent on leisure activities	Increasing	Decreasing
Frequency of vacations	Several short weeklong vacations	One 4–5-week vacation in August
Spending	Cannot leave the park empty-handed; gift giving is important	Gift giving is unnecessary
Acceptance of U.S. products	High	Low
Size of the park	Unimportant	Important
Attachment to Disney characters	Very high	Scorn American fairy-tale characters
Appeal of Disney entertainment	High	Low
Disneyland theme park	Symbol of new lifestyle	Seen as an American lifestyle
Tolerance for long queues	Very tolerant; used to crowds and lines	Intolerant
Acceptance of dress codes for workers	Very high, part of the culture is wearing uniforms	Very low; seen as an attack on individualism
Clean-cut grooming	Part of the culture	Attack on individualism
Politeness to strangers	Part of the culture	Not always
Enjoy being part of a team	Part of the culture	Not always
Follow instructions of one's superiors	Always	Sometimes question authority

special gifts to share them with the world. He envisioned a Magic Kingdom where these stories would come to life, and called it Disneyland. Now his dream returns to the lands that inspired it. Euro Disneyland is dedicated to the young, and the young at heart. . . with a hope that it will be a source of joy and inspiration for all the world.

The Walt Disney Company wanted to develop the commercial and residential property it had purchased surrounding the theme park and then sell off the properties while maintaining ownership and control over their commercial use. Real estate sales were expected to supply 22 percent of Euro Disney's revenue beginning in 1992 and climbing to 45 percent by 1995.

The revenue from land development was expected to help pay down Euro Disney's massive $3.5 billion debt. But when the park opened, Europe was in a recession, and it became obvious that Euro Disney had severely miscalculated the French real estate market, which was quite depressed. When Tokyo Disneyland opened in 1983, the Japanese economy was booming and the Japanese were spending a large portion of their discretionary income on leisure and entertainment. Tokyo Disneyland reaped the benefits. But in Europe, the recession caused people to cut back on leisure and entertainment. Euro Disney suffered. Disney miscalculated the impact of the recession in Europe.

Euro Disney also miscalculated how Europeans take vacations. Disney hoped that people would take one-week excursions to Euro Disney throughout the school the year. Instead, Europeans prefer to save up their vacation until August and take a 4–5-week vacation then. The cost of spending one week at Euro Disney was almost the same cost as renting a vacation home in Europe for a month. Once again, Euro Disney suffered from a loss of revenue. Disney expected labor costs to be about 13 percent of total revenue. Instead, it was 24 percent of total revenue in 1992 and climbed to 40 percent in 1993.

DISNEY'S INTEGRATED SERVICES

The Walt Disney Company's integrated services generate revenue from four sources: (1) admission to the theme parks and other attractions, (2) food, (3) shopping, and (4) accommodations. The company recognized the potential for increased profits through accommodations. In Disneyland and Walt Disney World, it allowed others to build moneymaking hotels around the theme parks. This was seen as a mistake that the company regretted. At Disneyland, the Walt Disney Company owned only 1,000 hotel rooms out of 20,000. At Walt Disney World, it owned only 5,700 hotel rooms out of a possible 70,000. The Walt Disney Company also generates revenue from the sale of cruise vacation packages and the rental of vacation club properties.

In Tokyo, the company believed that it again made a mistake by not investing heavily in accommodations. When the decision to build Tokyo Disneyland was made, the Walt Disney Company was worried about making a heavy investment in a new cultural environment. The decision was made to limit their financial risk with a small investment in accommodations surrounding the park. This put limitations on profit generation in exchange for a perceived reduction in risk and uncertainty. The risks were further reduced because the theme park was not owned or operated by the Walt Disney Company. It received a predefined royalty.

It is important to understand that an increase in daily park attendance does not necessarily translate into a significant increase in profits unless the average stay in a hotel is lengthened, which, in turn, would generate revenue from the high-profit-margin businesses, such as hotels, restaurants, and shops. Admission to the theme parks is not a high-margin business.

The Walt Disney Company took an overly optimistic position with Euro Disney, believing that what worked in Japan could be transplanted into Europe. The company did not want Euro Disney or any of its other theme parks to be seen as just theme parks. It wanted Euro Disney to be viewed as a vacation resort or a destination vacation where visitors would stay for four to five days or longer. The company wanted people to see Disney theme parks as a source of family- and adult-oriented high-quality entertainment. Therefore, included in the Euro Disney plans were a 27-hole golf course, 5,800 hotel rooms (more hotel rooms than in the city of Cannes), shopping malls, apartments, and vacation homes. The Walt Disney Company also planned on building a second theme park next to Euro Disney to house an MGM Film Tour site at a construction cost of $2.3 billion. There was even talk of building a third theme park by 2017. This would help fill the large number of hotel rooms. By the year 2017, Euro Disney, under the terms specified in its contract with the French government, was required to complete construction of a total of 18,200 hotel rooms at varying distances from the resort. All of this assumed that Euro Disney would be as popular as Disneyland, Walt Disney World, and Tokyo Disneyland. The Walt Disney Company believed that it had a firm grasp on the EEFs and set higher standards for Euro Disney than even the theme parks in the United States.

DISNEY UNIVERSITY

Just as in the United States, Euro Disney set up a Disney University to train approximately 20,000 employees and cast members who applied for jobs at Euro Disney. Training was designed to enforce the Disney culture as well as policies and procedures that had worked well for decades. The training had to be completed well before the park opened. Employees were expected to be bilingual or trilingual and were required to attend training sessions conducted by Disney University on behavior codes and how to talk to park guests. The company stressed that all visitors should be treated as guests rather than customers.

The Walt Disney Company also established rules related to facial hair (none was allowed), dress codes, covering of tattoos, limited jewelry and makeup, no highlighting or streaking of the hair, limitations on the size of fingernails, and the wearing of appropriate undergarments. The French saw this as an attack on their individual liberties.

Euro Disney and all other Disney theme parks had very strict rules sharing behind-the-scenes Disney information. Photography and filming were strictly forbidden in backstage areas. The edges of the parks were lined with ride buildings and foliage to hide areas that were not for the public to see. Numerous gates, separate from those for the public, allowed entrance into the park for cast members and parade cars. When gates around the park are opened, anything that could be seen through them is considered part of the Disney magic. Therefore, from the second the gates are opened, all of the crew must be in character and in place to "perform." Since the Euro Disney complex is so big, shuttle buses are needed to take cast members to different parts of the park via roads behind the parks.

ANTI-AMERICAN SENTIMENT GROWS

According to the literature, it appears that the Walt Disney Company understood the sociocultural and economic issues but perhaps did not pay them enough attention. The company did launch an aggressive public relations program targeting young children and government officials. Even with the community relations program, the company was still viewed as being insensitive to the French culture, the people's need for privacy, and individualism. French labor unions were opposed to the dress code. This created an anti-American climate that eventually led to deviations from the original plan for Euro Disney.

The Walt Disney Company addressed some EEFs concerning culture through behavior modification. Others required changes in the design of the attractions. The company wanted the park to show that many of the Disney characters had a European heritage. This was necessary because Euro Disney was competing against the historical architecture and sights of Paris. Simply changing the restaurant menus to serve more European food was not enough. When possible, attractions had to have a European flavor. For example, Snow White and the Seven Dwarfs were located in a Bavarian village. Cinderella was located in a French inn. Discoveryland featured story lines from the French author Jules Verne. Castles in the attractions closely resembled the architecture of castles in Europe.

Although the Walt Disney Company did the best it could to address the EEFs related to culture without altering its image and reputation, it could not adequately address the factors related to politics. For example, since many visitors were staying at Euro Disney just for one day, traffic through the town was heavy and the accompanying noise irritated many local residents. The communities surrounding Euro Disney were mainly farming communities that opposed the construction of the park. The French government had to step in to ease tension. Shortly after Euro Disney opened, French farmers used the park as a site for a protest and drove their tractors to the entrance and blocked it. This globally televised act of protest was aimed not at the Walt Disney Company but at the U.S. government, which had been demanding that French agricultural subsidies be cut. Also, Euro Disney faced several disputes with French labor unions that believed that the park was attacking the civil liberties of their union members.

Anti-American sentiment increased even more when tensions grew surrounding America's war in Iraq and France's refusal to back it. The number of American tourists visiting France dropped drastically, hurting the tourism industry, especially in Paris. One worker at the restaurant atop the Eiffel Tower noted that Spanish and Italian tourists had replaced American tourists. Other factors that affected tourism were the harsh weather in Europe, a series of transportation strikes, and the outbreak of severe acute respiratory syndrome (SARS) in Asia.

UNDERESTIMATING THE IMPACT OF CULTURE

It appeared that, even though the Walt Disney Company took steps to address almost all of the cultural differences, it had underestimated the magnitude of the differences in culture between the United States and Europe. The impact could be clearly seen through the EEFs, as shown in Table II.

TABLE II   IMPACT OF CULTURE ON THE EURO DISNEY ENTERPRISE ENVIRONMENTAL FACTORS

Enterprise Environmental Factor	Projected Impact	Actual Impact
Monopoly	Euro Disney would be a monopoly. Europeans would be willing to pay the above-market admission fees. Prices for Euro Disney were higher than at all other European theme parks and at the Disney theme parks in the United States.	It is difficult to define leisure and entertainment as a monopoly. Unlike necessities, such as water or electricity, which are often monopolies, most people can find other, less expensive forms of entertainment and leisure activities. Euro Disney functioned more as an oligopoly that has few suppliers with either similar or dissimilar products or substitute products.
Vacation resort	Europeans would see Euro Disney as a vacation resort and stay for four or five days or longer.	Europeans saw Euro Disney as a one-day short excursion. This meant that accommodations were not necessary, and Euro Disney was not seen as a resort.
Alcoholic beverages	The Walt Disney Company believed that the Europeans would accept the fact that no alcoholic beverages would be allowed in the theme park.	Europeans wanted wine and alcoholic beverages with their meals. The ban on alcohol demonstrated insensitivity toward the French culture. The French are the biggest consumers of wine worldwide. People therefore refused to eat in the theme parks. Some brought wine coolers in their cars and had tailgating parties. Also, cigarettes were not sold in the theme parks.
Integrated services	Europeans would stay four or five days at the theme park.	Europeans would stay just one day, spend the entire time on the rides, and spend very little time shopping. High admission fees also led to them spending less shopping. The actual revenues from shopping, food, accommodations, and admission fees was significantly below target levels.
Cost for a family of four	Based on data from Disney Tokyo, the cost for a family of four at Disney Tokyo was approximately $600 per day excluding accommodations. The Walt Disney Company assumed Europeans would pay the same costs.	Europeans believed that $280 per person per day was too much.

(continued)

TABLE II *(Continued)*

Enterprise Environmental Factor	Projected Impact	Actual Impact
Per-capita spending	Euro Disney assumed that per-capita spending in the park would be $33 per person.	Actual spending was revised down to $29 per person, significantly less than Disneyland and Walt Disney World in the U.S. and almost 50 percent less than Tokyo Disneyland.
Mealtime seating capacity	Based on Disneyland and Walt Disney World data, Americans would "graze" all day on snacks and fast foods. Europeans would do the same, thus implying that restaurant seating capacity at Euro Disney could duplicate other theme park seating.	Expecting 60,000 visitors a day, Disney built 29 restaurants capable of feeding 14,000 visitors each hour. However, Europeans appear to eat healthier than Americans. Most Europeans prefer to eat a healthy lunch at exactly 12:30 p.m. Most restaurants could not handle the number of customers arriving at the same time. Europeans are intolerant of long queues.
Park staffing	Euro Disney employees would accept the standards and codes that were set at other Disney theme parks.	Park employees and guests felt that they were being "Americanized." In the first nine months, 1,000 of the 10,000 workers quit.

Other mistakes were made:

- At various locations in the park, there were an insufficient number of rest rooms.
- Park staffing assumed that Friday would be a busy day and Monday a light day when, in fact, the reverse was true.
- Park management underestimated the success of the convention business and had to increase convention facilities.

MISSING THE TARGETS

On opening day, Euro Disney had expected that as many as 500,000 visitors and 90,000 cars might try to enter the park, even though the capacity of the park was estimated at slightly above 50,000 visitors. Approximately 50,000 visitors showed up, and only three out of every 10 visitors were native French. Attendance figures were disappointing. Some people argued that the low attendance was due to the Walt Disney Company's insensitivity to the French culture. Others believed it was due in part to economic conditions in Europe at that time. As the year progressed, Euro Disney lowered its daily projection from 60,000 to 25,000 visitors. Walt Disney stock plunged, eventually losing one-third of its value.

In the first two years of operations, Euro Disney's losses were estimated at $2 billion. Euro Disney also had a debt load of $3.5 billion with some interest payments as high as 11 percent. The 22 percent operating profit from land development, which was planned to pay down the debt, never materialized. Hotel occupancy was 55 percent rather than the expected 68 percent. Some hotels were shutting down for the winter season. Operating expenses had risen from an expected 60 percent of revenue to 69 percent of revenue. The MGM Film Studio theme park project was put on hold.

Although project management seemed to be successful with regard to construction of the park, the miscalculation of the impact of the EEFs was quite apparent. Euro Disney was seen as a project management success, Imagineering at its best, but possibly a business failure. The possible causes of failure were because the Walt Disney Company:

- Failed to recognize competitive leisure and entertainment offerings.
- Failed to recognize the sociocultural and economic issues.
- Had a wrong assessment of market conditions, which led to strategic and financial miscalculations.
- Took on an overly ambitious $3.5 billion debt load that was hard to pay off.
- Overdeveloped the property and land.
- Failed to recognize guest awareness of pricing.

There were also communication issues. Park executives were not returning phone calls to the media, resulting in a reputation for failed communication with the media.

Three interesting comments about the Walt Disney Company appeared in articles and newspapers. In one article, Euro Disney was viewed as "a cultural Chernobyl." In another article, a European banker stated that "Euro Disney is a good theme park married to a bankrupt real estate company and the two can't be divorced." A former Disney executive stated that during Euro Disney financing negotiations, "We were arrogant. It was like we're building the Taj Mahal and people will come—on our terms."

People were even attacking the name of the park. People outside of Europe often viewed "Euro" as being synonymous with fashion, glamour, and even high society. As Michael Eisner, Disney's CEO at the time, stated:

> As Americans, the word "Euro" is believed to mean glamorous or exciting. For Europeans it turned out to be a term they associated with business, currency, and commerce. Renaming the park "Disneyland Paris" was a way of identifying it with one of the most romantic and exciting cities in the world.

Changing the name was the Walt Disney Company's attempt to decommercialize the theme park. In addition to changing the name of the theme park in October 1994, the company took additional steps to overcome the impact of the EEFs. Previously, energetic visitors could cover all of the rides in about five hours. There were not enough attractions to convince people to stay overnight. Euro Disney eventually:

- Enhanced theme park areas, such as Frontierland, Space Mountain, and Animal Kingdom.
- Added new attractions, bringing the total number of attractions to 29. Some new additions were Zorro, Mary Poppins, Aladdin, Cinderella's Castle, Temple of Peril, and Nautilus.
- Stressed the European heritage of many of the Disney characters.
- Cut park admission prices by 33 percent.
- Cut hotel room costs by 33 percent.
- Offered discount prices for winter months.
- Offered cheaper meals in the hotels
- Allowed restaurants to serve wine and beer; however, the French never forgot that originally wine and beer were prohibited.
- Offered more foods from around the world.
- Changed its marketing and advertising strategy to include "California is only 20 miles from Paris" and "Fairy tales can come true."
- Lowered projected daily attendance from 60,000 to 25,000 people per day.
- Offered package deals that were affordable to everyone. However, this did not include the entrance fees to the park, which were still higher than in the United States.

The Walt Disney Company wanted to make people believe that, once they entered Disneyland Paris, they had escaped the real world. They would be in "a kingdom where dreams come true." To do this, the company had to recognize that the European culture was not the same as U.S. or Japanese culture. The company would not be able to "Americanize" some cultures.

DEBT RESTRUCTURING

In the fall of 1993, optimism and euphoria over the park came to an end and Euro Disney was in financial distress with a $3.5 billion debt load. If the Walt Disney Company pulled the plug on Euro Disney, there would have been a bankrupt theme park and a massive expanse of virtually worthless real estate. This would certainly blemish the company's image globally and could significantly hamper its plans for construction of other theme parks outside of the United States.

The Walt Disney Company developed a rescue plan for Euro Disney that initially was rejected by the French banks. The company fought back by imposing a deadline for agreement by March 31, 1994, and even threatened possible closure of Euro Disney if debt restructuring did not take place. Eisner believed that the French already had so much money invested in the park that they would be forced to restructure the debt. By mid-March, the Walt Disney Company's commitment to support Euro Disney had risen to $750 million. When the banks refused to consider the refinancing plan, Eisner announced in early March to the shareholders that the park might be closed by the end of that same month; the decision would be announced at the annual shareholders' meeting on March 15.

On March 14, the banks capitulated, fearing a huge financial loss if Euro Disney closed. A new preliminary deal was struck whereby Euro Disney's lead banks were required to contribute an additional $500 million. The aim was to cut the park's high-cost debt in half and make Euro Disney profitable by 1996, a date considered unrealistic by many analysts.

Part of the deal stated that the Walt Disney Company would spend about $750 million to buy 49 percent of the new rights offering that was estimated at $1.1 billion. The banks agreed to forgive 18 months of interest payments on the outstanding debt and to defer all principal payments for three years. The banks would also underwrite the remaining 51 percent of the rights offering. For its part, in addition to its $750 million, the Walt Disney Company agreed to eliminate for five years its lucrative management fees (3 percent of revenue), royalties on the sale of tickets (10 percent), and concession sales (5 percent). The company's management fees were approximately $450 million per year, regardless if Euro Disney lost money. Royalties would gradually be reintroduced at a lower level. The capital infusion was not well received by shareholders, even though they recognized that, if the park went into receivership, further expansion and the company's image could be hurt. Some believed that this debt restructuring was just a temporary bandage and that unfavorable changes in the economy or the EEFs could require future debt refinancing.

Prince Al-Waleed, nephew of King Fahd of Saudi Arabia, purchased 24 percent of Euro Disney S.C.A. for $500 million. After the restructuring, the Walt Disney Company's stake in Euro Disney S.C.A. dropped from 49 to 39 percent. The remaining 37 percent was held by a collection of more than 60 banks, mostly French, and individual shareholders, primarily from the European community.

The debt restructuring, which included debt payment forgiveness and deferral of some principal payments, was a desperately needed lifeline for Euro Disney and gave it some financial breathing room to change its marketing strategy and attract more visitors. By 1995, with debt refinancing and some theme park enhancements in place, Euro Disney had its first quarterly profit of $35.3 million. However, there was no guarantee that Euro Disney's financial headaches would completely disappear.

By 1996, attendance at Disneyland Paris was more than at the Louvre art museum, the Eiffel Tower, and Buckingham Palace. At the same time, Tokyo Disneyland was having remarkable attendance success. In 1999, Tokyo Disneyland had 17.5 million visitors, more than any other theme park worldwide.

WALT DISNEY STUDIOS PARK

A second theme park, the $2.3 billion MGM Film Studio Tour, was scheduled to open in 1996, although plans were canceled around mid-1992 due to the resort's financial crisis at the time. After the resort began to make a profit, these plans were revived on a much smaller scale. The new theme park included a history of films, including cinema, cartoons, and how films are made. The new budget was $600 million. The MGM Studio theme park was renamed Walt Disney Studios Park and opened on March 16, 2002. It was dedicated to show business, themed after movies, production, and behind-the-scenes events. In 2013, the park hosted approximately 4.4 million guests, making it the third-most visited amusement park in Europe and the twenty-first most-visited in the world, although it has the lowest attendance figures of all 11 Walt Disney theme parks. According to the Walt Disney Company's CEO, Michael Eisner:

> To all who enter this studio of dreams . . .welcome. Walt Disney Studios is dedicated to our timeless fascination and affection for cinema and television. Here we celebrate the art and the artistry of storytellers from Europe and around the world who create magic. May this special place stir our own memories of the past, and our dreams of the future.

The company had planned to open a third park in Disneyland Paris by 2017, but this plan was push back to 2030.

ANOTHER DEBT RESTRUCTURING

By 2000, Euro Disney's restructured debt load had risen to $2 billion. With the opening of the MGM Studio park, Disneyland Paris now included seven hotels, two convention centers, 68 restaurants, and 52 boutiques. But Europe's economy was struggling. The slowdown in the European travel and tourism industry had negatively affected Euro Disney's operations and cash flow. The company was strapped for cash. Once again, the word "bankruptcy" raised its ugly head. Euro Disney's financial difficulties forced it to focus on short-term cash flow rather than expansion, enhancing rides, and building new attractions.

In response to the cash flow situation, Euro Disney S.C.A.[2] initiated discussions with its lenders and the Walt Disney Company to obtain waivers of its fiscal 2003 loan covenants and to obtain supplemental financing to address Euro Disney's cash requirements. As a result of an agreement entered into on March 28, 2003, the Walt Disney Company did not charge Euro Disney royalties and

---

[2] Euro Disney S.C.A. is a subsidiary of the Walt Disney Company and the owner of Euro Disney Associés S.C.A., which operates Disneyland Paris. The name of the park itself was changed from Euro Disney Resort to Disneyland Paris in 1995.

management fees for the period from January 1, 2003, to September 30, 2003. Additionally, the Walt Disney Company agreed to allow Euro Disney to pay its royalties and management fees annually in arrears for fiscal 2004, instead of quarterly.

In fiscal 2005, Euro Disney S.C.A. completed a financial restructuring, which provided for an increase in capital and refinancing of its borrowings. Pursuant to the financial restructuring, the Walt Disney Company agreed to conditionally and unconditionally defer certain management fees and royalties and convert them into long-term subordinated debt and provide a new 10-year line of credit for liquidity needs.

Jeffrey Speed, the chief financial officer of Euro Disney, said that the modified agreement would provide "significant liquidity."

2007 TO 2013

By the end of 2007, Disneyland Paris had more than 14 million visitors for the year. The theme park had 54 attractions, 54 shops, and 68 themed restaurants. In 2008, Disneyland Resort Paris welcomed its 200 millionth guest since the opening in 1992. Exhibit III shows the attendance figures for the six years from 2008 to 2013.

A study done by the Inter-Ministerial Delegation reviewing the contribution of Disneyland Resort Paris to the French economy was released in time for the resort's twentieth anniversary in March 2012. It found that, despite the resort's financial hardships, it has generated "37 billion euros in tourism-related revenues over twenty years" and supports on average 55,000 jobs in France annually, and that one job at Disneyland Paris generates nearly three jobs elsewhere in France.[3]

TABLE III ATTENDANCE FIGURES FOR 2008 TO 2013

Year	Disneyland Europe Theme Park	Disney Studio Park
2008	12,688,000	2,612,000
2009	12,740,000	2,655,000
2010	10,500,000	4,500,000
2011	10,990,000	4,710,000
2012	11,500,000	4,800,000
2013	10,430,000	4,470,000
World Ranking	6	21

[3] Inter-Ministerial Delegation for the Euro Disney Project," Disneyland Paris.

In 2012, the Walt Disney Company announced that it would again refinance the debt of Disneyland Paris with a loan of $1.6 billion and a credit facility of $120 million. Disneyland Paris had not been profitable in 12 of the first 20 years of operations.

THE WALT DISNEY COMPANY'S 2013 10K REPORT

The following information was extracted from the Walt Disney Company's 2013 10K report:

> Parks and Resorts revenues increased 9%, or $1.2 billion, to $14.087 billion due to an increase of $1.1 billion at our domestic operations and an increase of $112 million at our international operations. Domestic theme park revenue was $11.394 billion and international theme park revenue was $2.693 billion.

Table IV shows some additional information relative to the theme parks and resorts. In the exhibit, numbers in parentheses show a decrease from the previous fiscal year.

OCTOBER 2014

For the year ending September 30, 2014, the revenue from Disneyland Paris had dropped by 3 percent from the previous year and loss was estimated between €110 and €120 million. The loss from the previous year was €78 million.

Some investors wanted the Walt Disney Company to "pull the plug" on the hemorrhaging of cash at Disneyland Paris and close the park. On the other end of the spectrum was a petition written in six languages and signed by more than 8,000 people. Titled "Save Disneyland Paris," the petition cited several problems that needed to be addressed at the theme park, including poor maintenance and upkeep of the grounds, a need for better food choices, and a need for newer and upgraded attractions. Upgrades were also needed at the Walt Disney Studios Park, where some believed that attractions should be added based on recent movies, such as *The Avengers* and *Iron Man 3*.

TABLE IV   2013 10K SUPPORTING DATA

		Domestic	International
Parks and Resorts	Attendance	4%	(2%)
	Per-capita guest spending	8%	4%
Hotels	Occupancy rate	79%	81%
	Available room nights (in thousands)	10,558	2,466
	Per-room guest spending	$267	$309

The Walt Disney Company understood that the survival of Disneyland Paris was based on repeat visitors. For that reason, it decided to provide Disneyland Paris with $1.3 billion (€1 billion) over 10 years for improvements to the theme park and Walt Disney Studios Park. In addition, the Walt Disney Studio would postpone principal payments on its debt until 2024.

FEBRUARY 2017

Disney's $1.3 billion cash infusion in 2014 was partially used to make improvements and renovations. Revenue began to rebound until the Paris terrorist attacks in 2015. For the fiscal year ending September 30, 2016, Disney Europe lost about $260 million.

On February 10, 2017, Disney announced that it would make its second cash infusion in three years. This time, Disney would invest $1.6 billion and buy out all other shareholders in Disney Europe. The money would also be used for improvements, new attractions, reducing debt, and increasing liquidity at the resort.[4]

CONCLUSIONS

Improper assumptions about EEFs can wreak havoc on a project. Not all impacts created by EEFs can be controlled or managed effectively. The Walt Disney Company did what was expected of a company of its stature to correct the impact of EEFs while protecting its name, image, and reputation. Although the factors may not have had a direct impact on the way the theme park projects were managed, especially the quality and aesthetic aspects, the factors can and did have a direct bearing on how people define the success and failure of a project.

It is important to understand that theme parks must grow continuously. They must add more rides, update existing attractions, and improve other aesthetic elements as necessary. Doing this requires capital, which often makes it difficult to pay down a large debt load.

The Walt Disney Company demonstrated "Imagineering" at its best, not only in the design and construction of the theme park attractions but also in the way that it handled necessary changes due to cultural issues. Disneyland Paris is an American theme park. The company maintained its brand name and image. Cultural issues can never be resolved in a manner where everyone is 100 percent pleased. But if I am ever in such a conflicting position, I would want Disney in my corner.

---

[4] www.latimes.com/business/la-fi-euro-disney-20170210-story.html.

QUESTIONS

1. Did the Walt Disney Company spend enough time and effort initially to understand the impact of the enterprise environmental factors as related to culture?
2. What steps did the company take to address the cultural issues?
3. How should the Walt Disney Company defend itself from French labor unions that argued that what was being taught to the 20,000 workers at Disney University was a violation of their individual rights?
4. What lessons were learned from Tokyo Disneyland?
5. What could the Walt Disney Company do in the first two years of operations at Euro Disney to get a higher occupancy rate?
6. Do you believe that the company would have allowed Euro Disney to close in 1994 because of its financial headaches?
7. Can executives or project managers ever really control enterprise environmental factors?
8. How can we prevent last-minute scope changes caused by executive meddling, assuming the changes will have a major impact on cost?
9. Euro Disney had three major debt restructurings between 1993 and 2013. Why was each debt restructuring necessary? What was the driving force that caused each debt restructuring?
10. Should enterprise environmental factors be tracked the same way that we track and report budgets and schedules?

Disney (D): The Globalization of Disneyland

In the late 1970s, the Walt Disney Company made the decision to begin expanding internationally. Tokyo Disneyland was to be the first Disney theme park built outside of the United States. Although the Walt Disney Company understood the enterprise environmental factors surrounding Disneyland and Walt Disney World, there were unknowns with opening a theme park in Tokyo. First, Japan has a winter season that could impact attendance. Second, Disney was unsure as to whether the Japanese would embrace the Disney characters. Having an American theme park in the middle of Japan was seen as a risk.

Although several globalization options were available to the Walt Disney Company, only three of the options are considered in this case study. Each option requires some sort of contractual agreement, and each type of agreement is impacted by the assumptions and associated risks made concerning the enterprise environmental factors. First, the company could assume the cost for constructing a theme park wholly owned by the Walt Disney Company. The cost of doing this would require expenditures in the billions of dollars. The company would have to work directly with foreign governments, labor unions, and stakeholders.

While it could be done, the risks and costs involved in doing this were considered prohibitive, especially for the first theme park built outside of the United States. Therefore, the other two options were licensing agreements and joint ventures.

LICENSING AGREEMENTS

A licensing agreement is a legal contract between two parties, known as the licensor and the licensee. In a typical licensing agreement, the licensor, such as the Walt Disney Company, grants the licensee the right to produce and sell goods, apply a brand name or trademark, or use patented technology owned by the licensor. Legal restrictions often mandate that the licensee must be a company in the host's country that is willing to accept such an arrangement with the licensor. In exchange, the licensee usually submits to a series of conditions regarding the use of the licensor's property and agrees to make payments known as royalties.

Licensing agreements cover a wide range of well-known situations. For example, a retailer in the theme park might reach agreement with the Walt Disney Company to develop, produce, and sell merchandise bearing Disney characters. Or a construction company might license proprietary theme park design technology from the Walt Disney Company to gain a competitive edge rather than expending the time and money trying to develop its own technology. Or a greeting card company might reach agreement with the Walt Disney Company to produce a line of greeting cards bearing the images of popular Disney animated characters.

One of the most important elements of a licensing agreement covers the financial arrangement between the two parties. Payments from the licensee to the licensor usually take the form of guaranteed minimum payments and royalties on sales. Royalties typically range from 6 to 10 percent, depending on the specific property involved and the licensee's level of experience and sophistication. Not all licensors require guarantees, although some experts recommend that licensors get as much compensation up front as possible. In some cases, licensors use guarantees as the basis for renewing a licensing agreement. If the licensee meets the minimum sales figures or theme park attendance figures, the contract is renewed; otherwise, the licensor has the option of discontinuing the relationship.

Another important element of a licensing agreement establishes the time frame of the deal. Many licensors insist on a strict market release date for products licensed to outside manufacturers. This strict release date also applies to the time needed to construct the theme park. After all, it is not in the licensor's best interest to grant a license to a company that cannot build the theme park in a timely manner or that never markets the products. The licensing agreement also includes provisions about the length of the contract, renewal options, and termination conditions.

Another common element of licensing agreements covers which party maintains control of copyrights, patents, or trademarks. Many contracts also include a provision about territorial rights, or who manages distribution in various parts of the country or the world. In addition to the various clauses inserted into agreements to protect the licensor, some licensees may add their own requirements. They may insist on a guarantee that the licensor owns the rights to the property, for example, or they may insert a clause prohibiting the licensor from competing directly with the licensed property in certain markets.

There are advantages as well as disadvantages with licensing agreements. The primary advantage is that the licensing agreement can limit a company's financial exposure. It is entirely possible that the Walt Disney Company may not have to provide any financial support for the construction of the new theme park. It would provide expertise in the design, construction, and management of the park and its attractions. It can demand that all attractions be identical to those at Disneyland and Walt Disney World. The theme park could be an exact duplicate of those two parks.

In exchange, the Walt Disney Company would receive royalty payments based on admission fees and the sale of food, beverages, and merchandise. The company may also receive royalty payments for the use of Disney characters, including use in themed hotels. The Walt Disney Company can also collect a percentage of sponsorship fees. The company would receive royalty payments regardless of whether the theme park lost money.

The disadvantages are limitations on profitability and possibly future opportunities. If the theme park is highly profitable, the Walt Disney Company would receive only royalty payments and would not share in the profitability. All profits may stay with the licensee. The licensee may demand that the Walt Disney Company not be allowed to enter certain markets that could be seen as competitors for the new theme park. This could limit the company's ability for future expansion in foreign markets. Since the Walt Disney Company would not be managing the theme park under the licensing agreement, it could face a loss of quality control that affects its image and reputation. Licensing agreements, therefore, can minimize risk for licensors and maximize risk for licensees.

JOINT VENTURES

A joint venture is a business agreement in which the parties agree to develop, for a finite time, a new entity and new assets by contributing shared equity. The theme park would be shared ownership. Both parties may exercise control over the enterprise and consequently share revenues, expenses, and assets.

A joint venture takes place when two parties come together to take on one project. In a joint venture, both parties are equally invested in the project in terms of money, time, and effort to build on the original concept. Although joint ventures are generally small projects, major corporations also use this method to diversify or expand their business globally. A joint venture can ensure the success of smaller projects for those that are just starting in the business world or for established corporations. Since the cost of starting new projects is generally high, a joint venture allows both parties to share the burden of the project's start-up cost as well as the resulting profits.

Since money is involved in a joint venture, it is necessary to have a strategic plan in place. In short, both parties must be committed to focusing on the future of the partnership rather than just the immediate returns. For example, it may take years before a theme park arrives at the desired yearly attendance figures. Ultimately, both short-term and long-term successes are important. In order to achieve these successes, honesty, integrity, and communication within the joint venture are necessary.

With joint ventures, there can be a dominant partner and participation of the public. There may also be cases where the public shareholding is substantial but the founding partners retain their identity. In such cases, local governments may provide some funding in the way of loans or tax incentives in the hope of creating jobs.

Further consideration relates to starting a new legal entity on foreign soil. The licensor may have to abide by legal requirements in the host country related to ownership of shares of stock in the new venture, the use of local labor, abiding by local union contracts, how procurement will be performed, and restrictions on land development. Such an enterprise is sometimes called an incorporated joint venture and can include technology contracts (rights related to know-how, patents, trademarks, brand use agreements, and copyright), technical services, and assisted-supply arrangements.

Joint ventures are profit and risk maximization strategies for licensors and risk minimization strategies for licensees. A joint venture does not preclude the licensor (such as the Walt Disney Company) from collecting royalties. However, it does require that both licensor and licensee make significant financial contributions. The result is usually a much larger project than each party could afford if each had to do it alone. The main disadvantage is that licensor and licensee can have different opinions on critical decisions.

TOKYO DISNEYLAND

The Walt Disney Company's partner for Tokyo Disneyland was be the Oriental Land Company. It had to decide whether the partnership should be based on a joint venture or a licensing agreement. Unsure about how the EEFs would affect the acceptance of the theme park and the fact that this was the Walt Disney Company's first theme park outside of the United States, it opted for the risk minimization strategy of a licensing agreement. Under this agreement, Tokyo Disneyland was not partially or wholly owned by the Walt Disney Company. With the licensing agreement, the company would receive royalty payments of 10 percent of the admission fees and 5 percent of the sales on food, beverages, and merchandise.

It receives its royalty payments even if Tokyo Disneyland loses money. The Walt Disney Company did have a small investment in the theme park ($3.5 million), which amounted to 0.42 percent of the initial construction cost. Because it opted for the risk minimization licensing agreement, the Walt Disney Company decided not to invest heavily in land development surrounding the theme park.

In April 1979, the first basic contract for the construction of Disneyland in Tokyo was signed. Japanese engineers and architects flocked to California to tour Disneyland and prepare to construct the new operating dreamland in Tokyo. Just one year later, construction of the park began. Hundreds of media reporters covered the story, indicative of the high expectations for the park. Although the building process was successful, the final cost of Tokyo Disneyland was almost double the estimated budget, costing ¥180 billion rather than the projected ¥100 billion Nevertheless, Tokyo Disneyland has been a constant source of pride since opening day over 30 years ago.

With only a few exceptions, Tokyo Disneyland features the same attractions found in Disneyland and Walt Disney World's Magic Kingdom. It was the first Disney theme park to be built outside the United States, and it opened on April 15, 1983. The park was constructed by Walt Disney Imagineering in the same style as Disneyland in California and Magic Kingdom in Florida.

There are seven themed areas in the park: the World Bazaar; the four classic Disney lands: Adventureland, Westernland, Fantasyland, and Tomorrowland; and two mini-lands: Critter Country and Mickey's Toontown. Many of the games and rides in these areas mirror those in the original Disneyland, as they are based on American Disney films and fantasies. Fantasyland includes Peter Pan's Flight, Snow White's Scary Adventures, Dumbo the Flying Elephant, and others based on classic Disney films and characters. The park is noted for its extensive open spaces, to accommodate the large crowds that visit the park.

The day the park opened in 1983, attendance was 13,200 visitors. On August 13 of the same year, more than 93,000 people visited the park. This one-day attendance record surpassed that of all other Disney theme parks to that time. Three years later, Tokyo Disneyland broke the record again with a one-day attendance of 111,500 people.

EURO DISNEY (DISNEYLAND PARIS)

In Tokyo Disneyland's first year of operations, 1983, it was evident that the park would be a success. In 1984, the Walt Disney Company made the decision to build a second foreign theme park, this time in Europe. The company wanted to build a state-of-the-art theme park, a decision that eventually led to "budget breaker" scope changes, and wanted it to open by 1992. Many of the changes were last-minute changes made by Michael Eisner, Disney's CEO.

Using the first year's results from Tokyo Disneyland, highly optimistic financial projections were established for Euro Disney based on the expectation of 11 million visitors the first year and 16 million visitors yearly after the turn of the 21st century.

The Walt Disney Company viewed the Euro Disney theme park as a potentially profitable revenue generator for decades since the company would have a leisure and entertainment monopoly in Europe. Unlike Tokyo Disneyland, the Walt Disney Company opted for a joint venture that was designed around a profit maximization strategy. The company would receive 10 percent royalties on admissions, 5 percent royalties on food, beverage and merchandise sales, a management fee equivalent to 3 percent of revenue, a licensing fee for using the Disney name and characters, 5 percent of gross revenues of themed hotels, and 49 percent of the profits. The Walt Disney Company also received royalties from companies that invested in and endorsed specific rides. If Euro Disney was as successful as Tokyo Disneyland, the Walt Disney Company could receive more than $1 billion in royalties and profit sharing each year.

Part of the decision to use a joint venture relationship was because the Walt Disney Company realized that it had made serious mistakes in Disneyland, Walt Disney World, and Tokyo Disneyland by not investing heavily in property development surrounding the theme parks. To maximize its profits, the Walt Disney Company agreed to build 18,200 hotel rooms surrounding Euro Disney by 2017.

WALT DISNEY STUDIOS PARK

The Walt Disney Company recognized that, if you want people to come back to theme parks again and again, you must add new attractions, build adjacent theme parks on closely related topics, or do both. The $2.3 billion MGM Film Studio Tour was scheduled to open in 1996, although these plans were canceled around mid-1992 due to the resort's financial crisis. After the resort began to make a profit, the plans were revived on a much smaller scale. The new theme park included a history of films, including cinema, cartoons, and how films are made. The new budget was $600 million. The MGM Studio theme park was renamed Walt Disney Studios Park and opened on March 16, 2002. It was dedicated to show business, themed after movies, production, and behind-the-scenes. Just like Euro Disney, this park was part of the original joint venture relationship rather than just a licensing agreement.

In 2013, the park hosted approximately 4.4 million guests, making it the third-most visited amusement park in Europe and the 21st-most visited in the world though it had the lowest attendance figures of all eleven Walt Disney theme parks.

TABLE I   TOKYO DISNEYLAND ATTENDANCE: 1983 TO 1997

Year	Attendance
1983	9,933,000
1984	10,013,000
1985	10,675,000
1986	10,665,000
1987	11,975,000
1988	13,382,000
1989	14,752,000
1990	15,876,000
1991	16,139,000
1992	15,815,000
1993	16,030,000
1994	15,509,000
1995	16,986,000
1996	17,368,000
1997	16,686,000

TOKYO DISNEYSEA

In 1997, Tokyo Disneyland recognized the need for a second theme park because attendance was leveling off, as indicated in Table I.

Between 75 and 80 percent of the visitors to Tokyo Disneyland were repeat visitors. Even with new attractions being built each year, there was apprehension that people might not return to the park after two or three visits. Furthermore, a drop in attendance by as much as 4 percent over the next four years was expected. A new theme park would be needed.

The concepts and designs for a Disney sea park had been in development at Walt Disney Imagineering for more than 20 years. However, the Walt Disney Company recommended that the new theme park be similar to the Walt Disney Studios Park that was in the planning stages for Euro Disney. Disney's partner in Japan, the Oriental Land Company, believed that the Japanese would not be as enamored with moviemaking as were Americans and Europeans. Instead, the decision was to build a Tokyo DisneySea for about $3.5 billion, based on recognition of the Japanese love for the sea. Unlike Tokyo Disneyland, the new park would be more adult-themed, including faster, scarier rides and shows designed for an older audience.

The success of Tokyo Disneyland made it quite apparent that the Walt Disney Company would have been better off financially had it chosen a joint venture rather than a licensing agreement. However, a sea park was not the same as a Disneyland park. The Walt Disney Company believed that the Tokyo DisneySea did have some risks. The Oriental Land Company believed that the Tokyo DisneySea could be just as successful as Tokyo Disneyland, but coming up with $3.5 billion was very risky. The Oriental Land Company would have preferred to minimize its risks by having a joint venture but eventually negotiated a licensing agreement for the $3.5 billion theme park. The debt for Tokyo Disneyland had been paid off in three years after the park opened. The Oriental Land Company believed that the Tokyo DisneySea could also be paid off in a reasonably short period of time.

In 2013, Tokyo Disneyland hosted 17.21 million visitors, moving its ranking to the world's second-most visited theme park, surpassing Disneyland in California but falling behind the Magic Kingdom in Florida. However, as seen in Table II, the Tokyo DisneySea attracted 14.08 million visitors in 2013, making it the world's fourth-most visited them park. In 2013, a total of 132,549,000 visitors visited Disney theme parks.

FUTURE OF TOKYO DISNEYLAND

Since the park opened in 1983, Tokyo Disneyland has regularly been the most profitable Disney resort. By 1994, over 140 million people had entered through its gates (the population of Japan is only 127.6 million), and its popularity had increased. Just two years later, it employed 12,390 people, marking Tokyo Disneyland as the biggest workplace in Japan's diversionary outings. Although the attendance trend is similar to that of other Japanese theme parks, the revenue produced by Tokyo Disneyland is larger than all other national theme parks combined, thus greatly profiting the Japanese economy. Many speculate that Tokyo Disneyland is such an economic success due to timing and location; the theme

TABLE II   2013 ATTENDANCE FIGURES FOR SELECTED THEME PARKS

Theme Park	2013 Attendance	2013 Worldwide Ranking
Walt Disney World's Magic Kingdom, Orlando	18,588,000	1
Tokyo Disneyland	17,214,000	2
Disneyland, Anaheim	16,202,000	3
Tokyo DisneySea	14,084,000	4
Disneyland Paris	10,430,000	6
Disney's Animal Kingdom at Walt Disney World	10,198,000	7
Disney's Hollywood Studios, Walt Disney World	10,110,000	8
Ocean Park, Hong Kong	7,475,000	12
Hong Kong Disneyland	7,400,000	13
Walt Disney Studios, Disneyland Paris	4,470,000	21

park lies in a metropolitan area with a population of 30 million and opened at the height of a booming economy where hard-working citizens desired a fun escape from reality. One of the main goals of Tokyo Disneyland was to keep improving the park and move away from the restrictions of the Walt Disney Company. Japan had merged its national identity with Tokyo Disneyland ark by adding attractions with distinctly Japanese qualities. Cinderella's Castle displays the classic Disney character and story plot yet presents the story through the eyes of the Japanese. Meet the World, located in World Bazaar, shows true national identity and pride as it embodies Japanese history; instead of costumes, Meet the World characters wear the traditional Japanese kimono. Once nominated by Disney Legends, Masatomo Takahashi, the former president of the Oriental Land Company, states that this growth and development is one of the company's primary goals: "We must not just repeat what we receive from Disney. I am convinced that we must contribute to the cultural exchange between Japan and U.S.A."[1]

HONG KONG DISNEYLAND

In 1988–1989, negotiations began to bring the original Disneyland to Hong Kong. Hong Kong was recognized as an international finance center and the gateway to China. The Walt Disney Company recognized that, even though many countries and cities in Southeast Asia may be on the cutting edge of technology, they were not familiar with many of the Disney products, including comics of Disney characters such as Mickey Mouse. Because of the potential risk of limited brand awareness, marketing and advertising would be critical. Even with limited brand awareness, the Hong Kong government recognized significant benefits in the venture:

- Hong Kong Disneyland would attract millions of tourists a year, create thousands of jobs, enrich the quality of life, and enhance Hong Kong's international image.
- The world-class theme park had the potential to provide Hong Kong with a net economic benefit of billions of dollars over 40 years.
- It was estimated that attendance in the park's first year of operation would be over 5 million, gradually rising to around 10 million a year after 15 years.
- About 18,400 new jobs were expected to be created directly and indirectly on opening, rising to 35,800 over a 20-year period.
- Around 6,000 jobs were expected to be created during the construction of facilities for Phase I of Hong Kong Disneyland. In addition, some 10,000 jobs were expected to be created by the land reclamation and other infrastructure works funded by the government.

---

[1] From the Oriental Land Company website, www.olc.co.jp/en/50th/03.html. Accessed February 2017.

The benefits of the base case were based on several assumptions, including these:

- The park opens in 2005.
- The park's total attendance in its first year of operation is estimated at 5.2 million.
- The park gradually reaches full annual capacity of 10 million total visitors after 15 years.
- Nearly all employees at Hong Kong Disneyland would be from Hong Kong. About 40 Disney employees from around the world would manage the park initially. But eventually about 35 local employees would be trained to take up these management duties.
- Disney would provide master planning, project management expertise, real estate development, design of the attractions, and other such support activities.
- Staff training for key personnel would take place in Hong Kong and the United States. In the United States, trainees would receive hands-on experience at existing Disney theme parks.
- In Hong Kong, the company would develop suitable training packages for a wide spectrum of Hong Kong Disneyland employees. A Disney University would be established as part of this process.
- Hong Kong Disneyland would attract 3.4 million incoming tourists from outside Hong Kong in Year 1, rising to 7.3 million after 15 years.

Hong Kong Disneyland is located on reclaimed land in Penny's Bay, Lantau Island. It is the first theme park built in Hong Kong, and is owned and managed by the Hong Kong International Theme Parks. Unlike Disneyland Paris, the Walt Disney Company preferred to be actively involved in park management rather than just being an investor. As part of a negotiated joint venture agreement, the government contributed $2.9 billion to build the park and Walt Disney Company contributed $314 million.

Risks

Despite the Walt Disney Company's experience with other theme parks, there were several risks in regard to the Hong Kong project. Some of these risks that emanated from EEFs included:

- The Chinese people's willingness to accept an American theme park.
- The Chinese culture.
- Potential cost overruns that could require that the Walt Disney Company provide additional financial support.

- Weather conditions.
- Uncertain market conditions.
- Hong Kong had another theme park, Ocean Park, which had opened in 1977. Both parks could be competing for the same tourists.
- Political uncertainty.
- A change in the government's policy for acting as a financial partner.
- Legal barriers affecting the joint venture.
- Counterfeit products.

The park opened to visitors on September 12, 2005. The park consists of five themed areas: Main Street, U.S.A., Fantasyland, Adventureland, Tomorrowland, and Toy Story Land. Cast members speak in Cantonese, English, and Mandarin. Guide maps are printed in traditional and simplified Chinese as well as English, French, and Japanese.

The park has a daily capacity of 34,000 visitors—the fewest of all Disneyland parks. The park attracted 5.2 million visitors in its first year, below its target of 5.6 million. Visitor numbers fell 20 percent in the second year to 4 million, which led to criticism from local legislators. However, park attendance increased by 8 percent in the third year, attracting a total of 4.5 million visitors in 2007. In 2013, the park's attendance increased to 7.4 million visitors, making it 13th in world park attendance.

Feng Shui Culture

The Walt Disney Company learned an unpleasant lesson about the importance of culture and EEFs from the negative publicity following the launching of Disneyland Paris. The company was attacked as being insensitive to European, and especially French, culture. The Walt Disney Company attempted to avoid similar problems of cultural backlash by attempting to incorporate Chinese culture, customs, and traditions when designing and building the resort, including adherence to the rules of feng shui. Feng shui is a local culture where numbers, colors, and images can represent good luck as well as bad luck. Buildings and structures must face in certain directions depending on their surroundings. There must be a balance between the elements of earth, wood, and fire. For instance, a bend was put in a walkway near the Hong Kong Disneyland Resort entrance so good *qi* (*chi*) energy would not flow into the South China Sea. Lakes, streams, and waterfalls were strategically placed around the theme park to signify the accumulation of wealth and good fortune.

The Walt Disney Company hired a feng shui expert to assist with designing the park and the attractions to focus on bringing the largest amount of good luck. The company was taking no chances with even the smallest details. Some of the feng shui features that were implemented are listed next.

- September 12 is considered as a lucky day for opening a business. Hong Kong Disneyland was officially opened on September 12, 2005.
- Various earthly elements important in feng shui, such as wood, fire, earth, metal, and water, were carefully balanced throughout the resort . For example, projections of a rolling fire in one restaurant bar enhance the fire element at that location, while fire is prohibited in other areas.
- Hong Kong Disneyland's main gate and entrance was positioned in a north–south direction for good luck. Another landscaped area was designed east of the theme park to ensure this north–south positioning, also enhanced by large entry portals to the area.
- Hong Kong Disneyland was carefully positioned on Lantau Island in Penny's Bay among hills and sea for the best luck. The lucky feng shui hill formations in the area include "white tiger" and "green dragon."
- The actual park entrance was modified to maximize energy and guest flow in order to help the park's success.
- Individual attraction entrances inside the Disney Park have been positioned for good luck as well.
- Large rocks are placed throughout Hong Kong Disneyland because they represent stability in feng shui. Two boulders have been placed within the park, and each Disney hotel in the resort has a feng shui rock in its entrance and courtyard or pool areas. The boulders also prevent good fortune from flowing away from the theme park or hotels.
- Water features play an important role in the Hong Kong Disneyland landscaping because they are extremely beneficial in feng shui. Lakes, ponds, and streams are placed throughout the park to encourage good luck, fortune, and wealth for the resort. A large fountain featuring classic Disney characters welcomes guests at the entrance to the park and to provide good luck (and for the taking of pictures).
- The Hong Kong Disneyland Hotel and the Disney's Hollywood Hotel were built in carefully selected locations with water nearby in a southwest direction to maximize prosperity from feng shui.
- The Hong Kong Disneyland Resort hotels have views of the waterfront onto the ocean and South China Sea. This provides good feng shui.
- The main ballroom at the Disneyland Hotel at the Hong Kong Disneyland Resort is 888 square meters, because 888 is a number representing wealth.
- The elevators at the Hong Kong Disneyland Resort do not have the number four, and no building (including the resort hotels) has a fourth floor. The number four is considered unlucky in the Chinese culture because, when pronounced, it sounds like the Chinese word for death.
- Red is an extremely lucky color in Chinese culture, so it is seen frequently throughout the park, especially on the buildings on Main Street, U.S.A.
- No clocks are sold at Hong Kong Disneyland stores because in Chinese the phrase "giving clock" sounds like "going to a funeral."

- No green hats are sold in Hong Kong Disneyland stores because it is said in Chinese culture that a man wears green to indicate that his spouse has cheated on him.

CRITICISMS

Overcrowding

Just before the grand opening, the park was criticized for overestimating the daily capacity limit. The problem became apparent on the charity preview day on September 4, 2005, when 30,000 locals visited the park. The event turned out to be a disaster, as there were too many guests. Wait times at fast food outlets were at least 45 minutes, and wait times at rides were up to two hours.

Although the park's shareholders and the Hong Kong government pressured the park to lower the capacity, the park insisted on keeping the limit, agreeing to relieve the capacity problem only by extending the opening time by one hour and introducing more weekday discounts. However, according to park officials, local visitors tended to stay in the park for more than nine hours per visit; thus, the adaptations would do little to solve the problem.

During Chinese New Year 2006, many visitors arrived at the park in the morning bearing valid tickets but were refused entry, because the park was already at full capacity. Disgruntled visitors attempted to force their way into the park by climbing over the barrier gates. Management was forced to revise the ticketing policy and designated future periods close to Chinese public holidays as special days during which admission would be allowed only with a date-specific ticket.

Initially, there were only 22 attractions, fewer than any other theme park. In July 2009, an agreement was reached between the Hong Kong government and Disney to add 20 more attractions. The Walt Disney Company would invest $450 million in the expansion and provide a loan to the theme park.

Fingerprinting

As at other Disney theme parks, visitors to Hong Kong Disneyland have their finger biometrics scanned at the entry gate. Visitors are not warned of the policy beforehand. Fingerprinting is done of all visitors older than 11 years of age and is used to associate a ticket with the person using it. The company claims that the surface of a guest's finger does not contain sufficient information to re-create a fingerprint image. Nonetheless, forensic specialists note that the data collected are more than adequate to establish a positive identification.

Public Relations

The Walt Disney Company initially refused to release attendance figures after media reports surfaced saying the park's attendance numbers might be lower than expected. The company finally declared on November 24, 2005, that the park had over 1 million guests during its first two months of operation.

In response to negative publicity locally and to boost visitor numbers, Hong Kong Disneyland offered $50 discounts for admission tickets to holders of Hong Kong identification cards in the period before Christmas 2005. Also, from March to June 2006, the park offered Hong Kong identification card holders the opportunity to purchase a two-day admission ticket for the price of a single-day ticket.

OCEAN PARK HONG KONG

Ocean Park Hong Kong opened in 1977. At that time, the park had a monopoly on theme park entertainment in Hong Kong since it was the only theme park. As it was the only park and was owned by the government, over the years, many attractions became outdated. It was not under any pressure to add more attractions and grow. When a deal was reached in 1999 to bring Disneyland to Hong Kong, it sounded like a death sentence for Ocean Park Hong Kong because it did not have the financial strength of the Walt Disney Company.

Initially, Ocean Park seemed to have lost its identity. But Ocean Park's strength was the fact that it seen as an educational park rather than an entertainment park. The park had an aquarium and live animals as well as some attractions. Its ticket prices were significantly lower than the proposed admission fees to Hong Kong Disneyland.

Rather than risk park closure, a reengineering effort was initiated. A subway line was built to the park, and the Chinese government gave the park a pair of pandas, bringing the total to four pandas. Additional hotels were built. The government also acted as guarantor for a loan to the park. The park successfully fended off the threat of Hong Kong Disneyland and hosted foreign visitors who wanted to visit both parks. In 2012, Ocean Park was the winner of the prestigious Applause Award; it was the first ever Asian attraction to be recognized as the best theme park in the world. In 2013, Ocean Park's attendance surpassed that of Hong Kong Disneyland.

FUTURE GLOBALIZATION

The future of the Walt Disney Company may focus on vacation resorts surrounding theme parks. But to get people to the theme parks, the company must get young children acquainted or hooked on Disney characters. In China, the company is getting children acquainted with its brand name at an early age. The Walt Disney Company operates dozens of English-language schools throughout China, where Disney characters and stories are used as teaching aids.

The Walt Disney Company opened Shanghai Disneyland in 2015. It is three times the size of Hong Kong Disneyland and cost $5.5 billion. Two additional theme parks will be attached to Shanghai Disneyland sometime in the future.

The park was financed 30 percent with debt and 70 percent with equity. The Walt Disney Company has a 43 percent stake in the joint venture; with the remaining 57 percent controlled by the state-run holding company Shanghai Shendi Group, which is a consortium of three companies owned by the Shanghai government.

It is expected that the Walt Disney Company will continue its globalization efforts and expand elsewhere over the next several decades.

QUESTIONS

1. What is the fundamental difference between a licensing agreement and a joint venture as related to the Disney theme parks?
2. Why did the Walt Disney Company opt for a licensing agreement for Tokyo Disneyland?
3. Why did the Walt Disney Company opt for a joint venture agreement with Euro Disneyland?
4. Does the size of the theme park have a bearing on whether a licensing agreement or a joint venture should be selected?
5. What is the difference between a theme park and a vacation resort?
6. If the goal is a vacation resort, should the Walt Disney Company negotiate a licensing agreement or a joint venture?
7. Why was it necessary to build the Walt Disney Studios Park as part of Euro Disneyland?
8. Why was it necessary to construct Tokyo DisneySea?
9. For the agreement with Tokyo DisneySea, would the Walt Disney Company have preferred a licensing agreement or a joint venture?
10. What did the Walt Disney Company see as the risks with Hong Kong Disneyland?
11. What is the feng shui culture?

Disney (E): Ocean Park Hong Kong: Competing against Disney

Ocean Park Hong Kong opened in 1977. The park had a monopoly on theme park entertainment in Hong Kong since it was the only theme park. As the only game in town, and owned by the government, it had many outdated attractions and was under no pressure to add more attractions and grow. But when a deal was reached in 1999 to bring Disneyland to Hong Kong, it sounded like a death sentence for Hong Kong Ocean Park because they did not have the financial strength of Disney. A decision had to be made on whether to compete with the Walt Disney Company. And if the park opted to compete, the question was "How?" and "How much money would be necessary?" If it chose complacency, there was a significant risk that Ocean Park might not exist in the future.

ENTERPRISE ENVIRONMENTAL FACTORS

Most project managers are assigned to projects shortly after the project is approved and possibly after the business case has been established. Project managers must have knowledge of the EEFs associated with project execution, but unfortunately these factors may not be described in the business case. Typical EEFs are listed next:

- State of the economy
- Present and future legislation
- Politics
- Consumer behavior
- Influence of labor unions
- Competitive forces
- Culture[1]

Other important EEFs address the traditional marketing and sales elements:

- Products to be offered
- Markets to be served
- Investment

Although other elements could be considered, these three EEFs are very important to this case study.

HISTORY

Ocean Park Hong Kong opened in January 1977. It was constructed with HK$150 million (approximately US$20 million) funded by the Hong Kong Jockey Club. The land was provided free by the Hong Kong government. Between 1982 and 1984, the Jockey Club allocated a further HK$240 million (approximately US$31 million) to developing facilities at Tai Shue Wan and thrill rides at the Summit.

Ocean Park ceased to be a subsidiary of the Jockey Club on July 1, 1987, becoming its own statutory body, with a government-appointed board. The Jockey Club established a HK$200 million (approximately US$26 million) trust to ensure the park's continued development. At present, Ocean Park is managed by the Ocean Park Corporation, a financially independent, nonprofit organization.

Even though some funding was available for continuous improvement efforts, Ocean Park's growth was quite slow. This was due to two factors: It had a monopoly on this type of entertainment in Hong Kong, and government-controlled enterprises tend to move slower than those in the private sector. The government was not even sure what the park should be—an Asian Sea World, a theme park like Disneyland, or a smaller park similar to Six Flags?

---

[1] Several enterprise environmental factors could have been listed here. These are the factors that impacted Euro Disney. See case study Disney (C): Disney Theme Parks and Enterprise Environmental Factors.

Part of Ocean Park's identity crisis and resulting complacency was attributed to economic conditions. The Asian financial crisis of 1997–1998 impacted consumer spending and thus park attendance. In the four-year period from 1999 to 2002, the park had lost money, and its survival was at stake. To make matters worse, the outbreak of severe acute respiratory syndrome (SARS) in 2003 had a serious impact on tourism. Attendance dropped almost 10 percent, from 3.4 million visitors in 2002 to 2.9 million visitors in 2003.[2] Ocean Park had a 75 percent decrease in profits, from US$2 million in 2002 to US$0.5 million in 2003.[2] Even though Ocean Park was somewhat profitable, and adding in the fact that Hong Kong Disneyland would open in just two years, many believed that Ocean Park was on life support and doomed to close.

THE DECISION TO COMPETE WITH DISNEY

Rather than throwing in the towel, Ocean Park Hong Kong decided to compete. Plans were prepared for the park's redevelopment. Using the EEFs of products to be offered, markets to be served, and investment, we can assess Ocean Park Hong Kong's decision.

With regard to products offered, Ocean Park Hong Kong could not compete directly with the Walt Disney Company. Rather, Ocean Park had to establish its own identity without using attractions that are similar to those of Hong Kong Disneyland.

Disneyland deals with storytelling, castles, cartoons, animation, and imagination. People visiting Disney theme parks are made to believe that they have escaped the real world and are living out part of a fantasy with characters such as Mickey Mouse, Winnie the Pooh, and Cinderella. Ocean Park Hong Kong would focus on real-world attractions, including live animals such as pandas, dolphins, and sea lions. The park also would have ride attractions, making it a marine-based theme park. Its new identity would include animals, the ocean, the environment, conservation, and, most of all, education.

Ocean Park would serve the same markets as it did previously plus have the advantage of having new customers that came to see Disney also visit Ocean Park Each park would be a feeder system for the other park. Ocean Park's development plan would include up to 80 attractions. Ocean Park believed that people who had visited Disneyland and Walt Disney World in the United States would be disappointed with Hong Kong Disneyland because of its small size and would then visit Ocean Park.

With regard to investment, in March 2005, Ocean Park Hong Kong announced a HK$5.5 billion (US$705 million) master redevelopment plan financed by bank loans to build the park into the world's best marine-based attraction. The goal

---

[2] Paul Wiseman, "Ocean Park Takes on Hong Kong Disneyland," *USA Today*, June 16, 2007.

was to make the park into a world-class, must-see landmark that would further strengthen Hong Kong as a premier theme park attraction. The park would double the number of animal and ride attractions from 35 to over 80 and would firmly establish itself as a tourist destination. The groundbreaking took place in November 2006, and the project was completed in six years over eight phases. Attractions opened under the master redevelopment plan include SkyFair, Amazing Asian Animals, Ocean Express, Sea Life Carousel, The Flash, Aqua City, the Rainforest, Thrill Mountain, and Polar Adventure.

Today, Ocean Park Hong Kong, known as just Ocean Park, is a marine mammal park, oceanarium, animal theme park, and amusement park, situated in Wong Chuk Hang and Nam Long Shan in the Southern District of Hong Kong, China. It is one of two large theme parks in Hong Kong, along with Hong Kong Disneyland.

Ocean Park has a wide array of attractions and rides, including four roller coasters, and animal exhibits with different themes, such as a giant panda habitat, a jellyfish and Chinese sturgeon aquarium, and a world-class aquarium featuring the world's largest aquarium dome, which displays more than 5,000 fish. Between 1979 and 1997, Ocean Park was most famous for its signature killer whale, Miss Hoi Wai.

Besides being an amusement park, Ocean Park is committed to merging entertainment and education while inspiring lifelong learning and conservation advocacy. This is done by operating observatories, laboratories, an education department, and the Ocean Park Conservation Foundation, Hong Kong (OPCFHK), a foundation that advocates, facilitates, and participates in the conservation of wildlife and habitats, with an emphasis on Asia, through research and education. In 2011–2012, the foundation funded 42 conservation projects, covering 27 species in 10 Asian countries for a total of HK$5 million.

Ocean Park was the first institution in the world to have success in artificial insemination of bottlenose dolphins and has developed numerous new breeds of goldfish.

ANIMALS

Ocean Park first gained accreditation from the Association of Zoos and Aquariums in 2002. In 2013, Ocean Park gained accreditation for a third successive five-year term, making it the only animal facility outside of the Americas to earn this industry recognition and validation of superior animal care, which meets or exceeds world standards, as established by the association.

The park's commitment to take full advantage of its unique collection of insects, fishes, birds, and marine mammals for scientific research has also been given a boost. With the increasing success of the park's breeding programs, births of rare shark species, bottlenose dolphins, sea lions, sea horses, penguins, anacondas, red-handed tamarins, pygmy marmosets, and different species of sea jellies have been recorded. Endangered birds and butterflies are also being hatched and reared at Ocean Park.

Animal on display at the park include giant pandas, dolphins, Chinese sturgeons, red pandas, Pacific walruses, spotted seals, southern rockhopper penguins, king penguins, gentoo penguins, Chinese giant salamanders, kinkajous, Sichuan golden snub-nosed monkeys, and orcas.

CONSERVATION

Ocean Park has directed much effort into education and research about animal conservation. It established the Ocean Park Conservation Foundation in 1993 and the Hong Kong Society for Panda Conservation in 1999. In July 2005, the two merged to form the Ocean Park Conservation Foundation, Hong Kong (OPCFHK), a registered charitable nongovernmental organization. With the ambition to advocate, facilitate, and participate in the conservation of wildlife and habitats, OPCFHK has provided a total of HK$9 million to 90 local and overseas projects since 2005, including various research projects on dolphins, horseshoe crabs, porpoises, giant pandas, snakes, and birds in various Asian countries.

Since 2006, OPCFHK has been collaborating with the Agriculture, Fisheries, and Conservation Department to handle cetacean stranding cases within Hong Kong waters. After the 2008 earthquake in Sichuan, OPCFHK established a Giant Panda Base Rebuilding Fund and donated equipment to affected nature reserves. In 2011–2012, the foundation funded 42 conservation projects, covering 27 species, in 10 Asian countries for a total of HK$5 million—all record highs.

Ocean Park has also facilitated learning through education programs throughout the years. The park established the Ocean Park Academy in 2004 to dedicate further efforts in education. Through the academy, the park runs educational tours for schoolchildren and workshops for teachers from the Hong Kong Institute for Education. Every year, the park offers over 35 core courses for around 46,000 students on six big topics: giant pandas and red pandas, dolphins and sea lions, birds, fishes, plants, and mechanical rides.

The Marine Mammal Breeding and Research Centre set up by Ocean Park houses nine dolphins and conducts research on dolphin breeding. The center is divided into six separate zones and provides behavioral training and basic husbandry to dolphins. It also plays a part in research work on the echolocation capabilities of dolphins.

To promote the idea of conservation to public, the official website of Ocean Park now features a "Conservation" session, which discusses the importance of conservation and some current conservation issues related to daily life. It also offers funny facts about some wildlife species as well as environmental threats and conservation.

GET CLOSER TO THE ANIMALS

Ocean Park runs a series of programs called "Get Closer to the Animals" that enable visitors to have close encounters with its resident animals. Its wildlife encounter programs run the gamut from hands-on experiences, such as swimming with dolphins at the Dolphin Encounter, to learning to be a panda keeper through the Honorary Panda Keeper Program. Visitors who wish to come face-to-face with fish join Grand Aquarium Scuba Diving for a journey in the Grand Aquarium (diver's certificate required). Nighttime in the Ocean's Depths offers a chance to camp inside the Grand Aquarium, spending a night viewing the underwater world. Visitors can also join tours like the Amazing Animals Ed-Venture, Grand Aquarium Ed-Venture, Polar Ed-venture, and Rainforest Ed-venture, which take groups behind the scenes at these facilities. Through the Polar Adventure, the Penguin Encounter, the Seal Encounter, and Honorary Polar Animal Keeper, people can meet with polar animals up close.

OTHER UNIQUE EXPERIENCES

The park can also be hired for birthday parties, wedding celebrations, and evening company outings. The park also offers various corporate training programs.

QUESTIONS RELATED TO ENTERPRISE ENVIRONMENTAL FACTORS

1. Why didn't Ocean Park Hong Kong grow like other theme parks?
2. How did Ocean Park differentiate itself from Hong Kong Disneyland?
3. Was Hong Kong Disneyland seen as a threat to Ocean Park before the expansion of Ocean Park?
4. Was Hong Kong Disneyland seen as a threat to Ocean Park after the expansion of Ocean Park Hong Kong?
5. Was Ocean Park a success?

Part 19

INDUSTRY SPECIFIC: THE OLYMPIC GAMES

It may seem glamorous to manage projects related to the Olympic Games. But one must always be careful what one wishes for because the grass is not always greener on the other side. All projects come with headaches. Most of us understand the headaches and the accompanying risks for our own industry and the types of projects we manage. But on some projects, such as those involving the Olympic Games, the enterprise environmental factors can be significantly different from what we have dealt with in the past.

Olympics (A): Would You Want to Manage Projects for the City Hosting the Olympic Games?

Today's Olympic Games are watched by more than two-thirds of the world's population. Managing projects for such an event carries with it prestige and certainly looks good on a resume. Unfortunately, the environment in which the project managers must perform may very well be different from anything they have dealt with in the past.

Olympic events are a blend of sports, politics, and big business. Perhaps sports, politics, and business should be the triple constraints on each project rather than the traditional time, cost, and scope constraints. Governance is provided by numerous stakeholders that may have political or hidden agendas. History has shown us that the final cost of some of the Games has been between five to ten times their original budgets. The challenges, and accompanying risks, are quite complex. If you fail as the Games' project manager, two-thirds of the world's population may know your name. You may be identified as the person whose projects were responsible for putting the host city on the brink of bankruptcy. Are you willing to accept this risk?

This case study describes the environment surrounding the Olympics Games, including some history. At the end of the case study are questions worded in a similar format to some of the questions many of you encountered when you sat for the Project Management Institute's Project Management Professional (PMP)®

Certification Exam. After reading the case study and thinking about the questions, ask yourself one more question; "Would you want to manage projects for the city hosting the Olympic Games?"

UNDERSTANDING THE OLYMPIC GAMES

For a little more than two weeks every other year, we watch the heroics of some of the greatest athletes in the world as they compete at the Summer and Winter Olympic Games. Over 13,000 athletes from more than 200 countries compete in more than 30 different sports and nearly 400 events. Many of the athletes have prepared for years for an event that is measured in seconds or minutes. For some athletes, the Olympics and its media exposure provide the chance to attain national and sometimes international fame.

But although the Olympic Games themselves last slightly over two weeks, many people do not realize the complexities of preparing for such events or the activities that follow after the Games have ended. This is where project management plays an important role. The preparation time for the Olympic Games can be more than a decade prior to the opening ceremonies, and business-related activities that are part of the Games can go on for years after the Games have ended. Perhaps project managers should receive gold, silver, and bronze medals, the same as athletes receive, for making sure that the Olympic Games take place as dictated by the project plans and the financial baselines.

PUBLIC-SECTOR PROJECT MANAGEMENT

Preparing for the Olympic Games mandates the completion of a multitude of projects that are managed by private-sector firms but within the public sector environment. The Olympics requires a close working relationship between the public and private sectors. Unfortunately, there are significant differences between project management in the public and private sectors. Some of the differences specifically related to the public sector when compared to the private sector include the following:

- There are significantly more stakeholders in the public sector.
- Getting all of the stakeholders to agree on goals and objectives may be difficult.
- Objectives are often established based on hidden political agendas.
- Political adversaries with their own hidden agendas may demand to be treated as active stakeholders.

Part of the section "Public-Sector Project Management" has been adapted from David W. Wirick, *Public-Sector Project Management* (Hoboken, NJ: John Wiley & Sons, 2009), pp. 8–10, 18–19.

- Political stakeholders can change as a result of elections and this could lead to changes in the projects' objectives and financial expectations
- There is extensive coverage by the news media.
- Project managers may need support from agencies that are not part of the project team.
- Private-sector project managers may not be able to fire underperforming public-sector team members.
- Public-sector team members may see the assignment to the project as a secondary job that may not impact their performance reviews.
- Regardless of how the private-sector project manager desires to manage the project, public-sector team members may have to follow inflexible government policies and rules that often create more work.
- The success of public-sector projects may not be measured until well into the future.
- The definition of public-sector success is significantly more complex than private sector projects and there may be an abundance of constraints and critical success factors.
- Since politicians in the public sector do not like to hear bad news, the number of metrics used to measure project performance is minimized.
- The funding for cost overruns may be paid for by future generations.
- Public-sector project managers and team members must follow the chain of command for information they need and support for decision making.

There are many reasons why public-sector projects fail, and many of these reasons have been identified more than once in past Olympic Games. They include:

- Inability to identify the key public-sector stakeholders.
- Optimistic schedules are developed with no contingency plans for late deliverables.
- Insufficient or unqualified resources.
- Not enough time devoted to up-front project planning.
- Constantly changing priorities due to political agendas.
- Inability to get an agreed-on prioritization of activities.
- Poor risk management practices for fear of exposing the seriousness of the risks.
- No revalidation of assumptions and constraints.
- Lack of repeatable project management processes.
- Inexperienced project managers.
- Failing to benefit from, or capture, lessons learned and best practices from other projects.

To understand the Olympic Games and how project management can be affected, it is important to understand the environment surrounding the Olympic Games. Five questions need to be answered. The answers to these questions have a direct impact on how projects and programs will be managed and the trade-offs that must be made.

1. How long is the life cycle for hosting the Olympic Games?
2. Are the Olympic Games a sports, business, or political event?
3. Why do cities or countries want to host the Olympic Games?
4. Where does the money come from to finance the Olympic Games?
5. How do we measure the success of the Olympic Games?

THE LIFE CYCLE FOR OLYMPIC GAMES

The typical life cycle for the Olympic Games is more than a decade long. But for some cities, such as Montreal, which hosted the 1976 Olympic Games, the life cycle ran for more than 40 years because it was 30 years after the Games had ended before the city was able to pay off its debt.

The life cycle for the Olympic Games is shown in Figure I. The life cycle begins with Phase I, which takes place between 11 and 9 years prior to the start of the Games. During this two-year period, the National Olympic Committee (NOC) in each country must determine which city or cities, if any, are interested in hosting the Olympic Games in 11 years.

Usually the NOC has already prepared a list of which cities would have the necessary infrastructure and possibly financial resources to host an Olympics. More than one city can be invited to submit bids to the NOC, but the NOC in each country submits only one city to the International Olympic Committee (IOC) at the end of this phase. The decision to bid to host the Games is more than just national pride; it is also seen by many as an investment in the city's future.

Each city interested in hosting the Games must prepare a feasibility and submit it to the NOC for evaluation. Phase I can last for up to two years because a typical feasibility study can be time-consuming and the preparation cost may

FIGURE I Life-cycle phases for the Olympic Games

Note: T= the year of the Olympic Games

be as much as $10 million or even more,[1] which is one of the reasons why some city governments that have financial issues will not consider acting as a host and paying for a feasibility study with little chance of recovering the cost downstream. City governments that are interested in hosting the Games may use their own internal resources to prepare the feasibility study or partner with private-sector firms that have experience in estimating various infrastructure costs. Improper estimating of infrastructure costs can turn a potentially highly successful sports event into a financial disaster. Project managers may not be invited to participate in the preparation of the bid or review the information prior to submittal to the NOC.

Several issues must be considered in Phase I:

- The final decision and approval of the information in the feasibility study may be made by politicians who may have political agendas, such as reelection opportunities, and therefore little concern about erroneous assumptions and/or unrealistic costs.
- The commission that is formed to create the bid always forecasts a financial windfall but fails to consider the downside aftereffects, such as cost overruns, poor land usage, and underutilized facilities.
- Sports boosters who participate in preparing the bid may exaggerate the benefits for hosting the Games and, at the same time, underestimate the infrastructure costs.
- Construction companies that participate in preparing the bid may underestimate the costs to guarantee getting the contracts, while knowing full well the costly scope changes will be necessary after contract award.
- When political agendas influence the feasibility study, backup or supporting information is at a minimum.
- Assumptions that were made are not always submitted as part of the feasibility study.
- Projections must be made for funding sources for the next nine years preceding the start of the Games.
- Projections on escalation factors over the next nine years must be made. This includes projections on cost of:
  - Inflation
  - Security
  - Housing for athletes, NOC officials, and tourists
  - Construction of sports facilities
  - Transportation
  - Land development

---

[1] Chicago's failed bid for the 2016 Olympic Games was estimated to cost $100 million.

History has shown that many of the bids prepared by the cities are just best guesses based on a political agenda and with faulty assumptions. Bids are often lowballed to increase the chances of winning. Some politicians blatantly lie to voters by stating that no public funds will be used for the venues. Then the same politicians sign the bid book submitted to the IOC, which states that public funds will be needed for the venues. This is one of the reasons why some cities do not release their bid information for perusal by the general public: fear of rejection and uproar by the voters. The history of Olympic financial disasters for host cities has left voters leery about having the Olympics in their city for fear of a heavy debt load, increased or new taxes, and a city overrun with "white elephants" when venues are abandoned after the Games.

In 1978, six years prior to the Los Angeles Olympic Games, a vote was held by the Los Angeles citizenry to prevent the use of public funds for the 1984 Olympics. This was understandable because the majority of host cities between 1932 and 1984 had financial issues after the Games were over. Denver, Colorado, won the rights to host the 1976 Winter Olympics. But in 1972, four years earlier, Colorado voters rejected the use of public funding for the Games. The Games were then moved to Innsbruck, Austria.

In December 2014, Boston submitted a bid to host the 2024 Olympic Games. Boston was in contention with Los Angeles, San Francisco, and Washington, DC, as host. Boston submitted a bid of $4.5 billion and stated that it would make maximum usage of the existing infrastructure and facilities at local universities. But a group called "No Boston Olympics" questioned the validity of the bid and the accompanying assumptions, and asked for a state ballot question to prevent Boston from hosting the Games. Letters appeared on the Internet and in the media in support of "No Boston Olympics." One such letter by Jonathan Kamens said:

> I have lived in Boston since 1989. I have been a homeowner here, in Brighton, since 1997. My wife and I have five children, all of whom were born in Boston, and two of whom are in Boston public schools. We are Bostonians. And we absolutely, positively, do not want Boston to host the Olympics in 2024.
>
> We do not have enough space. We do not have enough roadway capacity. We do not have enough public transportation capacity. We do not have enough housing capacity. We do not need to spend, literally, billions of dollars constructing facilities which will end up abandoned and unused after the party is over. We do not need years of construction disruption that will make the Big Dig look like a toddler digging in a sandbox.
>
> Boston has nothing to prove; we are already a world-class city. Hosting the Olympics will in the end make us less so rather than more.
>
> One cannot ignore the fact that the International Olympic Committee and Boston do have one thing in common: corrupt governance. It is impossible to ignore the fact that the head of the Boston Olympics exploratory committee,

John Fish, is the CEO of a construction company that would earn billions in revenue were the Olympics to be hosted here. It is impossible to ignore the fact that our mayor is beholden to the unions, whose members would also benefit financially from a Boston Olympics.

The problems I, and many others, foresee are hardly unique to Boston. In their current form, the Olympics don't benefit ANY city, state, or country in which they are hosted. They have become bloated and corrupt, and this sickness will only begin to heal when the IOC can no longer find any city willing to host the Olympics in their current form.

For the good of Boston and the good of the Olympics, I implore you to oppose the effort to bring them to Boston in 2024.[2]

A similar situation occurred in London when voters realized that the cost of the 2012 Olympics would be closer to $14.6 billion rather than the original estimate of $4.4 billion. In December 2011, an anti-Olympics poster competition was held.[3] One such poster from the competition appears in Figure II.

Once the bids are submitted to the NOC, the NOC evaluates each bid, selects one city to represent the NOC, and prepares the final bid package or bid book to be sent to the IOC. This occurs in Phase II, which is nine to seven years prior to the start of the Olympic Games. During this two-year period, the total cost to evaluate each city's bid and prepare the final bid book could reach $100 million.

WE DON'T WANT YOUR

OLYMPICS HERE

FIGURE II Anti-Olympics Poster

[2] http://blog.kamens.us/2014/07/07/olympics-in-boston-just-say-no/
[3] http://www.blowe.org.uk/2012/01/anti-olympics-poster-competition.html

One of the first things the NOC must do is to determine why a city, or the country, wants to host the Games. There are three common reasons:

1. Improve the infrastructure
2. Further develop the city
3. Build sports facilities

History has shown that those cities that focus on long-term benefits have a better chance of success. This includes constructing facilities that enhance urban life and have a positive impact on economic health long after the Games are over.

Cities can have more than one reason to host the Games. Some examples from previous Olympic Games are identified in Table I. There can also be political reasons for wanting to host the Games, such as improving a location's image or reputation on worldwide in order to attract new business. Regardless of the reasons, the cities need funding to fulfill their needs.

The NOC must be sure that whatever bid it submits conforms to Olympic standards. The NOC must validate the costs in each city's bid package, the assumptions made, projections on revenue sources, willingness to provide funding guarantees, cash flows, consideration of escalation factors, infrastructure requirements, and the need for new construction. The NOC can review post-Olympic reports from past Olympic Games for guidance. The NOC then submits its recommendation to the IOC.

Several factors may influence the outcome of Phase II:

- Politicians and the NOC are often too quick to believe what they have been told by the city offering to host the Games.
- The information provided by the cities may be in summary format, with no explanation of how the costs were actually derived.
- The assumptions made may or may not be documented.
- The assumptions, even if easily identified, are not tested for accuracy.
- Consideration of financial risks, if done at all, is done poorly.
- Cities argue that budgets cannot be confirmed until the projects are over, thus laying the groundwork for downstream cost overruns.

TABLE I   REASONS FOR WANTING TO HOST THE OLYMPIC GAMES

Improve Infrastructure	Develop the City	Build Sports Facilities
Beijing	Munich	Munich
Barcelona	London	Barcelona
Seoul	Athens	Seoul
	Montreal	Montreal
		Sydney

At the end of Phase II, each NOC must decide whether to submit a bid to the IOC for hosting the Games. The IOC then selects and announces the host city. This occurs approximately seven years prior to the opening ceremonies of the Olympic Games.

Phase III is the most critical phase and where true costs become visible. There are four primary activities associated with Phase III:

1. Construction of sports facilities
2. Construction and enhancements to the city's infrastructure
3. Fund raising
4. Establishing governance

Typical projects, including the infrastructure needs, are:

- Housing for athletes, officials, and tourists
- Communications systems and information technology (IT)
- Facilities for the media and the press
- Railways, stations, bridges, tunnels, interchanges, roads, and airports
- Power stations
- Traffic and crowd management
- Security
- Ticketing, marketing, and merchandising
- Finding and coordinating volunteers

Hosting an Olympics will incur expenses in the billions of dollars. The model that the IOC uses is that the IOC will provide a small percentage of the total costs and the rest must be provided by the city or country hosting the Games. The IOC wants guarantees from the city and/or country that sufficient funding will be made available for the Games. If the IOC sees an anti-Olympics movement and voter resistance to acting as a host, the IOC may remove that city from contention.

The greatest costs are usually attributed to the construction of sports facilities. If the national government does not participate in sports facility construction, then the city may have to take on a significant amount of debt, which most often must be repaid by raising taxes. This is why some cities that do not have existing sports facilities balk against hosting the Games.

In addition to sports facilities, there must be housing for the athletes in the form of an Olympic Village. If new construction is required, then the Olympic Village may end up suffering from many of the headaches from new construction projects. The construction of the Olympic Village for the 2016 Rio de Janeiro Olympic Games was still not fully completed when the athletes began arriving two weeks before the Games were set to begin. There were complaints about plumbing, electrical problems, puddles of water around electrical wiring, toilets that did not flush properly, water that poured down walls, and some flooding.

Although many believed that, from the outside, Rio's Olympic Village was one of the most beautiful of any Olympic Village, the inside of the facilities told a slightly different story. Kitty Chiller, head of the Australian delegation, had been at four previous Olympics and told reporters, "I have never experienced a village in this state—or lack of state—of readiness at this point in time."[4]

Many of the countries competing in Rio used hotels and other accommodations for their athletes until last-minute repairs could be made to the Olympic Village. Some delegations even hired their own tradesmen to make the needed repairs.

"There are some adjustments that we are dealing with and will be resolved in a short while," said Carlos Nuzman, the president of the organizing committee. "Every Olympic Village, because of their magnitude, needs some adjustments until it becomes perfect. The important thing is that everything will be resolved before the Games, without disturbing the athletes."[5]

Within the next two weeks, almost all of the athletes moved into Rio's Olympic Village, and the repairs were made. However, this does show the pressure placed on organizing committees for the Games, given the fact that the start date for the Games cannot change.

Infrastructure costs can be just as expensive as new facilities construction. Fortunately, many infrastructure projects are paid for by the national government and may appear as a cost to prepare for the Games. In most circumstances, infrastructure costs are the responsibility of the national government anyway, and would have been paid even without the Olympic Games. Under federal law in the United States, for example, the government will provide support for the Olympic Games when they are held in this country. The support is for security and some infrastructure projects, such as reconstructing highways that may experience heavy traffic, improving interchanges, constructing access roads to get to the venues, installing transit systems, and other capital improvements. The government does not provide any funding for the actual running of the Games.

In the 2002 Olympic Games in Salt Lake City, more than $600 million was provided by the federal government mainly for highway improvements. For the Vancouver Games in 2010, the Canadian government spent $1.9 billion for a rapid transit system from the airport to downtown Vancouver and another $900 million to build a Vancouver Convention Center that was later used by the Games' broadcasters. One of the reasons why the Atlanta and Los Angeles Olympic Games were so successful was because most of the infrastructure already existed. Minimal infrastructure funding was necessary.

---

[4] www.msn.com/en-us/sports/olympics/australia-team-wont-move-into-unfinished-athletes-village/ar-BBuKzqE?OCID=CALHeader

[5] www.msn.com/en-us/sports/olympics/blocked-toilets-leaking-pipes-and-exposed-wiring-at-the-olympic-village/ar-BBuKQTh?li=BBnb7Kz&OCID=CALHeader

Once Phase III begins, the governance structure for decision making is reasonably well defined. This is shown in Figure III. The IOC is the ultimate decision-making body and is responsible for choosing the host city for the Games. The IOC also determines the Olympic program of events, consisting of the sports to be contested. The NOC reports to the IOC and must comply with the Olympic Charter. Once a city is selected, an Organizing Committee for the Olympic Games (OCOG) is formed. Most often, the committee will have representation from both the public and the private sector. The initials of the city often appear before OCOG; for example, LA OCOG would be the Los Angeles Organizing Committee for the Olympic Games. The OCOG reports to both the NOC and the IOC.

Most Olympic Games are treated as programs, where a "program" is regarded as a grouping of projects. Therefore, the Games may have both program managers and project managers. The majority of the program and project managers report to the OCOG. However, if the national government is playing an active role in funding the Olympic Games, there could be a significant number of program and project managers reporting to both the NOC and the OCOG.

There are issues associated with Phase III. They include:

- Stakeholder engagement and transparency is complex because of the number of stakeholders.
- Determining the honesty of the stakeholders may be difficult.
- Politicians often do not want to hear bad news because of the unfavorable impact it can have on their reelection. Information may be suppressed.
- Project managers in both the public and the private sector may then prepare two sets of reports: All good news goes vertically up the chain of command and is filtered; bad news travels laterally and stays within the project teams.

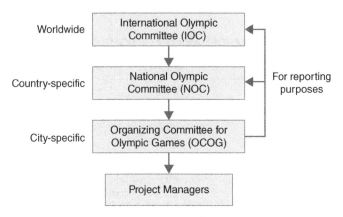

FIGURE III Governance structure

- Rigorous planning rarely happens.
- Budget and schedule estimates often are unrealistic.
- Almost all trade-offs result in cost increases.
- Politicians often succumb to pressure from contractors to work overtime to maintain schedules, thus causing budget increases. Sometimes this is just a ploy by contractors to earn more money, who know that replacing them at this late date may be impossible.
- Project teams will be highly diverse.
- It may be impossible to determine whether resources are qualified and if they are being used efficiently.
- Inspirational project leadership will be necessary.

The success of the Olympic Games is measured in Phase IV. Success can be defined as short-term success, long-term success, or both. Short-term success is often measured by the host's profit as a result of conducting the Games. This can be calculated immediately after the Games are over. Did the income to the city exceed the expenses to run the Games? Table II illustrates some short-term successes.

Short-term profitability can be misleading, especially if the host had to incur debt to run the Games. For example, Barcelona showed a profit of $8.6 million but was in the red for $6.1 billion; the Spanish government incurred $4 billion in debt and the remaining $2.1 billion was assigned to provincial governments. The Nagano (Japan) Olympics of 1998 showed a $28 million cash surplus but they still had a debt of $11 billion.

There are also long-term success measures, which are made after the crowds have vanished. These include enhancements to the city's image and reputation, bringing in new industry, additional tourism, investments in the city, the ability to sell or lease land that was developed, and the ability to sell or lease sports venues

TABLE II   SHORT-TERM PROFITABILITY OF SOME OLYMPIC GAMES

Host City	Year	Profit
Los Angeles	1984	$517.3 million
Seoul	1988	$1 billion
Barcelona	1992	$8.6 million
Atlanta	1996	$15.4 million
Beijing	2008	$190.3 million
London	2012	$91.3 million

With regard to the numbers above, there are several reports and studies that "estimate" profit and loss. They all seem to differ a bit. The numbers in the last column below were taken from: http://thechive.com/2016/08/18/which-host-city-profited-the-most-from-the-olympics-20-photos/

to professional teams. The long-term benefits may take years to be recognized, which is why there are question marks in Phase IV of Figure I.

Some benefits may not be quantifiable, but they can be significant; that is, they can stir national pride and psychological satisfaction. The Beijing Games, which some believe were the greatest Olympic Games ever, showed the world that China was now a major political and economic force. The Barcelona Games rebranded Barcelona as one of Europe's most popular tourist destinations. According to one poll, Barcelona went from 11th to fourth in European rankings for best cities to do business in. Other Games provided benefits from infrastructure projects that reduced traffic congestion, enlarged roads, built new subway systems, made airport improvements, and enforced better environmental safety procedures.

CITIES THAT HAVE LOST MONEY

There have been several cities that have lost money as a result of holding the Olympics, as shown in Table III.

The cost overruns from some of the past Olympic Games have put some of the host cities, as listed in Table IV, on the brink of bankruptcy.

There are three common reasons why some cities have had financial disasters. They were:

1. Unforeseen spending
2. Lack of long-term planning
3. Inability to utilize the venues after the Olympic Games

Atlanta is considered a success story. Using public funds, Atlanta constructed two arenas for the Games. These two arenas are now being used by the Atlanta Braves and Falcons. The Olympic Village is used as dormitories for Georgia Tech University students. This was part of the planning in preparation for the Atlanta Olympics.

But for many cities, hosting the Games was a losing proposition. If the venues are left idle after the Games are over, they become white elephants where

TABLE III   LOSSES FROM SOME PREVIOUS OLYMPIC CITIES

Host City	Year	Loss
Montreal	1976	$990 million
Sydney	2000	$2.1 billion
Athens	2004	$15 billion

TABLE IV   HOST CITIES THAT EXPERIENCED FINANCIAL DISTRESS

Year	Host City	Olympic Season
1976	Montreal	Summer
1980	Lake Placid	Winter
1998	Nagano	Winter
2000	Sydney	Summer
2004	Athens	Summer
2006	Turin	Winter
2010	Vancouver	Winter

maintenance and operations costs rise well above any fees that may be collected from users. The Sydney OCOG estimated that part of its $2.2 billion loss was due to the yearly $30 million cost to maintain the 90,000 seat Olympic Stadium. In 2005, Athens had spent $125 million for maintenance on two idle Olympic soccer field stadiums.

On May 12, 1970, extensive lobbying and diplomacy by Montreal mayor Jean Drapeau paid off when Montreal was awarded the 1976 Olympic Games over strong bids from Moscow and Los Angeles. Drapeau believed that the Games could be conducted for $250 million, with a small percentage of the cost coming from the IOC and the rest coming from the city of Montreal. The Games would therefore be self-funded because the government of Canada and Quebec Province refused to incur any financial liability. Speaking at a 1973 press conference, Drapeau stated: "Montreal can no more have a deficit than a man can have a baby."[6] His words later came back to haunt him and the citizens of Montreal.

Problems occurred right from the start. Labor union strikes caused construction delays and increased the costs. Many of the problems lasted until the opening-day ceremonies. The infield grass was laid in the Olympic Stadium the morning of the opening ceremonies. As costs began to escalate, Montreal converted the big "O" (which stood for the Olympics) to the big "OWE." In 1976, Montreal issued a tobacco tax to help pay the $990 million deficit. Paying off the deficit took 30 years. Including interest payments on the debt, Montreal's total cost was about $2.6 billion by 2006. However, the Montreal Games were considered a success by the viewers.

On September 5, 1997, Athens was awarded the 2004 Olympics, beating out Rome, Cape Town, Buenos Aires, and Stockholm. Greece was so busy celebrating its victory that it forgot the fact that it was the smallest country to host the modern

---

[6] Cooper Rollow, "No Deficit at Montreal Olympics, Vows Mayor." *Chicago Tribune*, February 11, 1973, p. 3 section 3.

Olympics. Thus the impact on the country's economy would be tremendous. Only a well-prepared plan and carefully orchestrated execution within a collaborative working environment would save the country from humiliation in front of the whole world.

Unfortunately, the Athens Olympic Games of 2004 is remembered as one of the greatest financial disasters in Olympic history. The decision to bid on the Games was more of an emotional one than a practical one. The motto that Athens selected for Games was "Welcome Home." To focus on the purity of the Games, Athens decided to avoid large-scale commercialization and minimize the number of Olympic sponsors. Athens also removed almost 10,000 billboards promoting the Olympic Games. The lack of revenue from commercialization had a devastating effect when the initial budget of $1.6 billion went out of control. Athens had underestimated the funds needed to develop its infrastructure for accommodations, security, environmental issues, traffic, and transportation. It also failed to give enough attention to other issues, such as IT, ticketing, volunteers, broadcasting and the press, crowd management, guest services, mascots, the Olympic Torch Relay, marketing, and merchandise sales.

Athens eventually incurred a $15 billion cash shortage. Some of the reasons for the financial disaster were attributed to:

- Lack of planning
- Irresponsible spending
- Nontransparent procurement methods
- Gross underestimation of infrastructure costs
- Missing a significant number of deadlines, thus creating fear of having to cancel the Games
- Adding in overtime to the budget and having to hire more workers

During the preparation for the Games, when the costs began to escalate, two chief financial officers were fired, and the politicians who promoted the Games lost bids for reelection.

The $15 billion deficit was about 5 percent of Greece's gross domestic product and amounted to almost $60,000 per Greek household. Taxpayers would have to share in the shortfall. Some people argue that the 2004 Olympic Games were a major factor leading up to Greece's current financial woes. As of 2016, 21 out of the 22 venues created for the Athens Olympic Games were rotting and unused. The three causes to financial disaster mentioned earlier—unforeseen spending, lack of long-term planning, and inability to utilize the venues after the Olympic Games—all affected Athens.

The grandeur of hosting the Games disappears quickly when the crowds vanish, venues are idle but still in need of maintenance and upkeep, and the host city must pay off a large debt by taxing its citizenry.

OLYMPIC GAMES REVENUE SOURCES

There are numerous sources of revenue for the Olympic Games. Some of the sources are listed next.

- Television rights
- Sponsorship fees
- Licensing fees
- Ticket sales
- Olympic lotteries
- Olympic commemorative coins
- Postage stamps depicting Olympic events
- Parking
- Leasing or selling venues after the Olympics
- Leasing or selling developed land

While the Olympic Games are seen by most people as the greatest sports event in the world, the Games still are business similar to most professional sports. Income must exceed expenses. Therefore, commercialization is a necessity to generate some funds. Fatal mistakes at the Athens Olympic Games were the failure to recognize that the Games are now a business, the importance of commercialization of the Games, and how much revenue could be generated through commercialization.

The 2008 Olympic Games were held in Beijing. The cost of the Games was estimated at $44 billion, and the Games were government sponsored. There was a very heavy focus on commercialization to raise revenue. There was also an emphasis on lowering construction costs without sacrificing quality. The Beijing Olympic Games netted a profit of $146 million and showed the benefits of aggressive commercialization.

The importance of commercialization was first recognized with the 1984 Olympic Games in Los Angeles, when city residents voted against using public funds for the Games. The Los Angeles Organizing Committee for the Olympic Games (LA OCOG) needed the Games to be self-financed, mainly through television rights, sponsorship fees, and ticket sales. The committee recognized the need for an Olympic sponsorship program, which eventually brought in 64 sponsors where each paid up to $4 million and/or provided in-kind products/services. The first two sponsors were the Coca-Cola Company and Anheuser Busch. The committee also instituted a licensee program giving manufacturers the right to sell products containing the Olympic logo in exchange for a royalty fee. Because the Games were in the summer, they were able to use vacant university residence halls rather than constructing a large Olympic Village. The committee arranged for 30,000 volunteers, which was the largest ever for an Olympic event at that time.

Commemorative coins have been part of the Olympic Games since 1952. The LA OCOG believed that it could generate $200 million or more in revenue if the coins were also legal tender. The original plan was to produce a set of 29 coins depicting various Olympic activities. Because the coins would be legal tender, Congress had to approve the minting of the coins. Congress eventually approved the minting of a set of three coins: two silver and one gold. The plan was to mint up to 50 million silver coins and 2 million gold coins, with the Secretary of the Treasury to determine the designs. Sales of the coins would include a surcharge of $10 for each silver coin and $50 for each gold coin. The proceeds from the surcharge would be split evenly between the United States Olympic Committee (USOC) and the LA OCOG. The manufacturing and marketing of the coins would be the responsibility of the U.S. Treasury. The LA OCOG had no role in the Olympic coin program except to receive a check from the Treasury every month. The retail price of the coins was set at $32 for the silver coin (face value: $1) and $352 for the gold coin (face value: $10).

The LA OCOG raised $150 million in corporate sponsorships, $286.8 million in television rights, and $150 million in ticket sales. Taxpayers bore none of the burden. The Los Angeles Olympic Games generated a profit of $517 million and brought more than $3.3 billion in benefits to Southern California after the Games were over.

CHANGING THE OLYMPIC IDEOLOGY

After the success of the 1984 Olympic Games, the IOC altered its ideology toward expanding the Games through the use of corporate sponsorship and the sale of television rights. By controlling the rights at the IOC level rather than the NOC level, the IOC would become financially independent. The Olympics were now becoming more of a business than a sports event.

The IOC sought to gain control of these sponsorship rights by establishing The Olympic Program (TOP) in 1985. This created an Olympic brand. Membership in TOP was, and is, very exclusive and expensive. A four-year membership costs $50 million. Members of TOP receive exclusive global advertising rights for their product category and use of the Olympic symbol (i.e., the interlocking rings) in their publications and advertisements. [7]

Sponsors demonstrated a willingness to come up with hundreds of millions of dollars in advertisements because, in little more than two weeks, their ads would

---

[7] The Olympic symbol, better known as the Olympic rings, consists of five interlocking rings and represents the unity of the five inhabited continents—Africa, Americas, Asia, Europe, and Australia (and the South Pacific). The colored version of the rings—blue, yellow, black, green, and red—over a white field forms the Olympic flag. These colors were chosen because every nation had at least one of these colors on its national flag.

be televised to two-thirds of the world's population. Membership in TOP included such companies as:

- The Coca-Cola Company
- McDonald's
- General Electric
- Manulife
- Omega
- Johnson & Johnson
- Panasonic
- Kodak
- Samsung
- Visa
- Lenovo
- Atos

The IOC has been criticized for allowing certain sponsors, especially those advertising junk food, to be part of TOP. For example, some groups have complained that, with the number of people around the world that are obese or overweight, promoting Coca-Cola and feasting at McDonald's should not be encouraged. Also, promoting tobacco, candy, and beer and alcoholic products to young audiences viewing the Games was seen as bad. At the 2012 Games in London, Cadbury was an Olympic partner, and its candy products were the official treats of the Games. Although the IOC does support public health and encourages healthy lifestyles, it also believes that there are gray areas when it comes to accepting sponsors.

There are levels of sponsorship. First, there are the worldwide sponsors that are part of TOP and partner directly with the IOC. There are also domestic partners and sponsors that may negotiate directly with each OCOG and may contribute goods and services rather than cash. Domestic partners and sponsors in all participating countries usually are sources of revenue for travel expenses to get their athletes to the Games. There are also exclusive suppliers that work with the OCOGs. Project managers may be asked to make changes to some of the projects due to the demands of the partners and sponsors. But perhaps the biggest headache may be getting the suppliers to keep their promises in a timely manner.

TOP also includes control of broadcast rights. The sale of broadcast rights enabled the IOC to increase the exposure of the Olympic Games worldwide, thereby generating more interest, which in turn created more appeal to advertisers' time on television. This cycle allowed the IOC to charge ever-increasing fees for those rights. For example, CBS paid $375 million for the rights of the 1998 Nagano Games, while NBC spent $3.5 billion for the broadcast rights of all the Olympic Games from 2000 to 2012. Broadcast rights to various Olympic Games are shown in Table V.

TABLE V   COST OF THE WORLDWIDE BROADCAST RIGHTS

Year	Host City	Worldwide Broadcast Rights
1960	Rome	$1.2 million
1964	Tokyo	$1.6 million
1968	Mexico	$9.75 million
1972	Munich	$17.8 million
1976	Montreal	$35 million
1980	Moscow	$100 million
1984	Los Angeles	$287 million
1988	Seoul	$403 million
1992	Barcelona	$635 million
1996	Atlanta	$930 million
2000	Sydney	$1.33 billion
2004	Athens	$1.5 billion
2008	Beijing	$2 billion
2012	London	$3.9 billion

Source: Stephen A. Greyser and Vadim Kogan, "NBC and the 2012 London Olympics: Unexpected Success," Harvard Business School working paper 14-028, September 15, 2013.

IOC HEADACHES

Although the IOC wanted to expand using TOP, the need for a much heavier focus on commercialization created headaches. Governments were making large financial commitments to the Games while the IOC appeared to be doing nothing to limit the lavish spending by host cities. As the cost of the Games increased, more and more countries were, and still are, refusing to submit bids to host the Games. To protect its public image about the increasingly large amount of money being spent to host the Games, the IOC has issued a reform called Olympic Agenda 2020, which is designed to reduce the cost of bidding for and hosting the games. The reform recommends the maximum use of existing and temporary facilities. However, the IOC has the right to add new sports to the Games to increase revenue from ticket sales, sponsorship fees, and TV rights.

The sale of the Olympic brand has been controversial because it also encourages spending. The argument is that the Games have become indistinguishable from any other commercialized sporting spectacle. Commercialization means that the Games now feature competition between businesses as well as athletes.

Parts of the section "IOC Headaches" have been adapted from these sources from *Wikipedia, The Free Encyclopedia*: "Olympic Games," https://en.wikipedia.org/w/index.php?title=Olympic_Games& oldid=758294281; and "Cost of the Olympic Games," https://en.wikipedia.org/w/index.php?title=Cost_ of_the_Olympic_Games&oldid=756062845. A source for additional information is *Wikipedia, The Free Encyclopedia*, "List of Olympic Games Scandals and Controversies," https://en.wikipedia.org/w/ index.php?title=List_of_Olympic_Games_scandals_and_controversies&oldid=756549622

Specific criticism was leveled at the IOC for market saturation during the 1996 Atlanta and 2000 Sydney Games. The cities were awash in corporations and merchants attempting to sell Olympic-related wares. The IOC indicated that it would address this issue to prevent overmarketing at future Games. Another criticism is that the Games are funded by host cities and national governments; the IOC incurs none of the cost yet controls all the rights and profits from the Olympic symbols. The IOC also takes a percentage of all sponsorship and broadcast income. Nevertheless, some host cities continue to compete ardently for the right to host the Games, even though there is no certainty that they will recover their investments.

Viewership of the Games also became a headache. Viewership increased exponentially from the 1960s until the end of the century due to the use of satellites to broadcast live television worldwide in 1964 and the introduction of color television in 1968. Global audience estimates for the 1968 Mexico City Games was 600 million; at the Los Angeles Games of 1984, the audience numbers had increased to 900 million; and that number swelled to 3.5 billion by the 1992 Summer Olympics in Barcelona. However, at the 2000 Summer Games in Sydney, NBC drew the lowest ratings for any Summer or Winter Olympics since 1968. This was attributed to two factors: increased competition from cable channels and the Internet, where results and video were displayed in real time. Television companies were still relying on tape-delayed content, which was becoming outdated in the information era. A drop in ratings meant that television studios had to give away free advertising time. With such high costs charged to broadcast the Games, the added pressure of the Internet, and increased competition from cable, the television lobby demanded concessions from the IOC to boost ratings.

After each Olympic Games are completed, the IOC reviews what occurred and usually recommends a number of changes to the Olympic program. At the Summer Games, the gymnastics competition was expanded from seven to nine nights, and a Champions Gala was added to draw greater interest. The IOC also expanded the swimming and diving programs, both popular sports with a broad base of television viewers. Finally, the American television lobby was able to dictate when certain events were held so that they could be broadcast live during prime time in the United States. The results of these efforts were mixed: Ratings for the 2006 Winter Games were significantly lower than those for the 2002 Games, while there was a sharp increase in viewership for the 2008 Summer Olympics, and the 2012 Summer Games became the most watched event in U.S. television history.

The IOC tried to maintain the Olympic ideals and conduct the Games free of political intervention. But as the TV audiences grew, the Olympic movement had to contend with political events including boycotts, protests, walkouts, terrorism, wars, human rights issues, and political instability. Some cities hosted the Games to show the world the merits of their political system. The Games evolved from a sporting event, then into a big business, and now a political tool used by governments where publicity and media exposure are more important than an environment for sporting competitiveness. The events surrounding the Games have often overshadowed the achievements of the Games themselves.

As TOP focused on television rights and commercialization, there was a growing fear that the IOC would not maintain its ethical standards and would try to cover up doping allegations and poor judging by some officials. If the public loses confidence in the integrity of the officials in the sports, they would also lose confidence in the Olympic movement.

Bribery, gift giving, and gift taking were also becoming an issue, as was demonstrated in the Salt Lake City Winter Games.[8] As the stakes grew higher for cities wanting to host the Games, even members of the IOC were found guilty of accepting bribes and gifts from cities that wanted the members' votes for selecting their city to host the Games.

PROJECT COMPLEXITY

The Olympics are the world's greatest sporting events, but they are still plagued by the fact that they are political programs and highly complex. Most people simply cannot imagine the complexity of some of these programs and accompanying projects, and the fact that it may take years for them to be completed. The IOC awards locations seven years in advance to give the host country sufficient time to build the necessary infrastructure and manage project complexity.

To understand the magnitude of some of the Olympic projects, look at IT as an example. Consider the following comments about the Games in Sochi, Russia, by Patrick Adiba, chief executive officer of Olympics and major events at Atos, a European IT services company and official IT partner for the Olympic Games:

> The technology infrastructure [for IT] is composed of 400 servers, 1,000 security network devices, and 5,600 computers that are responsible for things such as providing real-time Olympics information to 9,500 accredited broadcasters and members of the media. As well, it is delivering competition results to a global audience in less than a second; processing and activating accreditation badges for 200,000 "members of the Olympic Family"; and collecting and processing data for all of the more than 5,500 athletes taking part in the Games.[9]

Atos had been preparing for the Sochi Olympics for more than seven years. Because of the threat of terrorism and potential hackers, the company ran 100,000 hours of systems testing involving more than 9,000 different scenarios.

What we see on television is the end result of a massive undertaking. We might measure success by the quality of the performances of the athletes, the

---

[8] Jere Longman, "Leaders of Salt Lake Olympic Bid Are Indicted in Bribery Scandal." *New York Times*, July 21, 2000. Available at www.nytimes.com/2000/07/21/sports/olympics-leaders-of-salt-lake-olympic-bid-are-indicted-in-bribery-scandal.html.

[9] http://news.cnet.com/8301-10797_3-57618516-235/in-sochi-the-olympics-size-job-of-running-olympics-it-infrastructure/

glitter of the events such as the opening and closing ceremonies, the ability to see all of the Games in real time or recorded, and other such factors. Unfortunately, from a project management perspective, there is a significant difference between project success and project management success.[10] Although the end result may satisfy the millions of worldwide viewers and the host country, the way the project was managed could have been a partial or almost complete disaster.

THE CHALLENGES FROM THE TRIPLE CONSTRAINTS

Today's *PMBOK® Guide* focuses on competing constraints. Projects may have as many as 10 or more constraints—all with different degrees of importance—but let us just consider the traditional triple constraints of time, cost, and scope. Generally speaking, if you need to hold one constraint fixed, the other two may have to be changed. As an example, if the scope of the project is fixed, then we must consider trade-offs on time and cost. But the date for the Olympics cannot change. Trade-offs on time can have severe consequences. Therefore, trade-offs can be made only on cost and scope. However, there are limitations on how much scope can be altered. We cannot tell countries they must reduce their number of athletes by 20 percent because of a lack of housing nor can we tell them we are eliminating swimming events because sufficient funding does not exist to construct a swimming venue. Time waits for no Olympics project manager; it is fixed. The only solution is to pay more or do less. Therefore, the natural expectation is that costs will be allowed to escalate even though the budget was considered as fixed by the politicians. Cost overruns have occurred in many Olympics Games. In order to contain some of the cost overrun during the 2004 Athens Olympics, no roof was constructed over the swimming venue. When working with a fixed timetable, it is imperative that the scope be defined as early as possible. This will provide more opportunities for meaningful trade-offs.

PROJECT MANAGEMENT AND THE ENTERPRISE ENVIRONMENTAL FACTORS

Numerous enterprise environmental factors can impact the way that Olympic projects are planned and managed and how decisions are made. They include:

- Governance committee politicians who make decisions in their own best interest rather than in the best interest of the projects related to the Games.
- Politicians from the more than 200 countries attending the Games asking for "bells and whistles" to be included in certain projects for their personal whims.

---

[10] The reader should understand that, from this point forth, the use of the word "project" refers to both projects and programs.

- Working with budgets that cannot be confirmed until after the Games are over.
- Making financial projections on sponsorship fees, licensee royalties, and ticket sales revenue seven years before the Games take place.
- Being at the mercy of budgetary changes due to sudden economic downturns.
- Having certain countries boycott the Games for political reasons.
- Adhering to the religious requirements of certain countries.
- Worrying about human rights issues and sex discrimination.
- Having to deal with people involved in corruption, kickbacks, and embezzlement.
- Worrying about security issues, violence, and possible terrorism.

Each Olympics has had its problems. To understand how enterprise environmental factors and assumptions can change over the seven years between receiving the go-ahead to host the Games and the actual start of the Games, consider what happened at the 2016 Olympic Games in Rio de Janeiro.

- In 2009, Brazil was awarded the 2016 Summer Olympic Games. At the time, oil, which was one of Brazil's and Rio de Janeiro's primary revenue sources, had reached more than $80/barrel and Brazil's economy was robust. Oil eventually nosedived into the $20s per barrel, and by 2016, Brazil was struggling with its worst recession since the 1930s. The country's economy shrank by 5.4 percent in the first quarter of 2016.
- The Olympics budget had to struggle with last-minute cuts totaling $500 million. The cuts resulted in less seating for spectators at some venues and fewer amenities at the Olympic Village.
- The Rio Olympics were plagued by financial concerns that could lead to bankruptcy, forcing Rio's governor to declare a state of emergency and to ask the state for a cash infusion of $850 million since the province of Rio, and not the state, was responsible for funding the Olympics.
- By 2016, Brazil was in the midst of political upheaval with the impeachment of its president, Dilma Rousseff. Her successor, interim president Michel Temer, had a rough start to his presidency when, in just two months, three of his cabinet members, including the tourism minister, resigned amid allegations of their alleged involvement in the massive corruption scandal at Brazil's state-run oil company, Petrobras. Critics were now calling for Temer's impeachment.
- The threat of terrorism and security for the athletes and tourists was critical. Rio had one of the highest crime rates, 54 murders per 100,000 people annually, compared to the rate in Western Europe: one murder per 100,000 people annually. Rio used 85,000 security personnel for the

Games, which included 38,000 from the armed forces. This was twice the number used in the 2014 Sochi Olympics and eight times the number in the 2012 London Olympics. Unfortunately, by the end of June 2016, the city and state police had not been paid for overtime work for more than six months. At Rio's Galeano International Airport, more than 100 emergency services workers that included police and firefighters protested for their unpaid wages and held up a banner for visitors to see, stating: "Welcome to hell. Police and firefighters don't get paid; whoever comes to Rio de Janeiro will not be safe." The majority of the cash infusion was to be used to pay these back wages, as well as for security for the Games.

● Another part of the cash infusion was to complete the metro rail extension that would connect the Olympic facilities to the city center. This project was expected to be completed just four days before the Olympics were set to begin.

● The Zika virus, which is linked to microcephaly, a condition causing babies to be born with abnormally small heads, was a major concern that could reduce the expected number of athletes and tourists. In February 2016, there were 16,000 cases of Zika in just one week. More than 150 prominent doctors, scientists, and professors from around the world published an open letter calling for the Games to be moved or postponed.[11]

OTHER OLYMPIC ISSUES

Olympic events are political projects where politicians usually have a high expectation that by providing support for the Games, their reputation or image will increase as will their chances for reelection. Enterprise environmental factors must be considered right from the start, when the bid to host the Olympics is being prepared. To gain political support to host the Olympics, politicians establish a fixed cost for the event based on assumptions that are often unrealistic and not revealed. Enterprise environmental factors may or may not be considered at this time. Although government decision makers have a good grasp of how government financing works, they often fail to understand the basis for the budget and end up believing what they are told by subordinates or contractors hoping to win lucrative contracts. They trust the brief summary they have before them, which usually does not explain of how the numbers were derived, the assumptions made, the enterprise environmental factors, the financial risks, and any impact from future escalation factors on construction costs, security needed, housing, and transportation. Project managers are then asked to manage projects based on partial or unsubstantiated information.

---

[11] Max Kutner, "150 Doctors, Academics, Call for Olympics to Move because of Zika." *Newsweek*, May 27, 2016. Available at www.newsweek.com/rio-olympics-zika-open-letter-doctors-464381/.

The assumptions could include the availability of qualified labor, cost-of-living increases over the next seven years before the event itself, and a minimal escalation rate for raw material procurement. Accurate cost projections for work three months from now may be possible, but projections for seven years from now may be troublesome. More often than not, the people creating the bid have had little or no exposure to project management, and professional project managers may not be invited to participate in the establishment of the bid.

Although the budget may appear to be fixed for each project needed for the Games, the project manager must understand the impact of the enterprise environmental factors. First, budgets are established on the premise that funding will be available. If economic conditions change over the seven years prior to the Games, expected funding may be reduced or even withdrawn by potential sponsors. If fund raising brings in less than the necessary funding, trade-offs will be necessary. Project managers may succumb to political pressure in the order of performing the trade-offs. If more than optimal funding is available, project managers may be asked to enlarge their project, such as adding more seating to a stadium to generate more ticket revenue. This causes problems if construction has already started.

In general, budgets cannot be confirmed until the program is over. At least that is what politicians tell us. But, in reality, once the budget is established and publicized, politicians do not want to hear any bad news that might put their reelection campaign in jeopardy during the next seven years. When bad news occurs, politicians try to downplay their ties with a public-sector project that may be seen as a failure because they may serve as ammunition for opposing political parties and have an impact during elections. Project managers may be asked to withhold bad news.

In previous Olympics, on the average, the final cost has been three times the initial estimate. Voters know this and wonder where the funds for the overruns will come from. If voters are against hosting the Olympics because public funds are being used and taxes may be raised, then the project manager may be working in a hostile environment. It is entirely possible that some people may want the project to fail.

Embezzlement and kickbacks are also issues, even if the project manager is not involved. People may believe that the project manager is somehow involved in such financial irregularities. This could lead to an unfavorable working environment where the media places everything the project manager does under a microscope.

CAPTURING LESSONS LEARNED

As the funding needed to support the Games grows, so does the need to capture best practices and lessons learned from previous Olympic Games. Repeating mistakes can be costly.

Perhaps the most important lesson learned from the Montreal Olympics in 1976, Athens in 2004, Salt Lake City in 2010, and London in 2012 was the need for a collaborative working environment. Open and meaningful communications are essential.

When politicians create a culture where they refuse to hear bad news, the result is a breakdown in communications between politicians and various Olympic stakeholder groups. The end result is often a slowdown on projects because problem resolutions are beyond the project manager's level of authority. Action items can then remain in the system forever without receiving proper attention. In addition, an important lesson learned from Athens was the necessity for open and honest communications among stakeholders, including government stakeholders. For the 2010 Olympics in Salt Lake City, governance offices were created that used project charters. Authority and decision-making responsibilities were decentralized to various teams in order to keep the program moving. Information transmitted between the public and private sector was in language that everyone could understand, so people could make informed decisions based on evidence rather than guesses. This accelerated the handling of action items.

Several other lessons related to project management can be learned from past Olympics.

- There must be transparency on all assumptions made and how the costs in the bid were derived.
- Critical issues, including bad news, should be brought to the surface for resolution rather than being buried.
- There are limited opportunities for trade-offs other than an escalation of costs.
- Given that economic conditions can change over the life cycle of the Olympics, continuous revalidation of assumptions is necessary.
- Because of the length of the life cycle, risk management activities should focus on mitigating future damages.
- The sharing of information is critical.
- There must be a clear understanding of each party's role, responsibility, and decision-making authority. This includes governance personnel as well.
- Stakeholders and governance personnel must understand project management and how their actions and decisions can impact the project performance.
- Project managers must know how to deal with fraudulent activities, collusion, kickbacks, embezzlement, and influence peddling. Project managers may not be able to control these without support from above.

Best practices and lessons learned are now being captured in all Olympic events. Rather than call it a best practices library (as is common in industry), it is referred to as Project Management Legacy Learning. After the London Olympics, seminars were taught about lessons learned.[12] These seminars are expected to continue after each Olympics.

FUTURE OF THE OLYMPIC GAMES

Although the future of the Olympic Games is not in doubt, the process of selecting cities to host the Games may change significantly. Infrastructure costs are rising and have been a financial headache for host cities. Some experts believe that Japan's cost for hosting the 2020 Olympics will exceed $30 billion unless some existing structures can be used.[13] The IOC's decisions in the future may be based on using what cities have rather than what needs to be built. Very few cities have the infrastructure for all of the necessary Olympic venues, which may be the primary reason why fewer cities are competing to host the Games. Hosting the Games is a marketing bonanza for the host city and can generate significant revenue and notoriety provided that the city plans for the disposition of the venues after the Olympics are over.

Current thinking centers on reusable venues where some cities will be allowed to host the Games more than once in a reasonable time period. Although there will be a maintenance cost to keep up some of the venues between Olympics, the cost may be recoverable over multiple Games. The maintenance cost could be in the millions of dollars each year per venue.

Here are some of the ideas being considered regarding reusable venues.

- Greece will be the permanent host of the Games every four years since Greece was the birthplace of the Games. This may alienate other cities wishing to host the Games but would keep the cost of hosting the Games at a minimum.
- Three or four cities with existing infrastructure would be the permanent host of the Games, and the Games would rotate among those cities.
- The venues for the Games would be distributed over multiple cities every four years so that no single city would be overwhelmed. This would require multiple Olympic Villages and could create problems for advertisers, commercialization of the games, sponsorship rights, and broadcasting rights.
- The city awarded the rights to host the Games would be granted hosting rights twice, now and either eight or 12 years from now.

---

[12] Learning Legacy project, http://learninglegacy.independent.gov.uk/

[13] www.boston.com/sports/olympics/2016/09/29/expert-panel-warns-tokyo-olympics-cost-could-top-30-billion

Although each of the above ideas has both advantages and disadvantages, it shows that consideration is being given to the infrastructure costs necessary to host the Olympics.

QUESTIONS

For each question, assume that you are a project manager working for a private-sector contractor and are managing more than one project to prepare for hosting the Olympic Games. There are several possible alternatives to many of these questions, but your job is to think about what you would do.

1. You read over the statement of work for the project and discover that it was based on erroneous assumptions. Your cost baseline is insufficient to get the work done. What should you do next?

2. Some of the political stakeholders that are part of the OCOG appear to have hidden agendas and are now changing your project's requirements. This will cause a cost overrun. You are not sure if the changes are necessary. What should you do next?

3. Your project has a small management reserve built into the price. Your project sponsor, who happens to be part of the OCOG, wants to use the management reserve for adding in bells and whistles on another project where he is the sponsor. What should you do next, and whom can you go to for assistance?

4. Some of your team members appear to slow down their work on purpose in order to justify overtime. This will be costly if it is prolonged. What should you do next?

5. The OCOG slashes your budget and uses the money to put out fires on other projects that you are not managing. You realize that you cannot complete your efforts within the reduced budget. What should you do next? To whom should you talk?

6. A sponsor that is part of TOP has asked you to make structural changes to your project so that the sponsor's company logo can displayed more prominently. This will delay your project a bit and cause a cost overrun. The OCOG tells you to make the changes but within the existing cost baseline. What should you do next, and whom should you discuss this with?

7. Some domestic sponsors have asked you to make changes that will create a conflict with the requirements imposed by many TOP sponsors. What should you do next?

8. The contractors you hired are demanding more money in order to finish their work. You do not believe that they need additional funding. What should you do next, and whom should you discuss this with?

9. Several actions items require an agreement be reached by the stakeholders that reside in the OCOG. Their inability to come to an agreement is slowing down your project. What should you do next, and where can you get assistance?

10. You are a strong believer in the Project Management Code of Conduct established by PMI. One of the senior members of the OCOG is the owner of a company that has the responsibility for managing several of the projects. The budgets for these

projects have increased beyond belief, and some of the people working on these projects have told you that the increases are because of embezzlement and kick-backs. What should you do next, given that these are not your projects?

11. The OCOG has asked you to make a presentation to the media as to why your project is behind schedule and over budget. Should this be part of your job as a project manager?

12. The funding provided by the national and city governments for infrastructure projects was not sufficient for all of the work that needed to be done. You were told that government funding would cover all of the infrastructure costs. Since your projects directly interface with some of the infrastructure projects, you are fearful that part of your budget may be redistributed to these projects. What should you do next?

13. In the middle of your project, you have been informed that you now have a new sponsor who resides in the OCOG. The new sponsor may not know any of the critical issues you faced thus far and may change some of the requirements. What should you do next?

14. The project that you are working on could become a white elephant after the Games are over. If you make certain structural and design changes now, at an additional cost of $100 million, you can lower the yearly maintenance and operations costs by $15 million, should no buyers be found for the venue. What should you do next?

15. Now it is time to ask yourself one more question: "Do you want to manage projects related to hosting the Olympic Games?"

Olympics (B): The Olympics, Project Management, and PMI's Code of Ethics and Professional Conduct

Project managers are expected to abide by the Code of Ethics and Professional Conduct established by the Project Management Institute (PMI). But sometimes doing so is difficult. There are situations where executives in the company you work for or external stakeholders may act in an immoral or unethical manner for either personal or company gain. Although you, as a project manager, may not personally act in this unprofessional manner, you may be drawn into it by decisions made by your company's hierarchy. What if you like the company you work for and do not want to leave? What if you had no knowledge that this was happening until well into the project? What if you are asked to be part of this while managing a project? What if this is forced on you by the stakeholders after the project begins?

SOCHI OLYMPIC GAMES

The 2014 Winter Olympics in Sochi, Russia, had an initial budget of $12 billion, and the organizers had expected a cash surplus of over $300 million after the Games were over. The $12 billion estimate was significantly more than the $8 billion spent in the 2010 Winter Olympics in Vancouver, Canada. The final cost of the Sochi Olympics eventually ballooned to an estimated $51 billion and surpassed the $44 billion spent in the Beijing Summer Olympics in 2008, which had been the most expensive Olympic Games in history. Generally speaking, Summer Olympics should cost more than Winter Olympics because of the factors shown in Table I.

TABLE I   COMPARISON OF BEIJING SUMMER AND SOCHI WINTER OLYMPIC GAMES

Factors	Beijing Summer Games	Sochi Winter Games
Athletes	10,942	2,873
Events	302	98
Venues (sports disciplines)	28	15

There were political reasons behind the high cost of the Beijing Games. China wanted to show the world that it was now a political and economic power. The Olympics Games were one of the means by which China achieved its goal. Vladimir Putin hoped that, by hosting the Winter Olympics in 2014 and the World Cup in 2018, Russia would demonstrate to the world his nation's political and economic power.

There was a valid reason why the costs escalated at Sochi. Russia wanted Sochi to become the premier year-round resort area for years to come. This required that significant expenditures be made to improve the infrastructure in the region, specifically telecommunications, electric power generation, and transportation such as roads, railways, stations, bridges, tunnels, interchanges, and airports. Other infrastructure requirements were needed specifically to support the Games, such as housing for the athletes and for IOC officials and guests, facilities for the media and the press, and security. Generally, it is not a good idea to include infrastructure costs as part of the costs for running the Games because many such costs are expenditures that would most likely have been spent on development of the region even if the Games were not held. Infrastructure costs are long-term investments. In Sochi, only 13 of the 424 facilities built for the Olympics were sports-related.

The contracts for the infrastructure development, especially construction-related contracts, provided some people with the opportunity for embezzlement, kickbacks, graft, and other forms of corruption. There appeared to be an absence of fair competition in the awarding some contracts. Sochi was not the only Olympic Games to have these issues. The 2002 Salt Lake City Winter Olympics also suffered from corruption and mismanagement, as did other Olympic Games. Mitt Romney was brought in to clean up the mess in the Salt Lake City Olympics.

COST ESCALATIONS AT SOCHI

As work at Sochi began, the escalation of the costs provided ammunition for Putin's adversaries, who claimed that a good portion of the corruption was at the hands of government officials.

The opposition figures Leonid Martynyuk and former Deputy Prime Minister Boris Nemtsov claimed in a May report that up to $30 billion of the $51 billion budget had gone missing in "kickbacks and embezzlement" to close associates of Putin, claiming the Games had turned into a "monstrous scam.[1]

There was also criticism of the construction of a road that was deemed an infrastructure necessity, but many believed would not have been needed if it were not for the Games.

An 18-mile road between Sochi, where events such as hockey, speed skating and figure skating will be held, and the mountain sports cluster of Krasnaya Polyana [for skiing and snowboarding events] has become a symbol of the huge cost increases, spiraling to a reported $8.6 billion. "You could have paved this road with five million tons of gold or caviar and the price would have been the same," Nemtsov said in an interview with the RBK television channel in July.[2]

The massive construction requirements for the infrastructure improvements and the need to have the opening ceremonies on time required tens of thousands of construction workers, including 16,000 from outside Russia. Many of the non-Russian workers felt that they were being exploited.

A Human Rights Watch report accused firms contracted to build venues including the Central Olympic Stadium, the main Olympic village, and the main media centre of cheating workers out of wages and requiring them to work 12-hour shifts with few days off. The companies were also accused of confiscating passports and work permits, apparently to coerce employees to remain in exploitative jobs.[3]

For the Russian government, the Olympic Games were a sign of national prestige and pride. For Putin, the Games would be part of his legacy. Therefore, because of the importance of the Sochi Games, no one was afraid to let costs escalate and ask for more money for contracts.

It is also necessary to understand who was receiving the lucrative contracts.

Two kinds of private business interests are involved in Sochi: companies hired by state-owned corporations to carry out specific work and those who came on as investors, taking responsibility for various projects and putting up at least some of their own money. Among the first group, according to the Nemtsov and Martynyuk report and opposition magazine *New Times*, no one has gotten more money from Sochi than brothers Arkady and Boris Rotenberg, childhood

[1] Owen Gibson, "Sochi 2014: The Costliest Olympics yet But Where Has All the Money Gone?" www.theguardian.com/sport/blog/2013/oct/09/sochi-2014-olympics-money-corruption.

[2] Ibid.

[3] Ibid.

friends of Putin's from St. Petersburg who have become wealthy industrialists over the past decade. They received 21 contracts, according to the magazine, worth around $7 billion—more than the total cost of the Vancouver Olympics and around 14 percent of all spending for the Sochi Games.[4]

The private investors helping fund Olympic construction are most likely motivated less by the pursuit of large profits than a tacit understanding that under Putin they have certain obligations to the Kremlin and the nation at large. "They got a call with a voice saying, 'There exists the opinion that you should build this or that [project],'" says Sergei Aleksashenko, a former deputy chair of Russia's Central Bank now a fellow at Georgetown University.[5]

Both the Russian Government officials awarding the contracts and the contractors understood the rules of the game. Contractors wanted to make a profit, often an excessive one. Officials awarding the contracts wanted kickbacks. Many contracts were artificially inflated, without any justification, to satisfy both parties. Other contracts were allowed to increase after contract award, knowing that additional funding would be provided. Examples of how some of the costs had increased are shown in Table II.

Some Russian government officials have been charged with criminal activities. On February 6, 2013, Putin visited the ski jump facility in Krasnaya Polyana. Although the original cost of the venue was $40 million, the cost had risen to $267 million. Akhmed Bilalov was the head of the state-owned company responsible for construction of the ski jump venue. Bilalov was also vice president of the Russian Olympic Committee. The day after Putin's visit, Bilalov was removed from both positions and criminal charges were brought against him for fraud. He and his brother immediately left the country for London. The people working on the project stated that the proper geological tests were not done initially, leading to significant cost increases and constantly changing plans. The ski jump situation is an example of how big business, improper planning, and lax oversight leads to inefficient spending and possible abuse.

TABLE II   EXAMPLES OF VENUE COST ESCALATIONS

Venue	Final Cost	Times Original Cost
Ski jump	$267 million	6 times
Stadium	$700 million	14 times
Speed skating	$226 million	7 times
Bobsled	$76.5 million	1.6 times

[4] Joshua Yaffa, "The Waste and Corruption of Vladimir Putin's 2014 Winter Olympics," www.bloomberg.com/bw/articles/2014-01-02/the-2014-winter-olympics-in-sochi-cost-51-billion.
[5] Ibid.

Other issues related to corruption at the Sochi Games include:

- Russian taxpayers picked up 96 percent of the tab.
- The Olympic Stadium was two and a half times more expensive than similar stadiums in Europe; its cost may have gone up 14-fold in seven years.
- Three of Putin's old friends were awarded contracts totaling $15 billion.
- A Mafia-linked businessman with ties to Putin's friends built the Shayba hockey arena for $33 million over market price.
- Prime Minister Dmitry Medvedev's ski instructor was awarded contracts totaling $2.5 billion.
- A construction company owned by Siberian politicians with no experience building sports arenas spent $500 million on projects up to 2.3 times over market price.
- The local governor spent $15 million in Olympic funds on a helicopter for himself.
- A power station built at nearly double market price could not supply Sochi with enough electricity for the Olympics.
- The speed skating arena was seven times over budget and $130 million over market price.
- The Iceberg Skating Palace cost more than double the equivalent arena built for the 2006 Olympics in Turin.
- The state company in charge of the Olympics had four directors in six years and was the subject of numerous criminal investigations.
- Putin spent Russian Orthodox Christmas in a new $15 million church paid for out of the Olympic budget.[6]

EPILOGUE

Large, international sports events, such as the FIFA World Cup and the Olympic Games, are prone to corruption. Situations such as those that happened at Sochi probably occurred at several other Olympic Games but were not as publicized. But at Sochi, despite the cost overruns, schedule slippages, corruption, doping scandals that followed the Games, and complaints of shoddy workmanship, both the athletes and the worldwide audience appeared to be enormously satisfied with the Games. Putin achieved his goal. However, most residents of Sochi have seen very few economic benefits as a result of hosting the Games. Many of the problems that existed before the Sochi Olympics, such as poor transportation, potential for flooding, and electricity concerns, still remain unresolved.

---

[6] 6 Max Seddon, "16 Eye-Popping Examples of Alleged Corruption at the Sochi Olympics," www .buzzfeed.com/maxseddon/eye-popping-excerpts-from-a-report-alleging-corruption-at#.dbP8jg8Vne.

QUESTIONS

Considering the PMI Code of Ethics and Professional Conduct, answer the following questions. For each question, assume you are a project manager employed by a company in the private sector. Your company has one or more contracts for venues at the Olympic Games.

1. You have been assigned to manage one project and believe that there may be some corruption that could directly involve you in the future. As an example, an IOC official asks you to help her scalp tickets to some of the events with the promise that she will share some of the profits with you. What should you do?
2. You have been assigned to manage one project. Partway through the project, your senior management informs you that you must request (without any real justification) more money from the OCOG because one member of the OCOG wants a larger kickback. What should you do?
3. Partway through your project, one government stakeholder tells you that the government will pay you only 75 percent of the remaining cost; the other 25 percent will be withheld as a kickback. What should you do assuming that you need the remaining 25 percent to complete the project?
4. Partway through the project, your senior management tells you to sacrifice quality to keep costs down to compensate for the corruption. What should you do?

Olympics (C): Would you Want to Manage Projects for the Feeding of Athletes in the Olympic Village?

Imagine that you are responsible for feeding not just one Olympic athlete, or even all of the athletes from just one country, but every person in the Olympic Village for a little more than one month. This may include up to 15,000 athletes, as well as coaches and officials from 200 countries, and possibly more than 40,000 volunteers. Each athlete may have different dietary requirements. How do you start such a project? What people do you need as part of your project team? Are there specific skills your team should possess? How do you determine what food you will need and the quantity of each item?

PROJECT COMPLEXITY

The Olympic Village opens approximately two weeks prior to the actual start of the Olympic Games and remains open for approximately 33 days. During this period, the village must be capable of serving up to 60,000 meals per day, for a total of more than 1.5 million meals. There can be more than 500 different types of meals served each day. Exhibit I shows the food quantities needed at the London 2012 Olympic Games to serve 23,900 athletes and team officials.

TABLE I   FOOD QUANTITIES FOR THE LONDON 2012 OLYMPIC VILLAGE

Food	Quantity
Bread	25,000 loaves
Potatoes	260 tons
Seafood	92 tons
Poultry	35 tons
Meat	112 tons
Milk	19,800 gallons
Eggs	21 tons
Cheese	24 tons
Fruits and vegetables	370 tons

Source: Adapted from http://britishfood.about.com/od/introtobritishfood/a/Food-Facts-London-Olympic-Games-2012.htm.

FEEDING THE OLYMPIC ATHLETE

To stay at the top of their game, Olympic athletes maintain rigorous workout schedules. Athletes could spend years practicing for an Olympic event that lasts for seconds or minutes. Accompanying both the competitive events and the workouts is a regimented diet often defining what each athlete should ingest in training sessions prior to the competition, during the competition, during the cool-down or recovery period following the competition, and in the evening. Training and workout diets are often higher in calories, protein, vitamins, and minerals than competition diets. For some Olympians, postworkout and postcompetition dieting is just as important as training diets because grabbing any food that is available after a training session could be damaging if it is high in salt or sugar content. A typical Olympian can burn 800 calories during competition, and it may take four to eight hours to replace those calories. As Allen Tran, the head chef for the United States ski teams stated, "Training cannot be optimal if the fuel isn't optimal."[1] What is critical for each sport is the ratio of the nutrients. For strength-based sports, for example, emphasis is on protein. For endurance races, the ratio of carbohydrates is important.

Some Olympians have extreme eating habits, which may include 16 bananas every 24 hours to maintain potassium levels, 50 pieces of sushi after training, or special liquids.[2] A typical male adult consumes about 2,000 to 2,800 calories a day whereas Olympians may consume between 6,000 to 12,000 calories each day, based on the sporting event.

[1] Bill Pennington, "A U.S. Team Chef Shows His Own Competitive Spirit in Sochi." *New York Times*, February 8, 2014. Available at www.nytimes.com/2014/02/09/sports/olympics/us-ski-teams-can-now-order-in.html.

[2] http://www.edition.cnn.com/2012/08/03/sport/olympics-nutrition-phelps-blake/index.html

Food fuels the body, and each athlete demands a particular kind of sustenance. For instance, a 250-pound bobsledder may require proteins like steak to compete, while an endurance athlete like a cross-country skier needs carbs to go the distance.[3]

It seems that there is virtually no end to the power and stamina of Olympic athletes, which is due in part to the detailed guidance they get from experts about the right amount and type of food they need. . . . The science of fueling athletes for elite sports goes way beyond caloric intake to include percentages of carbohydrates versus proteins, quantities of fluids, and the timing of meals and snacks.[4]

Some athletes intentionally gain weight during workouts to reduce the chance of injury or illness but then slim down for the actual competition. For most Olympians, these extreme eating habits are their normal routine. Some athletes eat only two large meals daily whereas other athletes may eat four meals a day, each separated by four hours, and each meal is designed to satisfy different time-of-day dietary needs.

Dining at the Olympic Village can easily become a culinary trap. With more than 500 different menus covering the complete gamut of the health spectrum, athletes have an opportunity to feast on delicacies from other countries regardless of the dietary requirements they need to hone their bodies to perfection in their sport. All of the food in the Olympic Village is free to village residents.

Because of the extreme workouts and eating habits, Olympic athletes often supplement their dietary intake with vitamins. The athletes themselves are responsible for verifying that what goes into their mouths are not substances that are banned by the IOC or other sports governing authorities.

THE OLYMPIC VILLAGE

Olympians know the quality of the food in their own country. But there is always some concern over the quality of food provided by the city or country hosting the Olympics. The Olympic Games have a reputation of providing high-quality food for all of the athletes at the Olympic Village. When the athletes stray from the village, the quality of the food and the ingredients is unknown, and water may be undrinkable.

Because of terrorism threats, food goes from suppliers/distributors to warehouses to be security checked, leaving nothing to chance. Local government hygiene bureaus are used to validate quality. Executive chefs with work experience

---

[3] Kiri Tannenbaum, "Gold Medal Meals: What the Olympic Athletes Eat at the Vancouver Winter Games," www.delish.com, February 1, 2010. Available at www.delish.com/restaurants/a1274/winter-olympics-menu/

[4] Barbara Bronson Gray, "The Care and Feeding of Olympic Athletes." *HealthDay News*, July 27, 2012. Available at health.usnews.com/health-news/news/articles/2012/07/27/the-care-and-feeding-of-olympic-athletes?offset=100.

at previous Olympic Games supervise meal preparation. Registered dieticians and translators are in dining halls 12 hours a day to answer questions. Some registered dieticians have graduate degrees in sports nutrition.

The dining halls in a typical Olympic Village may serve a "World Menu" of more than 550 recipes. Special dietary and religious requirements are addressed as well. Pictograms appear at each food station to help athletes identify the foods being served. There are also kiosks that have translators and dictionaries to assist in identifying the nutritional content of the foods. Computers and printers are attached to the kiosks. For some Olympians, the hardest part of maintaining a regimented nutritional diet is finding the time to consume the needed calories and nutriments. The dining hall can prepare boxed lunches/meals transported in refrigerated trucks to the competition venues.

SUPPORTING SUSTAINABILITY

When London was bidding to host the 2012 Olympic Games, it promised that it would be the most sustainable Games ever. Specifically, it promised to support the consumption of local, seasonal, and organic produce. To understand the magnitude of this endeavor, consider that 1.3 million meals for the construction workers for the four-year period preceding the Games and then 14 million meals over 60 days for those people attending the Games, including visitors, would have to be prepared.

The responsibility for food for the construction workers was with the Olympic Delivery Authority (ODA). The ODA would be working with as many as 12 contractors, each of which would be responsible for its own catering. The ODA was expected to promote sustainable foods even though it had no such internal procurement policy at that time.

The responsibility for food during the Games rests with the OCOG in the city hosting the Games. The London OCOG (LOCOG) wanted all food vendors to serve healthy, local, and freshly produced food. Promoting an Olympic legacy of support for sustainability efforts was of high importance.

The IOC controls TOP, the group of key sponsors for Olympic events. Under the IOC agreement, the TOP sponsors have extensive rights to operate outlets in the Olympic Village and the Olympic Park itself, but they do not necessarily utilize them. Two of the key sponsors for the Games were McDonald's and Coca-Cola. In Sydney, McDonald's operated seven restaurants at Olympic venues, including two serving athletes in the Olympic Village, two serving the media, and three catering to the public. A large variety of other food providers are also used, but McDonald's has an "exclusion" on branded foods and can prevent other outlets from selling hamburgers, fries, or other foods (such as egg rolls) that are seen as too similar to McDonald's products. This arrangement does not, however,

---

Material in the section "Supporting Sustainability" has been adapted from www.sustainweb.org/pdf/ Feeding_the_Olympics.pdf.

affect other food sold that is deemed to be dissimilar to McDonald's products. Coca-Cola has the right to put its company brand on all menu boards at all venues.

In past Olympics, there was criticism from the general public that McDonald's and Coca-Cola do not provide health-conscious products and should not be allowed to functional as food and beverage sponsors. The LOCOG wanted all caterers and key sponsors, including McDonald's and Coca-Cola, to actively take the lead in promoting the use of local, seasonal, healthy, and organic produce as well as promoting the link between healthy eating, sports, and well-being. The focus was on healthy living.

As the Olympics began to grow in both the number of athletic events and visitors, so did the need for effective catering. Complicating matters even further was the need to satisfy the dietary requirements of all athletes in the Olympic Village. The full extent of the impact on the catering budget was not realized until after the 1996 Atlanta Games, resulting a larger-than-expected budget. Two years after London won the right to host the 2012 Olympic Games, the budget had already quadrupled from its original estimates.[5]

Because of the complexity of food services and support for sustainability, the plan was for the LOCOG to work with individual "master caterers," many of whom had experience at previous Olympics, and they would then manage their subcontractors. At the Sydney Games, there were eight to 12 master caterers, working with 400 individual caterers[6] and 700 catering outlets.[7] One of the largest caterers worldwide is Aramark, which has catered at the Olympics on 13 previous other occasions. At the Athens Olympics, Aramark served 75,000 meals a day to coaches, volunteers, and athletes throughout the event period.[8]

FAIRTRADE CERTIFICATION

Food purchased for the Games has an international social impact on the farmers and distributors supplying the food from other countries. Table II provides some food facts from the 1996 Olympic Village in Atlanta. There must be generous helpings of fruits and vegetables to go along with meat and fish for lean protein, healthy fats, and whole grains. Not all host cities or even countries have the ability to provide all of the necessary food.

If imported foods adhere to ethical trade standards, such as those embodied by Fairtrade certification, the Games enhance the international solidarity that the sports activities already set out to demonstrate. Fairtrade agreements cover items

---

[5] BBC News website, "Olympic Budget Rises to £9.3 Billion," March 15, 2007. http://news.bbc.co.uk/2/hi/uk_politics/6453575.stm

[6] Presentation by David Payne, catering advisor to the IOC, organized by the LOCOG, April 12, 2007.

[7] Official report of the XXVII Olympiad. http://www.gamesinfo.com.au/postgames/en/pg001540.htm

[8] A. Lisante, 2004, "Fueling Olympians: Philly's Aramark Corp's 13th Olympic Catering Gig." http://foodmuseum.typepad.com/food_museum_blog/2004/08/fueling_olympia.html. Aramark ultimately decided not to participate in the 2016 games (www.philly.com/philly/business/20160816_Philly_s_Aramark_is_sitting_out_these_Olympic_Games.html).

TABLE II   FOOD FACTS FROM THE 1996 OLYMPIC VILLAGE IN ATLANTA FOR 33 DAYS

Food	Quantity
Water	550,000 gallons
Milk	70,000 gallons
Pasta	52,000 pounds (dry weight)
Rice	34,000 pounds (dry weight)
Beef/Lamb	280,000 pounds
Poultry	150,000 pounds
Cheese	90,000 pounds
Eggs	576,000 fresh eggs
Margarine	32,800 pounds
Butter	30,000 pounds
Rolls	20,000 rolls
Apples	750,000 apples
Peaches	226,000 peaches
Strawberries	23,342 pints
Tomatoes	17,988 pounds
Asparagus	15,498 pounds
Melons	15,500 melons
Bean sprouts	2,800 pounds
Raisins	800 pounds
Lettuce	9,300 heads
Parsley	10,827 bunches

Source: http://btc.montana.edu/olympics/nutrition/questions01.html

such as coffee, tea, fruit, juice, sugar, chocolate, and other food products that may include imported ingredients. McDonald's already offers Fairtrade-certified coffee in 600 outlets in the United States.

THE COST OF FOOD

Although food in the Olympic Village is free to athletes, coaches, and team officials, visitors to the Olympic events are often at the mercy of the food vendors in the Olympic Park. At the London Olympics, visitors to the Olympic Park felt that the food was a bit expensive. Tables III and IV show the cost of various beverages and snacks at the 2012 London Olympics.

TABLE III   COST OF BEVERAGES

Beverage	Cost, US$
Beer	$11.30
Coca-Cola	$3.60
Water	$2.80
Tea	$3.13
Coffee	$4.00
Wine	$7.50

Source: Adapted from http://www.telegraph.co.uk/sport/olympics/9299883/London-2012-Olympics-small-bottle-of-beer-at-Games-will-cost-4.20-as-organisers-reveal-food-and-drink-prices.html.

TABLE IV   COST OF SNACKS

Snack	Cost, US$
Chicken Burrito	$10.15
Pasta with Chicken and Mushroom Sauce	$10.15
10-inch Pizza	$11.75
Fish & Chips	$13.30
Hotdog	$9.23
2 Small Bags of Chips	$4.70
Muffin and Cookie	$7.82

THE SUSTAINABILITY FOOD LEGACY

As stated previously, when London was bidding to host the 2012 Olympic Games, it promised that it would be the most sustainable Games ever. The goal was to establish a sustainability legacy, not just for the Olympic Games but for society in general. Part of the sustainability requirements established by the LOCOG included:

● The Food for Life targets of 75 percent unprocessed, 50 percent local, and 30 percent organic food should be set as a minimum standard for catering contracts during both the construction phase and during the Games themselves.

Material in the section "The Sustainability Food Legacy" has been adapted from www.sustainweb.org/pdf/Feeding_the_Olympics.pdf.

- Food outlets should be encouraged to use 100 percent UK vegetables and 80 percent UK seasonal fruit.
- 65 percent of the food sold should be vegetarian or vegan, with meat used sparingly in meat-based dishes; 100 percent of meat and dairy products should be organic and from the UK.
- Only fish from certified Marine Stewardship Council sources should be used.
- All tea, coffee, chocolate, fruit, and juice (where imported) should be Fairtrade certified.
- There should be minimal food packaging, with all waste reused, recycled, or composted; 100 percent composting of organic waste; 100 percent reuse or recycling of packaging.
- Free drinking-water fountains should installed throughout all Olympic sites.
- All possible avenues that would allow local, small, and medium-size enterprises to participate in catering activities during the construction phase and the Games themselves should be vigorously pursued.
- Before and at the Games, there should be visible and engaging food marketing that inspires and informs the public on the merits of healthy eating and its role in sports; an understanding of seasonal, local, and organic produce available; and the benefits of various eating habits for the local and global environment. This should include high-profile athletes promoting healthy and sustainable food.
- All catering staff should be trained in preparing fresh and healthy dishes and communicating this to their present and future customers, which will provide a sustainable catering legacy in its own right.

The future of feeding Olympians will be heavily based on a partnership established among key caterers, suppliers, and sponsors. Emphasis will be on healthy living, understanding nutrition, and focusing on sustainability of the foods we ingest.

QUESTIONS

Answer all of the following questions with regard to the project for feeding Olympians at the Olympic Village.

1. Who has the ultimate responsibility for the integration management function for feeding the Olympic Athletes?
2. With regard to scope management, how is the statement of work defined? In other words, who determines which foods are needed, the amount of each food, any special preparation instructions, and so on?

3. What trade-offs, if any, are available with regard to time management?
4. What trade-offs, if any, are available with regard to cost management?
5. How is the quality management function handled?
6. With athletes from more than 200 countries, how is communications management handled?
7. Who is responsible for procurement management for the food?
8. How is the staffing function of human resource management provided?
9. What are the major risks facing the Olympians?

Olympics (D): Managing Health Risks for Some Olympic Venues

Projects for the Olympic Games have many of the same characteristics as other types of projects; specifically, they can be significantly impacted by enterprise environmental factors. However, the enterprise environmental factors may take a different form than most of us are familiar with and may impose more risks. This case study focuses on the enterprise environmental factors that can impact the health and well-being of the Olympic athletes during pre-Olympic practices and the actual competitive events.

Health and well-being for the athletes can be looked at in several ways. Since 1970, the threat of terrorism has existed at the Olympic Games and other sporting events. The threats included attacks during the competition as well as attacks on athletes, officials, crowds of spectators, and tourists. At the Sochi Olympics, the Russian government assigned more than 40,000 security personnel as well as military personnel and equipment to safeguard the Games. At the Rio de Janeiro Olympic Games, there were 83,000 security personnel.

The IOC, local OCOGs, and national governments in the host countries are committed to the health and well-being of participants and visitors. But sometimes the health of the participating athletes can be at risk because of the environment in which they must participate rather than due to terrorism. As examples: What are the risks to the athletes due to the Zika virus? What are the risks to the athletes in the aquatic venues if the Olympic waterways were found to be contaminated with bacteria and viruses? These are the risks that will be discussed.

THE ZIKA VIRUS

The Zika virus is a mosquito-borne virus that appears to be widespread in Brazil and has been linked to birth defects. Zika has been around for decades, but only recently have issues been raised about its connection to birth malformations and neurological symptoms. There is no vaccine or treatment as yet, so combating the outbreak is heavily focused on eradicating mosquito populations and preventing mosquito bites. This was a concern not only to the officials hosting the Olympic Games in Rio de Janeiro, but to the athletes and visitors to the Games.

According to the World Health Organization, pregnant women and women who might get pregnant are at the greatest risk for contracting the virus. But anyone can get the virus, become a carrier, and bring the virus back to their homeland. Some countries considered pulling out of the Olympic Games in Rio for this reason. Ultimately no teams pulled out, but some individual athletes did withdraw out of concerns over Zika.[1]

The final decision on whether an Olympic athlete should attend was left up to the individual athletes. The United States Olympic Committee hired infectious disease specialists to assist the athletes in determining the risks and in making their own decisions about whether compete in the Games.

Rio expected upwards of 500,000 tourists, including 200,000 from North America, as well as 10,000 Olympic athletes and accompanying coaches, trainers and officials. However, if someone is infected and then return to his or her own country that has a much warmer client, the chances of the virus spreading significantly increases.[2]

The Rio 2016 Games took place in Brazil's winter months of August and September, when the dryer, cooler weather could significantly reduce the presence of the mosquitoes that carry the virus. Additionally, the Brazilian Government undertook measures to control mosquitos. There were inspections of each Olympic venue for puddles of stagnant water that could be a breeding ground for mosquitos.

Olympic officials also had to contend with the psychological aspect of athletes' fear of the Zika virus. Athletes were issued mosquito repellant as the primary defense against the mosquitoes. They were also encouraged to keep their

---

[1] Chiara Palazzo, "Rio Olympics: Which Athletes have Withdrawn over Zika Fears?" *The Telegraph,* August 4, 2016. Available at www.telegraph.co.uk/sport/0/rio-olympics-which-athletes-have-with-drawn-over-zika-fears/.

[2] This case study was prepared during and shortly after the Games. The full extent of unsafe conditions and their long-term consequences may not be known, or may not be made public, until years after the Games.

arms and legs covered as much as possible and not to leave windows or doors open in the sleeping accommodations.

GUANABANA BAY

Guanabana Bay, Rio de Janeiro, and some of the 50 accompany waterways that feed into it were the site for the rowing, sailing, wind surfing, canoeing, and marathon and triathlon venues at the Summer Olympic Games in 2016. These events were held in water that is severely contaminated with human sewage and teeming with bacteria and viruses. The waterways also have large floating debris, anything from furniture to dead animals. Some of the waterways are also breeding grounds for the mosquito-borne Zika virus.

For more than three decades, untreated water from toilets and showers, as well as whatever waste people put down their sinks, has been deposited into the rivers and streams that feed into Guanabana Bay. More than 70 percent of the waste from the 12 million residents of Rio is untreated and flows into these waters. The contamination and waste appears not only in the water but also along the shores of the beaches.

Even prior to the beginning of the games, several of the athletes who trained in Guanabana Bay fell ill with diarrhea, fevers, and vomiting as a result of exposure to the bacteria and viruses in the water. Exposure to fecal matter can lead to hepatitis A, dysentery, cholera, respiratory problems, and other diseases. Although rare, heart and brain disease are also possible. A German athlete who competed in Rio at a pre-Olympic sailing event in preparation for the Games had to be treated at a Berlin hospital for MRSA, a flesh-eating bacteria.

The organizers of the Brazilian Olympic Games had promised in their 2009 bid to host the Games that 80 percent of the pollution pumped into the bay would be reduced by the time the Games started. As part of its Olympic project, Brazil promised to build eight treatment facilities to filter out much of the sewage and prevent tons of household trash from flowing into the bay. Only one had been built by the start of the Games.

Some believe that Rio's bid to host the Games was part of a plan to force the cleanup of the bay. But now, Brazilian officials say that, at current depollution investment rates and even if additional funding were available, at least 10 more years would be required to reduce the pollution levels. The cleanup goal turned into an embarrassing failure. Federal police and prosecutors are investigating whether crimes were committed, and there is concern about where billions of dollars in funds have gone that were earmarked to improve sewage services and clean Guanabara Bay since the early 1990s.

Brazil has attempted to clean up its waters for decades but has run into problems of mismanagement and alleged corruption, which also delayed construction of Olympic facilities. The run-up to the 2016 Games was so troubled, in so many ways, that Olympic officials reportedly met in secret to suggest that the Games be moved to London, according to a report in *The Independent*.[3] Brazilian Olympic officials stated that there was no such "Plan B."

For some athletes, the risk was worth taking for the chance to win a gold medal and achieve international fame and recognition. More than 10,000 athletes competed in Rio and almost 1,400 of them came in contact with water polluted with bacteria and viruses.

As early as 2014 studies indicated that some of the bacteria found in the bay is a "super bacteria" resistant to most medications. These findings were confirmed by two subsequent studies, one in 2015 and a second just weeks before the Games. The super bacteria can cause hard-to-treat urinary, gastrointestinal, and pulmonary and bloodstream infections along with meningitis. According to the Centers for Disease Control, these bacteria contribute to death in up to half of the patients infected.[4]

There was also the risk that athletes would not immediately fall ill by coming in contact with the bacteria, but become carriers and bring the resistant bacteria back to their own countries, thus possibly infecting others. The same was true for spectators at Olympics beach venues. Super bacteria showed up in 90 percent of the water samples at Flamengo Beach and in 10 percent of the samples at Copacabana Beach. The most popular beaches for tourists are Ipanema and Leblon, which tested positive for the super bacteria 50 and 60 percent of the time, respectively. The immediate risk to an individual's health because of the super bacteria depends on the state of his or her immune system. These bacteria are opportunistic microbes that can enter the body, lie dormant, and then attack at a later date when a person's immune system is compromised for other reasons.[5]

TESTING THE WATER

Two questions must be answered concerning the testing of the water:

1. What pathogens should be tested for, bacteria or viruses, or possibly both?
2. Where in the bay should the testing take place?

Pathogens are infectious agents such as bacteria, viruses, or parasites that can cause disease in their hosts. The Rio OCOG followed the World Health Organization's "Safe Recreational Water Environments" guidelines and conducted a regular

---

[3] Adapted from Alfie Crow, "Brazil Is Struggling to Get Ready for the 2016 Summer Olympics," May 19, 2014. http://www.sbnation.com/2014/5/19/5732490/summer-olympics-2016-brazil-issues-water-london

[4] "Studies Find 'Super Bacteria' in Rio's Olympic Venues, Top Beaches," http://onlineathens.com/health/2016-06-11/exclusive-studies-find-super-bacteria-rio%27s-olympic-venues-top-beaches

[5] Ibid.

testing program of microbial water quality. Although the test results showed compliance with the guidelines, the argument was that testing focused on the levels of bacteria rather than viruses, which pose the greater health risks. Water testing can show safe levels of fecal bacteria and at the same time show dangerous viruses.

An investigation conducted by the Associated Press (AP) found that the water in several locations was infested with disease-causing viruses that were as much as 1.7 million times the levels that would be considered hazardous at beaches in Southern California and cause them to be closed to the public.[6] Another AP analysis of a decade's worth of government data on Guanabara Bay and other waterways showed that sewage pollution indicators consistently spiked far above acceptable limits, even under Brazilian laws, which are far more lenient on pollution than laws in the United States or Europe.[7] With pollution levels this high, exposure is almost unavoidable and the chance of infection is high.

Since the majority of the waste enters the waterways close to the shore, it was believed that as you get farther offshore into the deep waters of the bay, the risks from the pathogens would dissipate. However, the number of viruses found over a kilometer from the shore in Guanabana Bay, at locations where sailors would compete at high speeds and get utterly drenched, were equal to those found along shorelines closer to the sewage sources. Some testing showed pathogen levels that were over 16 times the amount permitted under Brazilian law.[8] These results indicated that it would be safer if the competition took place deeper in the lagoon or bay. "Kristina Mena, a U.S. expert in risk assessment for waterborne viruses, examined the AP data and estimated that international athletes at all water venues had a 99 percent chance of infection if they ingested just three teaspoons of water—although whether a person would fall ill depended on immunity and other factors."[9]

The International Sailing Federation reported that just over 7 percent of the sailors who competed in a mid-August Olympic warm-up event in Guanabana Bay fell ill, but the federation had not conducted a full count of how many athletes got sick in the two weeks following the practice sessions, the approximate incubation period for many of the pathogens in the water.[10] Some athletes who competed

---

[6] Kathryn Blackhurst, "Rio Olympic Water is 'Basically Raw Sewage,' Finds AP Investigation," July 30, 2015. www.newsmax.com/TheWire/rio-olympics-water-raw/2015/07/30/id/659685/.

[7] Jenny Barchfield, "Brazil Won't Clean Up Water Pollution in Guanabara Bay by 2016 Olympics, Officials Say", May 18, 2014. http://www.huffingtonpost.com/2014/05/18/brazil-water-pollution-guanabara-bay-2016-olympics_n_5347766.html.

[8] Brad Brooks, "Sewage-Infested Waters In Rio Place Olympic Athletes at Risk," December 2, 2015. http://www.huffingtonpost.com/entry/tainted-water-rio-olympics_565eaddfe4b08e945fed65f7.

[9] "Water in Brazil Olympic Venues Dangerously Contaminated", June 30, 2015. www.cnbc.com/2015/07/30/.

[10] Vinod Sreeharsha, "Olympic Sailing in Rio Still Planned for Polluted Guanabara Bay," *New York Times*, October 28, 2015. www.nytimes.com/2015/10/29/sports/olympics/olympic-sailing-in-rio-still-planned-for-polluted-guanabara-bay.html.

in aquatic events took precautions by wearing special clothing or even masks. Austrian sailor David Hussl said he and his teammates took such precautions, washing their faces immediately with bottled water when they were splashed by waves and showering the minute they return to shore. And yet Hussl said he had fallen ill several times.[11]

CONCLUSIONS

Many of the athletes felt that it was unfair that they had worked so hard for years and made sacrifices for their country only to have their destiny based on variables that might change the Games' outcome. Despite the risks, most athletes were still expected to compete in this environment for the chance of winning Olympic gold. They accepted the risks, took precautions where appropriate, and resigned themselves to competing in potentially dangerous water. It may be years after the Games are over before we know the true health issues caused by exposure to the contaminated waterways.

QUESTIONS

1. If you are the project manager for organizing these events, can the health risks be quantified?
2. What could a project manager do to possibly mitigate the health risks?
3. What are the chances that the Rio OCOG might recommend that these events be removed from the Games because of the health risks?
4. Would the International Olympic Committee remove the events from the Rio Olympics?
5. If you had been an Olympic athlete, and this were your only chance to compete, would you have accepted the health risks?
6. What could the OCOG have done to help mitigate the risks?
7. What can athletes do to mitigate the risks?
8. From a project management perspective, was holding these events despite these health risks a moral or ethical violation of the PMI Code of Conduct and Professional Responsibility?

Many of these questions are hypothetical and should be used as the basis for classroom discussion. Without knowing the final outcome of the health risks that appeared in the literature prior to the start of the Games, we can only speculate what might have happened. There can be more than one correct answer to some of these questions based upon the assumptions you have made.

---

[11] "Filthy Rio de Janeiro Water a Threat at 2016 Olympics," *New York Times,* July 30, 2016. www .nytimes.com/2015/07/31/sports/olympics/filthy-rio-de-janeiro-water-a-threat-at-2016-olympics.html.

Part 20

INDUSTRY SPECIFIC: THE COMMERCIAL AIRCRAFT INDUSTRY

The cost to design and develop a long-distance commercial aircraft is measured in the billions of dollars. The life expectancy of the planes is measured in decades, and a project manager could spend an entire career on just one aircraft.

In 1970, Boeing launched the 747 aircraft with the expectation of selling 400 planes. It has since sold 1,523 planes and has orders for 51 more. The Boeing 747 had a monopoly on the long-distance market for almost 40 years. But with 4 billion people expected to travel by air by 2030, the competition between Boeing and Airbus has increased. The three most recent aircraft in the long- and medium-distance marketplace are shown in Table I.

TABLE I   THREE AIRCRAFT SERVING THE LONG- AND MEDIUM-DISTANCE MARKET

Plane	Orders	Delivered
Boeing 777	1,893	1,424
Boeing 787	1,161	445
Airbus A380	319	193

Aircraft manufacturers have been pressured to design new planes with more fuel-efficient engines, more advanced technologies, and customer expectations of more amenities. But as expected, with any new technologies come risks, especially safety risks. Safety risks cannot always be identified through simulation and ground testing. Only through the actual use of the commercial aircraft can safety issues be addressed and resolved. For the project managers, safety could very well be the most important constraint on their projects.

Philip Condit and the Boeing 777: From Design and Development to Production and Sales

Following his promotion to Boeing chief executive officer (CEO) in 1988, Frank Shrontz looked for ways to stretch and upgrade the Boeing 767—an eight-year-old wide-body twin jet—in order to meet Airbus competition. Airbus had just launched two new 300-seat wide-body models, the two-engine A330 and the four-engine A340. Boeing had no 300-seat jetliner in service, nor did the company plan to develop such a jet.

To find out whether Boeing's customers were interested in a double-decker 767, Philip Condit, Boeing executive vice president and future CEO (1996), met with United Airlines vice president Jim Guyette. Guyette rejected the idea outright, claiming that an upgraded 767 was no match to Airbus's new model transports. Instead, Guyette urged Boeing to develop a brand-new commercial jet, the most advanced airplane of its generation.[1] Shrontz had heard similar suggestions from other airline carriers. He reconsidered Boeing's options and decided to abandon the 767 idea in favor of a new aircraft program. In December 1989, he announced the 777 project and put Philip Condit in charge of its

---

[1] Eugene Rodgers, *Flying High: The Story of Boeing* (New York: Atlantic Monthly Press, 1996), pp. 415–416; Michael Dornheim, "777 Twinjet Will Grow to Replace 747–200," *Aviation Week & Space Technology*, June 3, 1991, p. 41.

This case was presented by Isaac Cohen, San Jose State University, at the 2000 North American Case Research Association (NACRA) workshop. Reprinted by permission from the *Case Research Journal*. Copyright 2000 by Isaac Cohen and the North American Case Research Association.

management. Boeing launched the 777 in 1990 and delivered the first jet in 1995. By February 2001, 325 B-777s were flying in major international and U.S. airlines.[2]

Condit faced a significant challenge in managing the 777 project. He wanted to create an airplane that was preferred by the airlines at a price that was truly competitive. He sought to attract airline customers as well as cut production costs, and he did so by introducing several innovations—both technological and managerial—in aircraft design, manufacturing, and assembly. He looked for ways to revitalize Boeing's outmoded engineering production system, and update Boeing's manufacturing strategies. And to achieve these goals, Condit made continual efforts to spread the 777 program-innovations companywide.

Looking back at the 777 program, this case focuses on Condit's efforts. Was the 777 project successful, and was it cost effective? Would the development of the 777 allow Boeing to diffuse the innovations in airplane design and production beyond the 777 program? Would the development of the 777's permit Boeing to revamp and modernize its aircraft manufacturing system? Would the making and selling of the 777 enhance Boeing's competitive position relative to Airbus, its only remaining rival?

THE AIRCRAFT INDUSTRY

Commercial aircraft manufacturing was an industry of enormous risks where failure was the norm, not the exception. The number of large commercial jet makers had been reduced from four in the early 1980s—Boeing, McDonnell Douglas, Airbus, and Lockheed—to two in late 1990s, turning the industry into a duopoly, and pitting the two survivors—Boeing and Airbus—one against the other.

One reason why aircraft manufacturers failed so often was the huge cost of product development. Developing a new jetliner required an up-front investment of up to $15 billion (2001 dollars), a lead time of five to six years from launch to first delivery, and the ability to sustain a negative cash flow throughout the development phase. Typically, to break even on an entirely new jetliner, aircraft manufacturers needed to sell a minimum of 300 to 400 planes and at least 50 planes per year.[3] Only a few commercial airplane programs had ever made money.

The price of an aircraft reflected its high development costs. New model prices were based on the average cost of producing 300 to 400 planes, not a single plane. Aircraft pricing embodied the principle of learning by doing, the so-called

---

[2] "Commercial Airplanes: Order and Delivery, Summary," http/www.boeing.com/commercial/orders/index.html. Retrieved on February 2, 2000.

[3] P. Donlon, "Boeing's Big Bet" (an interview with CEO Frank Shrontz), *Chief Executive* (November/December 1994): 42; Michael Dertouzos, Richard Lester, and Robert Solow, *Made in America: Regaining the Productive Edge* (New York: Harper Perennial, 1990), p. 203.

learning curve:[4] Workers steadily improved their skills during the assembly process, and as a result, labor cost fell as the number of planes produced rose.

The high and increasing cost of product development prompted aircraft manufacturers to utilize subcontracting as a risk-sharing strategy. For the 747, the 767, and the 777, the Boeing Company required subcontractors to share a substantial part of the airplane's development costs. Airbus did the same with its own latest models. Risk sharing subcontractors performed detailed design work and assembled major subsections of the new plane while airframe integrators (i.e., aircraft manufacturers) designed the aircraft, integrated its systems and equipment, assembled the entire plane, marketed it, and provided customer support for 20 to 30 years. Both the airframe integrators and their subcontractors were supplied by thousands of domestic and foreign aircraft components manufacturers.[5]

Neither Boeing nor Airbus, nor any other post–World War II commercial aircraft manufacturer, produced jet engines. A risky and costly venture, engine building had become a highly specialized business. Aircraft manufacturers worked closely with engine makers—General Electric, Pratt and Whitney, and Rolls-Royce—to set engine performance standards. In most cases, new airplanes were offered with a choice of engines. Over time, the technology of engine building had become so complex and demanding that it took longer to develop an engine than an aircraft. During the life of a jetliner, the price of the engines and their replacement parts was equal to the entire price of the airplane.[6]

A new model aircraft was normally designed around an engine, not the other way around. As engine performance improved, airframes were redesigned to exploit the engine's new capabilities. The most practical way to do so was to stretch the fuselage and add more seats in the cabin. Aircraft manufacturers deliberately designed flexibility into the airplane so that future engine improvements could facilitate later stretching. Hence the importance of the "family concept" in aircraft design, and hence the reason why aircraft manufacturers introduced families of planes made up of derivative jetliners built around a basic model, not single, standardized models.[7]

The commercial aircraft industry gained from technological innovations in two other industries. More than any other manufacturing industry, aircraft construction benefited from advances in material applications and electronics. The

---

[4] John Newhouse, *The Sporty Game* (New York: Knopf, 1982), p. 21, but see also pp. 10–20.

[5] David C. Mowery and Nathan Rosenberg, "The Commercial Aircraft Industry," in Richard R. Nelson, ed., *Government and Technological Progress: A Cross Industry Analysis* (New York: Pergamon Press, 1982), p. 116; Dertouzos et al., *Made in America*, p. 200.

[6] Dertouzos et al., *Made in America*, p. 200.

[7] Newhouse, *Sporty Game*, p. 188; Mowery and Rosenberg, "The Commercial Aircraft Industry," pp. 124–125.

development of metallic and nonmetallic composite materials played a key role in improving airframe and engine performance. On the one hand, composite materials that combined light weight and great strength were utilized by aircraft manufacturers; on the other, heat-resisting alloys that could tolerate temperatures of up to 3,000 degrees were used by engine makers. Similarly, advances in electronics revolutionized avionics. The increasing use of semiconductors by aircraft manufacturers facilitated the miniaturization of cockpit instruments, and more important, it enhanced the use of computers for aircraft communication, navigation, instrumentation, and testing.[8] The use of computers contributed, in addition, to the design, manufacture, and assembly of new model aircraft.

THE BOEING COMPANY

The history of the Boeing company can be divided into two distinct periods: the piston era and the jet age. In the piston era from the 1920s to the 1940s, Boeing was essentially a military contractor producing fighter aircraft and bombers. By the jet age, which began in the 1950s, Boeing had become the world's largest manufacturer of commercial aircraft, deriving most of its revenues from selling jetliners.

Boeing's first jet was the 707. The introduction of the 707 in 1958 represented a major breakthrough in the history of commercial aviation; it allowed Boeing to gain a critical technological lead over the Douglas Aircraft Company, its closer competitor. To benefit from government assistance in developing the 707, Boeing produced the first jet in two versions: a military tanker for the Air Force (k-135) and a commercial aircraft for the airlines (707–120). The company, however, did not recoup its own investment until 1964, six years after it delivered the first 707 and 12 years after it had launched the program. In the end, the 707 was quite profitable, selling 25 percent above its average cost.[9] Boeing retained the essential design of the 707 for all its subsequent narrow-body single-aisle models (the 727, 737, and 757), introducing incremental design improvements one at a time.[10] One reason why Boeing used shared design for future models was the constant pressure it faced to move down the learning curve and reduce overall development costs.

Boeing introduced the 747 in 1970. Its represented another breakthrough; the 747 wide-body design was one of a kind; it had no real competition anywhere in the industry. Boeing bet the entire company on the success of the 747, spending on the project almost as much as the company's total net worth in 1965, the year the

---

[8] Mowery and Rosenberg, "The Commercial Aircraft Industry," pp. 102–103, 126–128.

[9] John B. Rae, *Climb to Greatness: The American Aircraft Industry, 1920–1960* (Cambridge, Mass.: MIT Press, 1968), pp. 206–207; Rodgers, *Flying High*, pp. 197–198.

[10] Frank Spadaro, "A Transatlantic Perspective," *Design Quarterly* (Winter 1992): 23.

project started.[11] In the short run, the outcome was disastrous. As Boeing began delivering its 747s, it struggled to avoid bankruptcy. Cutbacks in orders as a result of a deep recession, coupled with production inefficiencies and escalating costs, created a severe cash shortage that pushed the company to the brink of disaster. As sales dropped, the 747's break-even point moved farther and farther into the future.

Yet, in the long run, the 747 program was a triumph. The Jumbo Jet had become Boeing's most profitable aircraft and the industry's most efficient jetliner. The plane helped Boeing solidify its position as the industry leader for years to come, leaving McDonnell Douglas far behind and forcing the Lockheed Corporation to exit the market. The new plane, furthermore, contributed to Boeing's manufacturing strategy in two ways. First, as Boeing increased its reliance on outsourcing, six major subcontractors fabricated 70 percent of the value of the 747 airplane,[12] thereby helping Boeing reduce the project's risks. Second, for the first time, Boeing applied the family concept in aircraft design to a wide-body jet, building the 747 with wings large enough to support a stretched fuselage with bigger engines and offering a variety of other modifications in the 747's basic design. The 747–400 (1989) is a case in point. In 1997, Boeing sold the stretched and upgraded 747–400 in three versions: a standard jet, a freighter, and a "combi" (a jetliner whose main cabin was divided between passenger and cargo compartments).[13]

Boeing developed other successful models. In 1969, Boeing introduced the 737, the company's narrow-body flagship, and in 1982, it put into service two additional jetliners, the 757 (narrow-body) and the 767 (wide-body). By the early 1990s, the 737, 757, and 767 were all selling at a profit. Following the introduction of the 777 in 1995, Boeing's families of planes included the 737 for short-range travel, the 757 and 767 for medium-range travel, and the 747 and 777 for medium- to long-range travel. (See Table II.)

In addition to building jetliners, Boeing also expanded its defense, space, and information businesses. In 1997, the Boeing Company took a strategic gamble, buying the McDonnell Douglas Company in a $14 billion stock deal. As a result of the merger, Boeing became the world's largest manufacturer of military aircraft, NASA'S largest supplier, and the Pentagon's second largest contractor (after Lockheed). Nevertheless, despite the growth in its defense and space businesses, Boeing still derived most of its revenues from selling jetliners. Commercial aircraft revenues accounted for 59 percent of Boeing's $49 billion sales in 1997 and 63 percent of Boeing's $56 billion sales in 1998.[14]

---

[11] Rodgers, *Flying High*, 279; Newhouse, *Sporty Game*, chap. 7.

[12] M. S. Hochmuth, "Aerospace," in Raymond Vernon, ed., *Big Business and the State* (Cambridge, Mass.: Harvard University Press, 1974), p. 149.

[13] Boeing Commercial Airplane Group, Announced Orders and Deliveries as of 12/31/97, Section A 1.

[14] The Boeing Company 1998 Annual Report, p. 76.

TABLE II   TOTAL NUMBER OF COMMERCIAL JETLINERS DELIVERED BY THE
BOEING COMPANY, 1958 TO 2001

Model	No. Delivered	First Delivery
B-707	1,010 (retired)	1958
B-727	1,831 (retired)	1963
B-737	3,901	1967
B-747	1,264	1970
B-757	953	1982
B-767	825	1982
B-777	325	1995
B-717	49	2000
Total:	**10,158**	

Note: McDonnell Douglas commercial jetliners (the MD-11, MD-80, and MD-90) are excluded.

Sources: Boeing Commercial Airplane Group, Announced Orders and Deliveries as of 12/31/97; The Boeing Company 1998 Annual Report, p. 35; "Commercial Airplanes: Order and Delivery Summary." http://www.Boeing com/commercial/orders/index.html. Retrieved March 20, 2001.

Following its merger with McDonnell, Boeing had one remaining rival: Airbus Industrie.[15] In 1997, Airbus booked 45 percent of the worldwide orders for commercial jetliners[16] and delivered close to one-third of the worldwide industry output. In 2000, Airbus shipped nearly two-fifths of the worldwide industry output. (See Table III.) Airbus's success was based on a strategy that combined cost leadership with technological leadership. First, Airbus distinguished itself from Boeing by incorporating the most advanced technologies into its planes. Second, Airbus managed to cut costs by utilizing a flexible, lean production manufacturing system that stood in a stark contrast to Boeing's mass production system.[17]

As Airbus prospered, the Boeing Company was struggling with rising costs, declining productivity, delays in deliveries, and production inefficiencies. Boeing Commercial Aircraft Group lost $1.8 billion in 1997 and barely generated any profits in 1998.[18] All through the 1990s, the Boeing Company looked for

[15] Formed in 1970 by several European aerospace firms, the Airbus Consortium had received generous assistance from the French, British, German, and Spanish governments for over two decades. In 1992, Airbus signed an agreement with Boeing that limited the amount of government funds each aircraft manufacturer could receive, and in 1995, at long last, Airbus became profitable. "Airbus 25 Years Old," *Le Figaro*, October 1997 (reprinted in English by Airbus Industrie); Rodgers, *Flying High*, chap. 12; Andy Reinhardt, Seanna Browder, and Ron Stodghill II, "Three Huge Hours in Seattle: How Boeing and McDonnell Cut the Biggest Deal in Aviation History," *Business Week*, December 30, 1996, p. 40.

[16] Charles Goldsmith, "Re-engineering, After Trailing Boeing for Years, Airbus Aims for 50% of the Market," *Wall Street Journal*, March 16, 1998.

[17] "Hubris at Airbus, Boeing Rebuild," *Economist* 28 (November 1998).

[18] The Boeing Company 1997 Annual Report, 19; The Boeing Company 1998 Annual Report, 51

TABLE III   MARKET SHARE OF SHIPMENTS OF COMMERCIAL AIRCRAFT, BOEING, MCDONNELL DOUGLAS (MD), AIRBUS, 1992 TO 2000

	1992	1993	1994	1995	1996	1997	1998	1999	2000
Boeing	61%	61%	63%	54%	55%	67%	71%	68%	61%
MD	17	14	9	13	13				
Airbus	22	25	28	33	32	33	29	32	39

Sources: Aerospace Facts and Figures, 1997–98, p. 34; *Wall Street Journal*, December 3, 1998, and January 12, 1999; The Boeing Company 1997 Annual Report, p. 19; data supplied by Mark Luginbill, Airbus Communication Director, November 16, 1998, February 1, 2000, and March 20, 2001).

ways to revitalize its outdated production manufacturing system and to introduce leading-edge technologies into its jetliners. The development and production of the 777, first conceived of in 1989, was an early step undertaken by Boeing managers to address both problems.

THE 777 PROGRAM

The 777 program was Boeing's single largest project since the completion of the 747. The total development cost of the 777 was estimated at $6.3 billion and the total number of employees assigned to the project peaked at nearly 10,000. The 777's twin engines were the largest and most powerful ever built (the diameter of the 777's engine equaled the 737's fuselage), the 777's construction required 132,000 uniquely engineered parts (compared to 70,000 for the 767), the 777's seat capacity was identical to that of the first 747 that had gone into service in 1970, and its manufacturer empty weight was 57 percent greater than the 767's. Building the 777 alongside the 747 and 767 at its Everett plant near Seattle, Washington, Boeing enlarged the plant to cover an area of 76 football fields.[19]

Boeing's financial position in 1990 was unusually strong. With a 21 percent rate of return on stockholder equity, a long-term debt of just 15 percent of capitalization, and a cash surplus of $3.6 billion, Boeing could gamble comfortably.[20] There was no need to bet the company on the new project, as had been the case with the 747, or to borrow heavily, as had been the case with the 767. Still, the decision

[19] Donlon, "Boeing's Big Bet," 40; John Mintz, "Betting It All on 777," *Washington Post*, March 26, 1995; James Woolsey, "777: A Program of New Concepts," *Air Transport World* (April 1991): 62; Jeremy Main, "Corporate Performance: Betting on the 21st Century Jet," *Fortune*, April 20, 1992, p. 104; James Woolsey, "Crossing New Transport Frontiers," *Air Transport World* (March 1991): 21; James Woolsey, "777: Boeing's New Large Twinjet," *Air Transport World* (April 1994): 23; Michael Dornheim, "Computerized Design System Allows Boeing to Skip Building 777 Mockup," *Aviation Week and Space Technology*, June 3, 1991, p. 51; Richard O'Lone, "Final Assembly of 777 Nears," *Aviation Week and Space Technology*, October 2, 1992, p. 48.

[20] Rodgers, *Flying High*, 42.

to develop the 777 was definitely risky; a failure of the new jet might have triggered an irreversible decline of the Boeing Company and threatened its future survival.

The decision to develop the 777 was based on market assessment—the estimated future needs of the airlines. Boeing market analysts forecast a 100 percent increase in the number of passenger miles traveled worldwide and a need for about 9,000 new commercial jets during the 14-year period 1991 to 2005. Of the total value of the jetliners needed in that time, Boeing analysts forecast a $260 billion market for wide-body jets smaller than the 747. An increasing number of these wide-body jets were expected to be larger than the 767.[21]

A CONSUMER-DRIVEN PRODUCT

To manage the risk of developing a new jetliner, aircraft manufacturers generally first sought to obtain a minimum number of firm orders from interested carriers and only then commit to the project. Boeing CEO Frank Shrontz had expected to obtain 100 initial orders for the 777 before asking the Boeing board to launch the project, but as a result of Boeing's financial strength and the increasing competitiveness of Airbus, Schrontz decided to seek the board's approval earlier, after securing only one customer: United Airlines. On October 12, 1990, United placed an order for thirty-four 777s and an option for an additional 34 aircraft. Two weeks later, Boeing's board of directors approved the project.[22] Negotiating the sale, Boeing and United drafted a handwritten agreement (signed by Philip Condit and Richard Albrecht, Boeing's executive vice presidents, and Jim Guyette, United's executive vice president) that granted United a larger role in designing the 777 than the role any airline had played before. The two companies pledged to cooperate closely in developing an aircraft with the "best dispatch reliability in the industry" and the "greatest customer appeal in the industry." The agreement stated: "We will endeavor to do it right the first time with the highest degree of professionalism" and with "candor, honesty, and respect." Asked to comment on the agreement, Philip Condit said: "We are going to listen to our customers and understand what they want. Everybody on the program has that attitude."[23] Gordon McKinzie, United's 777 program director, agreed: "In the past we'd get brochures on a new airplane and its options . . . wait four years for delivery, and hope we'd get what we ordered. This time Boeing really listened to us."[24]

---

[21] Woolsey, "Crossing New Transport Frontiers," 20; Main, "Corporate Performance," pp. 102–103.

[22] Rodgers, *Flying High*, 416, 420–424.

[23] Richard O'Lone and James McKenna, "Quality Assurance Role Was Factor in United's 777 Launch Order," *Aviation Week and Space Technology*, October 29, 1990, pp. 28–29; Woolsey, "Crossing New Transport Frontiers," 20.

[24] John Mintz, "Betting it All on 777," *Washington Post*, March 25, 1995 www.washingtonpost.com/archive/business/1995/03/26/betting-it-all-on-777/24d26681-bc52-4604-b8d6-520d24be9838/?utm_term=.9bd252589565.

Condit invited other airline carriers to participate in the design and development phase of the 777. Altogether, eight carriers from around the world (United, Delta, American, British Airways, Qantas, Japan Airlines, All Nippon Airways, and Japan Air System) sent full-time representatives to Seattle; British Airways alone assigned 75 people at one time. To facilitate interaction between its design engineers and representatives of the eight carriers, Boeing introduced an initiative called "Working Together." "If we have a problem," a British Airways production manager explained, "we go to the source—design engineers on the IPT [integrated product teams], not service engineer(s). One of the frustrations on the 747 was that we rarely got to talk to the engineers who were doing the work."[25]

"We have definitely influenced the design of the aircraft," a United 777 manager said, mentioning changes in the design of the wing panels that made it easier for airline mechanics to access the slats (slats, like flaps, increase lift on takeoffs and landings), and new cabin features that made the plane more attractive to passengers.[26] Of the 1,500 design features examined by representatives of the airlines, Boeing engineers modified 300. (See Figure I.) Among changes made by Boeing was a redesigned overhead bin that left more stand-up headroom for passengers (allowing a passenger six-foot-three inches tall to walk from aisle to aisle), "flattened" side walls that provided the window seat occupants more room, overhead bin doors that opened down and made it possible for shorter passengers to lift baggage into the compartment, redesigned reading lamps that enabled flight attendants to replace light bulbs (a task formerly performed by mechanics), and a computerized flight deck management system that adjusted cabin temperature, controlled the volume of the public address system, and monitored food and drink inventories.[27]

More important were changes in the interior configuration (layout plan) of the aircraft. To be able to reconfigure the plane quickly for different markets of varying travel ranges and passenger loads, Boeing's customers sought a flexible interior plan. On a standard commercial jet, kitchen galleys, closets, lavatories, and bars were all removable in the past but were limited to fixed positions when the interior floor structure was reinforced to accommodate the "wet" load. On the 777, by contrast, such components as galleys and lavatories could be positioned anywhere within several "flexible zones" designed into the cabin by the joint efforts of Boeing engineers and the airline representatives. Similarly, the flexible design of the 777's seat tracks made it possible for carriers to increase the number of seat combinations and reconfigure seating arrangements quickly. Flexible

---

[25] Quoted in Bill Sweetman, "As Smooth as Silk: 777 Customers Applaud the Aircraft's First 12 Months in Service," *Air Transport World* (August 1996): 71, but see also Woolsey, "777, Boeing's New Large Twinjet," 24, 27.

[26] Quoted in Main, 20,): New Large Twinje, 112.

[27] Rodgers, *Flying High*, 426; Spadaro, "A Transatlantic Perspective," 22; Polly Lane, "Boeing Used 777 to Make Production Changes," *Seattle Times*, May 7, 1995.

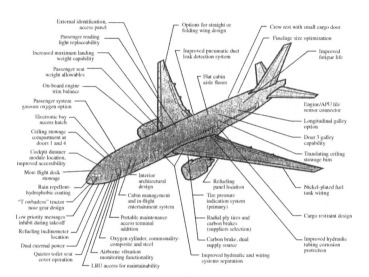

FIGURE I The 777: Selected design features proposed by Boeing airline customers and adapted by the Boeing Company

Source: The Boeing Company

configuration resulted, in turn, in significant cost savings; airlines no longer needed to take the aircraft out of service for an extended period to reconfigure the interior.[28]

The airline carriers also influenced the way in which Boeing designed the 777 cockpit. During the program definition phase, representatives of United Airlines, British Airways, and Qantas—three of Boeing's clients whose fleets included a large number of 747–400s—asked Boeing engineers to model the 777 cockpit on the plane. In response, Boeing introduced a shared 747/777 cockpit design that enabled its customers to use a single pool of pilots for both aircraft types at a significant cost savings.[29]

Additionally, the airline carriers urged Boeing to increase its use of avionics for in-flight entertainment. As a result, the 777 was equipped with a fully computerized cabin. Facing each seat on the 777 and placed on the back of the seat in front was a combined computer and video monitor that featured movies, video

---

[28] Spadaro, "A Transatlantic Perspective," 22; The Boeing Company, *Backgrounder: Pace Setting Design Value-Added Features Boost Boeing 777 Family*, May 15, 1998.

[29] Boeing, *Backgrounder: Pace Setting Design Value-Added Features Boost Boeing 777 Family*; Karl Sabbagh, *Twenty-First-Century Jet: The Making and Marketing of the Boeing 777* (New York: Scribner, 1996), p. 49.

programs, and interactive computer games. Passengers were also provided with a digital sound system comparable to the most advanced home stereo available, and a telephone. About 40 percent of the 777's total computer capacity was reserved for passengers in the cabin.[30]

The 777 was Boeing's first fly-by-wire (FBW) aircraft, an aircraft controlled by a pilot transmitting commands to the movable surfaces (rudder, flaps, etc.) electrically, not mechanically. Boeing installed a state-of-the-art FBW system on the 777 partly to satisfy its airline customers and partly to challenge Airbus leadership in flight control technology, a position Airbus had held since it introduced the world's first FBW aircraft, the A-320, in 1988.

Last, Boeing customers were invited to contribute to the design of the 777's engine. Both United Airlines and All Nippon Airlines assigned service engineers to work with representatives of Pratt & Whitney on problems associated with engine maintenance. Pratt & Whitney held three specially scheduled "airline conferences." At each conference, some 40 airline representatives clustered around a full-scale mock-up of the 777 engine and showed engineers gaps in the design, hard-to-reach points, visible but inaccessible parts, and accessible but invisible components. At the initial conference, Pratt & Whitney picked up 150 airline suggestions; at the second, 50; and at the third, 10 more suggestions.[31]

A GLOBALLY MANUFACTURED PRODUCT

Boeing contracted 12 international companies located in 10 countries and 18 U.S. companies located in 12 states to help manufacture the 777. Together, they supplied structural components as well as systems and equipment. Among the foreign suppliers were companies based in Japan, Britain, Australia, Italy, Korea, Brazil, Singapore, and Ireland; among the major U.S. subcontractors were the Grumman Corporation, Rockwell (later merged with Boeing), Honeywell, United Technologies, Bendix, and the Sunstrand Corporation. (See Table IV and Figure II.) Of all foreign participants, the Japanese played the largest role. A consortium made up of Fuji Heavy Industries, Kawasaki Heavy Industries, and Mitsubishi Heavy Industries had worked with Boeing on its wide-body models since the early days of the 747. Together, the three Japanese subcontractors produced 20 percent of the value of the 777's airframe (up from 15 percent of the 767s). A group of 250 Japanese engineers had spent a year in Seattle working on the 777 alongside Boeing engineers before most group members went back home to begin production. The fuselage was built in sections in Japan and then shipped to Boeing's huge plant at Everett, Washington, for assembly.[32]

---

[30] Sabbagh, *Twenty-First-Century Jet*, 264, 266.

[31] Ibid., pp. 131–132.

[32] Woolsey, "777: Boeing's New Large Twinjet," 23; Main, 20, 1992 New Large of t", p. 116.

TABLE IV   777 SUPPLIER CONTRACTS

U.S. Suppliers of Structural Components

Astech/MCI	Santa Ana, CA	Primary exhaust cowl assembly (plug and nozzle)
Grumman Aerospace	Bethpage, NY	Spoilers, inboard flaps
Kaman	Bloomfield, CT	Fixed training edge
Rockwell	Tulsa, OK	Floor beams, wing leading edge slats

International Suppliers of Structural Components

AeroSpace Technologies of Australia	Australia	Rudder
Alenia	Italy	Wing outboard flaps, radome
Embrace-Empresa Brasiera de Aeronautica	Brazil	Dorsal fin, wingtip assembly
Hawker de Havilland	Australia	Elevators
Korean Air	Korea	Flap support fairings, wingtip assembly
Menasco Aerospace/ Messier-Bugatti	Canada/France	Main and nose landing gears
Mitsubishi Heavy Industries, Kawasaki Heavy Industries, and Fuji Heavy Industries[a]	Japan	Fuselage panels and doors, wing center section wing-to-body fairing, and wing in-spar ribs
Short Brothers	Ireland	Nose landing gear doors
Singapore Aerospace Manufacturing	Singapore	Nose landing gear doors

U.S. Suppliers of Systems and Equipment

AlliedSignal Aerospace Company, AiResearch Divisions	Torrance, CA	Cabin pressure control system, air supply control system, integrated system controller, ram air turbine
Bendix Wheels and Garrett Divisions	South Bend, IN Phoenix/Tempe, AZ	Wheel and brakes Auxiliary power unit, air-driven unit
BFGoodrich	Troy, OH	Wheel and brakes
Dowly Aerospace	Los Angeles, CA	Thrust reverser actuator system
Eldec	Lynnwood, WA	Power supply electronics
E-Systems, Montek Division	Salt Lake City, UT	Stabilizer trim control module, secondary hydraulic brake, optional folding wingtip system
Honeywell	Phoenix, AZ Coon Rapid, MN	Airplane information management system, air data/inertial reference system

Rockwell, Collins Division	Cedar Rapids, IA	Autopilot flight director system, electronic library system displays
Sundstrand Corporation	Rockford, IL	Primary and backup electrical power systems
Teijin Seiki America	Redmond, WA	Power control units, actuator control electronics
United Technologies, Hamilton Standard Division	Windsor Lock, CT	Cabin air-conditioning and temperature control systems, ice protection system

International Suppliers of Systems and Equipment

| General Electric Company (GEC) Avionics | United Kingdom | Primary flight computers |
| Smiths Industries | United Kingdom | Integrated electrical management system, throttle control system actuator, fuel quantity indicating system |

*a*Program partners

Source: James Woolsey, "777, Boeing's New Large Twinjet," *Air Transport World* (April 1994): 24.

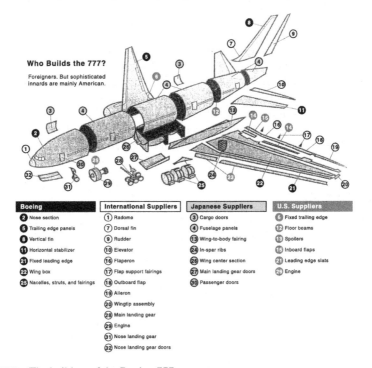

Who Builds the 777?

Foreigners. But sophisticated innards are mainly American.

Boeing	International Suppliers	Japanese Suppliers	U.S. Suppliers
② Nose section	① Radome	③ Cargo doors	⑥ Fixed trailing edge
⑤ Trailing edge panels	⑦ Dorsal fin	④ Fuselage panels	⑫ Floor beams
⑧ Vertical fin	⑨ Rudder	⑬ Wing-to-body fairing	⑮ Spoilers
⑪ Horizontal stabilizer	⑩ Elevator	㉔ In-spar ribs	⑯ Inboard flaps
㉑ Fixed leading edge	⑯ Flaperon	㉖ Wing center section	㉑ Leading edge slats
㉒ Wing box	⑰ Flap support fairings	㉗ Main landing gear doors	㉙ Engine
㉕ Nacelles, struts, and fairings	⑱ Outboard flap	㉚ Passenger doors	
	⑲ Aileron		
	⑳ Wingtip assembly		
	㉘ Main landing gear		
	㉙ Engine		
	㉛ Nose landing gear		
	㉜ Nose landing gear doors		

FIGURE II The builders of the Boeing 777

Source: Jeremy Main, "Corporate Performance: Betting on the 21st Century Jet," *Fortune*, April 20, 1992, p. 104.

Boeing used global subcontracting as a marketing tool as well. Sharing design work and production with overseas firms, Boeing required overseas carriers to buy the new aircraft. Again, Japan is a case in point. In return for the contract signed with the Mitsubishi, Fuji, and Kawasaki consortium—which was heavily subsidized by the Japanese government—Boeing sold forty-six 777 jetliners to three Japanese air carriers: All Nippon Airways, Japan Airlines, and Japan Air System.[33]

A FAMILY OF PLANES

From the outset, the design of the 777 was flexible enough to accommodate derivative jetliners. Because all derivatives of a given model shared maintenance, training, and operating procedures as well as replacement parts and components, and because such derivatives enabled carriers to serve different markets at lower costs, Boeing's clients were seeking a family of planes built around a basic model, not a single 777. Condit and his management team, accordingly, urged Boeing's engineers to incorporate the maximum flexibility into the design of the 777.

The 777's design flexibility helped Boeing manage the project's risks. Offering a family of planes based on a single design to accommodate future changes in customers' preferences, Boeing spread the 777 project's risks among a number of models all belonging to the same family.

The key to the 777's design efficiency was the wing. The 777 wings, exceptionally long and thin, were strong enough to support vastly enlarged models. The first model to go into service, the 777–200, had a 209-foot-long fuselage, was designed to carry 305 passengers in three class configurations, and had a travel range of 5,900 miles in its original version (1995) and up to 8,900 miles in its extended version (1997). The second model to be introduced (1998), the 777–300, had a stretched fuselage of 242 feet (10 feet longer than the 747), was configured for 379 passengers (three classes), and flew to destinations of up to 6,800 miles away. In all-tourist-class configuration, the stretched 777–300 could carry as many as 550 passengers.[34]

DIGITAL DESIGN

The 777 was the first Boeing jetliner designed entirely by computers. Historically, Boeing had designed new planes in two ways: paper drawings and full-size models called mock-ups. Paper drawings were two dimensional and therefore insufficient to account for the complex construction of the three dimensional airplane. Full-scale mock-ups served as a backup to drawings.

---

[33] John Mintz, "Betting It All on the 777: Making a New Jet on Which Its Future Rests, Boeing Remade Itself Too." *Washington Post*, March 26, 1995; Boeing Commercial Airplane Group, "777 Announced Order and Delivery Summary . . . as of 9/30/99."

[34] Rodgers, *Flying High*, 420–426; Woolsey, "777: Boeing's New Large Twinjet," 27, 31; "Leading Families of Passenger Jet Airplanes," Boeing Commercial Airplane Group, 1998.

Boeing engineers used three classes of mock-ups. Made up of plywood or foam, class 1 mock-ups were used to construct the plane's large components in three dimensions, refine the design of these components by carving into the wood or foam, and feed the results back into the drawings. Made partly of metal, class 2 mock-ups addressed more complex problems, such as the wiring and tubing of the airframe and the design of the machine tools necessary to cut and shape the large components. Class 3 mock-ups gave the engineers one final opportunity to refine the model and thereby reduce the need to keep on changing the design during the actual assembly process or after delivery.[35]

Despite the engineers' efforts, many parts and components did not fit together on the final assembly line but rather interfered with each other, that is, overlapped in space. The problem was both pervasive and costly. Boeing engineers needed to rework and realign all overlapping parts in order to join them together.

A partial solution to the problem was provided by the computer. In the last quarter of the 20th century, computer-aided design was used successfully in car manufacture, building construction, machine production, and several other industries; its application to commercial aircraft manufacturing came later, both in the United States and in Europe. Speaking of the 777, Dick Johnson, Boeing chief engineer for digital design, noted the "tremendous advantage" of computer application: "With mock-ups, the . . . engineer had three opportunities at three levels of detail to check his parts, and nothing in between. With CATIA [computer-aided three-dimensional interactive application] he can do it day in and day out over the whole development of the airplane."[36]

CATIA was a sophisticated computer program that Boeing bought from Dassault Aviation, a French fighter plane builder. IBM enhanced the program to improve image manipulation, supplied Boeing with eight of its largest mainframe computers, and connected the mainframes to 2,200 computer terminals that Boeing distributed among its 777 design teams. The software program showed on a screen exactly how parts and components fit together before the actual manufacturing process took place.[37]

A digital design system, CATIA had five distinctive advantages. First, it provided the engineers with 100 percent visualization, allowing them to rotate, zoom, and "interrogate" parts geometrically in order to spotlight interferences. Second, CATIA assigned a numerical value to each drawing on the screen and thereby helped engineers locate related drawings of parts and components, merge them together, and check for incompatibilities. Third, to help Boeing's customers service the 777, the digital design system created a computer-simulated human—a

---

[35] Sabbagh, *Twenty-First-Century Jet*, 58.

[36] Quoted in ibid., p. 63.

[37] Dornheim, "Computerized Design System Allows Boeing to Skip Building 777 Mockup," p. 50; Sabbagh, *Twenty-First-Century Jet*, p. 62.

CATIA figure playing the role of the service mechanic—who climbed into the three-dimensional images and showed the engineers whether parts were serviceable and entry accessible. Fourth, the use of CATIA by all 777 design teams in the United States, Japan, Europe, and elsewhere facilitated instantaneous communication between Boeing and its subcontractors and ensured the frequent updating of the design. And fifth, CATIA provided the 777 assembly-line workers with graphics that enhanced the narrative work instructions they received, showing explicitly on a screen how a given task should be performed.[38]

DESIGN-BUILD TEAMS

Teaming was another feature of the 777 program. About 30 integrated-level teams at the top and more than 230 design-build teams (DBTs) at the bottom worked together on the 777.[39] All team members were connected by CATIA. The integrated-level teams were organized around large sections of the aircraft; the DBTs, around small parts and components. In both cases, teams were cross-functional, as Philip Condit observed:

> If you go back . . . to earlier planes that Boeing built, the factory was on the bottom floor, and Engineering was on the upper floor. Both Manufacturing and Engineering went back and forth. When there was a problem in the factory, the engineer went down and looked at it. . . .
> With 10,000 people [working on the 777], that turns out to be really hard. So you start devising other tools to allow you to achieve that—the design-build team. You break the airplane down and bring Manufacturing, Tooling, Planning, Engineering, Finance, and Materials all together [in small teams].[40]

Under the design-build approach, many of the design decisions were driven by manufacturing concerns. As manufacturing specialists worked alongside engineers, engineers were less likely to design parts that were difficult to produce and needed to be redesigned. Similarly, under the design-build approach, customers' expectations as well as safety and weight considerations were all incorporated into the design of the aircraft; engineers no longer needed to "chain saw"[41] structural components and systems in order to replace parts that did not meet customers' expectations, were unsafe, or were too heavy.

The design of the 777's wing provides an example. The wing was divided into two integration-level teams, the leading-edge (the forward part of the wing)

---

[38] George Taninecz, "Blue Sky Meets Blue Sky," *Industry Week*, December 18, 1995, pp. 49–52; Paul Proctor, "Boeing Rolls Out 777 to Tentative Market," *Aviation Week and Space Technology*, April 11, 1994, pp. 36–39.

[39] Proctor, "Boeing Rolls Out 777 to Tentative Market," p. 37; Richard G. Olone, "777 Revolutionizes Boeing Aircraft Development Process," *Aviation Week and Space Technology*, June 3, 1991, p. 35.

[40] Quoted in Sabbagh, *Twenty-First-Century Jet*, 68–69.

[41] This was the phrase used by Boeing project managers working on the 777. See Sabbagh, *Twenty-First-Century Jet*, chap. 4.

and the trailing-edge (the back of the wing) team. Next, the trailing-edge team was further divided into 10 DBTs, each named after a piece of the wing's trailing edge. (See Exhibit I.) Membership in these DBTs extended to two groups of outsiders: representatives of the customer airlines and engineers employed by the foreign subcontractors. Made up of up to 20 members, each DBT decided its own mix of insiders and outsiders, and each was led by a team leader. Each DBT included representatives from six functional disciplines: engineering, manufacturing, materials, customer support, finance, and quality assurance. The DBTs met twice a week for two hours to hear reports from team members, discuss immediate goals and plans, divide responsibilities, set timelines, and take specific notes of all decisions taken.[42] Described by a Boeing official as little companies, the DBTs enjoyed a high degree of autonomy from management supervision; team members designed their own tools, developed their own manufacturing plans, and wrote their own contracts with the program management, specifying deliverables, resources, and schedules. John Monroe, a Boeing 777 senior project manager, remarked: "The team is totally responsible. We give them a lump of money to go and do th[eir] job. They decide whether to hire a lot of inexpensive people or to trade numbers for resources. It's unprecedented. We have some $100 million plus activities led by non-managers."[42]

EXHIBIT I. THE DBTS RESPONSIBLE FOR THE WING'S TRAILING EDGE

- Flap Supports Team
- Inboard Flap Team
- Outboard Flap Team
- Flaperon[a] Team
- Ailerona Team
- Inboard Fixed Wing and Gear Support Team
- Main Landing Gear Doors Team
- Spoilers[b] Team
- Fairings[c] Team

[a]Flaperons and ailerons are movable hinged sections of the trailing edge that help planes roll in flight. Flaperons are used at high speed; ailerons are used at low speed.
[b]Spoilers are the flat surfaces that lie on top of the trailing edge and are extended during landing to slow down the plane.
[c]Fairings are the smooth parts attached to the outline of the wing's trailing edge. They help reduce drag.

Source: Karl Sabbagh, *Twenty-First-Century Jet: The Making and Marketing of the Boeing 777* (New York: Scribners, 1996),

---

[42] *Main,* "Corporate Performance," p. 116; Sabbagh, *Twenty-First-Century Jet,* 69–73; Wolf L. Glende, "The Boeing 777: A Look Back," The Boeing Company, 1997, 4.

EMPLOYEES' EMPOWERMENT AND CULTURE

An additional aspect of the 777 program was the empowering of assembly-line workers. Boeing managers encouraged factory workers at all levels to speak up, offer suggestions, and participate in decision making. Boeing managers also paid attention to a variety of "human relations" problems faced by workers, problems ranging from child care and parking to occupational hazards and safety concerns.[43]

All employees entering the 777 program—managers, engineers, assembly line workers, and others—were expected to attend a special orientation session devoted to the themes of team work and quality control. Once a quarter, the entire "777 team" of up to 10,000 employees met offsite to hear briefings on the aircraft status. Dressed casually, the employees were urged to raise questions, voice complaints, and propose improvements. Under the 777 program, managers met frequently to discuss ways to promote communication with workers. Managers, for example, "fire fought" problems by bringing workers together and empowering them to offer solutions. In a typical *firefight* session, Boeing 777 project managers learned from assembly-line workers how to improve the process of wiring and tubing the airframe's interior: "staffing" fuselage sections with wires, ducts, tubs, and insulation materials before joining the sections together was easier than installing the interior parts all at once in a preassembled fuselage.[44]

Under the 777 program, in addition, Boeing assembly-line workers also were empowered to appeal management decisions. In a case involving middle managers, a group of Boeing machinists sought to replace a nonretractable jig (a large device used to hold parts) with a retractable one in order to simplify their jobs. Otherwise they had to carry heavy equipment loads up and down stairs. Again and again, their supervisors refused to implement the change. When the machinists eventually approached a factory manager, he inspected the jig personally and immediately ordered the change.[45]

Under the 777 program, work on the shop floor was ruled by the Bar Chart. A large display panel placed at different work areas, the Bar Chart listed the name of each worker, his or her daily job description, and the time available to complete specific tasks. Boeing had utilized the Bar Chart system as a management visibility system in the past, but only under the 777 program was the system fully

---

[43] O'Lone, "777 Revolutionizes Boeing Aircraft Development Process," p. 34

[44] O. Casey Corr, "Boeing's Future on the Line: Company's Betting its Fortunes Not Just on a New Jet, But on a New Way of Making Jets," *Seattle Times*, August 29, 1993. http://community.seattletimes. nwsource.com/archive/?date=19930829&slug=1718261; Lane, "Boeing Used 777 to Make Production Changes, Meet Desires of Its Customers. http://community.seattletimes.nwsource.com/archive/? date=19950507&slug=2119772.

[45] Casey Corr, sity.seattletimes.nwsource.com

computerized. The chart showed whether assembly-line workers were meeting or missing their production goals. Boeing industrial engineers estimated the time it would take to complete a given task and fed the information back to the system's computer. Workers ran a scanner across their ID badges and supplied the computer with the data necessary to log their job progress. Each employee "sold" his or her completed job to an inspector, and no job was declared acceptable unless "bought" by an inspector.[46]

LEADERSHIP AND MANAGEMENT STYLE

The team in charge of the 777 program was led by a group of five vice presidents, headed by Philip Condit, a gifted engineer who was described by one Wall Street analyst as "a cross between a grizzly bear and a teddy bear. Good people skills, but furious in the marketplace."[47] Each of the five vice presidents rose through the ranks, and each had a twenty-five to thirty years experience with Boeing. All were men.[48] During the 777 design phase, the five VPs met regularly every Tuesday morning in a small conference room at Boeing's headquarters in Seattle in what was called the "Muffin Meeting." There were no agendas drafted, no minutes drawn, no overhead projectors used, and no votes taken. The homemade muffins served during the meeting symbolized the informal tone of the forum. Few people outside the circle of five ever attended these weekly sessions. Acting as an informal chair, Condit led a freewheeling discussion of the 777 project, asking each VP to say anything he had on his mind.[49]

The weekly session reflected Boeing's sweeping new approach to management. Traditionally, Boeing had been a highly structured company governed by engineers. Its culture was secretive, formal, and stiff. Managers seldom interacted, sharing was rare, divisions kept to themselves, and engineers competed with each other. Under the 777 program, Boeing made serious efforts to abandon its secretive management style. Condit firmly believed that open communication among top executives, middle managers, and assembly-line workers was indispensable for improving morale and raising productivity. He urged employees to talk to each other and share information, and he used a variety of management tools to do so: information sheets, orientation sessions, question-and-answer sessions, leadership meetings, regular workers as well as middle managers. Condit introduced a three-way performance review procedure whereby managers were evaluated by

---

[46] O. Casey Corr, "Boeing's Future on the Line," and Lane, "Boeing Used 777 to Make Production Changes,"

[47] Quoted in Rodgers, *Flying High,* pp. 419–420.

[48] Sabbagh, *Twenty-First Century Jet,* p. 33.

[49] Ibid., p. 99.

their supervisors, their peers, and their subordinates.[50] Most important, Condit made teamwork the hallmark of the 777 project. In an address titled "Working Together: The 777 Story" and delivered in December 1992 to members of the Royal Aeronautics Society in London, Condit summed up his team approach:

> [T]eam building is . . . very difficult to do well but when it works the results are dramatic. Teaming fosters the excitement of a shared endeavor and creates an atmosphere that stimulates creativity and problem solving. But building team[s] . . . is hard work. It doesn't come naturally. Most of us are taught from an early age to compete and excel as individuals. Performance in school and performance on the job are usually measured by individual achievement. Sharing your ideas with others, or helping others to enhance their performance, is often viewed as contrary to one's self interest.
>
> This individualistic mentality has its place, but . . . it is no longer the most useful attitude for a workplace to possess in today's world. To create a high performance organization, you need employees who can work together in a way that promotes continual learning and the free flow of ideas and information.[51]

THE RESULTS OF THE 777 PROJECT

The 777 entered revenue service in June 1995. Since many of the features incorporated into the 777's design reflected suggestions made by the airline carriers, pilots, mechanics, and flight attendants were quite enthusiastic about the new jet. Three achievements of the program, in airplane interior, aircraft design, and aircraft manufacturing, stood out.

Configuration Flexibility

The 777 offered carriers enhanced configuration flexibility. A typical configuration change took only 72 hours on the 777 compared to three weeks in competing aircraft. In 1992, the Industrial Design Society of America granted Boeing its Excellence Award for building the 777 passenger cabin, honoring an airplane interior for the first time.[52]

Digital Design

The original goal of the program was to reduce "change, error, and rework" by 50 percent, but engineers building the first three 777s managed to reduce such modification by 60 percent to 90 percent. CATIA helped engineers identify more than 10,000 interferences that would have otherwise remained undetected until

---

[50] Dori Jones Young, "When the Going Gets Tough, Boeing Gets Touchy-Feely," *Business Week*, January 17, 1994, pp. 65–67; Jeremy Main, "Betting on the 21st Century Jet," *Fortune*, April 20, 1992, p. 117.

[51] Reprinted by The Boeing Company, Executive Communications, 1992.

[52] Boeing, *Backgrounder: Pace Setting Design Value-Added Features Boost Boeing 777 Family.*

assembly or until after delivery. The first 777 was only 0.023 inch short of perfect alignment, compared to as much as 0.5 inch on previous programs.[53] Assembly-line workers confirmed the beneficial effects of the digital design system. "The parts snap together like Lego blocks," said one mechanic.[54] Reducing the need for reengineering, replanning, retooling, and retrofitting, Boeing's innovative efforts were recognized yet again. In 1993, the Smithsonian Institution honored the Boeing 777 division with its Annual Computerworld Award for the manufacturing category.[55]

Empowerment

Boeing 777 assembly-line workers expressed a high level of job satisfaction under the new program. "It's a whole new world," a 14-year Boeing veteran mechanic said. "I even like going to work. It's bubbly. It's clean. Everyone has confidence."[56] "We never used to speak up," said another employee, "didn't dare. Now factory workers are treated better and are encouraged to offer ideas."[57] Although the Bar Chart system required Boeing 777 mechanics to work harder and faster as they moved down the learning curve, their principal union organization, the International Association of Machinists, was pleased with Boeing's new approach to labor–management relations. A union spokesman reported that under the 777 program, managers were more likely to treat problems as opportunities from which to learn rather than mistakes for which to blame. Under the 777 program, the union representative added, managers were more respectful of workers' rights under the collective bargaining agreement.[58]

UNRESOLVED PROBLEMS AND LESSONS LEARNED

Notwithstanding Boeing's success with the 777 project, the cost of the program was very high. Boeing did not publish figures pertaining to the total cost of CATIA, but a company official reported that, under the 777 program, the 3D digital design process required 60 percent more engineering resources than the older, 2D drawing-based design process. One reason for the high cost of using digital design was slow computing tools: CATIA's response time often lasted minutes. Another was the need to update the design software repeatedly. Boeing revised CATIA's design software four times between 1990 and 1996, making the system easier to learn and use. Still, CATIA continued to experience frequent software

---

[53] Taninecz, "Blue Sky Meets Blue Sky,", pp. 50–51; Woolsey, "777: Boeing's New Large Twinjet,"

[54] Boeing Rolls Out 777 to Tentative Market, p. 37.

[55] Boeing, *Backgrounder: Computing & Design/Build Process Help Develop the 777*, n.d.

[56] Casey Corr, "Boeing's Future on the Line."

[57] Lane, "Boeing Used 777 to Make Production Changes."

[58] Casey Corr, "Boeing's Future on the Line."

problems. Moreover, several of Boeing's outside suppliers were unable to utilize CATIA's digital data in their manufacturing process.[59]

Boeing faced training problems as well. One challenging problem, according to Ron Ostrowski, director of 777 engineering, was "to convert people's thinking from 2D to 3D. It took more time than we thought it would. I came from a paper world and now I am managing a digital program."[60] Converting people's thinking required what another manager called an "unending communication" coupled with training and retraining. Under the 777 program, Ostrowski recalled, "engineers had to learn to interact. Some couldn't, and they left. The young ones caught on" and stayed.[61]

Learning to work together was a challenge to managers too. Some managers were reluctant to embrace Condit's open management style, fearing a decline in their authority. Others were reluctant to share their mistakes with their superiors, fearing reprisals. Some other managers, realizing that the new approach would end many managerial jobs, resisted change when they could and did not pursue it wholeheartedly when they could not. Even top executives sometimes were uncomfortable with Boeing's open management style, believing that sharing information with employees was likely to help Boeing's competitors obtain confidential 777 data.[62]

Teamwork was another problem area. Working under pressure, some team members did not function well within teams and had to be moved. Others took advantage of their new freedom to offer suggestions but were disillusioned and frustrated when management either ignored these suggestions or did not act upon them. Managers experienced different team-related problems. In several cases, managers kept on meeting with their team members repeatedly until they arrived at a solution desired by their bosses. They were unwilling to challenge senior executives, nor did they trust Boeing's new approach to teaming. In other cases, managers distrusted the new digital technology. One engineering manager instructed his team members to draft paper drawings alongside CATIA's digital designs. When CATIA experienced a problem, he followed the drawing, ignoring the computerized design, and causing unnecessary and costly delays in his team's part of the project.[63]

Extending the 777 Revolution

Boeing's learning pains played a key role in the company's decision not to implement the 777 program companywide. Boeing officials recognized the importance

---

[59] Glende, "The Boeing 777: A Look Back," p. 10; *Air Transport World* (August 1996): 78.
[60] Woolsey, "777: Boeing's New Large Twinjet," 23.
[61] Mintz, "Betting It All on the 777."
[62] Lane, "Boeing Used 777 to Make Production Changes"; Rodgers, *Flying High,* 441.
[63] Lane, "Boeing Used 777 to Make Production Changes"; Rodgers, *Flying High,* 441–442.

of teamwork and CATIA in reducing change, error, and rework, but they also realized that teaming required frequent training, continuous reinforcement, and ongoing monitoring and that the use of CATIA was still too expensive, though its cost was going down. (In 1997, CATIA's "penalty" was down to 10 percent.) Three of Boeing's derivative programs, the 737 Next Generation, the 757–300, and the 767–400, had the option of implementing the 777's program innovations, and only one, the 737, did so, adopting a modified version of the 777's cross-functional teams.[64] Yet the 777's culture was spreading in other ways. Senior executives took broader roles as the 777 entered service, and their impact was felt through the company. Larry Olson, director of information systems for the 747/767/777 division, was a former 777 manager who believed that Boeing 777 employees "won't tolerate going back to the old ways." He expected to fill new positions on Boeing's next program—the 747X—with former 777 employees in their forties.[65] Philip Condit, Boeing CEO, implemented several of his own 777's innovations, intensifying the use of meeting among Boeing's managers and promoting the free flow of ideas throughout the company. Under Condit's leadership, all mid-level managers assigned to Boeing Commercial Airplane Group, about 60 people, met once a week to discuss costs, revenues, and production schedules, product by product. By the end of the meeting—which sometimes ran into the evening—each manager had to draft a detailed plan of action dealing with problems in his/her department.[66] Under Condit's leadership, more important, Boeing developed a new vision that grew out of the 777 project. Articulating the company's vision for the next two decades (1996–2016), Condit singled out "Customer satisfaction," "Team leadership," and "A participatory workplace" as Boeing's core corporate values.[67]

CONCLUSION: BOEING, AIRBUS, AND THE 777

Looking back at the 777 program 12 years after the launch and seven years after first delivery, it is now clear that Boeing produced the most successful commercial jetliner of its kind. Airbus launched the A330 and A340 in 1987, and McDonnell Douglas launched a new 300-seat wide-body jet in the mid-1980s, the three-engine MD11. Coming late to market, the Boeing 777 soon outsold both models. The 777 entered service in 1995, and within a year, Boeing delivered more than twice as many 777s as the number of MD11s delivered by McDonnell Douglas. In 1997, 1998, 1999, and 2001, Boeing delivered a larger number of 777s than the combined number of A330s and A340s delivered by Airbus. (See Table IV.)

---

[64] Glende, "The Boeing 777: A Look Back," 10.

[65] Sweetman, "As Smooth as Silk," 78.

[66] "A New Kind of Boeing," *Economist*, January 22, 2000, p. 63.

[67] "Vision 2016," The Boeing Company, 1997.

TABLE IV   TOTAL NUMBER OF MD11, A330, A340, AND 777 AIRPLANES DELIVERED BETWEEN 1995 AND 2001

	1995	1996	1997	1998	1999	2000	2001
McDonnell Douglas/ Boeing MD11	18	15	12	12	8	4	2
Airbus A330	30	10	14	23	44	43	35
Airbus A340	19	28	33	24	20	19	20
Boeing 777	13	32	59	74	83	55	61

Sources: For Airbus, Mark Luginbill, Airbus Communication Director, February 1, 2000, and March 11, 2002. For Boeing, The Boeing Company Annual Report, 1997, 1998, p. 35; "Commercial Airplanes: Order and Delivery, Summary." http//www.boeing.com/commercial/order/index.html. Retrieved February 2, 2000, and March 9, 2002.

A survey of nearly 6,000 European airline passengers who had flown both the 777 and the A330/A340 found that more than three out of four passengers preferred the 777.[68] In the end, a key element in the 777's triumph was its popularity with the traveling public.

---

[68] "Study: Passengers Voice Overwhelming Preference for Boeing 777." Press release, available at http://www.prnewswire.com/news-releases/study-passengers-voice-overwhelming-preference-for-boeing-777-74839487.html

Boeing 787 Dreamliner Battery Issues

THE SAFETY CONSTRAINT

When we discuss the triple constraint, we are generally referring to time, cost, and scope. But there are other constraints, and when human life is involved, safety becomes perhaps the most important constraint. There are many forms of safety. On information technology projects, safety protocols are installed to make sure that proprietary data is not compromised. Food and health care product manufacturers worry about product tampering and safety protection for consumers. Manufacturers worry about consumers using their products in a safe manner. Companies like Disney have safety as the number one constraint for rides and attractions at the theme parks. Most companies would rather allow projects to intentionally fail or be canceled before risking lawsuits over safety violations. This is particularly true if the there is a chance for loss of human life.

THE BOEING 787 DREAMLINER DECISION

The Boeing 787 "Dreamliner" is a long-range, mid-size wide-body, twin-engine jet airliner that can handle 242 to 335 passengers in typical three-class seating configurations. It is Boeing's most fuel-efficient airliner and is a pioneering

Adapted from "Boeing 787 Dreamliner battery problems," *Wikipedia*, https://en.wikipedia.org/wiki/Boeing_787_Dreamliner_battery_problems.

airliner with the use of composite materials (carbon fiber, aluminum, and titanium) as the primary material in the construction of its airframe and an electrical system using lithium-ion batteries. The 787 would reduce airline maintenance costs and replacement costs. The expectation was that the 787 would be 10 percent lower cost-per-seat mile than any other aircraft. The 787 was designed to be 20 percent more fuel efficient than the Boeing 767, which it was intended to replace.

In order to maximize shareholder value, Boeing decided to outsource 70 percent of the work on the 787 rather than the 35 to 50 percent outsourcing that was used on the 737 and 747 aircrafts. This was expected to shorten the development time from six to four years and lower development costs from $10 billion to $6 billion. The lowering of Boeing's assembly costs would spread significant financial risk to Boeing's suppliers, which were now responsible for more assembly work.

The longest-range 787 variant can fly 8,000 to 8,500 nautical miles, enough to cover the Los Angeles to Bangkok or New York City to Hong Kong routes. Its cruising airspeed is Mach 0.85, equivalent to 561 miles per hour at typical cruise altitudes. As of August 2016, 64 customers had ordered 1,161 of the Dreamliners, and Boeing had delivered 445 aircraft.

The airline industry spends significantly more than a decade and perhaps as much as $30 billion in designing a new commercial aircraft. But even in the design and manufacturing phases, safety issues and problems can still exist but remain hidden. The only real way to verify that safety issues have been addressed is through commercial use of the plane. Boeing had to push back the launch date of the 787 seven times, and the first few aircraft were delivered three years late. Boeing has reportedly spent $32 billion on the 787 program.

Boeing and Airbus may end up spending billions of dollars after the planes are put in use to resolve any and all safety issues. This is what consumers expect. And Boeing and Airbus comply, as they did with the battery problems on the 787 Dreamliner and the A380 as the case study will show.

INNOVATION PROBLEMS

In the Boeing 787 Dreamliner's first year of service (2011), at least four aircraft suffered from electrical system problems stemming from the lithium-ion batteries. Problems are common within the first year of any new aircraft design's life; it experienced the following problems during that first year:

November 1, 2011: Landing gear failed to deploy

July 23, 2012: Corrosion risk identified in an engine component

December 4, 2012: Leakage in fuel line connectors

December 4, 2012: A power generator failed

January 7, 2013: Smoke in the cockpit during an inspection

January 8, 2013: Faulty left wing surge tank vent

January 9, 2013: Indicator falsely reported brake problems

January 11, 2013: Engine oil leak

January 11, 2013: Crack developed on the cockpit wide screen

After a number of incidents including an electrical fire aboard an All Nippon Airways 787, and a similar fire found by maintenance workers on a landed Japan Airlines 787 at Boston's Logan International Airport, the United States Federal Aviation Administration (FAA) ordered a review of the design and manufacture of the Boeing 787 Dreamliner. following five incidents in five days involving the aircraft, mostly of which involved problems with the batteries and electrical systems. This was followed with a full grounding of the entire Boeing 787 fleet, the first such grounding since that of DC-10s following the American Airlines Flight 191 disaster in 1979.[1] It was reported that the plane had two major battery thermal runaway events in 100,000 flight hours, which substantially exceeded the 10 million flight hours predicted by Boeing, and had done so in a dangerous manner.[2]

In December 2012, Boeing CEO James McNerney told media outlets that the problems were no greater than those the company experienced with the introduction of other new models, such as the Boeing 777.[3] However, on January 7, 2013, a battery overheated and started a fire in an empty 787 operated by Japan Airlines (JAL) at Boston's Logan International Airport.[4] On January 9, United Airlines reported a problem in one of its six 787s with the wiring in the same area as the battery fire on JAL's airliner; subsequently, the U.S. National Transportation Safety Board opened a safety probe.[5]

On January 11, 2013, the FAA announced a comprehensive review of the 787's critical systems, including the design, manufacture, and assembly of the

---

[1] "Dreamliner: Boeing 787 Planes Grounded on Safety Fears," BBC News. www.bbc.com/news/business-21054089.

[2] Simon Hradecky, "Accident: ANA B787 Near Takamatsu on Jan 16th 2013, Battery Problem and Burning Smell on Board." *Aviation Herald.* http://avherald.com/h?article=45c377c5.

[3] "Boeing: Problems with 787 Dreamliner 'Normal,'" *Frequent Business Traveler*, December 16, 2012. www.frequentbusinesstraveler.com/2012/12/boeing-problems-with-787-dreamliner-normal/.

[4] "Fire Aboard Empty 787 Dreamliner Prompts Investigation," CNN, January 8, 2013, www.cnn.com/2013/01/07/travel/dreamliner-fire/; "Fresh Faults with Boeing Dreamliner Planes," BBC.com, January 8, 2013. www.bbc.com/news/business-20950287.

[5] "U.S. Opens Dreamliner Safety Probe," *Wall Street Journal*, January 9, 2013.

aircraft. U.S. Department of Transportation secretary Ray LaHood stated the administration was "looking for the root causes" behind the recent issues. The head of the FAA, Michael Huerta, said that so far nothing found "suggests [the 787] is not safe."[6] Japan's transport ministry also launched an investigation in response.[7]

On January 16, 2013, an All Nippon Airways (ANA) 787 made an emergency landing at Takamatsu Airport on Shikoku Island after the flight crew received a computer warning that there was smoke inside one of the electrical compartments.[8] ANA said that there was an error message in the cockpit citing a battery malfunction. Passengers and crew were evacuated using the emergency slides.[9] It was reported that there were no fire-suppression systems in the electrical compartments holding batteries, only smoke detectors.[10]

The oversight of U.S.-based aviation regulators into the 2007 safety approval and FAA certification of the 787 came under scrutiny, as a key U.S. Senate committee prepared for a hearing into the procedures of aviation safety certification However, an FAA spokesperson defended the group's 2007 safety certification of the 787 by saying, "The whole aviation system is designed so that if the worst case happens, there are systems in place to prevent that from interfering with other systems on the plane."[11]

On February 12, 2013, the *Wall Street Journal* reported: "Aviation safety investigators are examining whether the formation of microscopic structures known as dendrites inside the Boeing Co. 787's lithium-ion batteries played a role in twin incidents that prompted the fleet to be grounded nearly a month ago."[12]

On January 16, 2013, both major Japanese airlines, ANA and JAL, announced that they were voluntarily grounding or suspending flights for their fleets of 787s after multiple incidents involving different 787s, including emergency landings.

---

[6] "Boeing 787 Dreamliner to Be Investigated by US Authorities," *The Guardian*, January 11, 2013. www.theguardian.com/business/2013/jan/11/boeing-787-dreamliner-us-investigation.

[7] Anna Mukai, "Japan to Investigate Boeing 787 Fuel Leak as FAA Reviews," Bloomberg, January 15, 2013. www.bloomberg.com/news/articles/2013-01-15/japan-sets-up-team-to-probe-dreamliner-fuel-leak-as-faa-reviews.

[8] "Top Japan Airlines Ground Boeing 787s after Emergency," BBC, January 16, 2013. www.bbc.com/news/business-21038128.

[9] Ibid.

[10] Iain Thomson, "Boeing 787 Fleet Grounded Indefinitely as Investigators Stumped," *The Register*, January 25, 2013. www.theregister.co.uk/2013/01/25/boeing_787_ntsb_report/.

[11] "Boeing 787's Battery Woes Put US Approval under Scrutiny," *Business Standard*, February 22, 2013. www.business-standard.com/article/international/boeing-787-s-battery-woes-put-us-approval-under-scrutiny-113012300143_1.html.

[12] John Ostrower and Andy Pasztor, "Microscopic 'Dendrites' a Focus in Boeing Dreamliner Probe," *Wall Street Journal*, February 11, 2013. www.wsj.com/articles/SB10001424127887324880504782986735666960476.

These two carriers operated 24 of the 50 Dreamliners delivered to date.[13] The grounding was expected to cost ANA over $1.1 million a day.[14]

On January 16, 2013, the FAA issued an emergency airworthiness directive ordering all U.S.-based airlines to ground their Boeing 787s until yet-to-be-determined modifications were made to the electrical system to reduce the risk of the battery overheating or catching fire.[15] This was the first time that the FAA had grounded an airliner type since 1979. The FAA also announced plans to conduct an extensive review of the 787's critical systems. The focus of the review was on the safety of the lithium-ion batteries made of lithium cobalt oxide (LiCoO2). The 787 battery contract was signed in 2005[16] when LiCoO2 batteries were the only type of lithium aerospace battery available. Since that time, newer and safer[17] types, such as lithium iron phosphate(LiFePO), which provide less reaction energy during thermal runaway, have become available.[18] FAA approved a 787 battery in 2007 with nine "special conditions."[19] A battery approved by FAA (through Mobile Power Solutions) was made by Rose Electronics using Kokam cells,[20] but the batteries installed in the 787 were made by Yuasa.[21]

On January 20, the National Transportation Safety Board (NTSB) declared that overvoltage was not the cause of the Boston incident, as voltage did not exceed the battery limit of 32 V,[22] and the charging unit passed tests. The battery had signs of shortcircuiting and thermal runaway.[23] Despite this, on January 24, the NTSB

[13] Mayumi Negishi and Tim Kelly, "Japanese Airlines Ground Boeing 787s after Emergency Landing," Reuters, January 16, 2013, www.reuters.com/article/us-boeing-ana-idUSBRE90F01820130116; Justin McCurry, "787 Emergency Landing: Japan Grounds Entire Boeing Dreamliner Fleet," *The Guardian*, January 16, 2013.

[14] "Boeing Dreamliners Grounded Worldwide on Battery Checks," Reuters, January 17, 2013. www.theguardian.com/business/2013/jan/16/787-emergency-landing-grounds-787.

[15] Federal Aviation Administration, Press Release–FAA Statement, January 16, 2013. www.faa.gov/news/press_releases/news_story.cfm?newsId=14233.

[16] GS Yuasa, "Thales Selects GS Yuasa for Lithium Ion Battery System in Boeing's 787 Dreamliner." Press release, June 12, 2005. www.gsyuasa-lp.com/content/thales-selects-gs-yuasa-lithium-ion-battery-system-boeing%E2%80%99s-787-dreamliner.

[17] Brier Dudley, "Lithium-Ion Batteries Pack a Lot of Energy—and Challenges," *Seattle Times*, January 16, 2013.

[18] Per Erlien Dalløkken, "Her Er Dreamliner-Problemet" (in Norwegian), *Teknisk Ukeblad*, www.tu.no/artikler/her-er-dreamliner-problemet/233970, "Energy Storage Technologies—Lithium." Retrieved January 24, 2013.

[19] "Special Conditions: Boeing Model 787– 8 Airplane; Lithium Ion Battery Installation." FAA/*Federal Register*, October 11, 2007. "NM375 Special Conditions No. 25–359–SC"; Alwyn Scott and Mari Saito, "FAA Approval of Boeing 787 Battery under Scrutiny," NBC News/Reuters. Retrieved January 24, 2013.

[20] Supko /Iverson, "Li Battery Un Test Report Applicability," Nextgov, 2011. Retrieved January 23, 2013.

[21] Bob Brewin, "A 2006 Battery Fire Destroyed Boeing 787 Supplier's Facility," Nextgov, January 22, 2013.

[22] NTSB press release: "NTSB Provides Third Investigative Update on Boeing 787 Battery Fire in Boston." www.ntsb.gov/news/press-releases/Pages/PR20130120.aspx.

[23] NTSB, "NTSB Press Release," January 26, 2013.

announced that it had not yet pinpointed the cause of the Boston fire; the FAA would not allow U.S.-based Dreamliners to fly again until the problem was found and corrected. In a press briefing that day, NTSB Chairwoman Deborah Hersman said that the NTSB had found evidence of failure of multiple safety systems designed to prevent these battery problems, and stated that fire must never happen on an airplane.[24] The Japan Transport Safety Board (JTSB) said on January 23 that the battery in ANA jets in Japan reached a maximum voltage of 31 V (lower than the 32 V limit, which the Boston JAL 787 had reached), but had a sudden unexplained voltage drop to near zero.[25] All cells had signs of thermal damage before thermal runaway.[26] ANA and JAL had replaced several 787 batteries before the mishaps. As of January 29, 2013, JTSB approved the Yuasa factory quality control. The American NTSB continued to look for defects in the Boston battery.[27]

Industry experts disagreed on the consequences of the grounding: Airbus was confident that Boeing would resolve the issue[28] and that no airlines would switch plane type,[29] while other experts saw the problem as "costly"[30] and "could take upwards of a year" to fix[31]

The only U.S.-based airline then operating the Dreamliner was United Airlines, which had six.[32] Chile's Directorate General of Civil Aviation grounded LAN Air-

---

[24] Matthew Weld and Jad Mouwad, "Protracted Fire Inquiry Keeping 787 on Ground," *New York Times*, January 25, 2013.

[25] Sofia Mitra-Thakur, "Japan Says 787 Battery Was Not Overcharged," *Engineering & Technology*, January 23, 2013; Christopher Drew, Hiroko Tabuchi, and Jad Mouawad, "Boeing 787 Battery Was a Concern Before Failure," *New York Times*, January 30, 2013.

[26] Simon Hradecky, "ANA B788 Near Takamatsu on Jan 16th 2013, Battery Problem and Burning Smell on Board." *Aviation Herald*, February 5, 2013.

[27] Hiroko Tabuchi, "No Quality Problems Found at Battery Maker for 787," *New York Times*, January 28, 2013; Chris Cooper and Kiyotaka Matsuda, "Gs Yuasa Shares Surge as Japan Ends Company Inspections." *Businessweek*, January 28, 2013; NTSB press release, "NTSB Issues Sixth Update on JAL Boeing 787 Battery Fire Investigation," https://app.ntsb.gov/news/2013/130129b.html.

[28] "Airbus CEO 'Confident' Boeing Will Find Fix for 787," Bloomberg.com, online video, January 17, 2013. www.bloomberg.com/news/videos/b/6f5ce486-0820-4c1e-8313-a9efe3b80935, accessed February 2017.

[29] Robert Wall and Andrea Rothman, "Airbus Says A350 Design Is 'Lower Risk' Than Troubled 787," Bloomberg, January 17, 2013. www.bloomberg.com/news/articles/2013-01-17/airbus-says-a350-design-lower-risk-than-troubled-boeing-787. According to John Leahy, Airbus's sales chief, "I don't believe that anyone's going to switch from one airplane type to another because there's a maintenance issue," he said. "Boeing will get this sorted out" (quoted in Mary Jane Credeur, "Dreamliner's Lure Keeps Airline CEOs Wedded to Plane," www.bloomberg.com/news/articles/2013-01-18/dreamliner-s-promise-keeps-airline-ceos-wedded-to-troubled-plane).

[30] "'Big Cost' Seen for Boeing Dreamliner Grounding," Bloomberg.com, online video, January 17, 2013. www.bloomberg.com/news/videos/b/1a2745f9-f5b1-444a-8fc4-6e090a1ddd2b. Accessed February 2017.

[31] Martha C. White, "Is the Dreamliner Becoming a Financial Nightmare for Boeing?" *TIME*, January 17, 2013. http://business.time.com/2013/01/17/is-the-dreamliner-becoming-a-financial-nightmare-for-boeing/.

[32] "FAA Grounding All Boeing 787s," *Kiro Tv.* www.kiro7.com/news/faa-grounding-all-boeing-787s/246413556.

lines' three 787s.[33] The Indian Directorate General of Civil Aviation directed Air India to ground its six Dreamliners. The Japanese Transport Ministry made the ANA and JAL groundings official and indefinite following the FAA announcement.[34] The European Aviation Safety Agency had also followed the FAA's advice and grounded the only two European 787s, which were operated by LOT Polish Airlines.[35] Qatar Airways announced that it was grounding its five Dreamliners.[36] Ethiopian Air was the final operator to announce temporary groundings of its four Dreamliners.[37]

As of January 17, 2013, all 50 of the aircraft delivered had been grounded.[38] On January 18, Boeing announced that it was halting 787 deliveries until the battery problem was resolved.[39] On February 4, the FAA said it would permit Boeing to conduct test flights of 787 aircraft to gather additional data.[40]

On April 19, the FAA approved Boeing's new design for the Boeing 787 battery. This would allow the eight airlines that maintained a fleet of 787 planes to begin making repairs. The repairs would include a battery containment and venting system.[41] The new design would add more protection and would also increase the weight of the plane by more than 150 pounds. This was necessary to ensure safety. The repairs would cost $465,000 per plane. Boeing committed more than 300 people on 10 teams to make the repairs, which would take about five days per plane.[42]

ANA, which operates 17 Dreamliner jets, estimated that it was losing $868,300 per plane over a two-week period and would be talking with Boeing about compensation for losses. Other airlines were also expected to seek some compensation.

[33] "LAN suspende de forma temporal la operación de flota Boeing 787 Dreamliner," *La Tercera*, January 16, 2013. www.latercera.com/noticia/lan-suspende-de-forma-temporal-la-operacion-de-flota-boeing-787-dreamliner/.

[34] Anindya Upadhyay,"DGCA Directs Air India to Ground All Six Boeing Dreamliners on Safety Concerns," *Economic Times*, January 17, 2013. http://economictimes.indiatimes.com/industry/transportation/airlines-/-aviation/dgca-directs-air-india-to-ground-all-six-boeing-dreamliners-on-safety-concerns/articleshow/18056887.cms.

[35] "European Safety Agency to Ground 787 in Line with FAA," Reuters, January 16, 2013. www.reuters.com/article/boeing-787-easa-idUSL6N0AM0E020130117.

[36] "Qatar Airways Grounds Boeing Dreamliner Fleet," Reuters, January 14, 2013. www.reuters.com/article/us-qatar-boeing-idUSBRE90G0D020130117.

[37] "U.S., Others Ground Boeing Dreamliner Indefinitely," CNBC.com, January 17, 2013. www.cnbc.com/id/100385850.

[38] "U.S., Others Ground Boeing Dreamliner Indefinitely"; Jorn Madslien, "Boeing 787 Dreamliner: The Impact of Safety Concerns," www.BBC.com, January 17, 2013. www.bbc.com/news/business-21041265.

[39] "Dreamliner Crisis: Boeing Halts 787 Jet Deliveries," www.BBC.com, January 18, 2013. www.bbc.com/news/business-21095056.

[40] "FAA Approves Test Flights for Boeing 787," *Seattle Times*, February 7, 2013. www.seattletimes.com/seattle-news/faa-approves-test-flights-for-boeing-787/.

[41] "FAA Approves Fix for Boeing 787 Battery," *Los Angeles Times*, April 20, 2013.

[42] "ANA's Dreamliner Test Flight Seen as Step in Regaining Customers," *Bloomberg News*, April 27, 2013.

CONCLUSIONS

Boeing was successful in resolving the battery issues, and sales of the 787 are doing well. Both Airbus and Boeing understand the importance of customer confidence. If aircraft customers lose confidence in the aircraft manufacturer's ability to deliver a safe aircraft, the manufacturer will lose significant business. Aircrafts can have more than 100,000 components. There are more than 23,000 parts in the cabin area alone on the A380. Given that it takes at least 10 years and billions of dollars to design and test these planes, it is impossible to simulate all possible problems. Dry runs cannot simulate every possible scenario that could happen. The reliability of every part and every system can be proven only when the aircraft is in operations. The A380 has undergone more testing than any other jet. Yet, despite the testing, it may be some time until all of the problems are resolved. Because lives may be at stake, Airbus will be spending billions to correct all potential problems.

There will always be risks with the design and development of new aircraft. Typical risks include:

- *Innovation risks:* Dealing with new and unproven technologies
- *Outsourcing risks:* Expecting suppliers and partners to take on design and development risks
- *Tiered outsourcing risks:* Asking the suppliers and partners to manage and integrate the work of lower-tier suppliers
- *Offshore risks:* Having critical components manufactured far away from the final assembly plant
- *Communication by computer risks:* Expecting communication by computer to replace face-to-face communication
- *Labor relations risks:* Having critical decisions related to outsourcing made by executives without any input from the people doing the work
- *Disengaged C-suite risks:* Having executives who do not want to be involved in the day-to-day activities of designing a new plane
- *Project management skills risks:* Having a project team that lacks critical skills, such as in supply chain management.[43]

---

[43] Adapted from Steve Denning, "What Went Wrong at Boeing?" *Forbes*, January 21, 2013.

LESSONS LEARNED

There are several lessons learned from the Boeing 787 Program:

- A proper test campaign should be mandatory for new technology. In the case of 787 the number of tests needed to certify its batteries were not enough.
- Flaws in manufacturing, insufficient testing, and a poor understanding of an innovative battery all contributed to the grounding of Boeing's 787 fleet.[44]
- "We advanced the state of the art for testing lithium ion batteries," Mike Sinnett, Boeing's chief engineer for the 787 program, said at a hearing of the aviation subcommittee of the House Transportation Committee. The hearing was titled "Lessons Learned from the Boeing 787 Incidents."[45]
- Strict quality control on the products provided by the suppliers should be a mandatory requirement. Choice of the supplier based only on the quality is also a risk because of other factors.
- But whatever the outcome from correcting a problem, especially with so many lives at stake, the design and manufacture of new aircraft should be based solely on legitimate issues of cost and quality, and the selection process for suppliers should be transparent and untainted by other commercial or political concerns.
- "The greatest enemy of good aircraft is people who interfere with the freedom to shop for the highest quality."[46]
- Many of these problems were exacerbated by Boeing's decision to massively increase the percentage of parts it sourced from outside contractors. The wing tips were made in Korea, the cabin lighting in Germany, cargo doors in Sweden, escape slides in New Jersey, landing gear in France. The plan backfired. Outsourcing parts led to three years of delays. Parts did not fit together properly. Shims used to bridge small parts weren't attached correctly. Many aircraft had to have their tails reworked extensively. The company ended up buying some suppliers, to take business back in house. All new projects, especially ones as ambitious as the Dreamliner, face teething problems, but the 787's woes continued to mount. Unions blamed the company's reliance on outsourcing.[47]

(continued)

---

[44] Jad Mouawad, "Report on Boeing 787 Dreamliner Battery Flaws Finds Lapses at Multiple Points," New York Times, December 1, 2014.

[45] Ted Reed, "Boeing and FAA: What We Learned from 787 Problems," www.thestreet.com, June 12, 2013. www.thestreet.com/story/11948417/1/boeing-and-faa-what-we-learned-from-787-problems.html.

[46] James B. Stewart, "Japan's Role in Making Batteries for Boeing," New York Times, January 25, 2013.

[47] Dominic Rushe, "Why Boeing's 787 Dreamliner Was a Nightmare Waiting to Happen," The Guardian, January 18, 2013.

(continued)

- Careful control and the right attitude by the FAA. The FAA should maintain an intransigent attitude toward airplane makers by making sure that safety protocols are adhered to before awarding a certificate of air worthiness.
- The Dreamliner's problems are not just a Boeing issue. They are a lesson in the limits of outsourcing and the all-too-cozy relationships between regulators and the regulated that have caused problems across industries from automotive to food and financial services in recent years.[48]
- Consultant and former airline executive Robert Mann said Boeing's clout put pressure on the FAA to approve the Dreamliner speedily, despite its radical design and manufacturing process.
- The ultimate goal of designing a new aircraft should be to increase shareholder value.
- When the goal focuses on maximum shareholder value, especially in the short term, unnecessary risks often are taken and a good opportunity turns into a potential disaster.

QUESTIONS

1. Can safety be considered as a constraint on a project and considered to be at a higher priority than even time, cost, and scope?
2. Should the project manager on the 787 Program have the authority on how safety is defined, measured, and reported?
3. Other than the battery issues, what other types of safety issues do commercial aircraft manufacturers consider?
4. Are there safety standards that must be followed in the design of a new aircraft, or does the project team determine the safety protocols?
5. Why are aircraft purchasers worried about safety issues when they place orders for new planes?

---

[48] Ibid.

The Airbus
A380 Airplane

INTRODUCTION

The Airbus A380 airplane is a double-deck, wide-body, four-engine jet airliner produced by Airbus.[1] It is the first airliner able to carry almost 900 passengers. It has the most spacious interior of any jetliner ever built, the biggest wings, and the

---

[1] Airbus is the European aircraft-manufacturing consortium formed in 1970 as Airbus Industrie. It is now one of the world's top two commercial aircraft manufacturers, competing directly with the American Boeing Company and frequently dominating the jetliner market in orders, deliveries, or annual revenue. Its initial shareholders were the French company Aérospatiale (later Aerospatiale Matra, 1999) and the German company Deutsche Airbus (later DaimlerChrysler Aerospace Airbus, 1989), each owning a 50 percent share. Spain's Construcciones Aeronáuticas S.A. (CASA) joined in 1971 with a 4.2 percent share. Hawker Siddeley and other British companies were nationalized in 1977 into a single government conglomerate, British Aerospace (later BAE Systems, 1999), which joined Airbus with a 20 percent share in 1979. In 2000, all the partners except BAE Systems merged into EADS, which thus acquired an 80 percent share of Airbus (Eugene Rodgers, *Flying High: The Story of Boeing* [New York: Atlantic Monthly Press, 1996], pp. 415–416; Dornheim, "777 Twinjet Will Grow to Replace 747–200," p. 43). However, in September 2006, European Aeronautic Defence and Space Company (EADS) acquired BAE Systems' 20 percent share of Airbus, making it a 100 percent subsidiary of EADS. EADS is the major European aerospace company that builds commercial/military aircraft and helicopters, space systems, propulsion systems, missiles, and other defense products. In January 2014, EADS was reorganized as Airbus Group ("Commercial Airplanes: Order and Delivery, Summary," http/www.boeing.com/commercial/orders/index.html.). The group consists of the three business divisions: Airbus, Airbus Defence and Space, Airbus Helicopters.

greatest overall engine thrust. No civil airliner since the supersonic Concorde has aroused such passion, such controversy, and such fascination.[2] In addition to its remarkable engineering features, the A380 is very interesting from a management and didactic point of view, namely:

- Lessons learned from a project of this magnitude
- The different scenarios and strategies in the aircraft markets
- The ability of Airbus to compete with Boeing in every market segment
- The ability to create an icon capable of capturing the imagination of the world's travelers with its double-deck load
- The ability to relieve congestion at the world's busiest airports by taking up less space than two airliners it can replace

HISTORY AND MOTIVATION OF A380 PROJECT

In the preliminary design phase of a new aircraft, the first step is the "market survey."[3] This phase is important because it is closely connected with the success of the aircraft. The new design has to be better than what already exists in the market. In addition, it is essential to determine if the market is willing to accept the new aircraft; in other words, if a demand exists. The new aircraft has to have features that make it more appealing than its competitors, and the market should show promises of growth. Otherwise, a new design is unlikely to succeed. In this regard, it is interesting to look closer at the motivations that brought the A380 concept to market and what makes A380 different from and better than its competitors.

Considering the size of the project, the design of the A380 had a long gestation. In the summer of 1988, a small group of advanced project engineers of the Airbus New Product Development and Technology branch in Toulouse, France, started thinking about building a giant aircraft capable of carrying more than 800 passengers. The project, called "Ultra-High-Capacity Aircraft," was totally unknown to the rest of aircraft-manufacturing consortium, including the Airbus leadership. The leader of the project, Jean Roeder, the brilliant engineer behind the concept of the A330 and A340, believed that this project was something that Airbus simply had to do. For too long a time Boeing had benefitted from its 747 monopoly. "Airbus was making effort at this time to get 30 percent of the market, and we thought that this just would not be possible in the long term if we did not get a complete set of aircraft in our program," said Roeder.[4]

---

[2] Rodgers, *Flying High*, pp. 415–416; Dornheim, "777 Twinjet Will Grow to Replace 747–200," p. 43.

[3] Guy Norris, Mark Wagner, *Airbus A380: Superjumbo of the 21st Century.* St. Paul, MN: Zenith Press, 2005.

[4] Norris and Wagner, *Airbus 380*, p. 7.

In October 1988, Roeder asked for a meeting to discuss the new project with Airbus president Jean Pierson and chief operating officer Herbert Flosdorff. Roeder brought with him a model of the concept. "Pierson was clearly surprised; he did not expecting something so big," said Roeder. Immediately Pierson saw the possibilities but also the risk. Therefore, the idea remained a secret for another two years. Airbus started talking about the project within the company and began market research to assess if a market existed for an aircraft bigger than a 747.[5] DaimlerChrysler and British Aerospace, two partners in the Airbus consortium as it then was, pushed for cooperation with Boeing because they were worried about the risk of competing head on over such a big new aircraft. An earlier battle between the McDonnell Douglas DC-10 and Lockheed's L-1011 had weakened both firms and pushed Lockheed out of the commercial aviation business entirely. Boeing wanted to produce a plane substantially bigger than the 747 with the purpose of complementing rather than replacing it. However, the agreement did not seem reachable in the near term, and meanwhile Boeing continued to have the monopoly on the superjumbo market, as it could use its big profits from the 747 to cut the price of its other jets, like the 737, which were in a direct competition with Airbus. For this reason, in 1995, Airbus eventually decided to work alone. A year later, Jürgen Thomas, a veteran German engineer known as the father of the A380, was appointed to lead the project.

After a series of meetings with prospective customers,[6] Airbus became convinced that there was indeed a large market for a modern plane capable of carrying between 550 and 650 passengers up to 9,000 miles.[7] On December 19, 2000, the A380,[8] previously known as the A3XX and dubbed "The Flagship of the 21st Century," was launched commercially with 50 firm orders and 42 options from six major operators, including Emirates, Singapore Airlines, Qantas, Air France, Qatar Airways, and Korean Air. On April 27, 2005, at 10:29 a.m. in Toulouse, the Airbus A380 made its first flight. Finally on October 15, 2007, Singapore Airlines took delivery of the first Airbus A380–800. The aircraft, MSN-003,[9] entered service on October 25, flying between Singapore and Sydney.[10] Typical passenger seating is shown in Figure I.

---

[5] Guy Norris and Mark Wagner, *Airbus A380: Superjumbo of the 21st Century* (St. Paul, Minn.: Zenith Press, 2005).

[6] In this regard, it is important to point out the fundamental difference with the other markets. In markets with high technology and characterized by high costs, it is vital to produce something that will be of interest or, more explicitly, will be bought.

[7] "The Giant on the Runway," *The Economist*, October 11, 2007. www.economist.com/node/9944806.

[8] The A380 designation was a break from previous Airbus families, which had progressed sequentially from A300 to A340. It was chosen because the number 8 resembles the double-deck cross section, and is a lucky number in some Asian countries where the aircraft was being marketed (Norris and Wagner, *Airbus A380.*)

[9] "MSN" stands for "manufacturer's serial number," which is a unique number or code assigned to a unit. The first two MSNs of A380 have been retained by Airbus.

[10] Benét J. Wilson, "The Airbus A380: A History," airwaysnews.com, January 20, 2015.

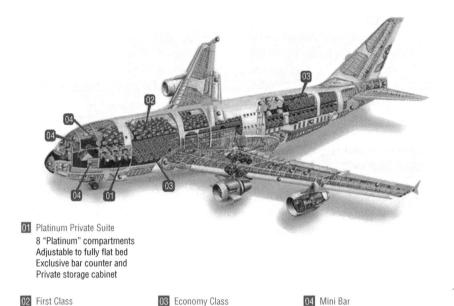

01 Platinum Private Suite
   8 "Platinum" compartments
   Adjustable to fully flat bed
   Exclusive bar counter and
   Private storage cabinet

02 First Class
   70 seats
   Adjustable seats can be
   laid flat
   All seats have aisle access

03 Economy Class
   428 seats with 76 in the upper
   deck and 352 in the main deck
   Adjustable seat headrest
   9" digital widescreen TV

04 Mini Bar
   passengers can sit on comfy
   chairs at the bar or laze on
   luxury sofas while having their
   favorite tipple delivered

FIGURE I   Airbus A380–800 superjumbo passengers placement seating

AIRBUS A380 VERSUS BOEING 747

Despite the rumors that the A380 was overweight and struggling to meet performance targets, Airbus delivered a remarkable plane. A380 is the "greenest" long-haul airliner, burning less than 1 gallon of fuel per passenger over 95 miles, fuel consumption comparable with that of a small turbo-diesel family car.[11] According to Airbus, the A380's cost per seat came out at 20 percent below that of the 747–400 offering the same cruise Mach number (Ma = 0.85). In addition, it has lower maintenance costs due to its 25 percent composite airframe and has a very advanced avionics.[12]

The Airbus A380 is the world's largest passenger jet, with a wingspan much wider than that of Boeing's 747–400.[13] (See Figure II.) The A380s upper deck extends along the entire length of the fuselage, with a width equivalent to a wide-body aircraft. This gives the A380–800's cabin 5,920 square feet of usable floor

[11] Norris and Wagner, *Airbus A380*.

[12] "The Giant on the Runway," *Economist*, October 11, 2007.

[13] Jad Mouawad, "Oversize Expectations for the Airbus A380," *New York Times*, August 9, 2014.

STANDARD PASSENGER CAPACITY

3-class configuration

747-400   416

A380   555

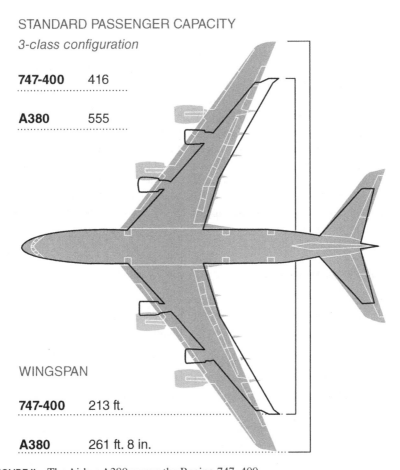

WINGSPAN

747-400   213 ft.

A380   261 ft. 8 in.

FIGURE II   The Airbus A380 versus the Boeing 747–400

space, 40 percent more than the next largest airliner, the Boeing 747–8, and provides seating for 555 people in a typical three-class configuration or up to 853 people in an all-economy-class configuration. Despite the dimensions, the A380 has a better field performance than the 747 due to its highly efficient wings. More in detail, for take-off and landing the A380–800 needs 2,990 meters and 2,103 meters respectively; this means 550 meters and 150 meters less than the 747–400. Finally, A380 has the quietest cabin in sky and a very smooth ride.[14]

---

[14] Royal Aeronautical Society, Hamburg Branch/DGLR/VDI/HAW and Axel Flaig, *Airbus A380: Solutions to the Aerodynamic Challenges of Designing the World's Largest Passenger Aircraft*, January 2008.

The Airbus A380 program has the sophistication of a space program. To deliver these performances, Airbus used several new technologies. For instance, the A380 is the first aircraft using computational fluid dynamics (CFD)[15] far more sophisticated than the previous-generation technology.[16] The CFD does not replace the wind tunnel test completely; the wind tunnel still is the central tool for design verification and performance determination of the realistic complete three-dimensional aircraft design, but by using the CFD, the number of experiments and the overall cost can be significantly reduced; the CFD enables engineers to design and preselect promising concepts before the wind tunnel is used.[17] The software used for A380 could simulate complex flow behavior in some areas, such as between the wing and the engine nacelle (engine enclosure), where drag problems can occur. In addition, the A380 is the first airliner to use a center wing box, which joints the wings to the fuselage, made mostly by carbon-fiber composite material. In this way it has been possible to save 1 ton from the overall wing box weight of 10 tons.[18]

AIRBUS A380 AMENITIES

Due to the size of the A380, all its major customers can offer passengers special amenities not found on other planes. Quite popular among the A380 airlines customers is the Onboard Lounge.[19] More specifically:

- The Emirates Airbus A380 offers to first-class and business class passengers the opportunity to enjoy in the lounge exclusive wines, champagne, cocktails, and beers plus hot and cold beverages served by a dedicated bartender.
- The Qatar Airways lounge area looks more like an executive club than an airplane. Business and first-class passengers can enjoy a range of five-star canapés, a fully stocked bar, leather seating, and fresh flowers, all of which provide an escape from the typical airplane environment.
- For first- and business class passengers of Etihad Airways who want to relax or socialize, there is "the Lobby." Inspired by intimate spaces found

---

[15] CFD is one of the tools (in addition to experimental and theoretical methods) available to solve and analyze problems that involve fluid flows. CFD relies on the numerical solution of Navier-Stokes equations, which are partial differential balance equations that describe how the velocity, pressure, temperature, and the like, of moving fluids are related.

[16] Bill Sweetman, "Jumbo Jumbo," *Popular Science* (November 2003), pp. 77–79.

[17] D. Reckzeh, *Aerodynamic Design of Airbus High-Lift Wings in a Multidisciplinary Environment*, European Congress on Computational Methods in Applied Sciences and Engineering, Jyväskylä (Finland), July 24–28, 2004.

[18] Sweetman, "Jumbo Jumbo."

[19] Katie Amey, "Mid-Air Cocktails, Five-Star Hors D'oeuvres and Live Sporting Events: the Most Luxurious On-Board Airline Bars in the Sky Revealed," *Daily Mail on Line*, October 10, 2014.

in boutique hotels, the Lobby is an in-flight, fully serviced lounge with a semicircular leather sofa and a large television screen.

● The A380 Korean Air airplanes are equipped with the "Celestial Bar," a relaxation space for first- and prestige class passengers. The chic interior design, exquisite lighting, and impressive sky lounge provide the perfect atmosphere. This is the spot to wind down and relax while sipping on specially crafted cocktails.[20]

However, only Emirates and Etihad can boast of having for first-class passengers private suites and shower spas. All suites are equipped with shoulder-height privacy doors that can be used to block out other passengers, making the suites really private rooms in the sky. Both airlines have showers furnished with spa toiletries; the difference is the shower-to-passenger ratio. Emirates has one shower for every seven passengers, while Etihad has "only" one shower for every nine passengers.[21] When the A380 was unveiled in 2005, the expectations about amenities were even greater; there was discussion of onboard shopping, full-service bars, lounges for surfing the Internet, a playroom to keep kids out of the way while their stressed-out parents relax at the spa, and even Las Vegas–style casinos.[22] Probably the realities of airline economics have kept such dreams grounded.

AIRBUS VERSUS BOEING: TWO DIFFERENT VISIONS

The decision of Airbus to build the A380 was based on the fact that air travel continued to grow by about 5 percent a year (see Figure III) and half of the world's 100 fastest-growing long-haul routes connected two big hubs, such as Hong Kong–London and New York–Tokyo.[23] Therefore, Airbus figured that the future of air travel belonged to big planes flying between major hubs.[24]

In contrast, Boeing came to radically different conclusions, arguing that the market for very large, long-haul planes had been fragmented by the increasing popularity and capability of so-called heavy twins, a new-generation airplane like its own 777 (introduced in 1995 and called "Triple Seven") with only two engines and able to carry more than 350 passengers in three classes flying for more than 10,000 miles. According to Boeing, aircraft like 747 and the A380 were now no more than "niche products." In other words, instead of the hub-and-spoke system, in which

---

[20] Korean Air website, "Arrive in Luxury—More Than Travel, It Is an Experience," www.Koreanair.com

[21] Loungebuddy blog entry, www.loungebuddy.com/blog/first-class-showdown-emirates-vs-etihad/.

[22] Bill Hoffman, "Even the Sky's No Limit—Biggest-Ever Jumbo Jet Has Store, Spa and Casino," *New York Post*, January 18, 2005. http://nypost.com/2005/01/18/even-the-skys-no-limit-biggest-ever-jumbo-jet-has-store-spa-casino/.

[23] "The Giant on the Runway."

[24] Mouawad, "Oversize Expectations for the Airbus A380."

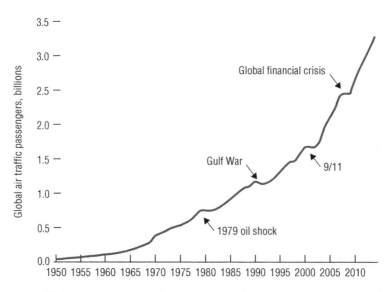

FIGURE III   Global air passenger traffic trend, 1950 to 2014

Source: International Air Transport Association

passengers fly in 747s or A380s to big hub airports and then take short-haul flights (spoke) to their final destination, Boeing maintained that passengers wanted the convenience of flying point to point. Boeing figured that traffic would move away from big hubs and toward secondary airports.[25] Therefore, while European engineers were developing the A380 plane, their counterparts at Boeing were working on alternative designs. Out of this effort came the 787 Dreamliner (introduced in 2011), with a carbon-composite fuselage, a host of electronic systems, and more efficient engines that could fly longer distances while consuming less jet fuel.[26]

When the Triple Seven entered service, airlines were worried that passengers would not accept the idea of flying across oceans on only two engines (even though such aircraft can manage on just one in an emergency). However, due to the reliability and power of high-bypass turbofan engines, people ceased to worry.[27] Recognizing the opportunity of this market, Airbus started to develop its own version of a long-range, fuel-efficient airplane, called the "A350-XWB,"[28] which entered service in 2015.[29]

---

[25] "The Giant on the Runway."

[26] Mouawad, "Oversize Expectations for the Airbus A380."

[27] "The Giant on the Runway."

[28] "XWB" stands for "eXtra Wide Body," due to the fuselage, which enables the widest seats of any jetliner in the same category, ensuring maximum comfort for passengers and crew.

[29] Mouawad, "Oversize Expectations for the Airbus A380."

PROGRAM MANAGER AND PROJECT MANAGER

So far we have discussed the decisions taken basically by top management. Nevertheless, when a project or concept is ready to be carried out and produced, everyone else in the company has an important role to play. With their passion and dedication, the engineers, technicians, workers, and others ensure that the project is realized in the best possible way. Two key figures play crucial roles at this state: the project manager and the program manager.

The project manager manages the operations of an individual project within the program. He or she coordinates time, budget, and resources and delegates tasks across the team. The project manager reports to the program manager on progress and changes made to the initial project plan.[30] The program manager has a greater breadth of responsibilities, oversees multiple project teams, and is held accountable for the overall outcome of the program. Numerous projects may feed into a program, and the program manager must monitor them all and understand how each contributes to program success.[31]

Figure IV shows where program and project managers are positioned in a typical organizational chart for the aerospace industry. More specifically, for each main component of Aircraft (fuselage, wings, etc.) there is an organizational structure like the one shown on the left side of the exhibit.

In this figure, the customer does not have a passive role (in other words, it does not receive only the final product); it can take an active role in the development of the entire airplane. It was Tim Clark, president of Emirates, the biggest customer for the A380, who came up with the idea to install two showers for first-class passengers. Airbus engineers thought the idea was crazy because it would require more fuel to carry the water for the showers. But Worth dismissed their objections, noting that the showers would immediately distinguish the plane from anything else in the air.[32]

The organizational chart displayed in Figure IV provides only a rough idea about the overall organization; in reality, the structure is much more complex. Specifically, many groups participate in the same task; therefore, there are several project managers. For this reason, it is vital that information and knowledge is exchanged among the teams, especially in a company like Airbus, where four countries, each with different cultures, must work together on large projects. If there is poor communication and coordination, the risk of serious problems is high.

In the autumn of 2006, the preassembled wiring harnesses, produced in the Airbus factory in Hamburg, Germany, were delivered to the assembly line in

---

[30] "The Giant on the Runway."

[31] Jeffrey Peisach and Timothy S. Kroecker, "Project Manager and Program Manager: What's the Difference?" *Defense AT&L* (July–August 2008).

[32] Mouawad, "Oversize Expectations for the Airbus A380."

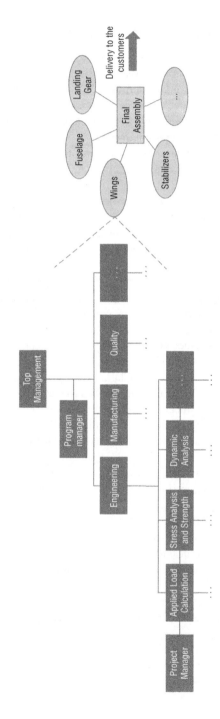

Figure IV Typical organizational chart in an airplane project

Toulouse. The workers discovered that the harnesses did not fit properly into the plane. The reason was that Hamburg had designed the wiring harnesses using an older version of CATIA software.[33] Unfortunately, there were issues of compatibility between the versions. This resulted in numerous design problems and inconsistencies. For example, wires that were manufactured to a specific length turned out to be too short. For an aircraft with more than 100,000 wires, this problem was costly to fix. Therefore, Airbus was forced to go back to the drawing board and redesign the wiring system. As a consequence, the entire project suffered a heavy setback, several months behind schedule, with an expected loss of $6.1 billion over the following four years.[34] Even though airlines that ordered the A380 did not cancel their orders, the reputation of Airbus was seriously compromised. Airbus was exposed to severe criticism and doubts on its structure and managements organization. Hans Weber, CEO of the Tecop International aviation consultant of San Diego, who had close contacts with the company's German operations, said: "The various Airbus locations had their own legacy software, methods, procedures, and Airbus never succeeded in unifying all those efforts."[35] Others, like Clark, criticized project time constraints that, according to him, were similar to smaller aircraft and did not take the greater complexity into account.[36] "Normally you need four to five years from the time you announce the launch of a new plane until delivery," said Jean-François Knepper, co-president of the European workers' committee at Airbus and a representative of the French union Force Ouvrière. "Airbus had never built a plane of this complexity before, and yet managers did not take the precaution of building more flexibility into the delivery schedule." Even Jürgen Thomas, the head of the large aircraft division at Airbus in the late 1990s, who some call the father of the A380, said: "It is very significant and very worrying, first there is the terrible financial loss, and secondly, you have the question of the confidence of the airlines." Nevertheless, on October 15, 2007, Airbus was able to deliver the first aircraft to Singapore Airlines.

Even though such kinds of mistakes seem inconceivable, unfortunately they are not rare. One of the most famous examples in aerospace history is the metric mishap that caused the loss of a NASA orbiter in the Mars mission of 1999. In that case, a $125 million Mars orbiter was lost because a Lockheed Martin engineering team used English units of measurement while the agency's team used the more conventional metric system for a key spacecraft operation. Specifically, after

---

[33] CATIA (computer-aided three-dimensional interactive application) is a professional computer-aided drafting program to aid in the creation, modification, analysis, or optimization of a design.

[34] Barry Shore, "Airbus 380," www.globalprojectstrategy.com

[35] Carol Matlack, "Airbus: First, Blame the Software," October 5, 2006. www.bloomberg.com35

[36] Nicola Clark, "The Airbus Saga: Crossed Wires and a Multibillion-Euro Delay," *The New York Times*, December 11, 2006. www.nytimes.com/2006/12/11/business/worldbusiness/11iht-airbus.3860198.html. A print version of this article also appeared in the *International Herald Tribune*, December 11, 2006.

a nearly 10-month journey to Mars, the probe fired its engine to push itself into orbit.[37] "People sometimes make errors," said laconic Edward Weiler, NASA's associate administrator for space science in a written statement. He added also: "The problem here was not the error, it was the failure of NASA's systems engineering, and the checks and balances in our processes to detect the error. That's why we lost the spacecraft."[38]

This is an important lesson learned that project managers must keep in mind: The success of one team is not a guarantee of success for the whole program if there is lack of coordination and communication among the teams. In addition, as suggested by Weiler, in a big and challenging project, there must exist fail-safe conditions that prevent a mistake made by a single team from going too far. Such problems can be avoided only through the checks and balances both within the team and among the teams and when work of one team is checked by other teams. Project managers are tasked with ensuring that such a complex structure is efficient.

The use of a new technology often can cause problems. At the beginning of 2013, several airlines, including Qantas and Lufthansa, found some cracks in the A380's wings. The cracks occurred because of a combination of materials and assembly issues as well as problems with the design process in adapting new materials into an existing design. They were caused by the interface between metal and carbon-fiber composite components inside the airliner's wings. As Tom Enders, CEO of Airbus Group, explained: "We thought we understood the properties of the materials and the interface between carbon fiber and metal and found out the wrong way we didn't know everything." He added also that Airbus did not have the right design controls in place to anticipate possible problems.[39] Airbus immediately applied a short-term fix to about a third of the A380s in service and planned longer-term repairs that required each plane to be grounded for eight weeks. That same year, Boeing faced a composite delamination manufacturing quality issue in the fuselage of its 787 Dreamliner.[40] Boeing said the issue was a "straightforward repair" and posed no "short-term safety concern"; those repairs would require only 10 to 14 days per plane and would not affect future deliveries.

Industry experts say it is not unusual for new planes to develop unforeseen faults in the first few years of flight. These faults normally are quickly identified

---

[37] Robin Lloyd, "Metric Mishap Caused Loss of NASA Orbiter," CNN.com, September 30, 1999. edition.CNN.com/TECH/Space.

[38] "NASA's Metric Confusion Caused Mars Orbiter Loss," CNN.com, September 30,1999. edition. CNN.Com/TECH/Space.

[39] Quoted in Ann R. Thryft, "Airbus Delays Repair of Cracked A380 Wings," June 19, 2012. www.designnews.com .

[40] Jon Ostrower, "Delamination Prompts Boeing to Inspect 787 Fleet," February 5, 2012. www.flightglobal.com.

and ironed out.[41] Thus, according to industry experts, nothing can be done to avoid such issues. Companies must have the right attitude and give immediate assistance to customers.

SALES DATA AND FUTURE OF THE A380

Table I summarizes all the airlines worldwide that have ordered the A380 as of August 31, 2016. As the figures show, the biggest customer for the A380 is clearly Emirates; it has signed almost half of all A380 contracts.

The overall backlog of 125 units looks to be shaky. The Dublin-based aircraft-leasing company Amedeo, even though it is confident of placing the world's largest jetliner with airlines,[42] has not yet placed the aircraft with any operators.[43] Virgin Atlantic has repeatedly delayed accepting delivery of the jet, and Richard Branson, the airline's billionaire founder, said even a decade and a half after the order was placed, that it must "still be determined" what role a double-decker would play. He added that though the six aircraft remain in Airbus's backlog, Virgin Atlantic has no obligation to take them; this is supported by the declaration that the A350 will be Virgin's "future flagship." Behind this decision there is the fact that the composite, twin-turbine A350 is less costly to run than the all-metal, four-engine A380, and also will be less of a challenge to fill as Virgin replaces its larger Boeing 747 jumbos.[44] For the same reason, Emirates airline president Tim Clark has been pushing Airbus to develop an A380neo ("neo" for new engine option) featuring more efficient engines. In this case, Emirates would be willing to increase its A380 fleet to 200, but Airbus would like additional customers before launching.[45] In addition, as Airbus CEO Fabrice Brégier said, if oil prices remain low, a more fuel-efficient option would be less attractive.[46] On this subject, consider these running costs per seat hour: for an A380, about $50; for 747–400, $90; and for 777–300ER, $44.[47]

---

[41] Tom Gardner, "Wing Cracks in £250million A380 Superjumbos that Grounded Entire Fleet Blamed on British Engineering," *Daily Mail on Line*, March 2, 2012.

[42] "Dublin Aircraft Lessor May Have Customers for A380s," *Irish Times*, March 23, 2015.

[43] Jens Flottau, "Airbus Halves A380 Production," *Aviation Week*, July 12, 2016. http://aviationweek.com/commercial-aviation/airbus-halves-a380-production.

[44] Andrea Rothman, "Virgin Atlantic A380s a Distant Dream After 'Flagship' A350 Deal," Bloomberg, July 11, 2016. www.bloomberg.com/news/articles/2016-07-11/virgin-atlantic-a380s-a-distant-dream-after-flagship-a350-deal.

[45] Flottau, "Airbus Halves A380 Production."

[46] Jens Flottau, "Emirates Would Buy 100 A380neos, Agrees to Amedeo Sale-Leaseback Deals," *Aviation Week*, January 25, 2015. http://aviationweek.com/commercial-aviation/emirates-would-buy-100-a380neos-agrees-amedeo-sale-leaseback-deals.

[47] AirInsight, "The A380s Future," November 20, 2015. https://airinsight.com/2015/11/20/the-a380s-future/.

TABLE I   A380 ORDERS AND DELIVERIES OF AIRLINE COMPANIES WORLDWIDE

A380 CUSTOMERS			
NAME	NATION	Orders	Deliveries
Emirates	United Arab Emirates	142	82
Singapore Airlines	Singapore	24	19
Amedeo	Ireland	20	
Qantas Airways	Australia	20	12
Lufthansa	Germany	14	14
Air France	France	12	10
British Airways	United Kingdom	12	12
Etihad Airways	United Arab Emirates	10	8
Korean Air	Korea, Republic of	10	10
Qatar Airways	Qatar	10	6
Undisclosed Buyers		10	
Asiana Airlines	Korea, Republic of	6	4
Malaysia Airlines Berhad	Malaysia	6	6
Thai Airways International	Thailand	6	6
Virgin Atlantic	United Kingdom	6	
China Southern Airlines Company	China	5	5
Air Accord	Bermuda	3	
All Nippon Airways	Japan	3	
Totals		319	194

Source: Airbus Press Center. Orders & Deliveries, August 31, 2016. www.airbus.com/presscentre

It is undeniable that A380s orders did not meet the aspirations of Airbus. Since 2000, as we have seen, the A380 has won 319 orders, less than a third of the 1,200 Airbus projected in the plane's first 20 years. Not a single U.S. airline has ordered it, and only five have been contracted for China, which will be the world's largest aviation market in 20 years.[48] For Airbus as a company, recovering the €25 billion ($28 billion) spent on the program will take many years, especially considering that the list price of $432.6 million is subject to discounts that airlines negotiate.[49]

---

[48] Andrea Rothman, "Airbus A380 Haunted by Feeble Orders Marks Decade in Skies," Bloomberg, April 27, 2015. www.bloomberg.com/news/articles/2015-04-26/airbus-a380-haunted-by-lack-of-orders-marks-decade-in-the-skies.

[49] Daniel Michaels, "Airbus Wants A380 Cost Cuts," *Wall Street Journal*, July 13, 2012.

The A380 demand has evaporated in recent years with the introduction of more nimble twin-engine jets like 787 and A350 that have a total order of 1,161 and 810, respectively. Some signs would seem to confirm the Airbus idea of the spoke-hub model,[50] but for the moment, it does not appear to be enough. In 2014, for instance, British Airways replaced three B777 flights between London and Los Angeles with two A380s. This resulted in 19 percent lower trip costs, 5 percent more premium seats, 7 percent fewer nonpremium seats, and 1 percent fewer seats overall while releasing a valuable pair of slots at each airport.[51] However, in general, airlines seem to prefer relatively smaller twin-engine jets that can fly nonstop to their ultimate destinations, bypassing large hubs. Airline managers are trained to be risk averse, and many believe that the A380 offers a lot of risk.[52]

At the Farnborough International Air Show in England on July 16–22, 2016, Brégier declared that the A380 was launched too early with respect to the market demand, maybe 10 years before the right time.[53] However, he is still convinced that airline companies will need such a big plane. More specifically, the Airbus Global Market Forecast for 2016 to 2035 anticipates that air traffic will grow at 4.5 percent annually. Therefore, an aircraft able to carry 500/600 people would be the best response for airline companies to reduce airport congestion, a fact that inevitably will become more critical in the future. In any case, to stay in line with the current A380s demand, Brégier announced that A380 production would be slashed from the current 2.5 aircraft per month to one per month by 2018.[54] A similar decision was earlier taken by Boeing: It would reduce, starting September 2016, the production rate of the B747 to 0.5 per month.[55] Airbus management is seeking to reduce production costs to allow the A380 to remain viable at lower production levels. In other words, the main goal is to keep the break-even rate of 27 deliveries achieved in 2015 also with at a lower rate, leaving the option to ramp up again when the market requires. On this subject, Brégier was quite clear: "With this prudent, proactive step we are establishing a new target for our industrial planning, meeting current commercial demand but keeping all our options open

---

[50] The *spoke-hub model* assumes that people will be willing to pay for the added luxury of flying from hub to hub, and then change planes to get to their final destination. Some people prefer to fly from point to point, nonstop, forego the luxury of added amenities, and pay a lower price for the ticket.

[51] Ian Goold, "Ainonline, Airbus, Airlines Happy with A380," AIN.online, July 13, 2014. www.ainonline.com/aviation-news/air-transport/2014-07-13/airbus-airlines-happy-a380.

[52] AirInsight, "The A380's Future," November 20, 2015. https://airinsight.com/2015/11/20/the-a380s-future/.

[53] "Airbus: 'l'A380 a été lancé trop tôt' (Brégier)," *Le Figaro*, March 18, 2016. www.lefigaro.fr/flash-eco/2016/03/18/97002-20160318FILWWW00074-airbus-l-a380-a-ete-lance-trop-tot-bregier.php.

[54] Guy Dutheil, "Faute de commandes, Airbus diminue de moitié la production de l'A380," *Le Monde Economie*, July 15, 2016. www.lemonde.fr/entreprises/article/2016/07/13/faute-de-commandes-airbus-diminue-de-moitie-la-production-de-l-a380_4968826_1656994.html.

[55] Jon Ostrower, "Boeing to Cut Production of 747s," *Wall Street Journal*, January 21, 2016.

to benefit from future A380 markets, which we consider. . .a given." In addition, he underlined his faith in the A380: "We are maintaining, innovating and investing in the A380, keeping the aircraft the favorite of passengers today and in the future. The A380 is here to stay!"[56]

CONCLUSIONS

All agree, friends and foes, that the A380 is phenomenal engineering achievement. However, since the beginning of its history, some have perceived the A380 as a magnificent cruise liner of the skies with unprecedented moneymaking potential; others see it as a huge white elephant with questionable economics and a slim chance of success.[57]

Only time and therefore the evolution/development of the airline market let us evaluate the A380 from a business standpoint. In this regard, it is wise to keep in mind what Sandy Morris of Jefferies Investment Bank reminds us: "The peak year for orders of the 747 came nearly 25 years after it first took to the skies. Airbus's A330 took 15 years to hit the heights and had no orders in 1994, six years after its launch."[58]

What we can say is that if Airbus had not chosen to develop such a plane, Boeing probably would have had a monopoly on the super-jumbo jet market with its 747. Boeing would have become stronger and therefore capable of investing more resources in the other segments. In economic terms, the opportunity cost[59] for Boeing would probably have been higher. This reasoning will be considered particularly valid if the growth of airlines market follows Airbus predictions.

An interesting aspect can be underlined in this project: the appeal and the interest that A380 has been able to create. Most people around the world are eager to fly on a A380, but why? Because it is the biggest civil plane in the world. Therefore, people are more excited about flying in one and receive much more pleasure by doing so. In this regard, the A380 is a kind of advertisement for Airbus. Its popularity among airline passengers around the world has grown: In 2015, the A380 was voted by *Global Traveler* readers as the best aircraft in the world. This

---

[56] Peggy Hollinger, "Airbus Slashes Production of A380 Superjumbo," *Financial Times*, July 12, 2016.

[57] Norris and Wagner, *Airbus A380*.

[58] Economist Redaction, "Airbus's Big Bet," *The Economist*, November 23, 2013.

[59] Applied to a business decision, the term *opportunity cost* refers to the profit that a company could have earned from its capital, equipment, and real estate if these assets had been used in a different way. Simply stated, an opportunity cost is the cost of a missed opportunity. Although opportunity costs are not generally considered by accountants—financial statements include only explicit costs, or actual outlays—they should be considered by managers. (Inc.com, "Opportunity Cost," www.inc.com/ency-clopedia/opportunity-cost.html).

achievement is particularly relevant considering that it marked the first time that honor went to an aircraft not made by Boeing.[60]

To take advantage of passengers' love for the A380, Airbus launched the website "I Fly A380" (www.iflya380.com). As Marc Fontaine, Airbus digital transformation officer, explained: "Booking systems today do not allow the passengers to easily choose their preferred aircraft and we decided to fill that gap by easing the access to the iconic A380 aircraft, for everybody. For the first time, a booking service puts the aircraft type as the criteria for flight selection. This seamless experience is a win for Airbus A380, it's a win for the airlines operators and it's a win for the passengers."[61] The A380 is passengers' favorite aircraft, and 60 percent of them are willing to make an extra effort to fly on this plane.[62]

QUESTIONS

1. What are the enterprise environmental factors that affect the success of the A380? How important are they?
2. What are the risks facing Airbus with the decision to launch the A380?
3. Airbus has several locations in four countries, and many of these locations must work together. What are the risks especially with regard to culture?
4. What are the risks that Airbus faces when it allows each potential buyer to customize its amenities?
5. How can oil prices impact the success of the A380?
6. What is the impact on project scheduling when customers delay or cancel their orders?
7. From a cost perspective, why is continuous production important for Airbus rather than periodic production?
8. What lessons were learned from the CATIA problem with the wiring and housing?
9. Do amenities and scope changes have a great impact on Airbus's hub-and-spoke concept or Boeing's point-to-point concept?

---

[60] Fox News Travel, "Boeing Loses Best Aircraft Award to Airbus A380 for the First Time Ever," December 14, 2015. www.foxnews.com/travel/2015/12/14/boeing-loses-best-aircraft-award-to-air-bus-a380-for-first-time-ever.html.

[61] Airbus Press Center, "Airbus Launches the 'I Fly A380' Website: Choose. Fly. Love A380," July 14, 2016. www.airbus.com/presscentre/pressreleases/press-release-detail/detail/airbus-launches-the-i-fly-a380-website-choose-fly-love-a380/

[62] Survey by Epinion: Independent agency surveying passengers arriving on A380 flights at London Heathrow, cited in "Flying on the iconic double deck is now a matter of choice, not chance!" Airbus press release, July 14, 2016. www.airbus.com/presscentre/pressreleases/press-release-detail/detail/airbus-launches-the-i-fly-a380-website-choose-fly-love-a380/.

Part 21

AGILE/SCRUM PROJECT MANAGEMENT

During the last two decades, executives have seen the benefits of project management and recognized that project management is now a corporate strategic competency. For that reason, they are now putting more trust in the hands of the project managers for both business and project-related decisions.

This trust has allowed for the development of more streamlined project management approaches, such as with agile and Scrum. These new methodologies function more as frameworks than rigid methodologies and allow project managers more freedom in customizing the project management approach to fit the needs of the customers. But with any new approach, there are always obstacles that must be resolved, especially when people are outside their comfort zones.

Agile (A): Understanding Implementation Risks

In the past decade, many information technology (IT) companies have changed their systems development life-cycle methodology to a more flexible framework approach, such as agile and Scrum. The basis for these flexible frameworks was the Agile Manifesto, introduced in February 2001, which had four values:

1. Individuals and actions over processes and tools
2. Working software over comprehensive documentation
3. Customer collaboration over contract negotiation
4. Responding to change over following a plan

The primary players in the framework included a Scrum Master—a product owner who represented the customer and the development team. The framework allowed for evolving scope and changes to be made while keeping the customer involved during the entire project. Work was broken down into sprints of two to four weeks, in which the team performed the tasks needed to be completed.

Without the formality of a rigid methodology, the need for very detailed up-front planning and expensive and often unnecessary documentation, projects could be aligned quickly to the customer's business model rather than the contractor's business model. Alignment of the framework to the way that the customer does business, combined with continuous and open communication with the customer, often was seen as the primary driver of project success, an increase in customer satisfaction, the creation of the desired business value the customer wanted, and repeat business.

The increase in the success rate of IT projects did not go unnoticed in other industries. But several questions needed to be addressed before these flexible methodologies could be applied to other types of projects and adopted in other industries.

- Will flexible methodologies work on large, complex projects?
- What if the work cannot be broken down into small sprints?
- What if the customer or business owner will not commit resources to the project?
- What if the customer does not want continuous communication over the life cycle of the project?
- What if the customer wants detailed up-front planning?
- What if the customer says that no scope changes will be authorized after project go-ahead?
- What if the customer will not allow for trade-offs on time, cost, and scope?
- What if the customer uses a rigid methodology that may be inflexible?
- What if your methodology cannot be aligned to the customer's methodology?
- What if during competitive bidding, the customer either does not recognize or understand the agile/Scrum approach or does not want it used on the project?
- What if the customer wants you to use its methodology on the project?
- What if you are one of several contractors on a project and you all must work together but each contractor has a different methodology?
- Can the use of a flexible methodology or framework prevent you from bidding on some contracts?
- Can your company maintain more than one methodology throughout the company based on the type of project being undertaken?

REMCO'S CHALLENGE

The executive staff at Remco Corporation was quite pleased with the one-day training program they attended on the benefits of using agile and Scrum on some of their projects. Remco provided products and services to both public and private sector clients, almost all of it through competitive bidding. IT was not required for any of the products and services Remco provided. Agile and Scrum had proven to be successful on internal IT projects, but there were some concerns as to whether the same approach could be used on non-IT-related projects for clients. There was also some concern as to whether clients would buy into the agile and Scrum approach.

Remco recognized the growth and acceptance of agile and Scrum as well as the fact that it might eventually impact its core business rather than just internal

IT. At the request of Remco's IT organization, a one-day training program was conducted just for the senior levels of management to introduce them to the benefits of using agile and Scrum and how their techniques could be applied elsewhere in the organization. The executives left the seminar feeling good about what they heard, but there was still some concern as to how this would be implemented across possibly the entire organization and what the risks were. There was also some concern as to how their clients might react and the impact this could have on how Remco does business.

THE NEED FOR FLEXIBILITY

Remco was like most other companies when it first recognized the need for project management for its products and services. Because executives were afraid that project managers would begin making decisions that were reserved for the executive levels of management, a rigid project management methodology was developed based on eight life-cycle phases:

1. Preliminary planning
2. Detail planning
3. Prototype development
4. Prototype testing
5. Production
6. Final testing and validation
7. Installation
8. Contractual closure

The methodology provided executives with standardization and control over how work would take place. It also created an abundance of paperwork.

As project management matured, executives gained more trust in project managers. Project managers were given the freedom to use only those parts of the standard methodology that were necessary. As an example, if the methodology required that a risk management plan be developed, the project manager could decide that the plan was unnecessary since this project was a very low risk. Project managers now had some degree of freedom, and the rigid methodology was slowly becoming a flexible methodology that could be easily adapted to a customer's business model.

Even with this added flexibility, there were still limits as to how much freedom would be placed in the hands of the project team. As shown in Figure I, the amount of overlap between Remco's methodology, the typical client's methodology, and the agile methodology was small. Remco realized that, if it used an agile project management approach, the overlap could increase significantly and lead to more business. But again, there are risks, and more trust would need to be given to the project teams.

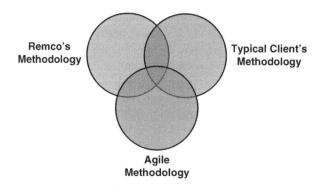

FIGURE I   Overlapping of methodologies

THE IMPORTANCE OF VALUE

Remco's project management community had spent quite a bit of time trying to convince senior management that the success of a project cannot be measured solely by meeting the triple constraint of time, cost, and scope. Rather, they argued that the true definition of success is when business value is created for the client, hopefully within the imposed constraints, and the client recognizes the value. Effective client-contractor communication, as identified in the Agile Manifesto, could make this happen.

The course reinforced the project managers' belief that measuring and understanding project value was important. Agile and Scrum was shown to be some of the best techniques to use when the value of the project's deliverables was of critical importance to the customers, whether they are internal or external customers.

Remco's project management experience was heavily based on the traditional waterfall approach where each phase of a project must be completed before the next phase begins. This creates a problem with measuring value as shown in Figure II. With the traditional waterfall approach, value is measured primarily at the end of the project. From a risk perspective, there exists a great deal of risk with this approach because there is no guarantee that the desired value will be there at the end of the project when value is measured, and there may have been no early warning indicators as to whether the desired value would be achieved.

With an agile approach, value is created in small increments as the project progresses, and the risk of not meeting the final business value desired is greatly reduced. This incremental approach also reduces the amount of time needed at the end of the project for testing and validation. When used on fast-changing projects, agile and Scrum methodologies often are considered to have built-in risk management functions. Though other methods of risk mitigation are still necessary, this additional benefit of risk mitigation was one of the driving forces for convincing

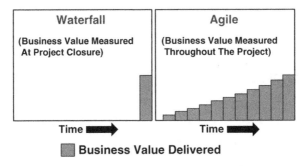

FIGURE II Creation of business value

executives to consider changing to an agile approach for managing projects. Agile and Scrum were seen as excellent ways to overcome the traditional risks of schedule slippages, cost overruns, and scope creep.

CUSTOMER INVOLVEMENT

For decades, project management training programs for public and private sectors recommended that customers should be kept as far away from the project as possible to avoid customer meddling. For this reason, most customers took a somewhat passive role because active involvement in the project could limit career advancement opportunities if the project were to fail.

Remco has some client involvement in projects for the private sector but virtually no involvement from public-sector government agencies. Public-sector organizations wanted to see extensive planning documentation as part of the competitive bidding process, an occasional status report during the life cycle of the project, and then the final deliverables. Any and all problems during the execution of the project were the responsibility of Remco for resolution with little or no client input.

Remco understood that one size does not fit all and that agile and Scrum might not be applicable to larger projects, where the traditional approach to project management using the waterfall methodology may be better. There would need to be an understanding as to when the waterfall approach was better than agile.

For agile and Scrum to work on some smaller projects, there would have to be total commitment from the customer and their management team throughout the life of the project. This would be difficult for organizations unfamiliar with agile and Scrum because it requires the customer to commit a dedicated resource for the life of the project. If the customer does not recognize the benefits of this, then it may be perceived as an additional expense incurred by awarding Remco the contract. This could have a serious impact on competitive bidding and procurement activities and make it difficult or impossible for Remco to win government contracts.

SCOPE CHANGES

With traditional project management, accompanied by well-defined requirements and a detailed project plan, scope changes were handled through the use of a change control board (CCB). It was anticipated that, with the amount of time and money spent initiating and planning the project, scope changes would be at a minimum. Unfortunately, this was not the case with most projects, especially large and complex ones. When scope change requests were initiated, it became a costly and time-consuming endeavor for the CCB to meet and write reports for each change request, even if the change request was not approved.

With agile and Scrum, accompanied by active customer involvement, frequent and cheap scope changes could be made, especially for projects with evolving requirements. The changes could be made in a timely manner without a serious impact on downstream work and with the ability to still provide the client with the desired business value.

STATUS REPORTING

All projects, whether agile or waterfall, go through the Project Management Institute's domain areas of initiation, planning, execution, monitoring and control, and closure. But the amount of time and effort expended in each domain area, and how frequently some parts can be repeated, can change. To make matters worse, government agencies often mandate standardized reporting documents that must be completed. Many of these are similar to Gantt charts and other scheduling techniques that take time to complete. Customers may not be pleased if they are told that the status now appears on a Scrum board along with stories.

Government agencies tend to use standardized contracting models; and stating in a proposal that the contractor will be using an agile or Scrum approach may violate their procurement policies and make Remco nonresponsive to the proposal statement of work.

MEETINGS

One of the concerns that Remco's executives had was the number of meetings needed for agile and Scrum and the number of participants in attendance. The time spent in meetings by the product owner and the Scrum Master, and in many cases the team itself, was seen as potentially unproductive hours that were increasing the overhead to the project. With the waterfall approach, meetings almost always resulted in numerous action items that often required months and additional meetings to resolve. In agile and Scrum, action items were kept to a minimum and were resolved quickly because the people on the team had the authority to make decisions and implement change. This also made it easier to create business value deliverables in a timely manner. There are also techniques available to minimize the time spent in meetings, such as creating an agenda and providing guidelines for how the meetings will be run.

Agile and Scrum work with self-governed teams made up of people with different backgrounds, beliefs, and work habits. Without a definitive leader in these meetings, there exists the opportunity for conflicts and poor decision making. Without effective training whereby each team understands that they are working together toward a common goal, chaos can reign. People must believe that group decisions made by the team are better than the individual decisions typical of the waterfall approach. Decision making becomes easier when people have not only technical competence but also an understanding of the entire project. Effective meetings inform team members early on that certain constraints may not be met, thus allowing them sufficient time to react. This requirement for more information may require significantly more metrics than are used in waterfall approaches. Sometimes executives may be invited to attend these meetings, especially if they have information surrounding enterprise environmental factors that may have an impact on the project.

PROJECT HEADCOUNT

In the waterfall approach, an exorbitant amount of time is spent in planning with the belief that a fully detailed plan must be prepared at project initiation and will be followed exactly and that a minimum number of resources will be required during project execution. Risk and unpredictability are then handled by continuous and costly detailed replanning and numerous meetings involving people who may understand very little about the project, thus requiring a catch-up time.

In the waterfall approach, especially during competitive bidding, the client may ask for backup or supporting data as to why project personnel are needed full time rather than part time. Some government agencies argue that too many full-time people is an overmanagement cost on the project.

With agile and Scrum teams, the scope of the project evolves as the project progresses, and planning is done continuously in small intervals. The success of this approach is based on the use of full-time people who are under no pressure from other projects competing for their services. The people on the project are often rotated through various project assignments; therefore, project knowledge is not in the hands of just a few. The team therefore can be self-directed, with the knowledge and authority to make most decisions with little input from external resources (unless, of course, critical issues arise). The result is rapid feedback of information, a capturing of best practices and lessons learned, and rapid decision making. Collaborative decision making involving stakeholders with diverse backgrounds is a strength of the agile approach. Once again, such an approach could be a procurement detriment if the client does not have a knowledge of agile and/or Scrum during competitive bidding activities.

Remco now seemed aware of many of the critical issues and had to decide about converting over to an agile approach. It would not be easy.

QUESTIONS

1. Given the issues in the case that Remco is facing, where should Remco begin?
2. What should Remco do if the customers will not commit resources to agile or Scrum projects?
3. How should Remco handle employee career development when employees realize that there are no formal positions on agile/Scrum projects and titles may be meaningless? What if employees feel that being assigned to an agile/Scrum team is not a career advancement opportunity?
4. How harmful might it be to an agile team if workers with critical skills are either reassigned to higher-priority projects or are asked to work on more than one project at a time?
5. How should you handle a situation where one employee will not follow the agile approach?
6. In meetings when there is no leader, how do you resolve personality issues that result in constant conflicts?
7. Can an agile methodology adapt to change faster than a waterfall methodology?
8. Will the concept of self-organized teams require Remco to treat conversion as a cultural challenge?
9. Can part of the company use agile and another part of the company use waterfall?

Agile (B): Project Management Mind-set

Jane had been a project manager for more than 15 years. All of her projects were executed using traditional project management practices. But now she was expected to manage projects using an agile approach rather than the traditional project management approach she was accustomed to. She was beginning to have reservations as to whether she could change how she worked as a project manager. This could have a serious impact on her career.

THE TRIPLE CONSTRAINTS

Jane believed that clear scope definition, sometimes on a microscopic level, had to be fully understood before a project could officially kick off. Sometimes as much as 30–35 percent of the project's labor dollars would be spent in scope definition and planning the project. Jane deemed the exorbitant amount of money spent planning the project a necessity to minimize downstream scope changes that could alter the cost and schedule baselines.

Senior management was adamant that all of the scope had to be completed. This meant that, even though senior management had established a target budget and scheduled end date, the project manager could change the time and cost targets based on the detailed scope definition. Time and cost had flexibility in order to meet the scope requirements.

With agile project management, Jane would have to work differently. Senior management was now establishing a budget and a scheduled completion date, neither of which were allowed to change, and management was now asking Jane how much scope she could deliver within the fixed budget and date.

PLANNING AND SCOPE CHANGES

Jane was accustomed to planning the entire project in detail. When scope changes were deemed necessary, senior management would more often than not allow the schedule to be extended and let the budget increase. This would now change.

Planning was now just high-level planning at the beginning of the project. The detailed planning was iterative and incremental on a stage-by-stage basis. At the end of each stage, detailed planning just for the next stage would begin. This made it quite clear to Jane that the expected outcome of the project would be an evolving solution.

COMMAND AND CONTROL

Over Jane's 15-year career, as she became more knowledgeable in project management, she became more of a doer than a pure manager. She would actively participate in the planning process and provide constant direction to her team. On some projects, she would perform all of the planning by herself.

With agile project management, Jane would participate in just the high-level planning, and the details would be provided by the team. This meant that Jane no longer had complete command and control and had to work with teams that were empowered to make day-to-day decisions to find the solution needed at the end of each stage. This also impacted project staffing; Jane needed to staff her projects with employees whose functional managers felt they could work well in an empowered environment.

Jane's primary role now would be working closely with the business manager and the client to validate that the solution was evolving. As project manager, Jane would get actively involved with the team only when exceptions happened that could require scope changes resulting in changes to the constraints.

RISK MANAGEMENT

With traditional project management that was reasonably predictable, risk management focused heavily on meeting the triple constraints of time, cost, and scope. But with agile project management, where the budget and schedule were fixed, the most critical risk was the creation of business value. However, since the work was being done iteratively and incrementally, business value was also measured iteratively and incrementally, thus lowering some of the risk on business value.

QUESTIONS

1. How easy would it be for Jane to use an agile project management approach from this point forth?
2. If Jane could change, how long would it take?
3. Are there some projects where Jane would still be required to use traditional project management?
4. Empowerment of teams is always an issue. How does Jane know whether the team can be trusted with empowerment?

Agile (C): Managing and Reporting Project Agility

Linda had been the head of her company's project management office (PMO) for the past five years. All projects were managed using traditional project management practices and a single project management methodology established by the PMO and updated continuously. But now, with her company's adoption of agile project management techniques on several of the projects, many of the activities in her PMO would change.

During the past decade, there had been a rapid growth in agile project management practices, not just in IT, but in other types of projects as well. Most of the principles of agile project management practices had provided beneficial results when applied to non-IT projects. While all of this sounds good, there were also some headaches that accompany the growth. Linda understood the changes that needed to be made regarding use of a different approach, but her biggest headache would be with metrics. Using just time, cost, and scope metrics would no longer be sufficient. What metrics would be needed to track and report agility?

There is an old adage in project management: "You cannot manage what you cannot measure." Therefore, to manage projects using agile techniques, Linda understood that she must establish metrics to confirm that the benefits are being realized and agile practices are being executed correctly. Fortunately, accompanying the growth in agile practices had been growth in metric measurement techniques. Today Linda believed that just about anything could be measured and there were good metrics for reporting performance.

METRIC MANIA

Linda was convinced that her biggest challenge would be to combat metric mania, the insatiable desire to create metrics for metrics' sake rather than for what is really needed. Linda knew that there were pros and cons to having both too many metrics and too few metrics. There is always confusion in what metrics to choose, as Linda found out.

The result of having too many metrics is that:

- We steal time from important work to measure and report these metrics.
- We provide too much data, and the stakeholders and decision makers find it difficult to determine what information is in fact important.
- We provide information that has little or no value.
- We end up wasting precious time doing the unimportant.
- Too many metrics can open the door for unnecessary questions from stakeholders and business owners and eventually lead to a micromanagement environment.

The result of having too few metrics is that:

- We may not be providing the right information for stakeholders to make informed decisions based on evidence and facts.
- Stakeholders may be confused as to what is really happening on the project until the project is over.
- Stakeholders and business owners may be misled as to the true status of the project. This could lower customer satisfaction and risk the loss of follow-on work with this client.

In traditional project management using waterfall charts, Linda's reporting had always been done around the metrics of time, cost, and scope. With the use of the earned value measurement system, the number of metrics soon increased to 12 to 15. Linda feared that when a new approach appears, such as agile and/or Scrum, metric mania would set in, and people would believe that they must create significantly more metrics than were actually needed.

Linda found numerous publications that discussed the 10 or 20 important metrics that needed to be reported in agile project management. Some publications stated that there may be as many as 50 different metrics, and there appeared to be some disagreements as to what metrics were really important. Initially, with the acceptance of new approaches, Linda felt that it might be necessary to allow metrics mania to exist until she could filter out those metrics that did not provide informational value. As companies become mature in using any new approach, the number of metrics reported is usually reduced.

Having a good metric selection process would ease the impact of metric mania and allay many of Linda's fears. Linda understood that the use of a PMO and the use of organizational metric owners who report solid to functional managers and dotted to the PMO would certainly help.

METRIC MANAGEMENT

Having a good metric management program can minimize the damage of metric mania but cannot always eliminate it. Linda established a four-step metric management program that would be supported by her PMO and used on all projects where agile concepts were required:

1. Metric identification
2. Metric selection
3. Metric measurement
4. Metric reporting

Linda then sent out a memo to her PMO team explaining her thoughts on metric management:

> Metric identification is the recognition of those metrics needed for fact-based or evidence-based decision making. Today's project management methodologies, such as agile and scrum, function more as frameworks or flexible methodologies than rigid methodologies. These flexible methodologies can be customized to a particular project's or business owner's needs. Our external clients would prefer to work with a contractor who has a flexible methodology that can be customized to the client's business model than to work with a rigid methodology that supports only our business model. Flexible methodologies that are adaptable to any situation are the future and we must be prepared.
>
> When dealing with flexible methodologies, the notion of standardized metrics disappears. Each project can have a set of metrics that are unique to that particular project. Metrics can also be unique to individual parts of a project. Since each stakeholder and business owner may have different needs and make different decisions in support of the project, it is advisable at the onset of a project to ask each stakeholder and business owner what metrics they would like to have reported. It is possible that each stakeholder on the same project will request a different set of metrics based on their own needs and the decisions they are expected to make.
>
> Metric selection is when you decide how many of the identified metrics are actually needed. Some companies maintain metric libraries and show the metric library to the stakeholders and business owners. It is not uncommon for the first response to be "I would like to have all of the metrics in the library reported on my project." This is when the devastating effects of metric mania set in. We must create a metric library, but do it the correct way.

Metric selection is the first step in resolving metric mania issues. Ground rules for metric selection might include the following:

- There is a cost involved to track, measure, and report metrics even if we use a dashboard reporting system rather than written reports.
- If the intent of a good project management approach such as agile or Scrum is to reduce or eliminate waste, then the number of metrics selected should be minimized.
- We should encourage viewers of metrics to select the metrics they need, not the metrics they want. There is a difference!
- Asking for metrics that seem nice to have but provide no informational value, especially for decision making, is an invitation to create waste. This is something we must avoid.

There is no point in selecting metrics that are difficult and costly to measure. We can be supported by the use of metric owners. These are people who track the use of one or more metrics on all applicable projects and seek out best practices for improvements in the way the metric is measured and reported. This information will be periodically provided to the PMO, and updates will then made to the metric library. Metric measurements can become costly when the measurement data needed does not come from our computerized information system.

Paperwork is the greatest detriment to most project managers. The future of project management practices is to create a paperless project management environment. This does not mean that we will be 100% paperless since some reporting is mandatory, but it does mean that we recognize that unnecessary paperwork is waste that needs to be eliminated. To do so, we will rely heavily on dashboard project performance reporting.

Each stakeholder or business owner can have dashboards that are customized with the metrics they selected. This customization may seem like an added expense, but we must also consider that, once the dashboards are designed, updating the information is relatively fast if we use Excel spreadsheets as the drivers for the data for the images on the dashboards. Although customized dashboard reporting may appear as an added expense initially, we must also consider the long-term impact it can have on business owner, stakeholder, and customer satisfaction.

We have been discussing that some flexible methodologies may have as many as 50 metrics. Dashboard reporting systems can force viewers to be selective in the metrics they wish to see on the dashboard. A typical dashboard screen has a limited amount of real estate, namely the space for usually only six to 10 metrics that are aesthetically pleasing to the eyes and can be easily read. Therefore, telling stakeholders or business owners that we wish to provide them with one and only one dashboard screen may force them to be selective in determining the metrics they actually need.

THE LOVE/HATE RELATIONSHIP

Linda wanted to make sure that her company did not end up the way other companies did with a love/hate relationship with metrics, especially metrics related to agility. She wanted metrics that could be used to shine a light on the accomplishments of the team by tracking performance, reporting the creation of business value, and identifying ways to reduce waste. Linda also wanted metrics that could be used to identify pain points, or are situations that bring displeasure to business owners, stakeholders, and clients. The team then would look for ways to reduce or eliminate the pain points.

The hate relationship occurs when metrics become a weapon used to enforce a certain behavior. While good metrics can drive the team to perform well, the same metrics can create a hate relationship if management uses them to pit one team against another. Another hate relationship occurs when the metrics are used as part of an employee's performance review. From her research, Linda discovered that this type of hate relationship arises when:

- Metrics are seen as the beginning of a pay-for-performance environment.
- Metrics are the results of more than one person's contribution, and it may be impossible to isolate individual contributions.
- Unfavorable metrics may be the result of circumstances beyond the employee's control.
- Employees may fudge or manipulate the numbers in the metrics to look good during performance reviews.
- The person doing the performance review does not understand that the true value of the metric may not be known and the stakeholders may not get a true representation of the project's status until some time in the future.
- Employees working on the same project may end up competing with one another rather than collaborating, and the project's results could end up being suboptimized.

LINDA'S CONCLUSIONS AND RECOMMENDATIONS

Linda believed that she just scratched the surface in the identification of metrics. Metrics would be a necessity with any and all project management approaches, including agile and Scrum. However, given the number of possible metrics that could be identified, she had to establish some guidelines to avoid metric mania conditions and love/hate relationships. Linda prepared the following list as a starting point:

- Select metrics that are needed rather what people think they might want without any justification
- Select metrics that may be useful to a multitude of stakeholders, clients, and business owners.

- Make sure the metrics provide evidence and facts that can be used for decision making.
- Make sure the metrics are used rather than just nice to have displayed.
- Do not select metrics where data collection will be time-consuming and costly.
- Do not select metrics that create waste.
- Do not use metrics where the sole purpose is for performance reviews and comparing one team against another.
- Make sure that the metrics selected will not demoralize the project teams.

QUESTIONS

1. Should metric identification and selection be a structured process?
2. Is it better to allow and even encourage metric mania to exist and then reduce the number of metrics or to let the number of metrics grow incrementally?
3. If you started small, what would be a reasonable number of metrics?
4. What metrics would be appropriate for agile project management, at least as a starting point?

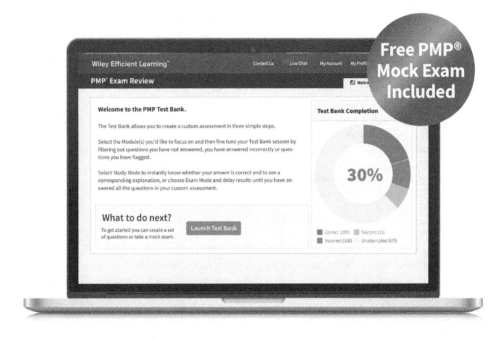

Index